New Trends in Coal Science

NATO ASI Series

Advanced Science Institutes Series

A Series presenting the results of activities sponsored by the NATO Science Committee, which aims at the dissemination of advanced scientific and technological knowledge, with a view to strengthening links between scientific communities.

The Series is published by an international board of publishers in conjunction with the NATO Scientific Affairs Division

A Life Sciences B Physics	Plenum Publishing Corporation London and New York
C Mathematical and Physical Sciences D Behavioural and Social Sciences E Applied Sciences	Kluwer Academic Publishers Dordrecht, Boston and London
F Computer and Systems Sciences G Ecological Sciences H Cell Biology	Springer-Verlag Berlin, Heidelberg, New York, London, Paris and Tokyo

Series C: Mathematical and Physical Sciences - Vol. 244

#18326529.

New Trends in Coal Science

edited by

Yuda Yürüm
Department of Chemistry,
Faculty of Engineering,
Hacettepe University, Ankara, Turkey

Kluwer Academic Publishers
Dordrecht / Boston / London

Published in cooperation with NATO Scientific Affairs Division

Proceedings of the NATO Advanced Study Institute on
New Trends in Coal Science
Datça, Turkey
August 23 – September 4, 1987

Library of Congress Cataloging in Publication Data

NATO Advanced Study Institute on New Trends in Coal Science (1987 : Datça, Turkey)
New trends in coal science : proceedings of the NATO Advanced Study Institute on New Trends in Coal Science, Datça, Turkey, August 23-September 4, 1987 / edited by Yuda Yürüm.
p. cm. -- (NATO ASI series. Series C, Mathematical and physical sciences ; no. 244)
Includes bibliographies and index.
ISBN 9027727902
1. Coal--Congresses. I. Yürüm, Yuda, 1946- . II. Title. III. Series.
TP325.N34 1987
662.6'2--dc19 88-23021
CIP

ISBN 90-277-2790-2

Published by Kluwer Academic Publishers,
P.O. Box 17, 3300 AA Dordrecht, The Netherlands.

Kluwer Academic Publishers incorporates the publishing programmes of
D. Reidel, Martinus Nijhoff, Dr W. Junk, and MTP Press.

Sold and distributed in the U.S.A. and Canada
by Kluwer Academic Publishers,
101 Philip Drive, Norwell, MA 02061, U.S.A.

In all other countries, sold and distributed
by Kluwer Academic Publishers Group,
P.O. Box 322, 3300 AH Dordrecht, The Netherlands.

Printed in The Netherlands

CONTENTS

FOREWORD

This volume contains the lectures presented at the Advanced Study Institute on "New Trends in Coal Science" which was held at Datça, Muğla, Turkey during August 23 - September 4, 1987. The book includes 23 chapters which were originally written for the meeting by some of the world's foremost investigators.

Chemists everywhere are carrying out exciting research that has important implications for the energy and fuels industries and for society in general. For the near future, coal resources will continue to be of great importance and science and technology of the highest order are needed to extend this fossil energy resource and to utilize it in an economical way that is also environmentally acceptable. These were the main purposes for the organization of this NATO ASI.

The Institute constituted two working weeks on structure and reactivity of coal and so is the book. Through the presentation of many specific recent results on structure and characterization of coal and its products the potential of new instrumental techniques is presented in the first part of the book. Finally the reactivity of coals at different conditions both in laboratory and industry is discussed. We hope that the volume will be of great use to research workers from academic and industrial background. In addition it could serve as a textbook for a graduate course on coal science and technology.

I would like to acknowledge the contribution of each of the authors in this book. Their research efforts have shed light on our understanding in coal science research. I look forward to their contributions for the advancement of this very challenging science.

I received a letter from Professor Peter H.Given in March informing me that his manuscript on "The Origin of Coal" was almost ready and he would be going to send it in April. On April 11, 1988 I received a letter from his secretary telling Professor Given had died on April 2, 1988. This was really an unexpected event for me. I knew that he had some health problems but I had not known his situation was that critical. Although Professor Given was in my list of lecturers of the NATO ASI, due to his health problems he had not been able to attend the Institute and his lectures were taken over by Dr.Gary Dyrkacz.

The loss of Professor Given will surely create a big emptiness in the field of coal science and I am confident that his contributions as corner stones in this field will guide the scientists for future prospects.

Many people contributed the success of the Institute on which this volume is based. I acknowledge the members of the Organizing Committee: Prof.Alec F.Gaines of Birkbeck College, Prof. John W.Larsen of Lehigh University and Dr.Francis P.Miknis of Western Research Institute and I take this occasion to thank them for their efforts before and during the Institute.

The Institute was generously sponsored by the Scientific Affairs Division of the NATO and their contribution is deeply acknowledged. In addition to the Scientific Affairs Division of the NATO, the Institute was supported by the Turkish Scientific and Technological Research Council (TÜBITAK) and Faculty of Engineering of Hacettepe University.

Finally, I acknowledge the efforts of the people who contributed to organizational aspects of the Institute, administrative matters and secretarial assistance. I include in this list Perla Yürüm, and my graduate students Murat Azık, Kerim Karabakan, Jale Özkısacık and Süleyman Tuncel. I should also convey my appreciation to the publishers themselves for their customary efficiency in bringing out this volume.

Ankara
April, 1988

YUDA YÜRÜM
Director
NATO ASI on
New Trends in Coal Science

LIST OF CONTRIBUTORS

1. Professor Peter H.Given †
   5 High Street
   S. Woodchester, Strout
   Glos. GL5 5EL
   England

2. Dr.Gary Dyrkacz
   Argonne National Laboratory
   Chemistry Division
   9700 South Cass Avenue
   Argonne, Illinois 60439
   U.S.A.

3. Professor John W.Larsen
   Lehigh University
   Department of Chemistry
   Bethlehem, Pennsylvania 18015
   U.S.A.

4. Dr.Francis P.Miknis
   Western Research Institute
   P.O.Box 3395
   University Station
   Laramie, Wyoming 82071
   U.S.A.

5. Professor Leon M.Stock
   University of Chicago
   Department of Chemistry
   5735 S. Ellis Avenue
   Chicago, Illinois 60637
   U.S.A.

6. Dr.Keith D.Bartle
   University of Leeds
   Department of Chemistry
   Leeds LS2 9JT
   United Kingdom

7. Dr.Leonidas Petrakis
Chevron Research Company
576 Standard Avenue
Richmond, California 94802
U.S.A.

8. Professor Alec F.Gaines
University of London
Birkbeck College
Malet Sreet
London WC1E 7HX
United Kingdom

9. Professor Jacob A.Moulijn
University of Amsterdam
Institute of Chemical Technology
Nieuwe Achtergracht 166
1018 WV Amsterdam
The Netherlands

10. Professor Larry L.Anderson
University of Utah
Department of Fuels Engineering
Salt Lake City, Utah 84117
U.S.A.

11. Professor Klaus J.Hüttinger
Universitat Karlsruhe
Kaiserstrasse 12
D-7500 Karlsruhe
Federal Republic of Germany

12. Professor Aral Olcay
University of Ankara
Faculty of Science
Department of Chemical Engineering
Ankara
Turkey

13. Professor Nevin Selçuk
Middle East Technical University
Department of Chemical Engineering
Ankara,
Turkey

PARTICIPANTS

Murat Azik	Hacettepe University, Department of Chemistry, Ankara, Turkey
Dr.Samih Bayrakceken	Ataturk University, Department of Chemistry, Erzurum, Turkey
Dr.Carlos G.Blanco	Instituto Nacional del Carbon, 33080 Oviedo, Spain
Esen Bolat	Yildiz University, Department of Chemical Engineering, Istanbul, Turkey
Dr.Kadim Ceylan	Inonu University, Department of Chemistry, Malatya, Turkey
Dr.Luigi Carvani	Eniricerche, 20097 S.Donato Milanese, Milano, Italy
Robert Davidson	International Energy Agency, Coal Research, London, England
Prof.Ekrem Ekinci	Istanbul Technical University, Department of Chemical Engineering, Istanbul, Turkey
E.Sibel Findik	Hacettepe University, Department of Chemistry, Ankara, Turkey
Prof.Jose L.Figueiredo	University of Porto, Department of Chemical Engineering, Porto-Codex, Portugal
Dr.Tracey Flynn	Heriot Watt University, Department of Chemistry, Edinburgh, Scotland
Rainer Hegerman	Universitat Karlsruhe, Institut fur Chemische Technik, Karlsruhe, Federal Republic of Germany
Dr.Freek Kapteijne	Universiteit van Amsterdam, Laboratorium voor Chemische Technologie, 1018 WV Amsterdam, Holland
A.Kerim Karabakan	Hacettepe University, Department of Chemistry, Ankara, Turkey
Nickolaos Lazaridis	Aristotelian University of Thessaloniki, Department of Chemical Technology, GR-54006, Thessalonika, Greece
Gernot Lietzke	Universitat Karlsruhe, Institut fur Chemische Technik, Karlsruhe, Federal Republic of Germany
Bernard Majchrowicz	Limburg University,Department of Chemistry, B-3610, Diepenbeek, Belgium
Dr.Ana Mastral	Consejo Superior de Investigacisones Cientificas, Instituto de Carboquimica, Zaragoza, Spain

Dr.Kostas Matis — Aristotelian University of Thessaloniki, Department of Chemical Technology, GR-54006, Thessalonika, Greece

Dr.Heinz J. Muhlen — Bergbau Forschung GmbH, 4300 Essen, Federal Republic of Germany

Meral Oguz — Middle East Technical University, Department of Enviromental Engineering, Ankara, Turkey

Mualla Oner — Yildiz University, Department of Chemical Engineering, Istanbul, Turkey

Jale Ozkisacik — Hacettepe University, Department of Chemistry, Ankara, Turkey

Prof.Antonio Palavra — Instituto Superior, Lisbon, Portugal

Maria.J.R.Pires — Instituto Superior Tecnico, Laboratorio de Tecnologia Quimica, 1000 Lisboa, Portugal

Dr.Ersan Putun — Anadolu University, Department of Chemical Engineering, Eskisehir, Turkey

Dr.K.B.Sharma — Geological Survey of Zambia, P.O.Box 50689, Lusaka, Zambia

Dr.Ola Raanes — SINTEF, Division of Metallurgy, N-7034 Trondheim-NTH, Norway

Dr.W.J.M.Schippers — Nederlandse Afval Fossilisatie, Postbus 85 5680 AB Best, Eindhoven, Holland

Fatma Sevin — Hacetepe University, Department of Chemistry, Ankara, Turkey

Prof.Tullio Sonino — Sonino and Associates, 16 Beeri Street, Rehovot 76352, Israel

Dr.Peter Tromp — Universiteit van Amsterdam, Laboratorium voor Chemische Technologie, 1018 WV Amsterdam, Holland

Suleyman Tuncel — Hacettepe University, Department of Chemical Engineering, Ankara, Turkey

Dr.John M.Vleeskens — Netherlands Energy Research Foundation, ECN, Petten, Holland

Dr.A.I.Zouboulis — Aristotelian University of Thessaloniki, Department of Chemical Technology, GR-54006, Thessalonika, Greece

THE ORIGIN OF COALS

Peter H. Given †
Professor Emeritus of Fuel Science,
Pennsylvania State University
College of Earth and Mineral Science
University Park, Pa. 16802, U.S.A.

ABSTRACT. Palaeobotanic evidence shows that coals are derived chiefly from the woody components of vascular plants, as their debris accumulates in a peat swamp or marsh. Coal formation has been a rare event in geological time, the principal eras in Europe and North America having been about 300 m. and 50-100m. years B.P. Major changes in plant anatomy took place between these two periods.

The results of a recent multi-disciplinary study of profiles of five modern peat deposits are reviewed. Statistical analyses of the line intensities found in Curie-point pyrolysis/mass spectrometry showed that changes in lignin and cellulose contents accounted for much of the variance in chemical composition at three sites but hydrocarbon and organic sulphur compounds seemed more important in accounting for variance at the others. With Py/GC/MS it was possible to detect characteristic pyrolysis products of α-cellulose in samples in which it could not confidently be identified by C 13 nmr or FTIR. A large number of phenols related to lignin were found. It was concluded that a wide variety of oxidized lignin molecules form the principal input to coalification, but with some contribution from polysaccharide derivatives, tannins and an unidentified highly aliphatic polymer. The meaning and significance of "gelification" of woody tissues in peat and young coals are discussed. There appears to be a remarkable correlation between preservation of features of anatomical and chemical structure on the one hand and particle size

Y. Yürüm (ed.), New Trends in Coal Science, 1–52.

of entities in the peat on the other. Variations in particle size are probably the source of huminite macerals. In fact, variations in the extent of tissue degradation are probably largely responsible for the need to include six huminite macerals in brown coal petrography. On deeper burial, huminites in brown coals are converted to what are known as vitrinites in bituminous coals.

PREFACE

As a result of the intelligent use of the powerful new instrumental techniques now available, major advances in our understanding of the origins of coal have been made very recently. An integrated account of these, in which attempts are made to achieve synthesis and a coherent narrative account, has not yet appeared in print. Accordingly, I propose in this paper to take the opportunity to offer such a synthesis. I have permitted myself a fair amount of speculation, to fill gaps in the narrative, and to indicate where more work is needed. The result has been greater length and the inclusion of more detail than many readers will require. However, I trust that at least parts will be useful to all readers, and the complete text to those engaged in research in this area. Also, the copious bibliography may be of some general utility.

I have felt encouraged to deal with the subject of coal origins in some depth by my conviction that it is of great importance to our understanding of coals as we now have them, and in particular of their highly heterogeneous character.

1. INTRODUCTION

Coal may be defined as an organic rock, consisting of an assembly of macerals, minerals and cations attached to the organic matter; macerals are derived from organs and tissues of vascular plants, variously altered by microbial and/or non-biological processes in peat swamps or marshes, or by purely thermal processes.

We know that macerals are derived from organs and tissues of the higher (vascular) plants because most lignites examined by transmitted light or by the fluorescence excited by UV or blue light illumination exhibit a considerable amount of altered though recognizable woody cellular structure, as well as the exines (outer walls) of spore or pollen grains, occasional resin ducts or other resin bodies and leaf cuticles, charcoal, etc.

In bituminous coals of Carboniferous age, it is not uncommon to find permineralized plant material or "coal balls". In these, the organic matter has been largely replaced by calcite, which excellently preserves fine details of the plant anatomy. Indeed, our knowledge of the anatomy of Carboniferous plants is mostly derived from the study of coal balls. Also, in bituminous coals, spore or pollen exines can nearly always be seen, and occasionally some remnants of the cell structure of woody tissues.

There are a few sites on earth where peat is still accumulating at the surface, but earlier formed peats have been buried, lie at some depth below it, and have been converted to coal. The most notable of these is in the Mahakam Delta in Indonesia, where, in a sediment thickness of some 4000 m., several peat strata and a large number of coal seams have accumulated, covering a range of rank and apparently all from the same plant genera as are now growing at the surface (Boudou, 1981; Monin et al., 1981; Dereppe et al., 1983).

2. ANCIENT PEAT AND ITS SETTING

A peat swamp is a waterlogged habitat of arborescent plants, and a marsh similarly a habitat of herbaceous plants (reeds, sedges, etc.). In a swamp or marsh there is often with time a net increase in the organic matter accumulated, at a rate of 0.1 - 2 mm per year (Grosse-Brauckman and Puffe, 1967; Given and Dickinson, 1975; Delaune and Smith, 1984). If estimates of the annual production of fresh plant biomass per unit area (for above-ground plant parts) are combined with the measured depth of a peat bed, its organic content and the radiocarbon age of its basal section, the percentage of plant productivity that is preserved as peat can be calculated to be only 5-10% (Given and Dickinson, 1975).

When a plant sheds above-ground organs, they lie on the surface for some time, where they are subject to extensive microbial degradation. Study of peats in the Everglades of Florida indicates that roots, rootlets and rhizomes are important sources of peat. Biomass production and destruction rates are, as far as I know, not available for the components of the rooting system. Since these components do not experience exposure in surface litter, the fraction preserved may be greater than is the case for aerial parts.

Deposits of petroleum have probably been accumulating more or less continuously for at least the last 10^9 years. Coals, as defined, could not have been formed more than about 350-400 million years ago, since vascular plants did not evolve till then. In fact, coal formation seems to require a rather rare set of conditions, since on any one continent it has occurred only in certain quite limited periods. Thus in North America and Europe all commercial deposits of coal were formed either in the Carboniferous (say 310-280 million years B.P.) or in the late Cretaceous and early Tertiary eras (say 40-90 million years B.P.). In Australia coals were formed chiefly in the earlier Mesozoic (270-180 million years B.P.), or (the Victorian brown coals) in the Tertiary (20-70 million years B.P.).

In the long intervals between episodes of formation of coal measures, a great deal of evolution within the plant kingdom has occurred (Stewart, 1983), and so the sources of the coal changed botanically, and probably chemically as well. This point will be taken up again later.

Much study has been made by geologists with the object of defining the kinds of geographic setting in which peat may accumulate and coals form. It is not proposed to discuss this matter here, except to say that low-lying basins in geological settings such that deep subsidence of strata is possible are obviously required. These may be inundated by saline, brackish or fresh water. The matter has been reviewed by Ferm (1974) and Weimer (1977) (see also Scott, In Appendix.

3. SOME REMARKS ON PLANT ANATOMY AND PHYSIOLOGY

In most of the trees common today, secondary xylem, popularly known as "wood", accounts for perhaps 90% of a cross-section. This does contain occasional strings of living cells, but consists mostly of dead cells (Stewart, 1983). Many modern plants are classified as Gymnosperms (e.g. conifers), a class that first emerged in primitive form during the later Carboniferous (ca. 300 m. years B.P.). The other principal modern class of plants is the Angiosperms (flowering plants). Of these, the arborescent forms evolved in the Cretaceous, perhaps 140 m. years ago (the dicotyledons), while the grasses, reeds and sedges (Angiospermous monocotyledons) did not become abundant until near the Cretaceous/Tertiary border, say 70 m. years ago. All three types of plant have contributed to coal formation. There are differences in anatomy and reproductive structures between them, but it is not necessary to discuss them here (for a summary, see Given et al., 1980).

However, a point needs to be made about monocotyledonous Angiosperms, that is, the reeds, sedges and grasses. Their above-ground organs are leafy and contain little lignified woody tissue. But studies of peat accumulations in Okefenokee Swamp (Georgia) and the

Table I

Stratigraphy of Upper Carboniferous Coals of N. American and Europe
based on Phillips and Peppers, 1984; Gignoux, 1955; and Trueman, 1954)

(reproduced, with permission, from Mudamburi and Given, 1985)

m. yrs. B.P.	British Units	Other European	Appalachian	Generalized U.S.		Stratigraphic Position, Coals of this study	Dominant plants
290							
	Stephanian	Stephanian	Monongahela	Upper	Pennsylvanian	Opdyke	tree ferns
			Connemaugh				
	Morganian	Westphalian D	Allegheny*	Middle		Ill. No. 5	Lycopods
		Westphalian C				Brookville	(Cordaites-peak)
	Ammanian	Westphalian B	Pottsville			Vitrinites V3-V7	Lycopods + seed ferns
320		Westphalian A		Lower			

*nearly equivalent to Carbondale in Illinois Basin

Everglades of Florida show that the root system of reeds and sedges is the principal contributor to the peat, and this system has not much less lignified tissue than the organs of arborescent plants. Indeed, Corvinus and Cohen (1980) concluded that peat derived from grassy plants contains greater contributions from the below-ground parts than peat from trees, and tends to contain more recognizable tissue.

Ferns, which are still extant, have an ancestry more ancient even than primitive Gymnosperms. In the middle Carboniferous (Westphalian A and B, in Europe, though A-D in Britain, or middle Pennsylvanian in North America: see Table I), the now extinct seed ferns, and Lycopods, dominated swamp vegetation. Later, as the climate of Europe and North America became drier (Pfefferkorn and Thomson, 1982; Phillips and Peppers, 1984; Phillips and Cecil, 1987) primitive Gymnosperms (Cordaitales) and the tree ferns became more important; tree ferns had evolved earlier, in the Devonian. Thus even within a single episode of coal formation, significant evolution occurred. In fact British coals of the Carboniferous formed some 10 million years earlier than the Carboniferous coals of the Appalachian and Interior provinces of North America, and were probably derived from different populations of plants (see Table I, and Mudamburi and Given, 1985)

All of these types of Carboniferous plants had different, even radically different, stem anatomy compared with fully evolved Gymnosperms and Angiosperms. The principal difference was that secondary xylem occupied a much smaller volume than in more highly evolved forms, while pith, cortex and sometimes other tissues occupied much greater volumes (Stewart, 1983; Given et al. 1980). Seed ferns (long since extinct) had strands of cells with lignified cell walls curving through the cortex and enclosing nests of living cells.

In the reproduction of those plants that form spores rather than seeds, infertile male and female embryo plants grow, and sperm somehow have to be transported from the male plant to the plant containing the egg. The probability of this occurring is low, and so the parent plant has to produce very large numbers of male and female spores to ensure reproductive success. In seed producers, fertilization of the ovum occurs already before the seed is shed. This is much more efficient, and hence far fewer pollen grains are needed compared with spores. Hence the generally much lower sporinite content of Cretaceous and Tertiary coals.

In chemical terms, the walls of parenchymatous (living) cells are composed chiefly of α-cellulose. The walls of cells in woody tissues are composed of perhaps six or more polymers, including α-cellulose, lignin, and as many as three hemicellulose polymers. The cells in the corky periderm ("bark") are woody in this sense, but also contain a waxy polymer called suberin. Furthermore, at least on young shoots, the periderm is covered with a thin film of the same substance that coats leaves (cutin).

As far as is known, these statements have always been true of vascular plants (with some reservations about suberin and cutin; see below). In this sense, therefore, one might say that all wood is made of the same substances; why should there be differences in behaviour during coalification? One might speculate that topochemistry could be important, that is, that the spatial relations of different types of cell wall and cell inclusions might influence the nature and extent of microbial action in peat. As we shall see, peats do differ in their chemistry according to the identity of the precursor plants.

A curious point arises with regard to the Lycopods. They had a quite small volume of secondary xylem and achieved the mechanical strength their tree habit required by having outside the xylem and cortex a massive layer of periderm (Stewart, 1983). The cell walls of the periderm of any arborescent plant consist of the lignocellulose complex as in wood, but part of the secondary wall is impregnated with the highly aliphatic, waxy polymer called suberin. In some coals a maceral type derived from the periderm cells can be recognized by their remnant woody structure of considerably lower reflectance than a vitrinite because of the aliphatic character of suberin. Yet Carboniferous coals derived from Lycopods appear to contain no suberinite (Teichmüller, 1982). Apparently the massive periderm did not contain suberin.

Of course, plants contain many classes of substance other than the polymers found in cell walls. They contain terpenoid resins but the composition and spatial distribution of these varies widely; these substances tend to polymerize to a solid when exposed to air, and have the function of sealing wounds against the entry of fungal spores. Mucilages and gums also have some function in sealing wounds. These are mostly water-soluble polysaccharide structures and certainly vary in different plants. All plants contain some hydrolysable or condensed tannins, which have the ability to complex with proteins, including enzymes; the content of these is very variable; they have some limited ability to prevent microbial decay (Exarchos and Given, 1975). Sterols are present also.

All of these factors could share responsibility for the differing response of plant organs *post mortem* to microbial decay. It has been known to mankind for a very long time that certain woods are preferable for some purposes: chestnut, eucalyptus and oak galls serve as sources of tannin, cedar boxes for storage of woollen goods, chestnut and willow for fences, elm for coffins, oak for ships' hulls. I doubt whether even today the biochemical or anatomical basis for such preferences are fully understood. Trees are different even though to a chemist they seem all much the same. The differences are one source of variance in coals.

4. PROCESSES IN MODERN PEATS

4.1. Introduction

It is common in the earth sciences to study processes occurring today as models for what may have happened in the remote past. This is the justification for studying the accumulation of peat as it takes place today. Much of modern peat is in bogs in northerly latitudes, and will probably not lead to coal formation (Given and Dickinson, 1975; Given, 1984). More extensive in the past were marshes and swamps, which, unlike bogs, have substantial inputs of nutrients in flowing surface waters and can be highly productive. In North America, Okefenokee Swamp in Georgia and the Everglades of Florida are good models to study. The latter has the advantage that some parts of it near the coast are saturated with saline water, as were some coal-forming swamps in the Carboniferous. Salinity is associated with high sulfur contents (Given and Miller, 1985).

4.2. Earlier Work

Peat as the precursor of coal has been studied for 70 years or so, but little progress was made in understanding peat-forming processes because of the lack of appropriate techniques and because of the pervasive but quite unwarranted assumption that plant polymers in peats were essentially unchanged from their condition in fresh plants. Even so, good work has been performed, though mostly on bog peats (reviewed by Given and Dickinson, 1975, and Given, 1975), which , as noted earlier, are not very sound models for use in studies of coal origins (but see Appendix).

Obviously, modern chromatographic methods, plus mass spectrometry where needed, must be used for analysis of volatile and/or soluble products from peat. Thus Dissanayake et al. (1982) used such methods to show a change of sterol content with depth in a Sri Lanka peat, resulting from a change in plant sources. Casagrande and Given (1974, 1980) used GLC analysis to show that acid hydrolysis of peats afforded yields of amino acids of 3-8% of the dry organic matter of peats, but also identified not only the 18-20 acids commonly present in proteins but also another 15 acids, many of which are known to be associated with the cell walls or metabolic processes of bacteria. They also found amino-sugars, most probably derived from cell walls of fungi. They concluded that if polypeptides from bacteria and fungi contribute to the organic matter of peats, it is likely that so also will the polysaccharides and lipids. In fact, Youtcheff et al. (1983) identified in solvent extracts and liquefaction products of three bituminous coals a series of pentacyclic triterpanes believed to originate from a substance common in bacterial membranes.

Lucas (1970), Given et al. (1973), and Lucas et al. (1988) found hydrolysates of two Florida peats to yield a distribution of sugars that was unexpected if the source were α-cellulose and hemicelluloses, and concluded that substantial contributions were being made by polysaccharides from bacterial slime or capsule layers. Exarchos and Given (1975) reported rapid destruction of C 14-labelled α-cellulose inserted into peats (at one site, the half-life of the insert was only 12 days). There appeared to be present in the peat a potential capability to destroy plant polysaccharides, which was incompletely realized in natural systems.

4.3. A Recent Interdisciplinary Study

One of the problems of the chemical study of peats is how to perform some preliminary fractionation designed to simplify subsequent more detailed analysis. Most authors have adopted procedures from soil biochemistry, which tend to be based on methods used in the fractionation of fresh plant tissue (Stevenson, 1982), accepting the assumptions that that course implies (e.g. Morita, 1980). In a recent partly published study, (Given et al. 1984; Ryan et al. 1987; Rhoads et al. 1987; Spackman et al. 1987), the author and eight colleagues in four different institutions have made an interdisciplinary investigation of peat profiles from four contrasting environments in the Florida Everglades and one in Okefenokee Swamp. The sites differed in salinity and the nature of the present plant cover.

It was decided to make a purely physical preliminary fractionation, which required no *a priori* assumptions. The peat samples from various depths were slurried in water, and size fractions were obtained by sieving the slurries on 20 mesh (180μm) and 80 mesh (85μm) sieves. The material retained on the sieves represented fragments of plant organs and tissues, which were examined microscopically (by transmission, by fluorescence, and between crossed polarizers; Spackman et al. 1988), and what passed through them appeared to be fine-grained humic matter. The mean fraction by weight retained on the sieves was 53±14% (standard deviation). The microscopic examination permitted identification of the principal plants and plant organs contributing to each level of a peat core, as well as a description of the alteration undergone by the tissues.

A selection of samples was subjected to C 13 nmr spectrometric analysis. Size fractions from 3-6 levels in each of the 5 cores were examined by Fourier Transform Infrared spectrometry and by various techniques based on Curie-point pyrolysis.

These pyrolysis techniques are not as well known to coal chemists as they should be. They are capable of yielding much detailed structural information about involatile organic substances and indeed provided a large part of our data on peats. Heating a ferromagnetic wire to the Curie point by induction is very rapid, and gives a precisely defined final temperature. Product analysis can be by on-line mass spectrometry with computerized multivariate interpretation of the spectra, or by gas chromatography with flame detector or mass spectrometry. Published studies of fresh, peatified and buried lignin and woods provide valuable reference data on which to base examinations of whole peats (Saiz-Jimenez and de Leeuw, 1984; Saiz- Jimenez et al., 1987; Boon et al., 1987). Reports of pyrolysis studies of bog peats may not be directly relevant to coal formation, but they have also contributed valuable reference data on the behaviour of cellulose and other polysaccharides (Boon et al., 1986; van Smeerdijk and Boon, 1987); similar information on microbial and algal polysaccharides is available (de Leeuw, 1986; Helleur et al., 1985). Extensive use of this background information has been made in investigations of peatified wood and coalified logs in brown coal seams (Boon et al., 1987; Philp et al., 1982). Also, it has proved possible to detect the transition from marsh to bog peat, on the basis of the change in the pyrolysis products of lignins and carbohydrates consequent upon the change in source vegetation (Halma et al., 1984).

In the study of peat to be discussed here each of the pyrolysis/MS runs was performed in quadruplicate on samples of each of the three particle size fractions from various levels of a core. Thus there were a large number of data points for each core, and so principal components and discriminant analyses were carried out. These enable the principal sources of variance in the data to be identified, and their trends with depth observed. Any one polymer in the peat will give a number of pyrolysis products, and the intensity of the mass spectral lines due to these will vary from sample to sample,

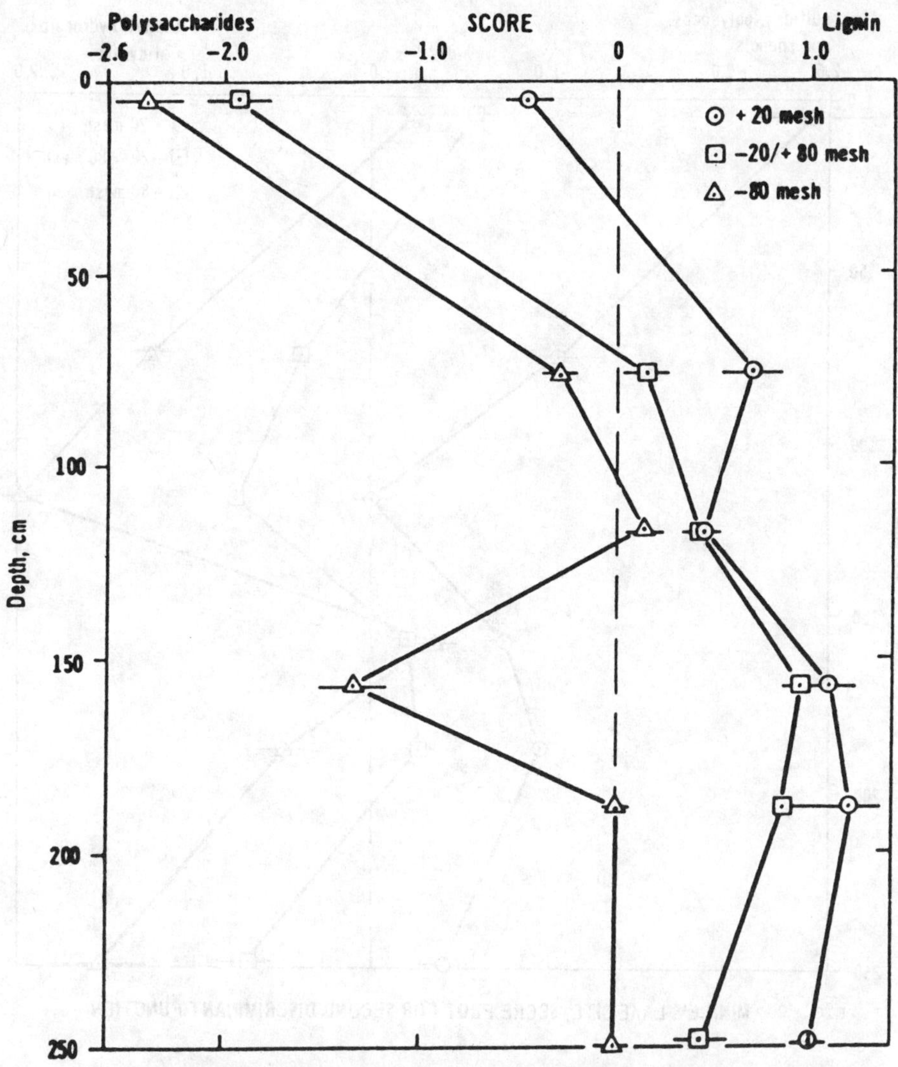

FIG. 1 MINNIE'S LAKE SITE, SCORE PLOT FOR FIRST DISCRIMINANT FUNCTION

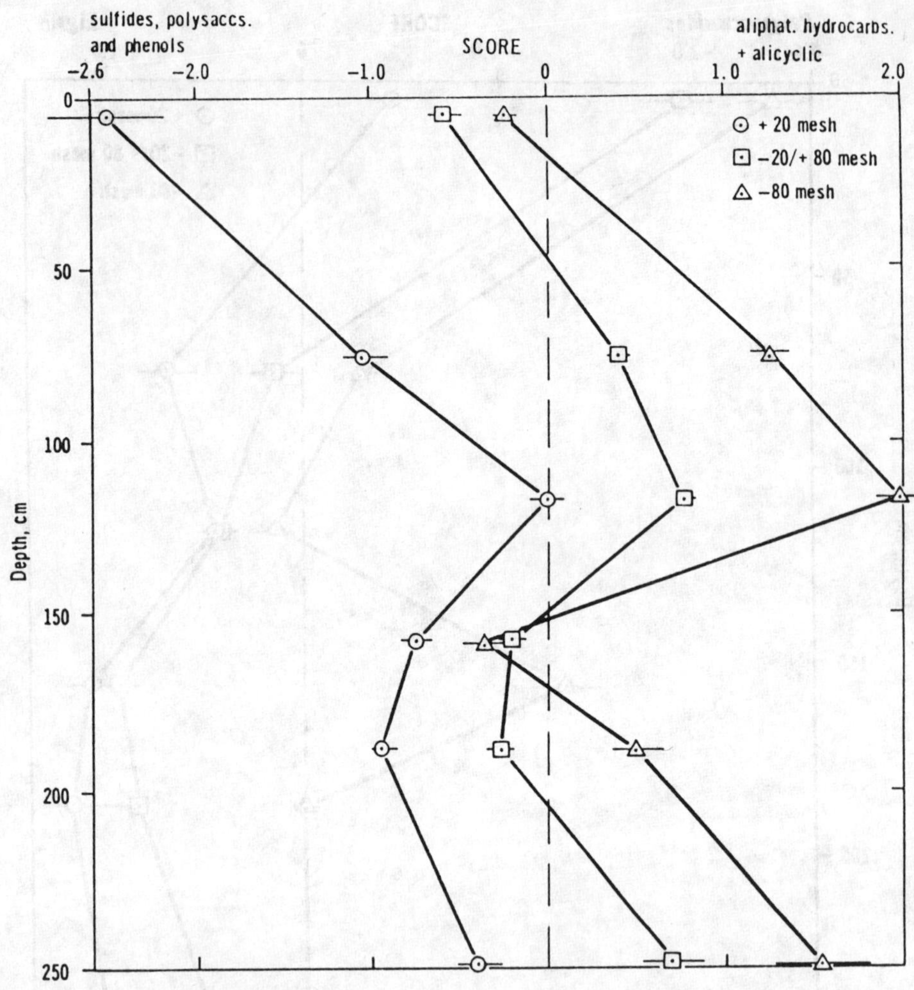

FIG. 2 MINNIE'S LAKE SITE, SCORE PLOT FOR SECOND DISCRIMINANT FUNCTION

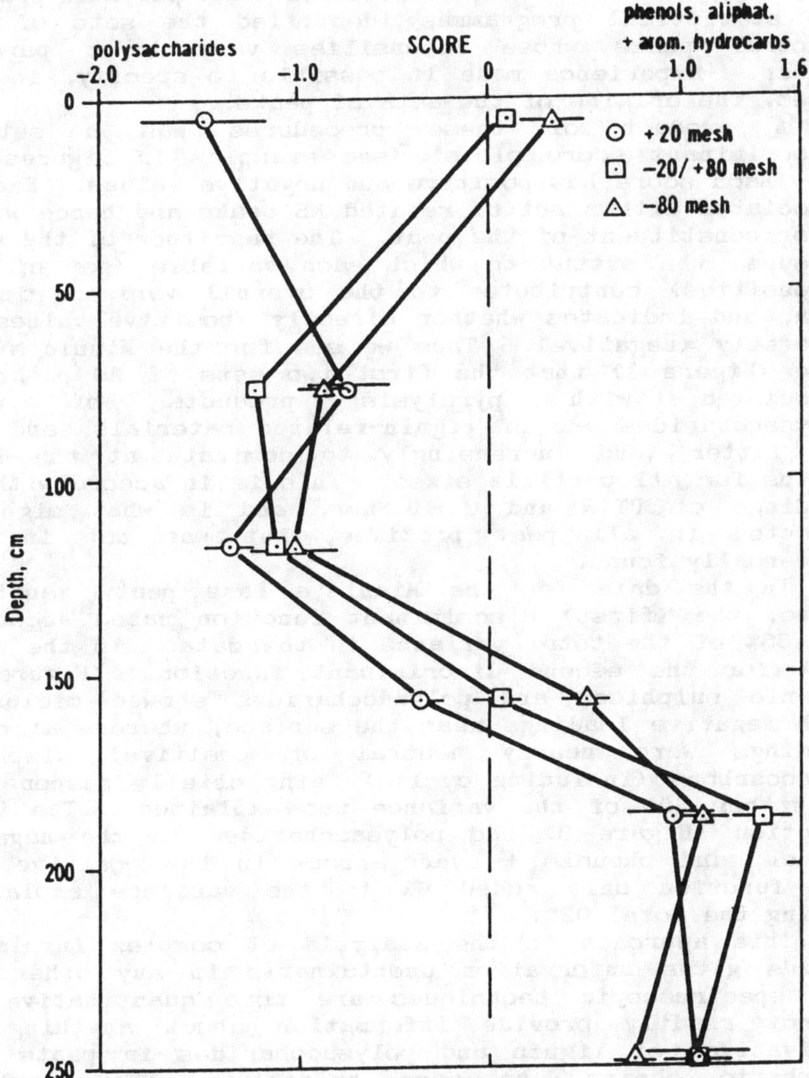

FIG. 3 **MINNIE'S LAKE SITE, SCORE PLOT FOR THIRD DISCRIMINANT FUNCTION**

according to the concentration of the polymer present. The statistical programmes identified the sets of mass spectral peaks whose intensities vary in a parallel manner. Experience made it possible to specify, in many cases, the origins of the sets of peaks.

A product of these procedures was a set of "discriminant score plots" (see examples in Figures 1 - 3). Each score has positive and negative values. Each is associated with a set of related MS peaks and hence with a major constituent of the peat. The magnitude of the score measures the extent to which each variable (set of peak intensities) contributes to the overall variance in the data, and indicates whether directly (positive values) or inversely (negative). Thus we see for the Minnie's Lake core (Figure 1) that the first two sets of MS peaks are associated with pyrolysis products of plant polysaccharides and of lignin-related materials, and that the latter tend increasingly to dominate at the lower depths for all particle sizes. This is in accord with the findings of FTIR and C 13 nmr, and is what might be expected in all peat profiles, but was not in fact universally found.

In the data for the Minnie's Lake peat, mentioned above, the (first) discriminant function named accounted for 63% of the total variance in the data. In the score plot for the second discriminant function (Figure 2), organic sulphides and polysaccharides showed medium to high negative loadings near the surface, whereas at depth loadings were nearly neutral or positive, aliphatic hydrocarbons (including cyclic) being chiefly responsible. A further 20% of the variance was explained. The third function (Figure 3) had polysaccharides in the negative branch, and phenols + hydrocarbons in the positive, but the function only added 9% to the variance explained, making the total 92%.

This approach to the analysis of complex involatile solids gives information unobtainable in any other way. The spectroscopic techniques are more quantitative but cannot readily provide information about anything but derivatives of lignin and polysaccharides in peats (and aliphatic chains, to some extent). The pyrolysis technique provides information on many compound classes whose formation explains significant amounts of the data matrix. Information on the score plots for the first three discriminant functions for each peat profile is assembled in Table II.

It is not necessary here to discuss Table II in any detail. It should be noted first that the Minnie's Lake site has fresh water, Rookery Branch is weakly brackish for part of the year, N. Harney and Joe River sites are strongly brackish, and Jewfish Key has full ocean

TABLE II

Substance Classes that Account for Significant Amounts of Variance Explained in the Score Plots for the First Three Discriminant Functions revealed for Peats by Py/MS

Site	Negative Arm	Positive Arm	Variance Explained % by each plot	cumulative
Minnie's Lake	1. polysac	lignin phenols	63	
	2. R_2S, polysac., phenols	ali. hcs.	20	92
	3. phenols, ali. + arom. hcs.	phenols, polysac., kets.	9	
Jewfish Key	1. alkanes, alkenes	phenols, R_2S	41	
	2. alk. phenols	R_2S	19	76
	3. phenols, arom. + ali. hcs.	R_2S, polysac.	16	
Rookery Branch	1. phenols	polysac.	54	
	2. R_2S, hcs.	phenols	24	93
	3. phenols, R_2S, N compds.	hcs., H_2S, Cl	15	
N. Harney River	1. hcs., N compds.	polysac., phenols	55	
	2. polysac.	lignin phenols	20	89
	3. H_2S, R_2S	lignin, polysacs.	14	
Joe River	1. arom. hc., N compds.	phenols, R_2S	34	
	2. alkyl phenols	R_2S	31	76
	3. polysacs., ali. hcs.	R_2S, phenols	11	

salinity. The vegetation is different at each site. Points of interest to be noted in Table II are:

1. The cumulative variance explained by each set of score plots is high, but the variance explained by the third discriminant function scores is small, and the compound assignments may not have much significance.
2. The variance explained by the scores for the various classes of compound varies widely from site to site: lignin and cellulose are not necessarily of overwhelming importance in explaining variance.
3. The saline peats have sulphur contents of 3-6% (dry basis), while the values for fresh and brackish peats is usually quite small. Even so, it is surprising how often sulphides appear in the discriminant functions.
4. N compounds occasionally appear, presumably arising from amino acids condensed with the plant polymers (Casagrande and Given, 1974, 1980).
5. Py/GC/MS experiments, C 13 NMR and FTIR showed the presence of an insoluble aliphatic polymer whose concentration tended to be higher in the fine-grained fraction of peat and to increase with depth. The aliphatic hydrocarbons in Table II were formed by pyrolysis of this material (the peat had been extracted with benzene/ethanol before pyrolysis).
6. In addition to the information in Table II, one can say that, in general, scores varied widely with depth, and from site to site. The trends with depth were often somewhat different for the material of different particle sizes.
7. The most important general conclusion is that in these five cores the chemistry of the peat can vary widely with salinity, the nature of the plants contributing to the peats, and the particle size of the material studied.

In Py/MS, one may in fact be able to analyse material from only about 50% of the solid. In Py/GC and Py/GC/MS the proportion is likely to be lower. On the other hand, the information obtainable is much more detailed (e.g. Saiz-Jimenez and de Leeuw, 1984). It was possible to apply these techniques involving GC to only a relatively small number of samples.

The finding of alkanes in the pyrolysis products has been mentioned. C 13 nmr showed that in the fine-grained fraction from the base of the Minnie's Lake profile 40% of the carbon was in alkane structures. Hatcher et al. (1982) have also found insoluble aliphatic material in peat. The source of this material is not known, but some possibilities will be mentioned in the following paper.

The characteristic pyrolysis products of polysaccharides are anhydro-sugars, such as levoglucosan from α-cellulose and starch, and various furan derivatives. Levoglucosan and furans characteristic of α-cellulose were found in all the samples examined, including the fine-grained fraction from the base of the Minnie's Lake profile that (from C 13 nmr) is so rich in methylene C. Other anhydrosugars were found, but not enough library spectra were available to permit identification of the parent polysaccharide.

The most interesting pyrolysis products were phenols, many of which bore obvious structural relations with lignin. The common lignin monomers, are 4-hydroxy-cinnamyl alcohols with 0, 1 or 2 methoxyl groups in the 3- or 3,5- positions. The relative amounts of these types of monomers differ as between the lignin of Gymnosperms, dicotyledonous and monocotyledonous Angiosperms. The pyrolysis products tended to contain representatives of all three types of structure, the distribution varying with depth. Factors that could affect distributions are transition from Gymnosperm to monocot. as the source of peat, which occurred at a certain depth in the Minnie's Lake profile, or the greater tendency to oxidation of the structures with three oxygen groups on each benzene ring. Philp et al. (1982) have shown that lignins from a Gymnosperm and a dicot. can be distinguished by Py/GC/MS.

4.4. General Conclusions from this Study on the Input to Coalification

Considering all five profiles and the results of all lines of investigation, we reach the following as among the more important conclusions:

1. Lignin is the most important, but certainly not the only, part of the input to coalification. Its skeletal integrity seems to be maintained well (pyrolysis), but a number of diverse changes, presumably peripheral and mostly oxidative, take place (broadening of FTIR bands), particularly at the lower depths and in the fine-grained fractions.
2. More or less intact α-cellulose is present in the surface peats (5-10%), but in a much lower ratio to lignin than in fresh wood. Cellulose tends to decrease with increasing depth and is either altered or mixed with other (bacterial?) polysaccharides. Nevertheless, some α-cellulose does persist even at depth; at least some polysaccharide is part of the input to coalification.
3. Alkane structures, organic sulphides, and N compounds condensed into insoluble form, are also part of the input to coalification.
4. The chemistry of peats varies with the nature of the contributing plants, salinity, and perhaps with environmental factors such as pH.
5. Contrary to expectations, the fine-grained fractions appeared to be essentially similar in structure to the

coarse; but they have suffered more peripheral alteration and loss of polysaccharides than the latter. Hence in any one peat sample there is a large number of lignin molecules that have suffered differing degrees of alteration. Similar comments could be made about polysaccharides. Thus the suite of derivatives of plant polymers from which coal is formed is a more complex mixture of materials than was present in fresh plants. This has important implications for ideas on the origin of huminite macerals.

These conclusions may have some general validity, since many levels in peat in five different environments were examined by a variety of techniques.

4.5. Other Modern Work

In concluding this discussion of recent peat studies, we should note some points from the literature that supplement the above. Casagrande and Given (1972) found that peats contained very little free amino acids, but considerable amounts in a form that yields amino acids on hydrolysis with 6N HCl. Ovsyannikova (1980) reports isolating several enzymes from peat: amylase, invertase, protease and polyphenol oxidase, which were partly free and partly complexed with humic matter.

Lukoshko et al. (1979, 1984) and Smychnik et al. (1979) found that the aqueous-dioxan-soluble part of the lignin in peat derived from reeds is degraded and depolymerized, but measurements by gel permeation indicate that some partly depolymerized lignin re-condenses to give fractions of higher molecular weight than were present in the fresh lignin. This provides an interesting rider to the account of lignin in peat given earlier.

Whether polymeric cellulose is or is not part of the input to coalification has been debated for many years, but until recently on the basis of unsatisfactory evidence. The δ C 13 values of a series of samples show slight enrichment in C 12 in the early stages of coalification, implying preferential loss of polysaccharide (Spiker and Hatcher, 1987). 160/180 ratios of peats, lignites and coals indicate, surpisingly, that cellulose is the principal source of O in coals (Dunbar and Wilson, 1983). Several authors (e.g. Hatcher et al., 1981) have found cellulose in various Tertiary coalified log samples, using C 13 nmr. Using this and other techniques, Wilson et al. (1987) have shown that fossil leaves in the Yallourn seam (late Miocene, 5-15 m.y.B.P.) contain some polysaccharide but this is quite different from cellulose.

Thus other workers confirm that at least a minor input to coalification is contributed by polysaccharides.

Bituminous coals are extensively cross-linked by various kinds of hydrogen bonds (Painter et al., 1987), which no doubt bridge O groups inherited from the organic matter of peat. Cohen et al. (1987) have identified certain components in 12 peats as huminitic macerals and measured the reflectance of each. Morita and Measures (1984) used a more elaborate approach, determining both fluorescence and excitation spectra, and also fluorescence decay times. The spectra were more like those of certain cellulose preparations rather than those of lignin or humic acid. Nevertheless, the fluorescence was attributed in part to a complex derived from cellulose and lignin derivates.

4.6. Comments on Microbial Activity in Peats

What information is available on the microbiology of peats was reviewed by Given and Dickinson (1975), but this was, and still is, a neglected area of study. What the coal geochemist wants to know, of course, is what the microorganisms are actually doing to the plant debris, and at what rate. Cultural studies, in which genera or physiological types are identified and counted, are not of much value for this purpose, since many bacteria, under poor conditions of growth, form spores, which, though viable, have a very low rate of metabolism and do almost nothing to substrates in their environment. There are indeed substantial numbers of microorganisms in Everglades peats, but it was possible to show that fungi and Actinomycete bacteria, though viable, were not actively attacking plant tissues at any significant rate (Dickinson et al., 1974; Given, 1972). Remnants of cells and hyphae of microorganisms have been detected in coals by electron microscopy (Taylor and Liu, 1987).

The literature on the decay of lignin by microorganisms has been extensively reviewed by Crawford (1981). A number of important points emerge from the review:
(i) Any one fungal rot or bacterium tends to attack preferentially either lignin or cellulose, (ii) even after extensive rotting woody structure persists, and the plant polymers are by no means fully removed from the system: "White-rotted lignins are still polymeric. White-rot attack on the lignin macromolecule probably does not involve extensive depolymerization prior to further decay, although some low-molecular weight lignin degradation fragments are lost to the growth medium during wood decay" (Crawford, 1981), (iii) the chemical changes undergone depend on the type of organism in question, but in general consist of the formation of carbonyl groups in the α-position to a benzene ring, Ar-CO-; production of α-β unsaturated aliphatic carboxylic acids, R-CH=CH.COOH; and

a minor amount of demethylation of methoxyl groups. These points justify some of the assumptions made above, and fill in some detail. The structural changes listed above are consistent with our FTIR and NMR observations.

Dehydrogenase enzyme contents, a measure of total respiratory activity, in the top metre or so of 10 Everglades peats have values comparable with those of fertile soils, indicating high microbiological activity. However, there is a sharp drop in activity in the lower parts of the profiles of fresh water peats, while activity is maintained at a modest level in the saline cores (Given et al., 1983). In the saline peats, the activity must include that of sulphate-reducing bacteria and of the other bacteria that produce the fatty acids the sulphate-reducers need as their fuel and carbon source; it is these bacteria that are ultimately responsible for the high pyrite and organic sulphur contents of saline peats (Cohen et al. 1985).

Fresh filter paper inserted into peats is rapidly destroyed, and so also, as noted above, is C 14-labelled cotton (Exarchos and Given, 1975). Thus viable cellulolytic organisms are present, but in the absence of fresh polymer their rate of growth is evidently slow. Glycosidic (hydrolysing) enzymes are highly specific: if a sugar unit is chemically altered in any way, the enzyme will have to search along the chain to find a linkage that it can recognize, thus reducing the rate of reaction. Phenol oxidase enzymes, which can depolymerize lignin and catalyse several other reactions, are less specific.

Bacteria generate a number of polysaccharides, cytochromes and porphyrin pigments, fatty acids, terpenoids and amino acids that are highly specific. Fungi produce many highly specific carotenoids. Well-directed searches for these might well be valuable means of assessing the role of microorganisms in peat formation.

The changes undergone by plant polymers are no doubt almost entirely catalyzed by the enzymes of microorganisms. It is an important part of coal geochemistry to know what physiological classes and genera are active in peats, and how limited is their activity. Physiological types present include glycosidic and phenol-oxidizing bacteria and fungi, chitinolytic, sulphate-reducing, nitrogen-fixing, photosynthetic, non-specific fermenting (e.g. *Clastridium*), sulphur-oxidizing, and denitrifying bacteria. One might be able to assess the activity of these by judicious isotope labelling of organisms or presumed substrate. (Culturing analyses are of little value for the purposes stated.)

Bacteria generate a number of polysaccharides, cytochromes and porphyrin pigments, fatty acids, terpenoids and amino acids that are highly specific.

Fungi produce many highly specific carotenoids. Well-directed searches for these might well be valuable means of assessing the role of microorganisms in peat formation.

5. LOSS OF CELLULAR STRUCTURE: GELIFICATION

5.1. The Problem: Aspects of Peat and Coal Heterogeneity

We now come to some complex and instructive phenomena, related to the processes by which cellular structures in peat and at the peat/brown coal transition are preserved, altered, or destroyed, and to the chemical consequences. It will be necessary first to discuss brown coals in some detail. A scheme showing interrelationships appears in Figure 4.

A minority of the material in some brown coal seams consists of "coalified logs" or "coal woods". These are seen as relatively massive blocks (>10 cm.) of substances with the texture and appearance of darkened and fibrous wood. As we shall see, the degree of preservation of plant anatomy and chemistry is variable, even within one block. The greater part of the seams that contain such blocks has no woody texture and has the usual coaly external appearance. However, under the microscope, some remnants of plant anatomy are usually to be seen in thin sections of brown coal, but in much of the material little or no cellular structure is evident, either because it has been degraded or because the organic material filling the lumina has the same optical characteristics as the cell walls.

5.2. Coalified Wood and Gelification

The above features of non-woody brown coal structure have long been believed to arise because of the incidence of a phenomenon known as gelification at a very early stage of coalification (Teichmüller, 1982). Coals appear to have been through some mysterious colloidal process in which the external woody texture, and much of the interior anatomy derived from the peat stage, are destroyed, and a continuous gel formed, which later solidifies to a coal-like solid. Until recently, the chemistry of this phenomenon had been little studied, and it is still not understood.

Spackman and Barghoorn (1966) found remarkably well preserved woody material in the Brandon lignite (Vermont, U.S.A.), enabling them to identify source genera and organs, and to describe the nature of the alteration seen in the woody cell walls. Some walls even showed the birefringence characteristic of α-cellulose; Siskov (1977) has also observed birefringence of cellulose in lignites as well as

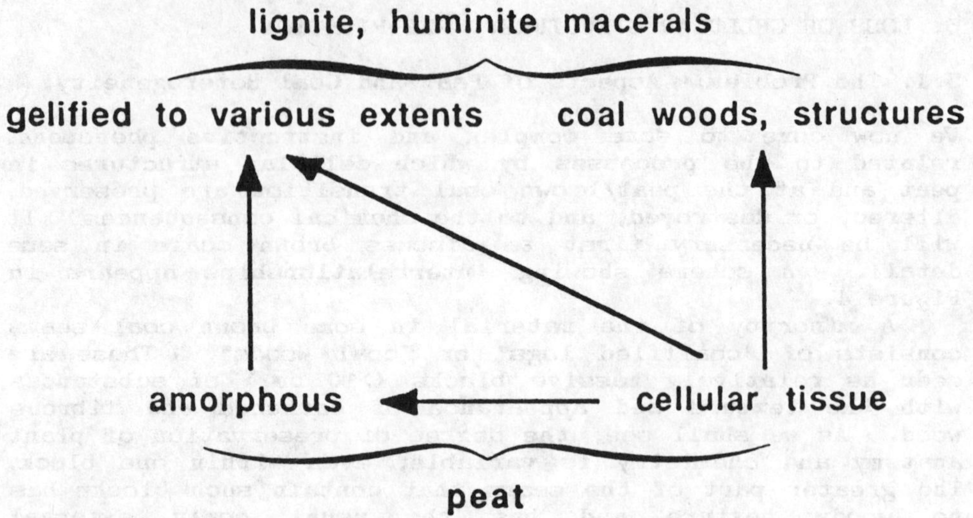

Fig. 4 Schematic Diagram illustrating aspects of Peat-Lignite Transition

peats. More recently Stout and Spackman (1983-5; see Stout, 1987) have found that this highly woody material accounts for roughly 10% of the thickness of the seam (see Stout, 1985; Stout and Bensley, 1987); The remainder of the seam, which Stout and Spackman referred to as "matrix", had the usual coal-like texture and appearance. Blocks of the highly woody material usually showed several layers or zones, in each of which the degree of preservation of tracheid and vessel cell walls differed somewhat. No significant chemical differences between layers could be detected by FTIR spectroscopy, except that the outermost gave a spectrum somewhat similar to that of the matrix, while the spectra of the other layers indicated that somewhat altered lignin was perhaps the principal constituent (Ryan and Given, 1985, unpublished). The spectra of matrix specimens showed little or no signs of the presence of lignin- or cellulose-related structures, and were quite different from those of logs.

Hatcher et al. (1981) reported lignin-like and even cellulose-like materials to be present in coalified logs in strata in Germany of Miocene-Eocene age (10-50 m. years B.P.), as evidenced by C 13 nmr spectra. Coalified logs are of relatively frequent occurrence in the thick brown coal seams of the Latrobe Valley, Victoria, Australia, and have been much investigated. They range in size from twigs less than a centimetre in diameter to large sections of stem several metres long (Russell, 1984). The Australian workers refer to "soft brown coal woods" or "fossilized woods" rather than coalified logs.

Coalified woods were compared with each other by earlier workers, but not usually with the matrix coal in which the woods were embedded. (However, Chaffee et al., 1984, found major differences in C 13 nmr between fossil wood and fully gelified coal). Members of a set of samples from one seam were described as displaying various degrees of gelification (the term seems to have been restricted to what is found in the woody samples). Pyrolysis/GC/MS, C 13 nmr and FTIR spectroscopy showed ungelified wood to contain altered lignin and cellulose. Samples of progressively higher degrees of gelification showed less recognizable cellular structure, less methoxyl and more carboxyl in the lignin, less or zero cellulose and higher aromaticity (Liu et al., 1982; Philp et al., 1982; Wilson et al., 1983; Russell, 1984; Russell and Barron, 1984; Chaffee et al. 1984). An important implication of this work and that of Hatcher et al. (1981) is that there is a correlation between chemistry and the degree of preservation of macroscopically visible woody texture.

5.3. Chemical Characteristics and Petrographic Composition

A classification of brown coal macerals by Teichmüller (1982) was adopted by the International Committee on Coal Petrography (1963, 1971, 1975). The degree of preservation of cellular structure is the main criterion on which the classification of huminites depends, and decreases downwards in the version of the scheme reproduced in Figure 5. The brown coal classification is intended to be read from left to right, and that for bituminous coals from right to left; broken lines indicate the relationships between the two schemes. Thus in a lignite the huminite maceral group contains a subgroup, humotellinite. This contains two macerals, textinite and ulminite; ulminite is further subdivided into texto-ulminite and eu-ulminite. All of the huminite macerals are derived from woody tissue except for corpocollinite (from various types of cell-fillings). Russell (1984) performed petrographic analyses on the soft brown coal woods, using the above system: increasing gelification was associatd with increasing contents of gelinite.

A number of lithotypes are distinguished in the normal coaly part of Victorian brown coals by colour and texture when partially dried out; they are described as pale, light, medium light, medium dark and dark (Verheyen and Johns, 1981; Chaffee et al., 1984). They do show some cellular structure in thin section, the content increasing in the series pale to dark. While the pale and light lithotypes contain up to 15% liptinites, the other lithotypes are largely or entirely huminitic; often with much attrinite, densinite, gelinite and some ulminite (Verheyen, 1982; Verheyen et al. 1984). The various lithotypes were evidently derived from different plant communities, and so most probably represent differing environments of deposition.

It may be suggested that, in general, the differences between the various huminitic macerals arise from processes in peat, which produce a variety of structured and amorphous particles, containing lignin and polysaccharide molecules chemically modified in a variety of ways. However, perhaps different macerals, with their different degrees of preservation of plant anatomy, in fact derive from different plants of varying resistance to decay. This may well be another important factor, but to assess its significance would require a careful anatomical study of structures preserved in humotelinitic and humodetrinitic macerals, which has rarely been made.

Another factor is suggested by Ting (1977), who discovered a silicified Palaeocene peat layer in a lignite seam in western North Dakota, in which fine details of anatomical structures were preserved. He concluded that

Fig. 5 Classification of Macerals in Coals[a]

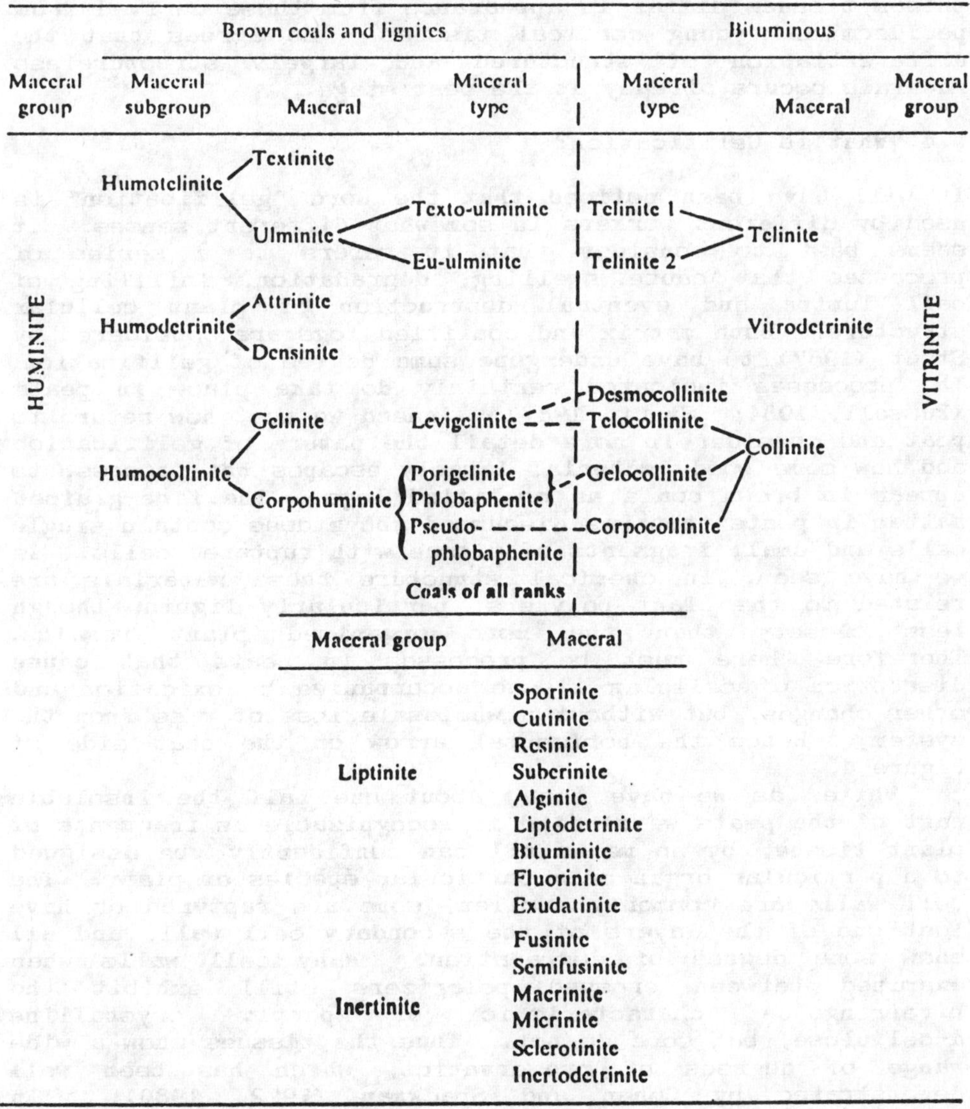

[a] International Committee for Coal Petrology (1963, 1971, 1975).

huminitic macerals from secondary xylem and the scleroids in phloem tissues differ in appearance from those derived from periderm and young cortical tissues. He agrees that the differentiation of structured and largely structureless macerals occurs already at the peat stage.

5.4. What is Gelification?

It will have been noticed that the word "gelification" is used by different workers in somewhat different senses. It seems best to consider that it refers to a series of processes that cause swelling, degradation, infilling of cell lumina and eventual destruction of plant cellular structure. Both matrix and coalified logs are considered by Stout (1987) to have undergone some degree of gelification. The processes indicated certainly do take place in peats (Russell, 1984; Stout, 1987). Indeed we must now return to peat and consider in more detail the nature of gelification and how some woody material largely escapes the process, to appear in brown coals as coalified logs. The fine-grained matter in peats that was discussed above does contain single cells and small fragments of tissue with ruptured cells. As we have seen, in chemical structure these materials are related to the plant polymers, particularly lignin, though less closely than the more organized plant remains. Therefore there must be processes in peats that cause disruption of cellular tissue accompanied by oxidation and other changes, but without a wholesale loss of mass from the system. Hence the horizontal arrow on the peat side of Figure 4.

While, as we have seen, about one half the insoluble part of the peats we studied is recognizable as fragments of plant tissue, by no means all can confidently be assigned to a particular organ or a particular species of plant. The cell walls are commonly swollen, some are ruptured or have lost one of the layers of the secondary cell wall, and all show some degree of degradation. Many cell walls when examined between crossed polarizers still exhibit the birefringence characteristic of partly crystalline α-cellulose, but some do not. Thus the tissues show a wide range of degrees of preservation, which has been well demonstrated by Cohen and Spackman (1972, 1980). In addition to the fragments of tissue already discussed, visual inspection of a peat core sliced in half longitudinally often reveals substantial pieces of "peatified" wood, which can be picked out with forceps. These extend the range of degrees of preservation (both botanical and chemical) to be observed. Such pieces were examined in detail and compared with corresponding pieces of fresh wood by Stout (1985, 1987) and by Stout and Bensley (1987), using a variety of techniques of optical microscopy.

The cell walls even of these well-preserved pieces of peatified wood show differing degrees of alteration and degradation, as indicated above. The authors described the processes resulting in these changes as gelification. Thus there is already in peat a range of variously altered tissues, as there is in the coalified logs discussed above.

But the logs in brown coal seams are often quite massive. Clearly a large piece of coalified wood can only have originated as a large piece of wood in the peat, not from the pieces of plant organs found abundantly in the Everglades and Okefenokee peats. Stout (1987), in the work referred to above, mostly relied on pieces of root or branch up to about a centimetre in diameter hand-picked from peat cores. In fact the coring equipment used (aluminium tube 7.5 cm. diam.) would not have been able to cut through large pieces of wood; if any were encountered, one would have to move the equipment to avoid them. During one coring operation in the coastal swamps of the Everglades, the coring tube hit something it could not penetrate. Subsequent excavation revealed a stump of red mangrove whose stem had evidently been broken off in a hurricane. The stump was some 25 cm. in diameter and the top was 40-50 cm. below the peat surface. The xylem wood was darkened but still strong and hard (the cambium, cortex and periderm layers had been removed). The radiocarbon age was "modern" (i.e. not more than 100 years old). The wood contained amino acids and carbohydrates of apparently bacterial origin. Spackman and Stout (see Stout, 1987) observed a few well-preserved logs in a wall of peat dug in archaeological excavations in central Florida. Thus to some extent the cores are biased samples in containing no large pieces of wood instead of, say, 10%. Observation in the field confirms that relatively few large pieces of stem or root are in fact buried in peat, whether or not recoverable by coring. Such pieces will present relatively little surface area to microbial attack, and this may be why, if they are deposited and buried (or if root, remain buried), they can survive for tens of millions of years with little change. The more common fate of plant organs is to be fragmented through fungal destruction of the adhesion between cells, yielding very small fragments of tissue. Nevertheless we can now infer that the rather uncommon well-preserved woody logs in brown coals are derived from the relatively rare massive logs in peats. Stout (1987) not only described the anatomical features of peatified logs but also studied their chemistry by Curie-point pyrolysis/gas chromatography/mass spectrometry (Boon et al., 1987); he found little evidence of alteration in the lignin, though rather more in the polysaccharides.

We can now see that a peat profile can contain material of particle size ranging from micrometers to metres.

Moreover, there appear to be strong correlations between chemistry and the preservation of cellular anatomy seen under the microscope. Thus if a piece of coalified wood is massive and excellently preserved from an anatomical point of view, it will exhibit marked chemical characteristics of somewhat altered lignin and cellulose. As the preservation becomes less good and the particle size smaller, the chemical characteristics become less clearly related to the plant polymers. It should be noted that the observations suggest that gelification involves transformations rather than destruction with substantial loss of mass: that is, gradual transformation of cell walls into some sort of gel, or an organic infilling of lumina, or fine particulates. These various processes are associated with, or accompanied by, a progresive series of chemical reactions, such as differential loss of the polymeric constituents, alterations of the non-depolymerized residues, addition of microbial cell constituents, and formation of humic acids. Bacteria and fungi are no doubt the principal agents of change.

5.5. Microbiological Problems

These concepts raise some problems related to microbial respiration and nutrition. In order to obtain energy and useful forms of carbon from a substrate, a microorganism usually catalyses complete or fairly complete combustion of the substrate. In obtaining energy from an insoluble polymer, a microorganism secretes extracellular enzymes whose function is to break off soluble monomer units, which can then diffuse in solution to and into the organism. In the case of polysaccharides, various enzymes hydrolyse the glycoside linkages between sugar units at the end and within polymer chains. Lignin is depolymerized progessively by phenol oxidase enzymes; the broken bond where a monomer unit has been removed will become an oxygenated functional group in the remaining polymer.

As a method of nutrition, the use of extracellular enzymes is inefficient: monomers may fail to diffuse to an organism before they undergo some further reaction, and carbohydrate oligomers may recombine or condense with phenols or amino acids to a new mixed polymer. It has been noted above that some partly degraded lignin molecules may re-condense to molecules of higher molecular weight than the orignal lignin molecule. The incidence of these reactions and side-reactions are the best speculative suggestion I can make at present as to how gelification can transform plant tissue into other solid forms without excessive loss of mass.

After burial of the peat under other sediments of increasing thickness, water will gradually be squeezed out, the temperature will rise and at some stage the sediment will be indurated into a rock. As long as plant polysaccharides persist in a form recognizable by bacterial enzymes, at least anaerobic alteration could continue, until the temperature reaches 60-70°C (at a depth of 1500-2000 m.?). If the strata are permeable and there is some circulation of ground water, even aerobic decay is possible, in which case lignin could also be altered. Thus continued bacterial gelification after burial of the peat stratum is feasible. Aerobic biodegradation is known to occur frequently in oil reservoir rocks as long as the temperature does not exceed about 60°C (Tissot and Welte, 1984).

5.6. Conclusion

The differential degradation of plant cell walls in peat and the excellent preservation, though still differential, of coalified logs in brown coals are interesting facts about the origins of coal that call for explanation. The coalified logs themselves provide clues to this explanation. These can only have originated from massive sections of plant organs in peat, and we have seen that such sections do exist in peats, though not accounting for a major fraction of the sediment. Nevertheless, they are very significant, because they extend by orders of magnitude the range of particle sizes found in peats. Two points should be noted: (i) it is the larger pieces of woody material in peat that preserve best the cellular anatomy, and (ii) there appear to be good correlations between preservation of anatomy and retention of the chemical structures of cellulose- and lignin-derived materials. Thus it is a reasonable hypothesis that the numerous wood-derived macerals in brown coals result from coalification because the original peat contained bodies of widely differing particle size and degrees of anatomical and chemical preservation.

The broadest range of anatomical preservation is to be found in young brown coals, which have not been as extensively studied as coals of higher rank, and not all brown coals contain coalified logs. We can only infer, therefore, that essentially all lignites and coals of higher rank have been subjected to gelification: it seems to be an essentially universal process in coal formation.

6. OTHER INPUTS TO COALIFICATION

6.1. Tannins

These are plant extractives that have long been known to have the ability to condense with the protein of skin, and have therefore been used in converting hides to leather. There are two classes.

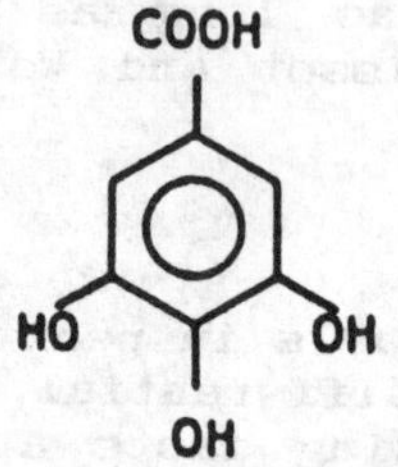

(sugars on OH's as ethers)
gallic acid derivatives,
"hydrolysable tannins"

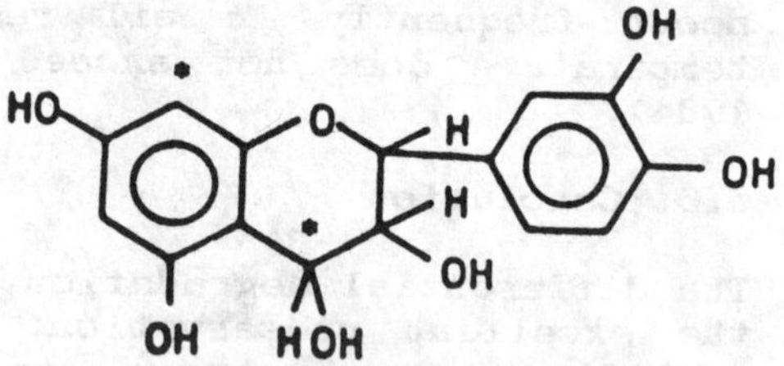

no sugars,
"condensed tannins"
* common sites of condensation polymerization

(reviewed by Haslam, 1975)

Both classes combine with and precipitate many proteins. As far as I know, the hydrolysable do not polymerize by themselves, but the condensed do polymerize in nature to insoluble red solids called phlobaphenes. Oligomeric condensed tannins occur in solution in the spongy parenchyma (living cells) of the leaves from a few plants, and polymerize to opaque cell fillings on senescence of the leaf. The bark cells of some trees contain insoluble solid polymeric tannins, which are often bright red or brown in colour. The soluble oligomeric condensed tannins precipitate a red solid naturally in cells of various organs of plants, and also on treatment with mineral acid (reviewed by Given, 1984).

Probably all vascular plants contain small quantities of one or other class of tannin, and a few contain enough to serve, or to have served, as a commercial source. The optically dense tannin cell fillings in some coalified tissues are described as the maceral phlobaphenite. No doubt because of the polyphenolic nature of phlobaphene, and the similarity of part of the structure to lignin, this maceral is assigned to the huminite group. However, it must be derived from the condensed tannins, and not the hydrolysable tannins as Teichmüller (1982) supposed. Pseudo-phlobaphenite in lignites and the supposedly derived corpocollinite in bituminous coals represent other kinds of optically dense cell filling, of poorly understood chemistry.

The mangrove trees from which peat in the Everglades coastal swamps is derived are rich in condensed tannins, which are partly preserved as insoluble cell fillings, and partly as humic acids (Chen, 1971; Cohen and Spackman, 1977).

6.2. Insoluble Aliphatic Polymers

It was recorded above that all of the Everglades and Okefenokee peats, after solvent extraction, yielded on Py/GC/MS homologous series of n.alkanes and alkenes, and it was inferred that these must be derived from an aliphatic polymer in the peat. FTIR and C 13 nmr spectroscopy confirmed that such a polymer is indeed present. Hatcher et al. (1982) have reported similar material in other peats. Possible biological origins of this polymer will be discussed in the companion paper.

Shadle et al. (1987) oxidized the asphaltenes from the liquefaction of three coals with trifluoroperoxyacetic acid, and found among the products a homologous series of long chain fatty acids. The authors concluded that these must have originated in long chain alkyl aromatics which had been part of the macromolecular network of the original coal. An interpretation of this observation is that the aliphatic polymer found in peats may end up as part of the macromolecular structure of the vitrinitic macerals. The microscopic examination of sections of the peat had revealed no discrete bodies that might have been composed of a highly aliphatic polymer, except for some cell walls of the green alga, *Botryococcus braunii*, in some Okefenokee profiles (Cohen, personal communication). Inasmuch as the concentration of the aliphatic polymer tended to be highest in the fine-grained fraction, the polymer may end up distributed within humocollinite macerals.

6.3. Substances of Low Molecular Weight

Resins are terpenoid substances, usually liquid. They often polymerize to a clear solid on exposure to air, and this is presumably the origin of resinite macerals, which often occur as blebs or fillings in resin ducts.

Peats contain a variety of substances of low molecular weight inherited from the parent plants and bacteria, such as amino acids (Casagrande and Given, 1974, 1980), hydrocarbons and fatty acids (Cooper, 1969; Perry et al. 1979), terpenes (e.g. Quirk et al., 1984), free phenols (Morita, 1973, 1975), and plant pigments including bacteriophaeophytin (Averina and Polikarpova, 1981). The subject was reviewed by Given and Dickinson (1975), and for the very deep Hula valley peat in Israel by Brenner et al. (1978).

All peats contain the ill-defined humic acids, the content often being in the range 10-20%. These are probably of quite low molecular weight but behave as colloidal polyelectrolytes in the presence of metal cations and water (Sipos et al. 1972). There are two principal modes of generation of humic acids: condensation of sugars and amino acids (the browning or Maillard reaction), and condensation of phenolic degradation products of lignin (and any other phenols present) with each other and some amino acids. In peats and mineral soils, the latter is most probably the more important (Flaig, 1966; Huc and Durand, 1974; Flaig et al. 1975; Schulten, 1987; a comprehensive review of the subject has been presented by Stevenson, 1982).

In the few cases where analyses have been made, the humic acids were found to be confined almost completely to the fine-grained (-85μm) fraction of peats (Given and Miller, 1970, unpublished). If this is generally true, the humic acids will after coalification be dispersed through the most highly gelified huminite macerals of brown coals (as in fact was found for Victorian brown coals by Verheyen and Johns, 1981).

As is well known, coals contain a complex mixture of substances of relatively low molecular weight which are extractable by solvents (e.g. Bodzek and Marzec, 1981; Radke et al., 1982; Youtcheff et al., 1982). These molecules include many "biological markers", that is, substances bearing clear structural relationships to compounds known to be present in vascular or algal plants or microorganisms (reviewed for coal by Given, 1984; Chaffee et al. 1986). These substances must be, or be derived from, molecules present in the original peat, where they may have been molecularly dispersed, present as colloidal micelles, or adsorbed on the cell walls of woody tissues. At any rate, they are part of the input to coalification, and one would suppose that after coalification could be occluded within almost any maceral. In fact it is macerals of the vitrinite group that yield the largest amount of products on solvent extraction of coals.

6.4. Inputs to Macerals other than the Huminite

This review has been limited largely to the fate of woody tissues in peat and coalification. The materials from which some liptinite (sporinite, cutinite, alginite, resinite) and inertinite (sclerotinite, some fusinites and semifusinites) macerals are derived are present in peats and are part of the input to coalification. They will be further discussed in the following paper.

7. INORGANIC CONSTITUENTS

It is not proposed to discuss here in any depth the inorganic constituents of coals, but their presence as part of coal should not be forgotten. Inorganics in lignites are of four different kinds (Miller and Given, 1986, 1987; Given and Miller, 1987; Karner et al., 1984, 1986; Tummavuori and Ako, 1980; Davis et al. 1984; Eskenazy, 1970): (i) cations held as salts by carboxyl groups (e.g. Ca, Mg, Na); (ii) cations, chiefly of trace elements, held as chelate coordination complexes on adjacent pairs of functional groups (e.g. Be, V); (iii) detrital minerals (principally clays and quartz) transported in flowing water from eroding rocks outside the peat basin; and (iv) authigenic minerals, formed *in situ* in the swamp or marsh (pyrite and calcite from bacterial activities, calcium oxalate from vascular plants, opaline silica from vascular plants, diatom tests and sponge spicules; Andrejko and Cohen, 1984).

Inorganics of all of these types enter the coalification process at the peat stage. Cations of group (i) are readily exchangeable with other ions present in flowing ground water, and so their composition could change after burial. Additional cations of group (ii) could similarly be added from ground water, though they will not readily undergo exchange reactions. After burial and induration, coals often develop cracks and cleats. In some areas, notably the Illinois Basin, post-burial pyrite may be deposited in cleats. The distribution of inorganics in any coal will depend on various factors, such as the nature of the rocks being eroded around the peat basin, and the local hydrogeology before and after burial. In fact the distributions are found to vary in the coal deposits of different coal basins (Glick and Davis, 1987; Given and Miller, 1987). Distributions of trace elements in peat deposits have often been related to contents in eroding rocks upstream from the deposit (Given and Dickinson, 1975).

As coals undergo metamorphism to fuels of higher rank, their functional groups are eliminated, carboxyl groups fairly rapidly, other groups more slowly. Thus inorganics of types (i) and (ii) will in course of time be released from the organic matter and either leached out or precipitated in some other form.

8. THE ORIGIN OF COALS

I feel that recent work on whole peats, peatified wood, and woody brown coals or coalified logs enables us to propose a fairly detailed scheme for the origin of vitrinitic material in coals, supported by evidence but partly hypothetical [since the revision of this manuscript was completed a thesis has appeared (Stout, 1987) which contributes much new and important experimental data and also proposes a detailed model for the origin of huminite macerals. See also Appendix].

Peats contain material of a continuum of particle sizes, ranging from large blocks of wood through centimetre-sized twigs or roots, sub-millimetre rootlets and root hairs, down to totally disrupted tissue such as single cells, cell fragments and fine-grained humic matter. All of these materials seem to consist chiefly of substances closely related to lignin, with lesser contributions from polysaccharides, tannins and other compounds. The plant polymers are chemically altered by oxidation and other reactions, and the finer the particle size, the greater are both the anatomical degradation and the degree of chemical alteration.

Laboratory studies show that microbiological rotting of wood depolymerizes only part of the starting material. The process tends to be self-limiting. That is, the rate of reaction eventually slows to negligible values while much plant material still survives in altered state. The new groups generated in the lignin polymer by the rotting process were reviewed above. The monomers produced in rotting are mostly burnt to carbon dioxide and water to provide energy for the microorganisms, but in part re-condense to humic acids.

The major input to coalification consists of under polymerized but chemically altered plant polymers, as summarized above. Thus, in terms of chemistry, a small number of kinds of molecule, the plant polymers, most probably give rise in peatification to large numbers of different molecules, sizeable groups of which have a common origin in lignin or one of the other plant polymers, and each group showing a spread of degrees of oxidation. The concept is that, for example, n essentially identical lignin molecules in fresh plant material give rise to n <u>different</u> lignin-related molecules in peat (see Figure 6). These differing but related molecules constitute the material of which the plant tissues of widely different size and degree of degradation are composed. It is therefore suggested that these variously degraded and chemically altered tissues are the precursors of the five or six huminite macerals that can at present be distinguished in brown coals, and exhibit differing degrees of preservation of cellular anatomy. There may in fact be many more than five or six macerals of huminitic character, but on accepted definitions we should not term these macerals, since they are not microscopically distinguishable by present techniques.

Thus I submit that the most important findings in this review are: (i) material of a very wide range of particle

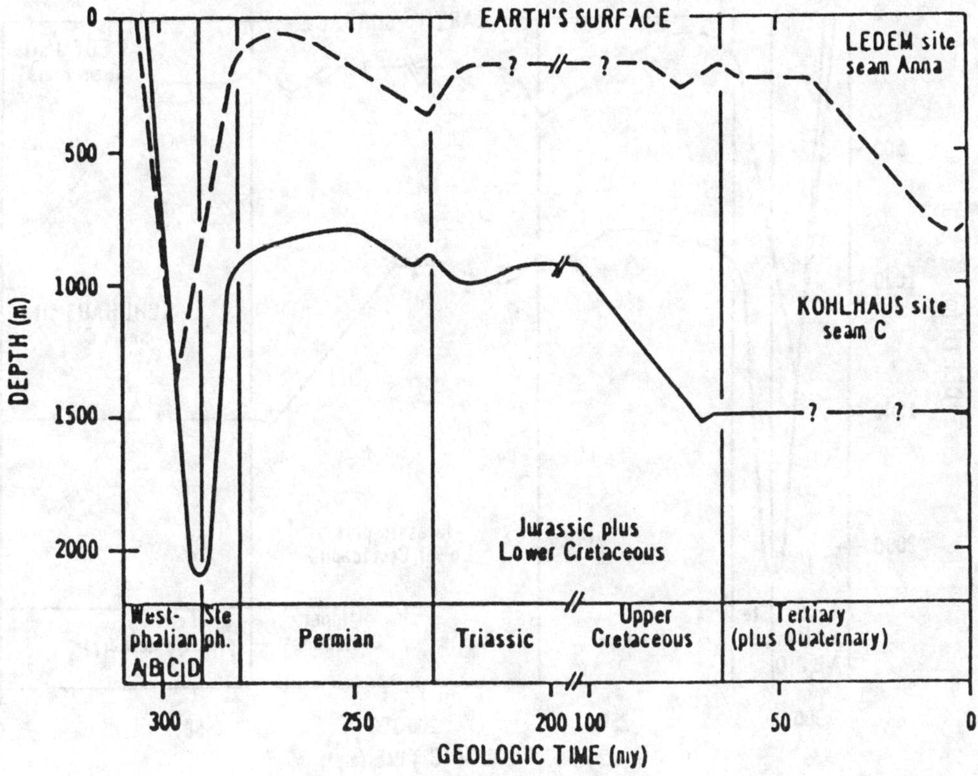

Fig. 6 Temperature history of two seams in the Ruhr, Germany
(after Radke *et al.*, 1982)

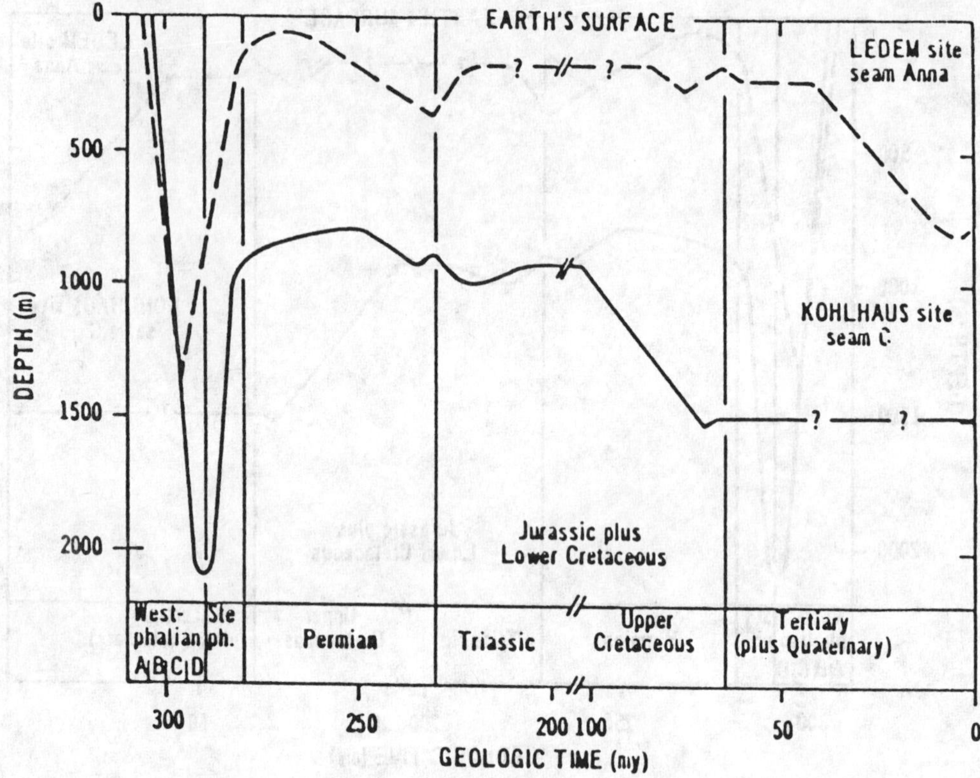

Fig. 7 Temperature history of two seams in the Ruhr, Germany (after Radke *et al.*, 1982)

sizes exist in peat, (ii) there is a strong correlation between degree of anatomical and chemical preservation of plant tissue on the one hand, and particle size on the other, (iii) materials of different particle sizes are the source of huminite macerals.

Any or all of these macerals may contain a second type of component, consisting of a complex array of relatively small molecules occluded within the primary cross-linked macromolecular component (see below, and Marzec, 1986). Some microbiological activity may continue after burial as long as the temperature does not exceed 60°C or so. It is known that both wood and α-cellulose yield a highly aromatic char if heated for several years in the absence of air at 200°C (Friedel et al., 1970). Presumably cellulose and other polysaccharides would aromatize (to phenols or hydrocarbons, not necessarily to a char) in some millions of years at some much lower temperature, and it may be assumed that this is the fate of any poly-saccharides surviving from peat. Lignin derivatives and other substances will tend to lose functional groups and hence form a variety of new cross links, both intra- and intermolecular. A result of cross-linking will be to create a number of partially closed cavities or cages from which it will be increasingly difficult to remove the relatively small molecules inherited from peat or broken off macromolecules during metamorphism. The temperature required to produce, or at least initiate, these changes might be in the range 40-70°C, or burial to depths of 800-1800 metres if the geothermal gradient has its average value of 30°C/1000 m. Not much is known about the depth of burial necessary to convert lignin-related substances into brown coal; most probably the above depths would suffice. Deeper burial and higher temperatures will yield bituminous coals. These contain a somewhat smaller suite of vitrinite macerals since differences in reflectance, colour and cellular structure tend to disappear as rank increases; it might be argued that descendants of the huminite macerals are still present, but cannot be distinguished by the techniques now available.

It should be realised that the depths of burial and temperatures experienced by coals after burial are far from being constant. Figure 7 shows a careful reconstruction of the temperature history of two coal seams in the Ruhr, Germany (from Radke et al., 1982). It will be seen that the highest temperatures were experienced for a very short time, and orogenesis then lifted the strata back towards the surface. Any attempt to construct a kinetic model must take account of highly non-isothermal kinetics.

ACKNOWLEDGEMENTS

I am grateful to Dr. William Spackman for a number of interesting discussions, his contributions to which constitute a significant part of this paper. He, with Dr. Nancy Ryan-Gray and Dr. Scott Stout, was responsible for the botanic studies of peat described above. The FTIR spectra of peats were obtained and interpreted by Dr. Carol Rhoads and Dr. P. C. Painter at Penn State. The C 13 nmr spectra and their interpretation were provided by Dr. R. J. Pugmire, University of Utah. The Py/MS, Py/GC and Py/GC/MS data were obtained by Dr. Nancy Ryan-Gray, working in the laboratories of, and supervised by, Dr. J. J. Boon, FOM Institute of Atomic and Molecular Physics, Amsterdam, and Dr. J. W. de Leeuw, Technological University of Delft, The Netherlands. I am very grateful to all of these persons for the use of their data.

REFERENCES

ANDREJKO M.J. and COHEN A.D. (1984) Contributions of ash and silica from
the major peat-producing plants in the Okefenokee swamp-marsh complex. In *The Okefenokee Swamp*, ed. A. D. Cohen, D. J. Casagrande, M. J. Andrejko and G. R. Best, Wetland Surveys, Los Alamos, N.M., pp. 575-592.

AVERINA N.G. and POLIKARPOVA N.N. (1981) Plant pigments in peat. *Vestsi*
Akad Navuk BSSR Ser. Khim. Navuk [1] 102-106. *Chem. Abs.* 1981, 95 no. 9689.

BODZEK D. and MARZEC A. (1981) Molecular components of coal and coal
structure. *Fuel*, 60, 47-51.

BOON J.J., DUPONT L., van der HAMMER, T. and de LEEUW J.W. (1986)
Characterization of a peat bog profile by Curie-point pyrolysis/mass spectrometry combined with multivariate analysis and by pyrolysis/gas chromatography/mass spectrometry. In *Peat and Water. Aspects of Water Retention and Dewatering in Peat*, ed. C. Fuchsmann, Elsevier, Amsterdam.

BOON J.J., van SMEERDIJK D. and STOUT S.A. (1987) Characterization of recent and fossil plant tissues by on-line pyrolysis/high resolution gas chromatography/LR and HR mass spectrometry and microscopic techniques. *Abstracts, 13th International Meeting on Organic Geochemistry*, pp. 286-287.

BOUDOU J.-P. (1981) Organic Diagenesis of deltaic sediments (Mahakam, Indonesia) (in French), D.Sc. Thesis, University of Orlèans, France

BRENNER S., KIAN R., AGRON N. and NISSENBAUM A. (1978) Hula Valley (Israel) peat: review of chemical and geochemical aspects. *Soil Sci.*, 125, 226-232.

CASAGRANDE D.J. amd GIVEN P.H. (1974), Geochemistry of Amino Acids in Some Florida Peat Accumulations. I. Experimental Approach and Total Amino Acid Concentrations,. *Geochim. Cosmochim. Acta*, , 38, 419-434.

CASAGRANDE D.J. and GIVEN P.H. (1980) Geochemistry of amino acids in some Florida peat accumulations. II. Amino acid distributions, *Geochim. Cosmochim. Acta*, 44, 1493-1507.

CHAFFEE A.L., JOHNS R.B., BAERKEN M.J., de LEEUW J.W., SCHENCK P.A. and BOON J. J. (1984) Chemical effects in gelification processes and lithotype formation in Victorian brown coal. In *Advances in Organic Geochemistry* 1983, ed. P.A. Schenck, J. W. de Leeuw and G.W.M. Lijmbach, *Org. Geochem.*, 6, 409-416.

CHAFFEE A.L., HOOVER D.S., JOHNS R.B. and SCHWEIGHARDT (1986) Biological markers extractable from coal. In *Biological Markers in the Sedimentary Record* ed. R. B. Johns, Elsevier, Amsterdam, pp. 311-345.

CHEN A. (1971) Flavonoid pigments in the red mangrove, *Rhizophora mangle L.* of the Florida Everglades and the peat derived from it. M.S. Thesis, Pennsylvania State University, 232 pp.

COHEN A.D. and SPACKMAN, W. (1972) Methods in peat petrology and their application to reconstruction of paleoenvironments. *Bull. Geol. Soc. Amer.*, 83, 129-141.

COHEN A.D. and SPACKMAN W. (1977) Phytogenic organic sediments and sedimentary environments in the Everglades-Mangrove complex: Part II. The origin, description and classification of the peats of southern Florida. *Palaeontographica*, 162, Part B, 71-114.

COHEN A.D. and SPACKMAN W. (1980) Phytogenic organic sediments and sedimentary environments in the Everglades-Mangrove Complex: Part III. The alteration of plant material in peats and the origin of coal macerals. *Palaeontographica*, 172, Part B, 125-149.

COHEN A.D., ANDREJKO M.J., SPACKMAN W. and CORVINUS D. (1984) Peat deposits of the Okefenokee Swamp. In *The Okefenokee Swamp*, ed. A. D. Cohen, D. J. Casagrande, M. J. Andrejko and G. R. Best, Wetland Surveys, Los Alamos, New Mexico, pp. 493-521.

COHEN A.D., SPACKMAN, W. and DOLSEN P. (1985) Occurrence and distribution of sulfur in peat-forming environments of southern Florida. *Int. J. Coal Geol.*, 4, 73-96.

COHEN A.D., RAYMOND R., ARCHULETA L.M. and MANN D.A. (1987) Preliminary study of the reflectance of huminite macerals in recent surface peats. *Org. Geochem.*, 11, 429-430.

COOPER W.J. (1971) Geochemistry of lipid components in peat-forming environments of the Florida Everglades, M.S. Thesis, Pennsylvania State University.

CORVINUS, D.A. and COHEN A.D. (1980) Microbotanical composition of Okefenokee peats. *Proc. 6th Internat. Peat Congress*, pp. 538-541.

CRAWFORD, R.L. (1981) *Lignin Biodegradation and Transformation*, John Wiley & Sons, New York, pp. 38-108.

DAVIS A., RUSSELL, S.J. RIMMER S.M. and YAEKEL J.D. (1984) Some genetic implications of silica and alumino-silicates in peat and coal. *Int. J. Coal Geol.*, 3, 293-314.

DELAUNE R.D. and SMITH C.J. (1984) Carbon cycle of peat in the Mississippi River deltaic plain. *Southeast Geol.*, 25, 61-68.

DEREPPE J.-M., BOUDOU J.-P., MOREAUX C. and DURAND B. (1983) Structural evolution of a sedimentologically homogeneous coal series as a function of carbon content by solid state C 13 nmr. *Fuel*, 62, 575-580.

DICKINSON C.H., WALLACE B. and GIVEN P.H. (1974) Microbial activity in Florida Everglades peats. *New Phytologist*, 73, 107-113.

DISSANAYAKE C.B., SENARATRE A. and GUNATILAKA A.A.A.L. (1982) Organic geochemical studies of the Muthurajawala peat deposit of Sri Lanka. *Org. Geochem.*, 4, 19-26.

DUNBAR J. and WILSON A.T. (1983) The use of oxygen-18/oxygen-16 ratios to study the formation and chemical origin of coal. *Geochim. Cosmochim. Acta*, 47, 1541-1543.

ESKENAZY G. (1970) Adsorption of beryllium on peat and coals. *Fuel*, 49, 61-67.

EXARCHOS C. and GIVEN P.H. (1977) Cell wall polymers of higher plants in peat formation: the role of microorganisms. In *Interdisciplinary Studies of Peat and Coal Origins*, eds. P. H. Given and A. D. Cohen, *Geol. Soc. Amer.*, Microform Publication No. 7, pp. 122-141.

FERM J.C. (1974) Carboniferous environmental models in eastern United States and their significance. *Geol. Soc. Amer.*, Special paper 148, pp. 79-95.

FLAIG, W. (1966) Chemistry of humic susbtances in relation to coalification. In *Coal Science*, ed. P. H. Given, Amer. Chem. Soc. Adv. Chem. Series No. 55, pp. 58-68.

FLAIG, W., BEUTELSPACHER H. and RIETZ E. (1975) Chemical composition and physical properties of humic substances. In *Soil Components*, ed. John E. Gieseking Vol. 1, Springer-Verlag, pp. 1-211.

FRIEDEL R.A., QUEISER J.A. and RETCOFSKY H.L. (1970) Coal-like substances from low-temperature pyrolysis at very long reaction times. *J. Phys. Chem.*, 74, 908-912.

GIGNOUX M. (1955) *Stratigraphic Geology* English translation of Fourth French edition of 1950 by Gwendolyn G. Woodford, W. H. Freeman Co., San Francisco, pp. 156-180.

GIVEN P.H. (1972) Biological aspects of the geochemistry of coal. In *Adv. Org. Geochem. 1971*, ed. H. R. von Gaertner and H. Wehner, pp. 69-92, Pergamon.

GIVEN P.H. (1975) Environmental organic chemistry of bogs, swamps and marshes. In *Environmental Organic Chemistry*, G. Eglinton, Senior Reporter, Chemical Society, London, pp. 55-80.

GIVEN P.H. (1984) An essay on the organic geochemistry of coal. In *Coal Science*, Vol. 3, ed. M. L. Gorbaty, J. W, Larsen and I. Wender, Academic Press, San Diego, pp. 63-252 and 339-341.

GIVEN P.H. and DICKINSON C.H. (1975) Biochemistry and microbiology of peats. In *Soil Biochemistry*, Vol. 3, ed. E. A. Paul and A. D. McLaren, Marcel Dekker, New York, pp. 123-212.

GIVEN P.H. and MILLER R.N. (1985) Distribution of forms of sulfur in saline peat cores. *Int. J. Coal Geol.*, 5, 397-409.

GIVEN P.H., and MILLER R.N. (1987) The association of major, minor and trace element profiles in four seams. III The trace elements in four lignites, and general discussion of the results of the whole study. *Geochim. Cosmochim. Acta*, 51, 1843-1853.

GIVEN P.H., CASAGRANDE D.J., IMBALZANO J.R. and LUCAS A.J. (1973) Biochemical aspects of early stages of coal formation. *Proc. Symp. Hydrogeochemistry and Biogeochemistry* (Tokyo 1970), Internat. Assoc. Geochem. Cosmochem., The Clarke Co., Washington, D.C., Vol. II, pp. 240-263.

GIVEN P.H., SPACKMAN W., DAVIS A. and JENKINS R.G. (1980) Some proved and unproved effects of coal geochemistry on liquefaction behavior with emphasis on U.S. coals. In *Coal Liquefaction Fundamentals*, Amer. Chem. Soc. Symp. Series No. 139, ed. D. Duayne Whitehurst, pp. 3-34.

GIVEN P.H., SPACKMAN W., IMBALZANO J.R., CASAGRANDE D.J., LUCAS A.J., COOPER W. and EXARCHOS C. (1983) Physico-chemical characteristic and levels of microbial activity in some Florida peat swamps. *Int. J. Coal Geol.*, **3**, 77-79.

GIVEN P.H., SPACKMAN W., PAINTER P.C., RHOADS C.A., RYAN N.J., ALEMANY L. and PUGMIRE R.J. (1984). The fate of cellulose and lignin in peats: an exploratory study of the input to coalification. In *Advances in Organic Geochemistry 1983*, ed. P.A. Schenck, J. W. de Leeuw and G.W.M. Lijmbach, *Org. Geochem.*, **6**, 399-407.

GLICK D.C. and DAVIS A. (1987) Variability in the inorganic element content of U.S. coals including results of cluster analysis. *Org. Geochem.*, **11**, 331-342.

GROSSE-BRAUCKMAN G. and PUFFE D. (1967) *Proc. 8th Internat. Congr. Soil Sci.*, Bucharest, 1964, **5**, 635.

HALMA G., VAN DAM D. HAVERKAMP J., WINDIG W. and MEUZELAAR H.L.C. (1984) Characterization of an oligotrophic-eutrophic peat sequence by pyrolysis-mass spectrometry and conventional analysis methods. *J. Anal. Appl. Pyrolysis*, **7**, 167-183.

HASLAM E. (1975) Natural proanthocyanadins. In *The Flavanoids* ed. J.B. Harborne, T. J. Mabry and H. Mabry, Academic Press, pp. 505-559.

HATCHER P.G., BREGER I.A. and EARL W.L. (1981) Nuclear magnetic resonance studies of ancient buried wood. I. Observations on the origin of coal to the brown coal stage. *Org. Geochem.*, **3**, 49-55.

HATCHER P.G., BREGER I.A., DENNIS L.W. and MACIEL G.E. (1982) Solid-state C 13 nmr of sedimentary humic substances: new revelations on their chemical composition. In *Aquatic and Terrestrial Humic Materials*, eds. R. F. Christman and E. Gjessing, Ann Arbor Science Publ., pp. 37-81.

HELLEUR R.J., HAYES E.R., CRAIGIE J.S. and McLACHLAN J.L. (1985) Characterization of polysaccharides of red-algae by pyrolysis-capillary chromatography. *J. Anal. Appl. Pyrol.*, **8**, 349-357.

HUC A.Y. and DURAND B.M. (1974) Study of humic acids and humin in recent sediments considered as precursors of kerogen. In *Advances in Organic Geochemistry 1973*, ed. B. Tissot and F. Bienner, Editions Technip, Paris, pp. 53-72.

INTERNATIONAL COMMITTEE for Coal Petrology (1963, 1971, 1975) *International Handbook of Coal Petrography*, Centre Nat. de Rech. Sci., Paris.

KARNER F.R., BENSON S.A., SCHOBERT H.H. and ROALDSON R.G. (1984) Geochemical variation of inorganic constituents in a North Dakota lignite. In *The Chemistry of Low-Rank Coals*. Ed. H. H. Schobert. Amer. Chem. Soc. Symp. Ser. 264, 175-194.

KARNER F.R., SCHOBERT H.H., FALCONE S.K. and BENSON S.A. (1986) Elemental distribution and association with inorganic and organic components in North Dakota lignites. In *Mineral Matter and Ash in Coal*, ed. K. S. Vorres, Amer. Chem. Soc. Symp. Ser. 301, 70-89.

de LEEUW J.W. (1986) Sedimentary lipids and polysaccharides as indicators for sources of input for microbial activity and short-term diagenesis. In *Organic marine Geochemistry*, ed. M. Sohn, Amer. Chem. Soc. Symp. Ser. pp. 33-61.

LIU S., TAYLOR G.H. and SHIBAOKA M. (1982) Biochemical gelification and the nature of some huminite macerals. In *Proc. Symp. Coal Resources: Origin, Exploration and Utilization in Australia*, ed. C. W. Mallet, Geol. Soc. Australia, Melbourne, pp. 145-152.

LUCAS A.J. (1970) Geochemistry of carbohydrates in some organic sediments of the Florida Everglades, Ph.D. Thesis, Pennsylvania State University, 263 pp.

LUCAS A.J., GIVEN P.H. and SPACKMAN W. (1988) Studies of peat as the input to coalification. I. Rationale and preliminary examination of polysaccharides in peats. *Int. J. Coal. Geol.*, in the press.

LUKOSHKO E.S., BAMBALOV N.N. and SMYCHNIK T.P. (1979) Change in lignin composition in the peat formation process. *Khim. Tverd. Topl.*, 3, 144-150, *Chem. Abs. 1979*, 91, no. 213648.

LUKOSHKO E.S., BAMBALOV N.N., KRUKOVSKAYA L.A. and SMYCHNIK T.P. (1984) An investigation of the composition of the lignins of peat-forming plants and the products of their decomposition. *Khim. Tverd Topl.*, 18 (1), 49-54. *Solid Fuel Chem.*, 18 (1), 43-48.

MARZEC A. (1986) Macromolecular and molecular model of coal structure. *Fuel Proc. Technol.*, 14, 39-46.

MILLER R.N. and GIVEN P.H. (1986) The association of major, minor and trace inorganic elements with lignites. I. Experimental approach and study of a North Dakota lignite. *Geochim. Cosmochim. Acta*, 50, 2033-2044.

MILLER R.N. and GIVEN P.H. (1987) The association of major, minor and trace inorganic elements with lignites. II. Minerals, and major and minor element profiles in four seams. *Geochim. Cosmochim. Acta*, 51, 1311-1322.

MONIN J.C., BOUDOU J.-P., DURAND B. and OUDIN J.L. (1981) Example of the enrichment of carbon-13 in coals in the process of coalification. *Fuel*, 60, 957-960.

MORITA H. (1973) Polyphenols in the benzene-ethanol extractives of an organic soil. *Geochim. Cosmochim. Acta*, 37, 1587-1591.

MORITA H. (1975) Polyphenols in the lime water extractives of peat. *Soil Sci.*, 120, 112-116.

MORITA H. (1980) Perspectives on carbohydrates as chemotaxonomic aids for peats. *Proc. 6th Internat. Peat Congress*, pp. 631-637.

MORITA H. and MEASURES R.M. (1984) (1984) Some observations on the laser fluorescence spectroscopy of peat. *Int. Symp. Peat Util. [Proc.]*, ed. C. H. Fuchsman and S. A. Spigarelli, pp. 391-403.

MUDAMBURI Z. and GIVEN P.H. (1985) Some structural features of coals of differing maceral distribution and geological history. *Org. Geochem.*, 87, 441-453.

OVSYANNIKOVA N.A. (1980) Isolation of protein from a protein-humus complex of peat. *Tr. VNIITP 1980*, 45, 142-145. (*Chem. Abs. 1981*, 95 No. 189817).

PAINTER P.C., SOBKOWIAK M. and YOUTCHEFF J. (1987) FT-IR study of hydrogen-bonding in coal. *Fuel*, 66, 973-978.

PERRY G.J., VOLKMAMN J.K. and JOHNS R.B. (1979) Fatty acids of bacterial origin in contemporary marine sediments. *Geochim. Cosmochim. Acta*, 43, 1715-1725.

PFEFFERKORN H.W. and THOMSON M.C. (1982) Changes in dominance patterns in upper carboniferous plant-fossil assemblages. *Geology*, 10, 641-644.

PHILLIPS T.L. and CECIL C.B., eds. (1985) Paleoclimate controls on coal resources of the Pennsylvanian System of North America (a symposium). *Int. J. Coal Geol.*, 5, 7-230.

PHILLIPS T.L. and PEPPERS R.A. (1984) Changing patterns of Pennsylvania coal-swamp vegetation and implications of climatic control on coal occurrence. *Int. J. Coal Geol.*, 3, 205-255.

PHILP R.P., RUSSELL N.J. GILBERT T.D. and FRIEDRICH J.M. (1982) Characterization of Victorian soft brown coal wood by microscopic techniques and Curie-point pyrolysis combined with gas chromatography-mass spectrometry. *J. Anal. Appl. Pyrol.*, 4, 143-161.

QUIRK N.M., WARDROPER A.M.K., WHEATLEY R.E. and MAXWELL J.R. (1984) Extended hopanoids in peat environment. *Chem. Geol.*, 42, 25-43.

RADKE M., WILLSCH H., LEYTHAEUSER D. and TEICHMüLLER M. (1982) Aromatic components of coal: relation of distribution pattern to rank. *Geochim. Cosmochim. Acta*, 46, 1831-1848.

RHOADS, C.A., PAINTER P.C. and GIVEN P.H. (1987) The use of Fourier Transform Infra-red spectrometry in the study of peat as the precursor to coal. *Proc. A.C.S. Symposium on Peat*, ed. D. J. Boron, *Int. J. Coal Geol.*, 8, [1/2], 69-83.

RUSSELL N.J. (1984) Gelification of Victorian Tertiary soft brown coal wood. I. Relationship between chemical composition and microscopic appearance and variation in the degree of gelification. *Int. J. Coal Geol.*, 4, 99-118.

RUSSELL N.J. and BARRON P.F. (1984) Gelification of Victorian Tertiary soft brown coal wood. II. Changes in chemical structure associated with variation in the degree of gelification. *Int. J. Coal Geol.*, 4, 119-142.

RYAN N.J, GIVEN P.H., BOON J.J. and de LEEUW J. W. (1987) Study of plant polymers in peats. Proc. A.C.S. Symposium on Peat, ed. D. J. Boron, *Int. J. Coal Geol.*, 8, [1/2], 85-98.

SAIZ-JIMENEZ C. and de LEEUW J.W. (1984) Pyrolysis/gas chromatography/
mass spectrometry of isolated, synthetic and degraded lignin. In *Advances in Organic Geochemistry 1983*, ed. P.A. Schenck, J. W. de Leeuw and G.W.M. Lijmbach, *Org. Geochem.*, 6, 417-422, Pergamon, Oxford.

SAIZ-JIMENEZ C., BOON J.J., HEDGES J.L., HESSELS J.K.C. and de LEEUW J.W.
(1987) Chemical characterization of recent and buried woods by analytical pyrolysis. Comparison of pyrolysis data with C 13 nmr and wet chemical data. *J. Anal. Appl. Pyrol.*, 11, 437-450.

SCHULTEN H.-R. (1987) Pyrolysis and soft ionization mass spectrometry of
aquatic/terrestrial humic substances and soils. *J. Anal. Appl. Pyrolysis*, 12, 149-186.

SCOTT A.C. ed. (1987) *Coal and Coal-Bearing Strata: Recent Advances*,
Blackwell Scientific Publications, Oxford (in particular, see R. S. Clymo, Rainwater-fed peat as a precursor of coal, pp. 17-23).

SHADLE L.J., NEILL P.H. and GIVEN P.H. (1987) The dependence of liquefaction
behavior on coal characteristics. 10. Structural characteristics of a set of high-sulphur coals and the asphaltenes derived from them. *Fuel*, submitted.

SIPOS S., SIPOS E., DEKANY I., SZANTO F. and LAKATOS B. (1972)
Investigation of humic acids and metal humates with analytical ultra-centrifuge. In *Proc. 4th Int. Peat Cong. (Otaniemi, Finland)*, 4, pp. 255-261.

SISKOV G. (1977) Optical diffraction of cellulose in peats and lignites in
linearly polarized light at small angles. *Freiberg Forschungh. C.*, C331, 167-73 (in German)

van SMEERDIJK D. and BOON J.J. (1987) Characterization of sub fossil
sphagnum leaf, rootlets of *Ericaceae* and their peat by pyrolysis/high resolution gas chromatography/mass spectrometry. *J. Anal. Appl. Pyrol.*, 11, 377-402.

SMYCHNIK T.P., LUKOSHKO E.S. and BAMBALOV N.N. (1979) Change in the molecular weight distribution of lignin in peat formation. *Vestsi Akad. Navuk BSSR Ser. Khim. Navuk*, [5], 114-116, *(Chem. Abs. 1980*, 92, No. 8018).

SPACKMAN W. and BARGHOORN E.S. (1966) Coalification of woody tissue as produced from a petrographic study of Brandon lignite. In *Coal Science*, ed. P. H. Given, *Adv. Chem. Ser. No. 55, Amer. Chem. Soc.*, pp. 695-706.

SPACKMAN W., RYAN N.J., RHOADS C.A. and GIVEN P.H. (1988) Studies of peat as the input to coalification. II. Sampling sites and preliminary fractionation. *Int. J. Coal Geol.*, in the press.

SPIKER E.C. and HATCHER P.G. (1987) The effects of early diagenesis on the chemical and stable isotope composition of wood. *Geochim. Cosmochim. Acta*, 51, 1385-1391.

STEWART W.N. (1983) *Palaeobotany and the Evolution of Plants*, Cambridge University Press, London.

STEVENSON F.J. (1982) *Humus Chemistry: Genesis, Composition, Reactions*, John Wiley & Sons, New York.

STOUT S.A. (1985) A Microscopic Study of the Fate of Secondary Xylem during Peatification and the Early Stages of Coal Formation. M.S. Thesis, Pennsylvania State University, 308 pp.

STOUT S.A. (1987) Tracing the microscopical and chemical origin of huminitic macerals in coal, Ph.D. thesis, Pennsylvania State University, 252 pp.

STOUT S.A. and BENSLEY D.F. (1987) Fluorescing macerals from wood precursors. *Int. J. Coal Geol.*, 7, 119-134.

STOUT S.A., BOON J.J. and SPACKMAN W. (1988) Molecular aspects of the peatification and early coalification of angiosperm and gymnosperm woods. *Geochim. Cosmochim. Acta*, 52, 405-414.

TAYLOR G.H. and LIU S.Y. (1987) Biodegradation in coals and other organic-rich rocks. *Fuel*, 66, 1269-1273.

TEICHMüLLER M. (1982) Origin of the petrographic constituents of coal. In *Stach's Textbook of Coal Petrology*, ed. E. Stach, M. Th. Mackowsky, M. Teichmüller, G. H. Taylor, D. Chandra and R. Teichmüller, translated by D. G. Murchison, Gebrüder Borntraeger, Berlin, pp. 219-294.

TING F.T.C. (1977) Microscopical investigation of the transformation (diagenesis) from peat to lignite. *J. Microsc. (Oxford)*, 109(1), 75-83.

TISSOT B.P. and WELTE D.H. (1984) *Petroleum Formation and Occurrence*, Springer-Verlag, Berlin, Second Edn.

TRUEMAN, Sir Arthur (1954) *The Coalfields of Great Britain*, Arnold, London, pp. 52-77, 167-198.

TUMMAVUORI J. and AHO M. (1980) Ion-exchange properties of peat. Part I. The adsorption of some divalent metal ions on the peat. *Suo*, 31, 45-51, *Chem. Abs. 1980*, 93, No. 156349.

VERHEYEN T.V. (1982) *Structural Characterization of Selected Australian Coals*, Ph.D. Thesis, University of Melbourne.

VERHEYEN T.V. and JOHNS R.B. (1981) Structural investigations of Australian coals I: A characterization of Victorian brown coal lithotypes and their kerogen and humic acid fractions by I.R. spectroscopy. *Geochim. Cosmochim. Acta*, 45, 1899-1908.

VERHEYEN T.V. and JOHNS R.B. (1982) Structural investigations of Australian coals III: A C 13 nmr study on the effects of variation in rank on coal humic acids. *Geochim. Cosmochim. Acta*, 46, 2061-2068.

VERHEYEN T.V., JOHNS R.B., BRYSON R.L., MACIEL G.E. and BLACKBURN D.T.
(1984) A spectroscopic investigation of the banding of lithotypes occurring in Victorian brown coal seams. *Fuel*, 63, 1629-1635.

WEIMER R.J. (1977) Stratigraphy and tectonics of western coals. In *Geology of Rocky Mountain Coal, A Symposium*. Ed. D. Keith Murray, *Colorado Geological Survey, Resource Series 1*, pp. 9-27.

WILSON M.A., VASSALLO A.U. and RUSSELL N.J. (1983a) Exploitation of relaxation in cross-polarization nuclear magnetic resonance spectroscopy of fossil fuels. *Org. Geochem.*, 5, 35-42.

WILSON M.A., VERHEYEN V., VASSAELLO A.M., HILL R.S. and PERRY G.J. (1987)
Selective loss of carbohydrate from plant remains during coalification. *Org. Geochem.*, *11*, 265-271.

YOUTCHEFF J.S., GIVEN P.H., BASET, Z. and SUNDARAM M. (1983) The mode of association of alkanes with coals. *Org. Geochem.*, 5, 157-164.

APPENDIX

Inevitably, significant new papers appear during the completion of any review, and one has to cut off citations somewhere. However, three specially important works appeared after the final revision of this paper was completed, and it has been felt necessary to note them briefly here.

The volume edited by Scott (1987) has a number of excellent chapters on the biology and ecology of peat deposits, as they relate to coal formation. The consensus of most (though not all) of the authors is that Carboniferous coals originated in subtropical climatic zones in raised bogs. Even the existence of subtropical raised bogs is a novel concept to me, though in fact some exemplars are to be found in the Far East today. The reasoning of these authors is that the clastic sediments in a marsh or swamp containing detritus washed in from surrounding rocks, will have a high mineral matter content, and coals formed from them by dewatering and compression will be impossibly rich in minerals. Therefore raised bogs with no input of water-borne nutrients are more likely sources.

Personally, I find this argument weak. It involves the common confusion that arises from referring to the amount of inorganics in peat or coal as the ash content, ignoring the fact that in peat much of the ash-forming constituents are cations bound to organic matter and fairly easily leached, rather than discrete mineral phases. However, the idea certainly cannot be rejected at the present time, and the papers contain some highly thought-provoking discussions.

The studies of Stout, Boon and Spackman on fresh, peatified and coalified woods have been referred to frequently in the main part of this article. Much of this work, including extensive use of Curie-point pyrolysis/MS and pyrolysis/GC/MS, has now appeared (Stout et al., 1988). I feel that this remarkable paper represents a major advance in studies of coal origins.

The value of the pyrolysis techniques is shown in an important paper comparing fresh and buried woods (Saiz-Jiminez et al., 1987).

THE NATURE AND ORIGINS OF COAL MACERALS

Peter H. Given †
(formerly) The Pennsylvania State University
College of Earth and Mineral Sciences
University Park, PA 16802

Gary R. Dyrkacz
Argonne National Laboratory
Chemistry Division
9700 South Cass Avenue
Argonne, Illinois 60439 USA

ABSTRACT. The definition and nature of coal macerals is presented in terms of some current research. Recent work on the nature of several precursors of the liptinite maceral groups are discussed and compared to the corresponding liptinite macerals. The importance of separating macerals is considered, and some recent results using the density gradient centrifugation method of maceral separation is briefly reviewed. It is demonstrated that vitrinites are complex mixtures, not simple macerals.

1. INTRODUCTION : WHAT IS A MACERAL?

(i) Definitions

The word "maceral" was introduced into coal science by Dr. Marie Stopes, who defined it as "...a microscopically discernible constituent of coal." This seems clear, but presents difficulties when one starts to think about it. For example, discernible by what microscopic technique? What are the constituents and why should one want to discern them?

Y. Yürüm (ed.), New Trends in Coal Science, 53–72.

(ii) Significance of Definitions

When one begins to consider what is known about macerals, it becomes clear that, conceptually, different macerals represent different kinds of stuff. Thus the vitrinite macerals are derived from the half dozen or so polymers that constitute the cell walls of woody tissue, which have been differentially degraded before burial; the different degrees and modes of degradation, and the resulting heterogeneity, were discussed in the previous paper. On the other hand, the maceral sporinite is derived from the outer walls of spores and pollen grains, which largely retain their original morphology; the outer walls, or exines, are composed of a single polymer, probably little altered in peat and the early stages of burial.

As a third example, consider the maceral suberinite. The corky periderm (inner and outer "bark") of trees contains ordinary thick-walled lignified cells, like those in secondary xylem ("wood"), except that one of the layers of the secondary wall (S_2) consists of a mixture of soluble ester waxes and an insoluble highly aliphatic polymer known as suberin, instead of the cellulose/hemicellulose/lignin mixture found in xylem. This suberin layer can be distinguished by optical microscopy only when differential staining techniques are used, but it is clearly seen, with a stratified appearance, in the electron microscope. It is this woody tissue with the suberin layer that is the presumed precursor of the suberinite maceral found in many Tertiary (40-70 m. years) and some Mesozoic (100-200 m. years) coals (Teichmüller, 1982, p. 268). Suberinite is distinguished in analysis presumably by a characteristic fluorescence color in blue light and by its low reflectance. Surely this is a maceral in a sense different from what is meant when the term is applied to vitrinites or sporinite. Spackman (personal communication) prefers to describe coal constituents that are clearly derived from plant organs as "phyterals."

An important review on the origins of macerals should be noted here: that by Dr. Marlies Teichmüller (the classification of macerals shown in Table 5 of the preceding paper is largely based on her work). This review is comprehensive and lavishly illustrated by photomicrographs, but with not as much organic geochemistry as some would wish.

(iii) The Heterogeneity of Maceral Groups

Most chemists who have some awareness of coal petrography seem content to suppose that coal is composed of three macerals: vitrinite, liptinite (or exinite) and inertinite. In fact, each of these classifications is a maceral group in which many constituents can be seen under the microscope. Chemists are disinclined to accept the fact that materials of different texture or shade of orange in transmitted light are necessarily different chemically. They would prefer to have a bottleful of each of the alleged constituents on which to make appropriate chemical tests. The idea that vitrinite is "homogeneous" is a popular misconception of many chemists, based on a term used by petrographers to describe the often featureless appearance of vitrinite examined in reflected light; this "homogeneity" is in fact an artifact of the way coal is prepared and examined by reflected light microscopy. Electron microscopy and oxidative etching experiments on coal samples show that vitrinite is rarely homogeneous. Etched samples can reveal a wealth of information about the original and degraded plant structures and give, at the very least, an idea of the physical heterogeneity of a sample (e.g., Kröger, 1964). Precisely how to make effective physical separations of macerals always has been a major experimental problem. The primary maceral groups do tend to have appreciably different mean densities and hardnesses, so that stage crushing, particle size separation, and differential flotation of coal in liquids of different densities can separate the groups from one another, as long as all maceral group concentrations are reasonably high.

This simple-minded component classification is, in fact, woefully inadequate. Table 1 (repeated for convenience from Table 5 of the preceding paper) represents a classification that can be strongly defended, although it cannot pretend to be exhaustive. The density fractionation outlined above certainly cannot afford the kind of resolution Table 1 calls for. The recently devised density gradient separation procedure (discussed below) is capable of resolving the maceral groups into several fractions, though at present the fractions cannot be equated to the macerals shown in Table 1.

As far as we know, no concentrates of individual huminite macerals have ever been prepared. There has been some preliminary density gradient separation of Victorian brown coal lithotypes, but not in quantity. (See Section 3). Taylor et al. (1981, 1982, 1983) studied the huminite macerals in the Yallourn brown coal from the Latrobe Valley in Australia using transmission electron microscopy. They concluded that the huminite macerals listed in Table 1 do, in fact, differ in their characteristics, and also that attrinite and densinite are themselves quite heterogeneous. Nomura (1982) compared the liquefaction of a number of Australian brown coals of differing petrographic make-up, and concluded that conversion was proportional to the content of densinite in the coal. He also found the highest yield of alkanes from densinite-rich samples, and believed that the alkanes had been physically occluded. Thus, some inferences can be drawn about huminite macerals without prior separation.

Moreover, and more important for present purposes, is the fact that present understanding of processes in peat provides very plausible reasons why a series of huminite macerals, differing in chemistry as well as the degree of preservation of cellular structure, is to be expected (see preceding paper). It is true that at present there is no established means of physically separating huminite macerals or of characterizing their chemistry in situ without physical separation. As we shall see, it may be possible to segregate some of the vitrinite macerals derived from the huminites, but at a cost. No doubt chemists will have to continue to use heterogeneous sets of maceral groups, or even whole coals. But at least they should clearly realize that any statement they make about a "vitrinite," and still more about a whole

Table 1 Classification of Macerals in Coals[a]

Brown coals and lignites				Bituminous		
Maceral group	Maceral subgroup	Maceral	Maceral type	Maceral type	Maceral	Maceral group
HUMINITE	Humotelinite	Textinite				VITRINITE
		Ulminite	Texto-ulminite	Telinite 1	Telinite	
			Eu-ulminite	Telinite 2		
	Humodetrinite	Attrinite			Vitrodetrinite	
		Densinite		Desmocollinite		
	Humocollinite	Gelinite	Levigelinite	Telocollinite	Collinite	
		Corpohuminite	Porigelinite, Phlobaphenite	Gelocollinite		
			Pseudo-phlobaphenite	Corpocollinite		

Coals of all ranks

Maceral group	Maceral
Liptinite	Sporinite
	Cutinite
	Resinite
	Suberinite
	Alginite
	Liptodetrinite
	Bituminite
	Fluorinite
	Exudatinite
Inertinite	Fusinite
	Semifusinite
	Macrinite
	Micrinite
	Sclerotinite
	Inertodetrinite

[a] International Committee for Coal Petrology (1963, 1971, 1975).

coal, actually represents an <u>average</u> property for a quite diverse set of materials. The next section emphasizes this point even more strongly.

(iv) Multicomponent Ideas of Maceral Make-up

Various experiments performed on whole coals lead to the conclusion that "coal" consists largely of a cross-linked, three-dimensional macromolecular network (Green and Larsen, 1984; Brenner, 1983, 1984). Since the coals studied were vitrinite-rich, the statement presumably refers to the mix of vitrinitic macerals present. In the light of the discussion in the preceding paper, this statement should be re-phrased as follows: "The vitrinitic macerals in a coal contain a mix of related but somewhat different cross-linked macromolecular network structures." At any rate, it should not be supposed that the vitrinitic material in any coal contains a single macromolecular network.

It is well known that a complex mixture of relatively small molecules can be extracted from coals with various solvents. The yields tend to be greater from vitrinitic macerals than from liptinitic (unless resinite is an important constituent) or inertinitic macerals (Given, 1984). Jurkiewicz et al. (1982) interpreted the results of pulsed free-electron decay ^{1}H nmr experiments on coals swollen with pyridine-d_5 as indicating that there were larger amounts of relatively small molecules in coals than could be readily extracted by solvents (see also Marzec, 1986). This view was largely confirmed for a set of 23 coals by Kamienski et al. (1987), using essentially the same method. However, they found three rather than two populations of protons having differing levels of rotational mobility. The protons of highest mobility were partly in molecules extracted by the pyridine-d_5 and partly in molecules still occluded in the coal. It was suggested that some of the protons of intermediate mobility might also be present in relatively small molecules which are clathrated into cavities or imperfections in the macromolecular network. By "clathrated" it is not the intention here to consider molecules as totally encaged and therefore totally unextractable. This clathration is to be regarded in a kinetic light, that is, some molecules are readily extractable, some only on long exposure or at elevated temperatures, and some are released so slowly at fairly low temperatures that they are not seen at all (except perhaps under liquefaction conditions). The quantitative interpretation of how protons of differing mobility are to be assigned to various types of molecule is not unambiguous; however, perhaps about 20-30% of any coal of up to about 86% C dmmf consists of the clathrated component.

The study of homologous series of long chain alkyl aromatics in a variety of extracts of one coal, by Singleton et al. (1987), using tandem MS analysis, confirms the concept that a given molecule or size of molecule clathrated in one coal can be found in cavities providing widely differing ease of escape.

It would seem that the clathrated component of coal must be regarded as part of a maceral or group of macerals. Its presence will affect the reflectance and density of the solid, and probably also its fluorescence. But chemical analyses, FTIR and ^{13}C nmr spectra, etc., of coals will represent weighted means for the composite of the two types of component.

Thus, vitrinitic macerals cannot be treated as entirely macromolecular, and we re-emphasize that the vitrinite group in any coal must be regarded as quite heterogeneous. As Given (1984) has already pointed out, trying to represent a heterogeneous mixture by a single average molecular structure is not a very useful exercise.

2. WHAT IS A COAL LITHOTYPE?

Coal lithotypes were first distinguished by Stopes and defined as "banded constituents of coals." They were characterized or recognized by different types or degree of lustré, in hand specimen, and by the type of fracture.

The word "lithotype" means rock type, a rock being an assembly of minerals. Spackman (personal communication) objects to vitrain being termed a lithotype, since it is not an assembly of macerals; moreover, fusain occurs in lenses rather than bands. However, clarain and durain are indeed banded assemblies of macerals.

The question could be asked: "Of what value is a classification based on lustre and fracture?" It could be of value in describing coal seams, if each lithotype represented a characteristic assembly of macerals, with similar assemblies being found in many coals. Clarain, as defined by Stopes, is the major material found in most seams. This material is a characteristic assembly of macerals, in the sense that it contains sporinite, micrinite, fusinite particles, and mineral grains dispersed in a structure-less matrix of vitrinitic macerals. The total vitrinite content of clarain is likely to be in the range of 50-95%.

Durain, as defined by Stopes, seems to have a very wide range of maceral distributions, if U.S. as well as European coals are considered. In European coals of Carboniferous age, it is useful to distinguish black durains, which commonly contain up to 70% liptinites (chiefly sporinite) from grey durains, in which micrinite may account for 40% or more of the lithotype. However, this distinction is meaningless with Carboniferous coals found in the U.S., and may or may not be useful elsewhere.

Thus, on balance, the lithotype classification is of limited value.

The related term "microlithotype" suffers from a similar cryptic definition problem, but it is defined in terms of the type and content of macerals composing a microlithotype band. To be defined as a microlithotype, a band must be at least 50 microns by 50 microns in size. For example, by definition, a clarite microlithotype can have a concentration of exinite ranging from 5% to 90% exinite. As bad as this definition system appears to be, there is one aspect in the concept of microlithotypes that may be significant: the microlithotypes represent different paleoenvironments. In many cases, it is not at all clear just what microlithotypes evolved from what types of environments. Nevertheless, the idea of a microlithotype recognizes the potential for a relationship between different environments and certain maceral assembleges. The concept of a microlithotype may then be necessary for a comprehensive understanding of the chemical variations of macerals in coal.

3. THE ORIGIN OF SOME MACERALS

Most coals are comprised primarily of huminite or vitrinite macerals. Therefore, a discussion of their origins is unavoidable in reviewing the origin of coals (see the preceding paper). In this section, there will be a discussion of some of the liptinites, since some interesting new results have appeared recently. Moreover, liptinites, in many cases, originate from fewer biopolymer precursors than do huminites or inertinites. Thus, these macerals may be more valuable in understanding the detailed chemical processes of coalification.

(i) Cutinite

The waxy cuticle on leaves, fruits and young shoots has an outer layer of soluble hydrocarbons and esters, on top of an insoluble polyester polymer of C_{16} and C_{18} -hydroxy fatty acids which is called cutin. Recently, another insoluble highly aliphatic polymer has been found in the cuticle of certain Angiosperm leaves and fruits (Nip et al, 1986) by pyrolysis/GC/MS studies. The detailed structure of this polymer is not yet known, nor how widely it is distributed. But together with cutin, the polymer could be a significant contributor to the cutinite maceral found in some coals.

Not very much detailed chemistry is really known about cutinite itself. The reason is that cutinite is relatively rare in most coals, and is therefore, not easily obtainable; or it is mixed with other liptinites. There are, however, several unusual coals which contain predominantly fossilized leaf cuticles. These "paper coals" consist of very high concentrations of finely laminated cuticles which can often be peeled away from each other. Typically, cutinites have a very low extractability in chloroform or benzene, and a high H/C ratio of from about 1.0 to 1.5 (Soos, I., 1966; Neavel and Miller, 1960).

Some recent preliminary work shows that the cutinite from both a Victorian brown coal and from an Indiana paper coal are quite different from modern cutin (Anderson et al., 1987). Both solid probe ^{13}C nmr and IR show no indications of the original ester linkages. Carboxylic acid and hydroxyl groups are present in both samples, but are more prevalent in the Victorian brown coal cutinite. Both cutinites are highly aliphatic, with an f_a of about 0.1 for the brown coal, and 0.2

for the Indiana coal. Allan (1975) found that oxidation of cutinite with permanganate produced monobasic acids that were almost exclusively C_{16} and C_{18}. The majority of materials found with this procedure were dibasic acids, from C_6 to C_{15} with C_8 predominating. Several benzene carboxylic acids were also produced; the most common was 1,2,4-benzene-tricarboxylic acid. The low extractability, coupled with the fact that cutinites retain their morphology, implies that the cuticles remain extensively cross-linked. However, the lack of ester groups suggests that the cross-links now holding the structure together have formed as a consequence of diagenesis, and probably were created before or during the time the ester groups were being lost.

(ii) Alginite and Bituminite

So-called boghead coals are known in various parts of the world. They are thought to be formed in ponds or small lakes deep enough to preclude the growth of vascular plants. Plant spores may be borne in by the wind, and leaves, twigs, etc., and may be washed in by surface water, but the principal source of organic matter is the algal matter in the water. Thus, boghead coals have algae as an important contributor.

It has been believed for a long time that the alginite macerals found in these coals were derived from the green colonial alga, *Botryococcus braunii* (Teichmüller, 1982). This organism is unusual in that it contains high concentrations of saturated and polyunsaturated hydrocarbons; the latter are known to form a rubber-like sheet when a lagoon in which the alga lived has dried out. Petrographers recognize alginite by the fossilized colonial morphology, and one wonders how liquid hydrocarbons inside the cells could preserve this external morphology. The cell walls of algae have been thought to be composed of polysaccharides, which are unlikely to be preserved. However, Largeau and his collaborators (1984, 1986) have recently showed that the cell walls of *B. braunii* contain an insoluble, resistant, highly aliphatic polymer. This polymer is, no doubt, able to preserve the cellular morphology as the alginite maceral. However, it is worth noting that direct comparison, by pyrolysis/GC/MS of the new cutin polymer discussed by Nip et al. and the algal cell wall polymer has shown that they are not identical.

Teichmüller (1982, p. 269-70) suggests that the bituminite maceral is derived from constituents of algae and/or bacteria. This maceral is said to be abundant in oil shales and oil source rocks, and to be common in sapropelic (subaquatic) coals. Bituminite occurs in the lignites of Germany and Japan, and in some bituminous coals. It has recently been studied in a number of U.S. coals (Given et al., 1985; Mudamburi and Given, 1985; Shadle et al., 1986; Nip et al, 1985). The "King Cannel" coal from southwest Utah proved to have 55% bituminite (Given et al, 1985), and five other bituminite-rich samples had somewhat less. Bituminite is evidently rich in hydrogen (8% or more?) and is highly aliphatic. The Utah coal also contained about 5% alginite, and it is felt that bituminite could very well represent disrupted or degraded colonial cell walls of *B. braunii*.

(iii) Sporinite

Spores and pollen grains have an outer wall or exine of "sporopollenin," an inner wall or intine of a polysaccharide mixture, and a cell membrane. The interior is largely triglycerides. Sporinite macerals are derived from the exine (hence the word "exinite"), and so one wants to know the chemical structure of sporopollenin. For some time, geochemists and biochemists have accepted the view of Brooks and Shaw (1978) that sporopollenin is an oxidative co-polymer of two carotenoid pigments, one of them esterified with palmitic acid. Thus, a major part of the structure of sporopollenin was thought to be isoprenoid, in which case one C atom in every five should be due to a methyl group.

The matter has recently been re-examined by Given, Ryan, Davidonis, Painter, and Pugmire (1985, unpublished), who used a new procedure due to Loewus et al. (1985) for isolating the exine from the intine and cell contents, which requires only very mild conditions. FTIR and ^{13}C nmr spectra make it clear that a predominately isoprenoid structure for this material is not feasible. In fact, sporopollenin appears to contain a highly aliphatic unbranched, or lightly branched, polymer, condensed or cross-linked with a lesser amount (30%) of a polysaccharide. Similar conclusions have been reached by Dyrkacz et al. (1987a), working with *Lycopodium clavatum* spores. Infrared, nmr, and oxidation with nitric acid, permanganate, or ruthenium tetroxide do not support an isoprenoid structure. In addition, nmr results have shown that the harsh acid treatment used in the isolation of sporopollenin can generate new sp^2 carbon centers. This latter observation throws under suspicion much of the earlier depolymerization work. Sporopollenin is probably not a single substance, and varies somewhat from plant to plant. Moreover, sporopollenin is defined as the insoluble residue left after organic solvent extraction, followed by treatment with base and strong acid. This broad phenomenological definition also complicates the study of sporopollenin. In addition, the conditions of pre-treatment used to obtain sporopollenin will alter one's definition of the material. At this point, we are more certain about what sporopollenin is not than what it is.

The chemistry of sporinite is even less well understood than that of sporopollenin. Typically, observable sporinites have H/C ratios of between 0.85 and 1.2, and O/C ratios of between 0.05 and 0.25. The "observable" qualification is necessary because as rank increases, the spores in coal become similar in reflectance to vitrinite; in fact, in the low volatile bituminous coal, they usually can no longer be seen. The fraction of aromatic carbon of most sporinites is lower than that of the vitrinites, and ranges from 0.4 - 0.6; conversely, vitrinites have values of 0.65 - 0.85. The few oxidation studies that have been done on sporinites also indicate a larger amount of aliphatic material, when compared with results for the vitrinite in the same coal sample (Allan, 1975; Winans, 1981). Recently, a sporinite has been subjected to $RuO_4/NaIO_4$ oxidation (Choi, 1986). This oxidation system is particularly useful for examining the aliphatic species in coal. Although benzene type ring structures are usually destroyed using this procedure, multi-ring aromatics yield aromatic oxidation products. The sporinite in a high volatile bituminous coal was found to contain a larger amount of longer chain alkyl diacids than the corresponding vitrinite. Interestingly, the type and content of aromatic carboxylic acids found in the sporinite and vitrinite were quite similar except a high amount of methyl substitution was found in the vitrinite. The implication from this study is that if one considers the alkyl groups as crosslinking agents for the aromatic moieties, then sporinite and vitrinite may be structurally alike; however, the sporinite has longer aliphatic crosslinking chains.

(iv) Sources of Alkyl Chains in Coals

Calkins has been interested in hydrocarbon chains in coals as possible commercial sources of light hydrocarbons, and observed unexpectedly large amounts produced by flash pyrolysis (Calkins et al., 1984). He therefore participated in a search for the origins of polymethylene moieties in coal (Calkins and Spackman, 1986). Various organs of a number of plants, several peats, some whole coals, and a few maceral concentrates were tested for their polymethylene content. The plant materials showed varying contents. The contents in coals were in the range 1-10%, with liptinite macerals having the highest amounts.

In light of this work, it seems of interest to list the entities in coals that contain long aliphatic chains. They include sporinite, cutinite, alginite, bituminite, and suberinite macerals. A homologous series of alkanes is largely clathrated in the macromolecular network of vitrinitic macerals (Youtcheff et al., 1981); this is also true for various homologous series of long chain alkyl aromatics (Baset et al., 1982; Mudamburi and Given, 1985b; Singleton et al., 1987). Peats contain an insoluble, highly aliphatic polymer (see preceding paper). Finally, trifluoroperoxyacetic acid oxidation of the asphaltenes from liquefaction of 26 bituminous coals yielded homologous series of long chain fatty acids, presumably derived from alkyl aromatic structures that had been part of the macromolecular network (Shadle et al., 1987).

4. PROPERTIES OF MACERAL CONCENTRATES OBTAINED BY DENSITY GRADIENT CENTRIFUGATION

The traditional approach to coal chemistry research, using entire whole coals, has resulted in misleading information and inherently complicates the explanation of results. What appears to be the obvious answer is that some form of separation of coal must be done. However, the effort devoted to separating macerals for chemical characterization of coal has been very minor, considering the voluminous literature existing on coal research.

One simple technique that has been used to obtain more pure maceral groups is handpicking, usually using coal samples that have an abundant concentration of the desired macerals. This requires that the desired macerals must be large enough, so they can be removed with a probe. It is very difficult to obtain pure macerals in this way, especially in the case of the smaller macerals. Another potential problem that has been ignored is the possibility that those macerals which normally are rare in most coals, but are concentrated in certain special coals used for handpicking, may not be representative of the more highly dispersed macerals. The fact that they are concentrated at all in coal samples suggests very special paleoenvironments. For example, cannel coals (with typically high spore contents) are formed in deep water environments that have quite different chemical conditions from a forest peat.

The densities of the three maceral groups are quite different, and most maceral separations use this property to separate the macerals. One finds that liptinites have the lowest density, followed by vitrinites, and then inertinites. Traditionally, macerals have been separated by sink-float techniques. This technique is operationally very simple, and consists of separating a coal at a specified density, removing the floats and sinks, and separating these at higher and lower densities respectively, until all the material has been separated. This technique, if properly used, can isolate very pure macerals, with the ability to liberate the macerals becoming the limiting factor in the final separation. Unfortunately, sink/float is a very time-consuming operation, often requiring multiple cycles at a single density to obtain a pure material. Dyrkacz et al. (1982, 1981) have advocated the use of surfactants to improve the separation. All maceral sink/float separations benefit from this, but the yield and purity of maceral fractions from single sink/float passes are still not always optimal, unless the coal concentration is very dilute.

Another density technique, density gradient centrifugation (DGC), has been used in recent years by Dyrkacz and co-workers to obtain quite pure maceral groups. The DGC techniques employed are similar to those used in the biosciences to separate cells and cellular components. There are several variations of this technique being used to separate the macerals, but they all have several features in common (Dyrkacz and Horwitz, 1982). In this procedure, the coal is first fine ground in a fluid energy mill to a top size of approximately 6 microns, and then chemically demineralized. The fine particle size is necessary to ensure that most of the macerals will be liberated from one another. Next, the coal is applied to an aqueous CsCl surfactant-containing density gradient, and, after centrifugation, the density-separated macerals can be isolated by fractionating the gradient. Unlike in a sink/float separation, the macerals literally appear as suspended bands within the centrifuge tube, because the density gradient encompasses the entire range of coal maceral densities (1.0 - 1.5g cm^{-3}).

Density gradient separation methods have several distinct advantages over the other available methods: 1.) For coal ranks ranging from brown coals to anthracites, the separation efficiency and

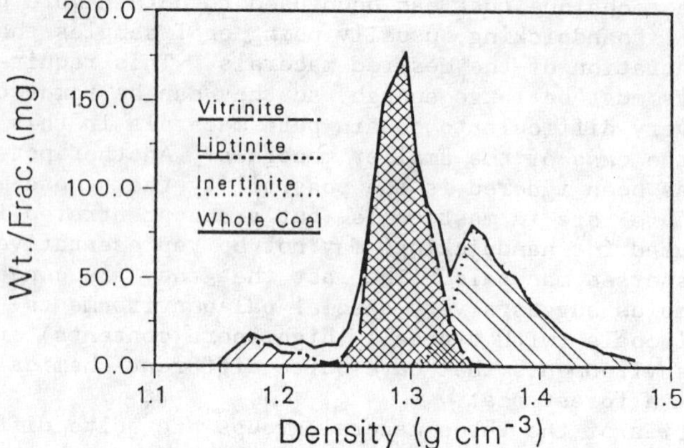

Figure 1. DGC separation of PSOC-732 (HVA bit., W. Virginia). Shaded distributions are those for monomaceral particles.

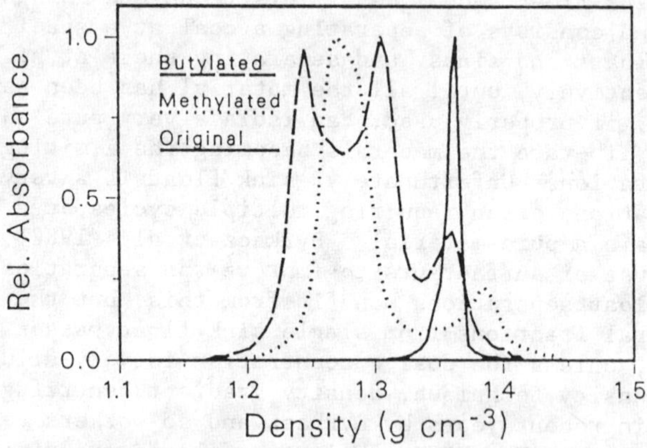

Figure 2. Comparison of alkylated and non-alkylated inertinite fractions from PSOC-732.

resolving power is very high. It would be possible to easily separate coal samples or macerals routinely with as little as 0.003g cm^{-3} density difference. Routine separations generate about forty usable fractions which would be comparable to doing forty separate sink/float cycles. Cross-contamination between density fractions is usually on the order of only a few percent. 2.) The technique immediately shows the entire density distribution of a coal. The effect of any changes made to a coal that would also affect the coal density can be examined in a few hours. Figure 1 is a typical DGC of a high volatile coal from West Virginia. The data are plotted as weight of coal found in each equi-volume density fraction versus the average CsCl density of that fraction. 3.) The overall density pattern itself provides some idea of the heterogeneity of the coal. The heterogeneity indicated by a density distribution could be due to both physical and chemical properties, because the densities are derived from a complex aqueous media. However, chemical heterogeneity appears to be the primary factor dictating the distributions. Thus, chemical heterogeneity parallels density heterogeneity. In this way, DGC can serve a dual role for both coal maceral separation and coal maceral characterization.

On the negative side, the price one pays for the high resolution density separation of macerals is that relatively small amounts of pure maceral groups are produced when compared to sink/float techniques. Sink/float coupled with DGC can be used to alleviate this problem to some extent, but at a cost in time for a separation. Secondly, the cost of the equipment can be relatively high; however, any good biochemistry department likely will have most of the necessary equipment and expertise for DGC procedures. A third problem is the fact that the fine grinding completely destroys the morphology of the material. Only the fluorescence or reflectivity of the particles remain for identification, which restricts final fraction analysis to maceral group identification only, though the initial analysis before fractionation is a useful guide.

In addition to the overall density distribution data shown in Figure 1, the maceral data obtained by petrographic analysis of selected density fractions are presented. These maceral data represents particles which contained under 10% contamination of any other maceral group. These particles are considered monomaceral or pure maceral particles. Two important facts emerge with this additional information. First, the maceral groups are all broad distribution bands, even for coals which contain greater than 95% vitrinite. Second, there are particle density overlaps between adjacent maceral groups, with vitrinite and inertinite exhibiting the highest degree of overlap (Dyrkacz, Bloomquist and Ruscic, 1984a). These results have been found to be true for over forty coals that have been separated. The natural overlap of the maceral groups places limitations on the use of any simple series of density separations to completely resolve maceral groups.

As has been previously mentioned, the broad character of the maceral bands is indicative of a broad range of chemistry being present. Support for this conclusion has been shown by elemental data analyses for a number of separated coals (Dyrkacz, Bloomquist and Ruscic, 1984a). Only maceral fractions that were at least 90% pure were used in this study. Except in the highest rank coals, the H/C ratio for these samples follows a monotonic decrease with increasing density, even within individual maceral groups. The O/C ratios follow a more complex pattern. The trends between maceral groups are clear: the O/C ratios decrease, in the order vitrinite>inertinite>exinite, for all the coals. Within individual maceral groups, for all the coal studied, no clear trends emerge.

More recently, Dyrkacz et al. (1987b) have used DGC to examine vitrinites obtained from three sets of high volatile U.S. bituminous coal microlithotypes. Each set represented microlithotypes physically separated by only a small vertical distance within the same seam. Indications from the literature (Leighton, 1959; Brown, Cook and Taylor, 1964; Binder, Duffy and Given, 1963; Lapo, 1978; Shiboaka, Stephans and Russell, 1979; Hutton and Cook, 1980) suggest that the vitrinite from different microlithotypes can have different chemical properties. Unfortunately, in all of the earlier studies the vitrinites were not completely isolated from the other macerals, and thus there was an element of uncertainty in the conclusions. The high maceral resolving power of DGC was indeed able to confirm and extend the earlier reports. The coals ranged from 0.0% to 22% liptinite. It was found that above approximately 10% liptinite concentration, the vitrinite density distributions showed a shift to lower density and corresponding higher H/C ratio than that found for most of the vitrinites with less than 10% liptinite levels. Moreover, the shift to higher H/C ratio was true of the entire series of vitrinite density distributions. Extraction of several of the vitrinite samples with benzene/methanol did not appreciably change the results. Infrared spectroscopy of the extracted residues indicated a greater amount of aliphatic C-H stretch for the high liptinite vitrinite microlithotypes. However, there was one microlithotype studied which did not fit the trends. A vitrinite with only 4.8% liptinite showed the same lower density pattern and higher H/C ratio as the high liptinite coals. The only significant difference was that the infrared results showed a very high methyl stretching mode.

Because each of the sets of microlithotypes were from the same channel sample, each set had the same catagenic history. The differences observed in the vitrinites must be related to the original paleoenvironmental conditions. These differences can still be seen in high volatile bituminous coals. In addition, even within the same variety of microlithotype, quite different chemical characteristics can occur. Whether these differences are due to the same environmental conditions, but are derived from different types of plant material or from the same type of plant material under different environmental conditions, is not clear. This work reinforces the comment made earlier about the dangers of assuming that any given vitrinite is "reagent grade".

Some recent DGC work on Victorian brown coal huminites suggests that they also show similar variations (Anderson et al., 1987).

Another recent extension of the DGC technique to attempt to understand coal and maceral heterogeneity made use of a "two-dimensional" approach (Choi, Dyrkacz and Stock, 1987). In this case, macerals were first separated in a density gradient. Selected maceral fractions were O-alkylated, and then again separated in a density gradient. The idea behind this approach was to take advantage of the chemical differences between maceral groups and individual macerals to promote further density resolution. The O-alkylation of maceral fractions from two coals with either butyl or methyl iodide and tetrabutylammonium hydroxide was used. Figure 2 shows data from a single narrow density inertinite fraction (>98% inertinite) of one of the coals. The butylated inertinite is particularly interesting, showing a broad new density pattern with three new bands. This was one of the more spectacular changes noted for this coal. Maceral density fractions of pure liptinite and pure vitrinite showed either broadened peaks or additional new minor peaks. A higher density inertinite peak was also butylated, and did not show the presence of the middle band. Only two bands were observed, with a density shift appropriate to the higher density of the fraction. In the case of the former inertinite fraction, the three bands were isolated. FTIR spectroscopy showed the presence of decreasing aliphatic C-H stretch with increasing density of the new bands, which would be expected if more butyl groups were being added. From the elemental data, the amount of alkyl groups added per 100 carbons for each band was 3.0, 1.7 and 0.3. The fact that three distinct bands appeared means that there must be three distinct chemical classes of particles. If this were not true, there would only be a shift in the overall density distribution from the original density. Also, the fact that one inertinite band seems to have been lost at a higher density suggests that the reactive particles overlap in the original density fraction, but very probably do not have the same density distribution. Even in the case of the other alkylated macerals, the broader bands imply that there is a small amount of increased density resolution due to alkylation. It is impossible to ascribe maceral names to the bands which have separated. The DGC technique can only allow classification into maceral groups, because of the loss of morphological information on fine grinding. The separated material could be different inertinite macerals as we recognize them in a coal sample, or it could be variations within one maceral, such as semi-fusinite.

This work may herald a new and more refined approach to understanding the chemical and physical heterogeneity of coal. Extension of this work to different chemical reactions, coupled with DGC, hopefully will offer a clearer idea of not only the heterogeneity of coal, but of coal macerals as well. If necessary, tandem cycles of different reactions and separations could be used in a cascade. This approach certainly will not lead to a quick understanding of coal, but as least its methodology has the potential for doing so, when compared to the shotgun approach that has been the mainstay of coal research up until now.

SUMMARY

What we have tried to do in this article is to show some of the problems that exist in understanding what coal macerals are. Not only is the comprehension of what coal is very difficult due to the natural structural complexity of coal itself, but this comprehension can also be hindered by a lack of realization of the true nature of coal macerals. Effective coal research requires a broad base of knowledge of many areas of natural science, not the least of which is the nature of the original materials that are incorporated into coal.

ACKNOWLEDGMENTS

Gary Dyrkacz wishes to acknowledge support for his work under the auspices of the Office of Basic Energy Sciences, Division of Chemical Sciences, U.S. Department of Energy, under contract number W-31-109-ENG-38.

REFERENCES

Allan, J. (1975), Natural and Artificial Diagenesis of Coal Macerals, Ph.D. Thesis, University of Newcastle Upon Tyne.

Anderson, K., Dyrkacz, G. R. and Johns, B. (1987), Work in progress.

Baset, Z. H., Pancirov, R. J. and Ashe, T. R. (1980), Organic Compounds in Coal: Structure and Origins, in Advances in Organic Geochemistry 1979, ed. A.G. Douglas and J. R. Maxwell, Pergamon, Oxford, 619-630.

Binder, G. R., Duffy, L. J. and Given, P. H. (1963), Selective Chemical Reactions for the Study of Coal Macerals, Amer. Chem. Soc. Preprint, Div. Fuel Chem., 7, 145-153.

Brenner, D. (1983c), In situ Microscopic Studies of the Solvent-Swelling of Polished Surfaces of Coal, Fuel, 62, 1347-1350.

Brenner, D. (1984), Microscopic in situ Studies of the Solvent-Induced Swelling of Thin Sections of Coal, Fuel, 63, 1324-1329 (Sept.).

Brooks, J. and Shaw, G. (1978), Sporopollenin: A Review of Its Chemistry, Palaeobiochemistry and Geochemistry, Grana, 17, 91-97.

Brown, H. R., Cook, A. C. and Taylor, G. H. (1964), Variations in the Properties of Vitrinite in Isometamorphic Coal, Fuel, **43**, 111-124.

Calkins, W. H. and Spackman, W. (1986), Tracing the Origins of Polymethylene Moieties in Coal, Int. J. Coal Geol., **6**, 1-19.

Calkins, W. H., Hovsepian, B. K., Dyrkacz, G. R., Bloomquist, C. A. A. and Ruscic, L. (1984), Coal Flash Pyrolysis: 4. Polymethylene Moieties in Coal Macerals, Fuel, **63**, 1226-1230.

Choi, C. Y. (1986), Investigation of the Chemical Structure and Reactivity of Bituminous Coal Macerals, Ph.D. Thesis, University of Chicago.

Choi, C. Y., Dyrkacz, G. R. and Stock, L. M. (1987), Density Separation of Alkylated Coal Macerals, Energy & Fuels, **1**, 280-286.

Dyrkacz, G. R., Hayatsu, R., Stock, L. M., and Botto, R. (1987a), Unpublished results.

Dyrkacz, G. R., Ruscic, L., Bloomquist, C. A. A. and Crelling, J. (1987b), In preparation.

Dyrkacz, G. R. and Horwitz, E. P. (1982), Separation of Coal Macerals, Fuel, **61**, 3-12.

Dyrkacz, G. R., Bloomquist, C. A. A. and Horwitz, E. P. (1981), Laboratory Scale Separation of Coal Macerals, Sep. Sci. Tech., **16**, 1571-1588.

Dyrkacz, G. R., Ruscic, L. and Bloomquist, C. A. A. (1984a), High-resolution Density Variations of Coal Macerals, Fuel, **63**, 1367-1373.

Dyrkacz, G. R., Ruscic, L. and Bloomquist, C. A. A. (1984b), Chemical Variations in Coal Macerals Separated by Density Gradient Centrifugation, Fuel, **63**, 1166-1173.

Given, P. H. (1984), An Essay on the Organic Geochemistry of Coal. In Coal Science, Vol. 3 ed., M. L. Gorbaty, J. W. Larsen and I. Wender, Academic Press, San Diego, pp. 63-252 and 339-341.

Given, P. H., Davis, A., Kuehn, D., Painter, P. C. and Spackman, W. (1985), A Multifaceted Study of a Cretaceous Coal with Algal Affinites. I. Provenance and Analyses of Coal Samples, Int. J. Coal Geol., **5**, 247-260.

Green, T. K. and Larsen, J. W. (1986), Coal Swelling in Binary Solvent Mixtures: Pyridine-Chlorobenzene and N,N-Dimethylaniline-Alcohol, Fuel, **63**, 1538-1542.

Hutton, A. C. and Cook, A. C. (1980), Influence of Alginite on the Reflectance of Vitrinite from Joadja, NSW, and Some Other Coals and Oil Shales Containing Alginite, Fuel, **59**, 711-714.

Jurkiewicz, A., Marzec, A. and Pislewski, N. (1982), Molecular Structure of Bituminous Coal Studied with Pulsed Nuclear Magnetic Resonance, Fuel, **61**, 647-650.

Kamienski, B., Pruski, M., Gerstein, B. C. and Given, P. H. (1987), Mobility of Hydrogen and Extractability in Coals: A Study by Pulsed NMR, Energy & Fuels, **1**, 45-50.

Kröger, C. (1964), Zur Struktur und Konstitution der Steinkohlen., Erdöl Kohle Erdgass Petrochemie, **17**, 802-812.

Lapo, A. V. (1978), Comparative characteristics of Vitrinites of Carboniferous Coals of the Ukraine and Jurassic Coals of Siberia, Fuel, **57**, 179-183.

Largeau, C., Casadevall, E., Kaduri, A. and Metzger, P. (1984), Formation of *Botryococcus*-Derived Kerogens--Comparative Study of Immature Torbanites and of the Extant Alga *Botryococcus braunii*. In Advances in Organic Geochemistry 1983, ed. P. A. Schenck, J. W. de Leeuw and G. W. M. Lijmbach, 327-332.

Largeau, C., Devenne, S., Casadevall, E., Kaduri, A. and Sellies, N. (1986), Pyrolysis of Immature Torbanite and of the Resistant Biopolymer (PRB A) Isolated from the Extant Alga *Botryococcus braunii*. Mechanism of formation and Structure of Torbanite. In Advances in Organic Geochemistry 1985, ed. D. Leythaeuser and R. Rullkötter, 1023-1032.

Leighton, L. H. (1959), The Variability of Vitrinites in British Coal Seams, Fuel, **38**, 155-164.

Loewus, F. A., Baldi, B. G., Franceschi, V. R., Meinert, L. D. and McCollum, J. J. (1985), Pollen Sporoplasts: Dissolution of Pollen Walls, Plant Physiol., **78**, 652-654.

Marzec, Anna (1986), Macromolecular and Molecular Model of Coal Structure, Fuel Proc. Technol., **14**, 39-46.

Mudamburi, Z. and Given, P. H. (1985a), Multifacetted Study of a Cretaceous Coal with Algal Affinites. II. Composition of Liquefaction Products, Org. Geochem., **8**, 221-232.

Mudamburi, Z. and Given, P. H. (1985b), Some Structural Features of Coals of Differing Maceral Distribution and Geological History, Org. Geochem., **8**, 441-453.

Neavel, R. C. and Miller, L. V. (1960), Properties of Cutinite, Fuel, **39**, 217-222.

Nip, M., de Leeuw, J. W., Schenck, P. A., Meuzelaar, H. L. C., Stout, S. A., Given, P. H. and Boon, J. J. (1985), Curie-point Mass Spectrometry, Curie-point Pyrolysis/Gas Chromatography/Mass Spectrometry and Fluorescence Microscopy as Analytical Tools for the Characterization of Two Uncommon Lignites, J. Anal. and Appl. Pyrolysis, In press.

Nip, M., Tegelaar, E. W., Brinkhuis, H., de Leeuw, J. W., Schenck, P. A. and Holloway, P. J. (1986), Analysis of Modern and Fossil Plant Cuticles by Curie Point Py-GC and Curie Point Py-GC-MS: Recognition of a New, Highly Aliphatic and Resistant Biopolymer. In Advances in Organic Geochemistry 1985, ed. D. Leythaeuser and J. Rullkötter, 769-778.

Nomura, M. (1982), A Study of Direct Liquefaction of Australian Brown Coal, Seisan To Gijutsu, **34** [4], 39-40; IEA Coal Abstracts 1983, 7 [8], 847.

Phillips, T. L. and Peppers, R. A. (1984), Changing Patterns of Pennsylvanian Coal Swamp Vegetation and Implications of Climatic Control on Coal Occurrence, Int. J. Coal Geol., **3**, 205-255.

Shibaoka, M., Stephans, F. and Russell, N. J. (1979), Microscopic Observations of the Swelling of a High Volatile Bituminous Coal in Response to Organic Solvents, Fuel, **58**, 515-522.

Shadle, L. J., Jones, A. D., Deno, N. C. and Given, P. H. (1986), Multifacetted Study of a Cretaceous Coal with Algal Affinites. III. Asphaltenes from Liquefaction and a Comparison with some Hydrogen-rich Texas Coals, Fuel, **65**, 611-620.

Shadle, L. J., Neill, P. H. and Given, P. H. (1987), The Dependence of Liquefaction Behavior on Coal Characteristics, 10. Structural Characteristics of a set of High-Sulphur Coals and the Asphaltenes Derived from them, Fuel, Submitted.

Singleton, K. E., Cooks, R. G., Wood, K. V., Rabinovich, A. and Given, P. H. (1987), Product Distributions and Their Relation with Coal Liquefaction Conditions, Fuel, **66**, 74-82.

Soos, I. (1966), Die Kennzahlen der Braunkohlen-Gemegteile. II. Kutinit, Acta Geologica Hung., **10**, 59-63.

Taylor, G. H., Shibaoka, M. and Liu, S. (1981), Vitrinite Macerals and Coal Utilization, In Proc. 1981 Int. Conf. Coal Sci., Düsseldorf, pp. 74-79.

Taylor, G. H., Shibaoka, M. and Liu, S. (1982), Characterization of Huminite Macerals, Fuel, **61**, 1197-1200.

Taylor, G. H., Liu, S. and Shibaoka, M. (1983), Huminite and Vitrinite Macerals at High Magnification, Proc. 1983 Int. Conf. Coal Sci., Pittsburgh, pp. 397-400.

Teichmüller, M. (1982), Origin of the Petrographic Consistuents of Coal. In Stach's Textbook of Coal Petrology, ed. E. Stach, M.-Th. Mackowski, M. Teichmuller, G. H. Taylor, D. Chandra and R. Teichmüller, translated by D. G. Murchison, Gebruder Borntraeger, Berlin, pp. 219-294.

Winans, R. E., Dyrkacz, G. R., McBeth, R. L., Scott, R. G. and Hayatsu, R. (1981), Characterization of Separated Coal Macerals, Proc. Int. Conf. Coal Sci., Düsseldorf, pp. 22-27.

Youtcheff, J. S., Given, P. H., Baset, Z. and Sundaram, M. (1983), The Mode of Association of Alkanes with Coals, Org. Geochem., 5, 157-164.

MACROMOLECULAR STRUCTURE OF COALS: STATUS AND OPPORTUNITIES

John W. Larsen
Lehigh University
Department of Chemistry
Bethlehem, Pennsylvania 18015

ABSTRACT. The cross-linked macromolecular structure of bituminous coals is described. Methods for its investigation and some of its properties are discussed.

1. INTRODUCTION

The aim of this review is to concisely summarize the state of knowledge of the macromolecular structure of coals, to point out problems with current models and areas in which our knowledge is incomplete, and to discuss the ways in which macromolecular structure ideas are being used in studies of coal reactivity. This is not an exhaustive review. Only those papers most directly relevant to the structural issues under consideration will be cited. This inevitably involves difficult choices and readers are cautioned that much good and useful work is not cited here. This article is also limited to coals and will not contain an introduction to or discussions of the relevant polymer chemistry. Numerous excellent texts exist and the reader who is puzzled by some aspects of the underlying polymer chemistry and physics is directed to these sources for enlightenment.[1,2]

Coals are extraordinarily complex heterogeneous substances, so complex that one must question the wisdom of applying to them models derived for homogeneous polymeric substances, especially quantitatively. First, it is necessary to point out that many very complex polymer systems are in commercial use, for example rubbers filled with carbon black and zinc oxide, and complaints about the application of quantitative macromolecular structural models to these systems are rare. Surely coals are much more heterogeneous and probably much more complex, but this certainly should not hinder the attempt to apply a number of structural models and treatments to coals. For without the attempt, we will never know what works and can be used. It is necessary to acknowledge explicitly that

Y. Yürüm (ed.), New Trends in Coal Science, 73–84.

coals are complex mixtures and that often we are looking at properties which are averaged over all of the components in that mixture. The test by which we shall judge the application of polymer models to coals is utility. Can the models be used to make predictions and can they provide understanding of phenomena in a clearer and more economical way than the alternatives?

Another issue is that of the essential correctness of the view that coals consist of macromolecular networks. Certainly the lignins from which much of coals are formed are themselves macromolecular networks and it seems likely that the coalification process results in alterations of this network, probably without degradation to a complex mixture of monomers. In their mechanical properties, in their thermodynamic properties, and in their reactions, coals behave like cross-linked macromolecular networks and so it is best to treat them as such and to apply the existing theories to this complex system as best we can.[3] If the theories are inadequate to deal with a substance as complex as coal, so be it. This is a comment on our theories, not on coal. Undoubtedly our approaches are enormously oversimplified. The complexity and heterogeneity of the material may be such that it is not worth progressing beyond rather simple models. The desires for simple, easily used predictive relationships and for very accurate structural models sometimes do conflict. It seems that in applying macromolecular structure models to coals, we are just now entering this stage of conflict and it will be interesting to see how things progress. While the complexity of the material and its variability will certainly impose constraints on the application of macromolecular structural models and calculations to coals, they should not prevent the attempt. Only experiment and study will reveal ultimately the limitations to which this approach is subject.

2. ORIGINS OF THE MACROMOLECULAR NETWORK STRUCTURE

One can find isolated statements that coal is polymeric and short references to the polymeric structure of coal throughout the early literature.[4] Some models of coal structure seem to have been macromolecular in concept without ever explicitly stating this. Credit for the origination of the idea that coals were macromolecular network structures and were best treated as such must go to the individual who argued explicitly for that structure and who first utilized that structural model in considerations of coal reactivity. That individual is D.W. Van Krevelen.[3] Early and still convincing arguments for the macromolecular network character of coals appear in his book. In a pair of early papers, he modeled the coalification process as the polymerization of a collection of monomers to give a cross-linked macromolecular system.[5,6] While the details of this process have since been questioned,[7] the fundamental notion behind these papers is one of the most important in all of coal chemistry. That notion is that coals are macromolecular network systems and the processes which they undergo must be considered as operating on macromolecular networks. The coalification process must be considered within the framework of macromolecular chemistry. It is a minor extension of this to realize that all reactions of coals must be considered within the framework of macromolecular chemistry. Many years have passed since Van Krevelen's original papers and this view is only now being aggressively developed.

Following Van Krevelin's work, a few papers appeared using classical polymer chemical techniques to probe the macromolecular structure. In particular, Sanada and Honda and Kirov used classical solvent swelling techniques to investigate cross-link density of coal macromolecular networks.[8,9] Again, one can criticize the details of the work in both papers, but the approaches and ideas are sound and bring credit to their originators.

After this auspicious beginning, the notion that coals were cross-linked macromolecular networks was largely abandoned. It was replaced by the idea that coals consisted of macromolecules which were associated by hydrogen bonds and other non-covalent interactions. The principal driving force behind much of these ideas was the apparent solubilization of coals using reactions which were thought not to cleave covalent bonds. In some cases, reactions such as the Sternberg reductive alkylation which were originally thought not to cleave significant numbers of σ bonds were later shown to significantly degrade the coal macromolecules.[10] In other cases, it was shown that "solutions" were actually colloidal suspensions of very high molecular weight materials.[11] In the early 1970's, a few workers began to argue that coals were cross-linked macromolecular networks and to carry out experiments based on this hypothesis.[12,13,14] While detractors remain, these ideas have demonstrated sufficient utility and provided enough new insight into the behavior of coals such that their acceptance has become quite wide spread.

3. STRUCTURE

3.1 Network Structure.

This section shall be arbitrarily divided into two subsections. In the first, I will address the structure of the macromolecular network itself, the insoluble portion of high vitrinite coals excluding mineral matter. This will ignore the structure of the extractable material. Of course, the coal extracts and the network are inextricably interrelated and interdependent. Much of what one learns about either is applicable to the other. For the sake of clarity, this interrelationship shall be temporarily ignored and we will proceed to consider first the network structure and then the structure of the extract and its relationship to the structure of the parent network.

In order to make the following discussion as clear as possible, I will describe the macromolecular structural model for bituminous coals in detail without providing evidence for the various assertions. That evidence will be found in the discussion that follows.

Coal structure consists of a set of clusters attached by groups containing bonds potentially capable of undergoing free rotation. The clusters are the aromatic and hydroaromatic systems whose overall size and arrangement was determined with X-ray measurements by Hirsch and which are prominent features of many coal structures such as those postulated by Heredy and Wender, Wiser, and Shinn.[15-18] The macromolecular structure model provides no information about any of the structural details of these clusters nor does it provide any information about the nature of the groups linking the clusters together, except to require that there be in principle free rotation about at least one of the bonds in the connecting linkage. Most of the clusters are divalent, that is covalently bonded to only two other clusters. If all were divalent, one would have a set of linear macromolecules which would be, in principle, soluble. Some portion of the clusters are tri- or higher valent. Their presence in the coal causes the whole system to be linked into one large molecule, one in which a chain of covalent bonds attaches any atom to all other atoms. Such divalent or higher clusters are known as branch points and we are especially interested in the quantity $\overline{M}_c$, the number average molecular weight of the chains between branch points. Branch points in the connecting links are possible and Stock has provided evidence for their existence.[19]

In addition to the covalent branch points, many branch points exist because of hydrogen bonding. A hydrogen bond from a donor to an acceptor within the same cluster is network inactive and will have no influence on network properties or behavior. However, a hydrogen bond between one cluster and another cluster will constitute a branch point and, if not broken, will affect $\overline{M}_c$. Strongly basic solvents can break coal-coal hydrogen bonds.[20] If $\overline{M}_c$ could be measured using standard solvent swelling techniques and strongly basic solvents, the results would be quite different from those obtained with solvents which do not break hydrogen bonds. Coals in which all the hydrogen bonds have been removed by hydroxyl derivation are expected to swell a great deal even in non-polar solvents and this is observed.[21]

The most fundamental feature of any cross-linked macromolecular network is the number average molecular weight between cross-links ($\bar{M}_C$) which can be obtained most directly from measurements of mechanical properties or from solvent swelling measurements. It is this last approach which has received the most attention. In this experiment, the expansion of the coal after it has dissolved a solvent is measured. If one knows the volume fraction of the polymer at equilibrium swelling(V) and the thermodynamics of interaction between the solvent and the coal, usually expressed as the Flory χ parameter, one can obtain values for $\bar{M}_C$. This treatment assumes a structural model for the coal, and three structural models have been used: those of Flory and Rehner,[1] of Kovac,[22] and of Peppas[23]. All of these assume that the mixing between the swelling solvent and the coal is random. Since coals contain many hydroxyl groups which are capable of hydrogen bonding, solvents which are hydrogen bond acceptors will not mix randomly with the coal and cannot be used in quantitative determinations of $\bar{M}_C$.

3.2 Solvent Swelling Studies

First, let us consider coal swelling experiments in which there are no hydrogen bond interactions. This has been accomplished in two ways: 1) solvents incapable of hydrogen bond formation have been used 2) the hydroxyl groups in coals have been derivatized in order to prevent hydrogen bonding. Mixtures of coals and non hydrogen bonding solvents have been shown to follow regular solution theory.[21] A plot of swelling versus solubility parameter of the solvent is a bell shape curve with a clear maximum from which the non polar solubility parameter of the coals can be established. This enables the interaction parameter to be calculated so that values for M_C can be derived from swelling measurements of coals in non polar solvents. The observed regular solution theory behavior for Bruceton coal is shown in Figure 1 and Figure 2 shows the values of M_C calculated from swelling results in a set of non-polar solvents for this same coal. The three curves in the figure refer to the native unextracted coal, the coal which has been extracted with pyridine, and coals in which all of the hydroxyl groups have been removed by acetylation. The Kovac treatment has been used and this results in a dependence of M_C on the molecular weight of the average cluster size of coal. The lines are experimental and to derive the M_C value for the different coals, one picks their favorite number average

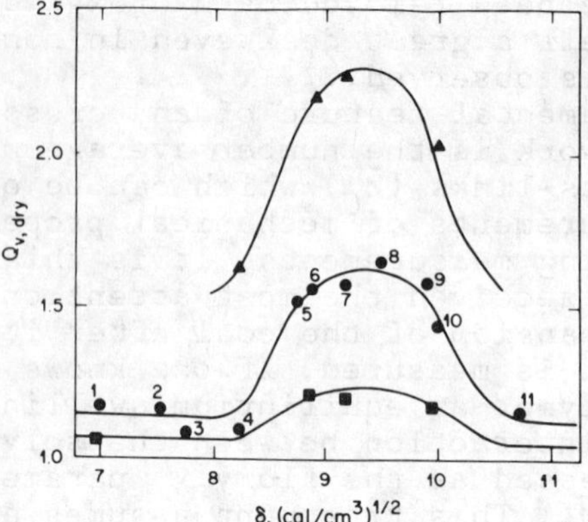

Figure 1. Swelling ratio ($Q_{v\text{-}dry}$) for Bruceton coal (■), pyridine-extracted Bruceton coal (●), and oxygen-acetylated, pyridine-extracted Bruceton coal (▲) as a function of the Hildebrand solubility parameter of the swelling solvent. The swelling solvents are (1) n-pentane, (2) n-heptane, (3) methylcyclohexane, (4) cyclohexane, (5) o-xylene, (6) toluene, (7) benzene, (8) tetralin, (9) naphthalene, (10) carbon disulfide, (11) biphenyl (from ref. 21).

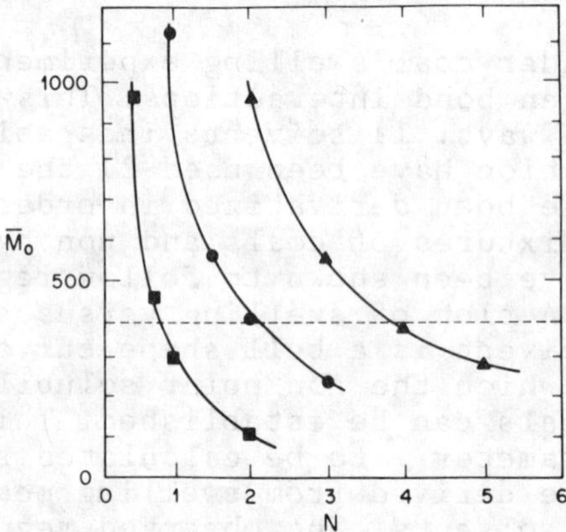

Figure 2. Relationship between the number average molecular weight of a "cluster" (M_o) and the number of clusters between branch points (N) for Bruceton coal (■), pyridine-extracted Bruceton coal (●), and oxygen-acetylated pyridine-extracted Bruceton coal (▲) from the Kovac equation (from ref. 21).

cluster molecular weight and finds the intercept of that weight with the desired line and reads the number of clusters from the x axis. Multiplication of the average cluster weight times the number of clusters gives the value for $\bar{M}_C$. The $\bar{M}_C$ values obtained are chemically reasonable. For the native coal, there is essentially one cross-link per cluster. Since non-polar solvents were used, this includes non-covalent as well as covalent branch points. This would be a very stiff hard material. In the pyridine extracted coal, the hydrogen bond population has been significantly reduced so the cross-link density is decreased. Finally, the acetylated coal shows the lowest cross-link density due to removal of hydrogen bonds and steric interference with other non-covalent interactions.

The swelling measurements discussed so far have all been volumetric measurements in which the net expansion of the coal when placed in liquid solvent is measured. It has been shown that coals swollen, dried and reswollen using the same solvent and this technique reach the same equilibrium swelling value[24]. The swelling is a reversible equilibrium process. However, Duda and Hisu have shown that the swelling of a pyridine extracted bituminous coal using non-polar solvent vapors is not reversible.[25] When the solvent is removed from the coal, it does not return to its original size and a second swelling gives a new and larger expansion of the coal as well as a faster rate of expansion. In addition, attempts to calculate $\bar{M}_c$ values from these vapor swelling data yield nonsense values. Two important conclusions can be drawn from this study. The first is that coal swelling can easily give a value which is not the thermodynamic equilibrium swelling. Second, removal of solvent from swollen coals may leave them in a different state from their original one. Coals are very complex networks and the motion of the coal molecular segments during the swelling process is slow and cruical. The system may approach equilibrium very slowly and be left in a different and non-reproducible state if solvent is removed before equilibrium is obtained.

3.3 Mechanical Properties

The value of $\bar{M}_c$ has a profound effect on the mechanical properties of coals. Coals as mined are glassy and strained.[26] Under the very high pressures experienced in coalification, they have adopted a configuration which is not in equilibrium at atmospheric pressure.[25,27] They are locked into this configuration by non-covalent interactions which serve as cross links. Each cluster is a branch point and coals are strained, glassy solids. If the hydrogen bonds are removed either by hydroxyl derivitization or addition of a basic solvent and if a plasticizer is present, the coal becomes rubbery.[26] This has been demonstrated by stress-strain measurements.[28] The mechanical properties are profoundly changed by a decrease in the number of non-covalent cross-links.

The transition from a glass to a rubber can be induced by heating. The temperature at which this occurs (T_g) is a very important parameter which has only recently been established for a number of coals. Peppas measured a thermal transition for several coals near 600° K using standard DSC techniques and assigned this to a glass to rubber transition.[29] This assignment was confirmed by work published by Lynch who used NMR to characterize the increase in proton mobility which occurs near 600 K when coals are heated.[30] Below T_g, there is limited mobility of chain segments. Diffusion rates are very low and cage effects are enormous. Above the transition temperature, diffusion rates increase by 10^2 to 10^4 as segmental mobility increases sharply. Coal above its T_g is a different material than coal below T_g and this has not been included in most thinking about coal chemistry. A discussion of its importance in coal pyrolysis will soon be available.[31]

The importance of lowering T_g for direct liquefaction is great. Poutsma has shown that anchoring bibenzyl leads to enhanced formation of polynuclear aromatic products when it is pyrolyzed.[32] If a coal is glassy, the groups it contains will be immobile and diffusion will be slow. Heating a glassy coal is expected to lead to formation of PNA products, as observed in direct liquefaction.[33] In a rubbery coal, both diffusion rates and mobility are greatly enhanced thus reducing the tendency to form PNA products. The effects of the physical state of coals on their reactions has been ignored and may be an area of great importance.

3.4 Coal Extraction

The state of the coal also affects extraction yields. To take a concrete example, it has long been known that more chloroform soluble material is removed from bituminous coals by pyridine than is by chloroform extraction.[34] In a coal in its glassy state, much soluble material will be trapped by slow diffusion rates and entanglements which are permanent due to low molecular mobility. If an extracting solvent does not induce a phase transition from glass to rubber, much soluble material may remain trapped in the coal network. Extraction with a solvent which induces the phase change will mobilize much of the formerly trapped material which dissolves and is removed. Thus, pyridine, whose presence induces a glass-to-rubber transition, will extract more chloroform-soluble material than chloroform will because the extract is no longer locked in the glassy coal. There is a conflict in the literature over whether pyridine lowers T_g for coals to room temperature or not.[27,28,29] There is agreement that it is lowered significantly. The state of a coal being extracted is important to the results of the extraction process.

The extracts themselves are directly related to the macromolecular network. In the discussion that follows, we shall ignore the highly aliphatic portion of the extracts. It does not have the potential to be easily incorporated into the network and is a spectator to the macromolecular structure changes occurring during coalification. It was early argued by van Krevelen that coals were formed by the condensation polymerization of a monomer soup that resulted from the degradation of deposited plant matter.[6] In their excellent book, Tissot and Welte have ascribed kerogen formation to such a polymerization process.[35] For coals, this view has been strongly rebutted by Given[36] and Larsen[7,37] has argued that the coalification of bituminous coals is a depolymerization. In either event, the coal extract comprises material which is being incorporated into the insoluble network or which is being derived from it. In both situations, there are well established relationships between the amount and the molecular weight distribution of the extracted material and between both of these quantities and $\bar{M}_c$ for the network.[1,37,38] The statistics of the polymerization or depolymerization process governs the form of the relationship.

3.5 Coalification Process

It has long been recognized that the molecular weight of bituminous coal extracts increased as the amount of extract increased.[3] This can only be true if the extracts are being produced by the depolymerization of a macromolecular network.[37,38] Visualize a network including a low molecular weight fragment and one of high molecular weight. On average, the large fragment will be bonded to the network at more sites than the small fragment. If bonds are broken at random, on average the small fragment will be released before the large one. The molecular weight of the soluble products will increase as the network depolymerization advances. This is as observed for bituminous coals.

These ideas are directly relevant to coal depolymerization kinetics and have been applied by Solomon through Monte Carlo calculations to coal pyrolysis.[39] Often, the rates of coal conversion are followed by measuring the increase in the solubility of the products and then these data are treated using kinetic laws derived for reactions of non-network molecules in solution. This may be a convenient way of presenting data, but the kinetics have no fundamental meaning. What is needed are measures of the rate of bond cleavage during coal conversion and the analyses of these data using the statistics of macromolecular networks. Such studies are underway in our laboratory.

This article cannot be properly called a review because too many important topics have been left out. A discussion of some topics of interest to the authors is a better description. Our purpose is to make clear that coal chemistry is best considered within the framework of macromolecular network chemistry. When that is done, many confusing issues disappear and many worthwhile experiments are revealed.

ACKNOWLEDGEMENTS. We are grateful to the following organizations for support of our work discussed here: U. S. Department of Energy, PSE & G Research Co., Exxon Research and Engineering Co., Gas Research Institute, and Exxon Education Foundation.

REFERENCES

1. Flory, P. J. Principles of Polymer Chemistry; Cornell Univ. Press: Ithaca 1953.

2. Treloar, L. R. G. The Physics of Rubber Elasticity; Clarendon Press: Oxford 1975.

3. Van Krevelen Coal; Elsevier: New York 1981.

4. Green, T.; Brenner, D.; Kovac, J.; Larsen, J. W. in Coal Structure Meyers, R. A. Ed., Academic Press: 1982.

5. Dormans, H. N. M.; Van Krevelen, D. W. Fuel **1960** 39 273-292.

6. Van Krevelen, D. W. Fuel, **1965** 44, 229-242.

7. Larsen, J. W.; Mohammadi, M.; Yiginsu, I.; Kovac, J. Geochim. et Cosmochim. Acta 1984, 48, 135-141.

8. Sanada, Y.; Honda, H. Fuel, **1966**, 45, 295.

9. Kirov, N. Y.; O'Shea, J. M.; Sergeant, G. D. Fuel, **1968** 47, 415.

10. Stock, L. M. in Coal Science, Vol 1, Gorbaty, M. L.; Larsen, J. W.; Wender, I. Eds., Academic Press:New York 1982.

11. Hombach, H. P. Erdol Kohle, Erdgas, Petrochim. Compendium **1975**, 74/275, 750. Larsen, J. W.; Lee, D. Fuel **1983**, 62, 918-923.

12. Larsen, J. W.; Kovac, J. in Organic Chemistry of Coal, Larsen, J. W. Ed., ACS Symp. Ser. 1978, 71, 36-49.

13. Lucht, L. M.; Peppas, N. A. in Chemistry and Physics of Coal Utilization, Cooper, B. R.; Petrakis, L. Eds., AIP Conf. Proc. 1981, 70, 28-48.

14. Berkowitz, N. Technol. Use Lignite 1982, 1, 414. Wachows, H.; Ignasiak, T.; Strausz, O. P.; Carson, D.; Ignasiak, B. Fuel **1986**, 65, 1081-1084.

15. Hirsch, P. B. Proc. Roy. Soc. (London) **1954**, A226, 143. Cartz, L.; Hirsch, P. B. Phil. Trans. Roy. Soc. (London) **1960**, 252, 557-602.

16. Heredy, L. A.; Wender, I. Am. Chem. Soc. Div. Fuel Chem. Preprints **1980**, 25(4), 38-45.

17. Wiser, W. H. EPRI Conf. on Coal Catalysis 1973, Palo Alto, CA.

18. Shinn, J. H. Fuel **1984**, 63, 1187-1196.

19. Stock, L. M.; Wang, S.-H. Fuel **1986**, 65, 1552-1562.

20. Larsen, J. W.; Baskar, A. J. Energy Fuels **1985**, 1, 230-232.

21. Larsen, J. W.; Green, T. K.; Kovac, J. J. Org. Chem. **1985**, 50, 4729-4735.

22. Kovac, J. Macromolecules **1978**, 11, 362.

23. Peppas, N. A.; Lucht, L. M. Chem. Eng. Commun. **1984**, 30, 291.

24. Larsen, J. W.; Lee, D.; Shawver, S. E. Fuel. Proc. Techn. **1986**, 12, 51-62.

25. Hsieh, S. T.; Duda, J. L. Fuel **1985**, 66, 170-178.

26. Brenner, D. Fuel **1985**, 64, 167-173.

27. Cody, G. W. Jr.; Larsen, J. W.; Siskin, M.; manuscript, in preparation.

28. Brenner, D. Am. Chem. Soc. Div. Fuel Chem. Preprints **1986**, 31(1), 17-24.

29. Lucht, L. M.; Larson, J. M.; Peppas, N. A. Energy Fuels **1987**, 1, 56-58.

30. Sakurovs, R.; Lynch, L. J.; Maher, T. P.; Banerjee, R.N., Energy Fuels **1987**, 1, 167-172.

31. Larsen, J. W. Fuel Proc. Techn. in press.

32. Buchanan, A. C. III; Dunstan, T. D. J.; Douglas, E. C.; Poutsma, M. L. J. Am. Chem. Soc. **1986**, 108, 7703-7715.

33. Whitehurst, D. D.; Mitchell, T. O.; Farcasiu, M. Coal Liquefaction; Academic Press: New York 1980.

34. Brown, H. R.; Waters, P. L. Fuel **1966**, 45, 17.

35. Tissot, B. P.; Welte, D. H. Petroleum Formation and Occurrence; Springer-Verlag: New York 1978.

36. Given, P. H. in Coal Science, Vol. 3; Gorbaty, M. L.; Larsen, J. W.; Wender, I. Eds., Academic Press: 1984.

37. Wei, Y.-C.; Larsen, J. W. Energy Fuels, submitted for publication.

38. Yan, J. F. Macromolecules **1981**, 14, 1438-1445. Yan, J. F., Johnson, D. C. J. Appl. Polym. Sci. **1981**, 26, 1623-1635.

39. Squire, K. R.; Solomon, P. R.; Carangelo, R. M.; DiTaranto, M. B., Fuel **1986**, 65, 833-843.

SOLVENT SWELLING OF COALS

Elaine M. Y. Quinga and John W. Larsen
Department of Chemistry
Lehigh University
Bethlehem, PA 18015

ABSTRACT. The literature on solvent swelling of coals up to the end of 1985 has been reviewed and discussed.

1. INTRODUCTION

This is a review of the literature dealing with the swelling of coals by solvents. When brought into contact with most organic solvents, coals absorb the solvent and swell. The amount of swelling depends on both the coal and the solvent. This document contains all of the coal swelling data which had been published in the refereed literature before the end of 1985. Our purpose is to present these data and to interpret them.

The necessary background for that interpretation is the physical chemistry of cross-linked polymers. It is only in this intellectual framework that the data can be rationalized. We cannot include an extensive discussion of that well-developed, active science area here. The necessary background can be found in standard texts and monographs[1-4].

More than 25 years ago, van Krevelen proposed that coals were three-dimensionally cross-linked macromolecular networks and successfully treated them as such[5]. Unaccountably, this idea was largely ignored for almost 20 years. Recently this suggestion has been extensively developed by several groups and is now generally, but not universally[6-9], accepted. It has changed the way chemists think about coals and led to new research and insights.

Bakelite, the polymer once used to make billiard balls, and rubber are two familiar materials which are cross-linked networks. Their properties are quite different and the structural features responsible for the differences are well and quantitatively understood. There exist a variety of experimental techniques for studying macromolecular networks and a well developed theoretical

Y. Yürüm (ed.), New Trends in Coal Science, 85–116.

framework to guide their application and evaluate their results. The experiments discussed in this brief review are a principal technique of modern polymer chemistry and the data obtained with coals are intelligible only within the framework of macromolecular structures. A brief, qualitative introduction to cross-linked networks will therefore be provided. More thorough and quantitative introductions to this topic are available[1-4].

Consider a set of linear macromolecules as shown in A below. They will, in principal, dissolve in a good solvent. B shows these chains linked together to form a three-dimensionally cross-linked network with the dots (•) indicating a covalent bond between chains. This network is insoluble in all solvents. It is one large molecule and will not dissolve. A rubber stopper is one large molecule, one obviously too large to dissolve. When contacted by a solvent for which it has an affinity, the solvent will dissolve <u>in the solid</u>, and the solid will expand to hold it. This experiment, done quantitatively, is capable of revealing a good deal about the nature of the macromolecular network. We discuss in this review the effects of swelling coals with solvents.

The two factors which control the amount by which a coal swells in a solvent are the magnitude of the solvent-coal interactions, usually expressed as the Flory interaction parameter χ, and the cross-link density of the coal, usually expressed as the number average molecular weight between branch points ($\bar{M}_c$). The more favorable the interactions between the solvent and the coal, the greater the amount of solvent present in the coal at equilibrium and the greater the swelling will be. The effect of $\bar{M}_c$ is more complex. The molecular chains in the cross-linked network shown in B are linked together in a three-dimensional array. The links between the chains, shown as black dots at chain intersections, are called branch points. $\bar{M}_c$ is the number average molecular weight of the chain segments between branch points. As the coal swells, these branch points move away from each other. The farther they can move, the more the coal can expand. Their limits of motion are set by the length of the chains connecting them. Therefore, the longer the chain segments, that is the larger $\bar{M}_c$, the more the coal can swell. The two factors which must be considered in any rationalization of coal swelling are the coal-solvent interactions (χ) and the cross-link density or $\bar{M}_c$ of the coal.

The size of $\bar{M}_c$ governs many of the mechanical properties of coals. A highly cross-linked (low $\bar{M}_c$) material will be stiffer, less flexible than a less cross-linked material. Mechanical properties such as shear moduli depend directly on $\bar{M}_c$. It is necessary to know M_c to understand a coal's mechanical properties.

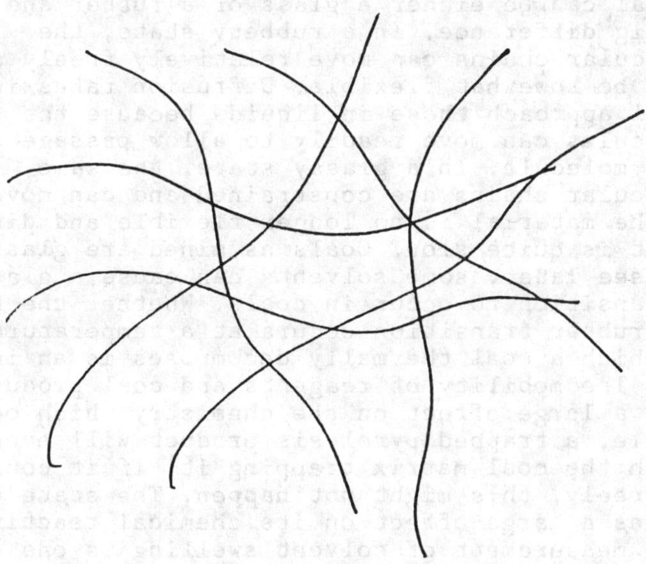

A. MIXTURE OF LINEAR MACROMOLECULES

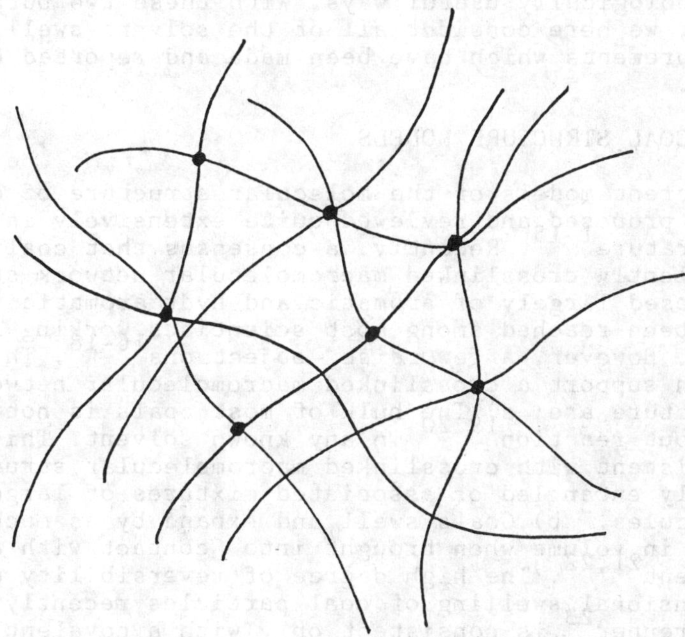

B. A CROSS-LINKED MACROMOLECULAR NETWORK

A coal can be either a glass or a rubber and this makes a big difference. In a rubbery state, the macromolecular chains can move relatively freely and the coal will be somewhat flexible. Diffusion rates in the solid will approach those in liquids because the macromolecules can move readily to allow passage of the diffusing molecule. In a glassy state, the same macromolecular chains are constrained and can move but little. The material is no longer flexible and diffusion through it is quite slow. Coals as mined are glasses. As we shall see later, some solvents can cause a glass-to-rubber transition to occur in coals. Whether the thermal glass to rubber transition occurs at a temperature below that at which a coal thermally decomposes is an important question. The mobility of reagents and coal products in coals has a large effect on the chemistry which occurs. For example, a trapped pyrolysis product will eventually react with the coal matrix trapping it. If it could diffuse freely, this might not happen. The state of the network has a large effect on its chemical reactivity.

The measurement of solvent swelling is one of the best methods for probing the macromolecular network structure of coals. Swelling coals also provides opportunities for modifying their properties in technologically useful ways. With these two purposes in mind, we here consider all of the solvent swelling measurements which have been made and reported before 1986.

2. COAL STRUCTURE MODELS

Different models of the molecular structure of coal have been proposed and reviewed quite extensively in the literature[1-15]. Recently, a consensus that coal is a covalently crosslinked macromolecular network structure composed largely of aromatic and hydroaromatic "clusters" has been reached among most scientists working in this area. However, a few raise objections[16-18]. The arguments which support a crosslinked macromolecular network structure are: a) The bulk of most coals is not soluble without reaction[19,20] in any known solvent. This is consistent with crosslinked macromolecular structure or a highly entangled or associated mixtures of large molecules. b) Coals swell and expand by as much as 2.5-fold in volume when brought into contact with a good solvent[21,22]. The high degree of reversibility of the dimensional swelling of coal particles recently observed by Brenner[23] is consistent only with a covalently crosslinked structure and rules out entanglements as the sole associative force[18-21]. (c) Coals are viscoelastic[24,25]. d) Hydrogenolysis increases the amount of soluble material in a coal and these new soluble materials are very similar to those naturally occurring in coal. Coals behave as if they are very large molecules being cleaved[5].

It is clear from the foregoing considerations that coals must be covalently crosslinked networks. Other non-covalent interactions also play important structural roles. Larsen, et al.[9,26] proposed that coals consist of a covalently bonded network of clusters which are extensively hydrogen bonded to each other and that the hydrogen bonding between clusters provide most of the crosslinks responsible for the brittle, rock-like character of bituminous coals. Peppas and Lucht[27,28] proposed a crosslinked macromolecular coal structure in which entanglements among the chains play a large role. Szeliga and Marzec proposed that coal macromolcules are bound together through electron-donor interactions[29,30].

3. EQUILIBRIUM SWELLING

Coals do not dissolve, rather they swell when they come in contact with a good solvent[21,22]. Dryden developed a classification of suitable solvents for coals based on his extensive studies on the extraction of low rank coals. The best solvents contain a nitrogen or oxygen atom possessing an available unshared pair of electrons[31]. This phenomenon is particularly striking with solvents containing nitrogen atoms such as ethylenediamine and pyridine. Solvents which extract large amounts of soluble material from coals are also good swelling solvents.

Equilibrium swelling is an important investigative tool into the thermodynamic interactions and physicochemical structure of coal. The degree of swelling of coal by solvents Q, (Q = swelling ratio) is the ratio of the volume of swollen coal to original coal volume at equilibrium with the swelling solvent. It may take many days to reach equilibrium due to slow diffusion rates. The initial gravimetric technique for measuring the expansion of coal due to swelling is quite simple but often takes a long time[32-36]. The coal is exposed to solvent vapors in a vacuum desiccator. The swelling ratio is the ratio of the mass of swollen coal after equilibration to the mass of dry coal corrected for the mass of the solvent which fills the coal pores and does not contribute to the swelling. The equilibration time often is several days. A more rapid and convenient way of measuring swelling of coal by solvents was developed by Dryden[10c] and used by Liotta[38] and Larsen[37]. The dry coal is placed in an 8 mm. o. d. Pyrex tube and is centrifuged for 5 min. in a Fisher centrifuge at 1725 rpm. The height of the coal is measured as h_1. After breaking up the column of packed coal, an

Table I. Coal Elemental Analysis.

	Coal Name	%C	%H	%O	%O+N	%O+N+S
1.	Big Brown Lignite	69.45	5.62	21.09		
2.	Wyodak (Rawhide)	70.57	5.01	21.62		
3.	North Dakota	69.16	5.31	23.93		
4.	Bruceton	82.57	5.20	8.84		
5.	Illinois No.6(1)	79.80	5.03	11.85		
6.	Illinois No.6(2)	76.35	5.71	10.38		
7.	Ashibetsu Bituminous	81.1	5.5			13.4
8.	Odaira Lignite	65.1	5.0			29.9
9.	Nakago	74.3	5.3			20.4
10.	Takamatsu	79.0	5.0			16.0
11.	Bibai	80.9	5.9			13.2
12.	Yubari I	84.9	6.2			8.9
13.	Yubari II	85.2	6.3			8.5
14.	Hashima	86.6	5.6			7.8
15.	Yatake	88.7	3.3			6.9
16.	Hongei	93.0	3.3			3.7
17.	Bituminous Coal hvB Coal	80.7	5.6		10.9	
18.	Furst Leopold	80.9	5.4		12.0	
19.	Hohenzollen	82.5	5.2		11.9	
20.	Heinitz	85.4	5.5		8.3	
21.	Heinitz vitrain	82.5	5.05		12.2	
22.	Heinitz durain	82.05	5.75		11.7	
23.	Heinitz fusain	63.15	2.7		32.9	
24.	Emscher Emil	86.5	5.1		7.25	
25.	Escheiler	88.6	4.7		5.7	
26.	Hebe (W.A.)	79.9	4.5	14.6		
27.	Greta (N.S.W.)	82.4	6.2	8.7		
28.	Bulli (N.S.W.)	88.2	5.1	4.7		
29.	Coal A (Bituminous)	82.32	5.51	5.89		
30.	Coal E (Anthracite)	91.29	2.91	2.10		
31.	Coke No.0	89.96	0.91	2.39		
32.	PSOC Code No. 418	69.94		21.46		
33.	" 791	72.25		20.17		
34.	" 414	72.99		19.25		
35.	" 211	77.08		14.68		
36.	" 207	79.82		12.23		
37.	" 402	82.48		9.00		
38.	" 341	86.01		5.29		
39.	" 1029	88.12		4.81		
40.	" 647	91.54		1.37		
41.	" 384	94.17		0.62		
42.	Bruceton	82.3	5.2	9.7		
43.	PSOC-219	82.1	5.7	10.9		
44.	PSOC-213	78.9	5.7	9.5		
45.	Illinois No. 6	76.2	5.8	11.5		
46.	Wyodak	73.3	6.3	18.3		

Table I (continued)

47. Big Brown	67.5	5.4	24.7
48. SRS 800	80.5	5.3	12.3
49. SRS 600	83.6	5.4	9.2
50. SRS 400	86.6	5.5	6.0
51. SRS 300	89.5	4.8	4.3

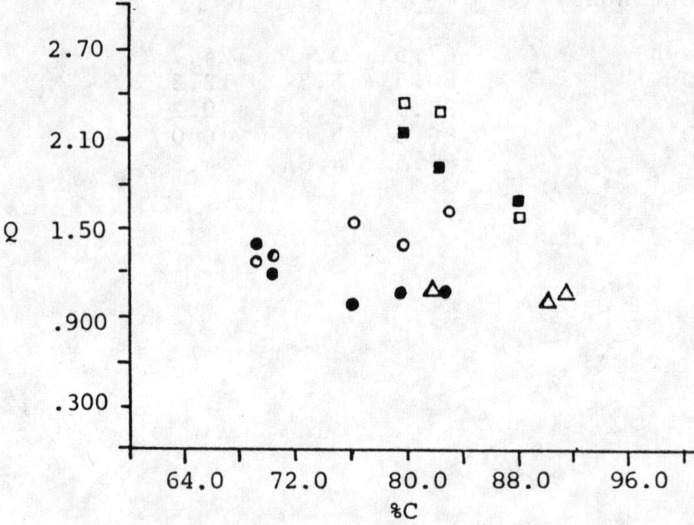

Fig. 1. Relation between equilibrium swelling and the carbon content of coals in non-hydrogen bonding solvents at room temperature.

□ Kirov et al. (benzene) ■ Kirov et al. (cyclohexane)
○ Green (benzene) ● Green (cyclohexane)
△ Moore (benzene)

excess amount of solvent (3-4 ml) is introduced and the tube is vigorously shaken to ensure thorough mixing. The coal is again centrifuged and the height is measured as h_2. The mixing and centrifuging is repeated until a constant height , h_2 , is obtained. The swelling ratio is $Q= h_2/h_1$. This technique does not require the corrections for pore filling required by gravimetric measurements. The two techniques have been directly compared and shown to be equivalent[37].

Swelling ratios of a number of coals using a variety of solvents are presented in this report. The elemental analyses of the coals are given in Table I. Coal samples 1-6[39] were pyridine extracted and dried in vacuum at 110°C overnight. Coals 7-16[22,40] were pulverized to pass through a 30 mesh Tyler sieve and the vitrain of specific gravity less than 1.30 was isolated by the float-and-sink method using mixtures of benzene and carbon tetrachloride. These vitrains were dried in vacuum for a few hours and stored in a calcium chloride desiccator. Coal 17[29] was milled and sieved and the <0.43 mm fraction was dried under vacuum at 170°C and kept under nitrogen. Coals 18-25[41] were ground to pass a sieve with 900 meshes per sq. cm. and dried to constant weight in an oven at 150°C. Coals 26-28[33] were Soxhlet extracted successively with DMF, pyridine, benzene, acetone and methanol. Coals 29-31[42] were ground to pass a 200 (I. M. M.) sieve and were heated in an oven at 150°C. Coals 32-41[43] were pyridine extracted and dried under vacuum at 60°C and stored in a nitrogen purged glove box. Coals 42-47[44] were pyridine extracted and dried in vacuum. Finally, coals 48-51[45] were impact-milled and sieved and pyridine extracted.

4. SWELLING IN NON-HYDROGEN BONDING SOLVENTS

The solvents are divided into two classes: non-hydrogen bonding and hydrogen bonding solvents. The swelling ratios of the coal samples in non-hydrogen bonding solvents (Table II) are less (<25-50%) than those in hydrogen bonding solvents (Table III). The data from Kirov et al.[33] for non-hydrogen bonding solvents are significantly higher than the others. These unusually high swelling ratios may mean that the Kirov coal samples are different from the other coal samples. The swelling ratios in benzene, toluene and cyclohexane are plotted in Figure 1 versus the carbon content of the different coal samples. The Kirov et al. data stand out as different from the rest of the coal samples. There are not enough data to decide whether Kirov's high values are due to differences in coal properties or experimental errors. There is no relationship between elemental composition and swelling in

Table II. Non-hydrogen bonding solvents and their swelling ratios.

Coal Name	Solvent	Q
Big Brown Lignite	Benzene	1.29
	Toluene	1.23
	Methyl Chloride	1.45
Wyodak(Rawhide)	Cyclohexane	1.23
	Toluene	1.320±.01
	Nitromethane	1.34
	Nitrobenzene	1.42±0.01
	Acetonitrile	1.27
North Dakota	Cyclohexane	1.32
	Toluene	1.18±0.03
	Benzene	1.42
	Nitrobenzene	1.67±0.06
	Acetonitrile	1.33±0.01
	Nitromethane	1.44
	n-Pentane	1.19
Bruceton	Cyclohexane	1.08±0.02
	Toluene	1.55±0.02
	Benzene	1.58±0.03
	Carbon Tetrachloride	1.46
	Nitromethane	1.43
	Nitrobenzene	1.62±0.06
	Acetonitrile	1.43±0.03
	Naphthalene	1.58
	n-Pentane	1.17
	Biphenyl	1.15
	o-Xylene	1.51
	Tetralin	1.65
	n-Heptane	1.16
Illinois No.6(1)	Pentane	1.14
	Cyclohexane	1.11±0.04
	Toluene	1.39±0.05
	Benzene	1.38±0.01
	Carbon Tetrachloride	1.34
	Nitrobenzene	1.67±0.06
	Acetonitrile	1.39±0.01
Illinois No.6(2)	Biphenyl	1.11±0.01
	Cyclohexane	1.00
	Benzene	1.52±0.02
	o-Xylene	1.41
	Tetralin	1.52±0.02
	n-Heptane	1.00
	Methylcyclohexane	1.03

Table II (continued)

Ashibetsu Bitum.	Benzene	1.53
Odaira Lignite	Benzene	1.05
Bituminous Coal	Benzene	1.0
hvB Coal	Nitrobenzene	1.1
	Acetonitrile	1.15
	Nitromethane	1.18
Hebe (W.A.)	Cyclohexane	2.13
	Xylene	2.32
	Toluene	2.32
	Benzene	2.32
	Carbon Tetrachloride	1.49
Greta (N.S.W.)	Cyclohexane	1.89
	Xylene	2.32
	Toluene	2.50
	Benzene	2.27
	Carbon Tetrachloride	2.22
	Nitrobenzene	2.44
Bulli (N.S.W.)	Cyclohexane	1.67
	Xylene	1.72
	Toluene	1.78
	Benzene	1.56
	Carbon Tetrachloride	1.64
	Nitrobenzene	1.62
Coal A (Bitum.)	Benzene	1.07
	Pentane	1.02
	Heptane	1.02
	Octane	1.00
	Nonan	1.00
	Decane	1.00
	Toluene	1.03
	Xylene	1.01
Coal E (Anthra.)	Benzene	1.06
Coke No.0	Benzene	1.01

Table III. Hydrogen bonding solvents and their swelling ratios.

Coal Name	Solvent	Q
Big Brown Lignite	Pyridine	2.03
	Methanol	1.57
	Ethanol	1.58
	Ethylenediamine	1.97
	2-Propanol	1.53
Wyodak(Rawhide)	1,4-Dioxane Pyridine	1.72±0.02
	Ethanol	2.12±0.06
	Methanol	1.30±0.06
	Ethanol	1.46±0.04
	Acetone	1.54
	2-Propanol	1.29±0.08
	Methylene Chloride	1.64
	Chloroform	1.74
North Dakota	1,4-Dioxane	1.78
	Pyridine	2.04
	Methanol	1.28±0.04
	Ethanol	1.36±02
	Acetone	1.44
	2-Propanol	1.45±0.06
	Methylene Chloride	1.94
	Chloroform	2.21
	Formamide	1.23
	Aniline	2.31
	2,6-Lutidene	1.79
	Chlorobenzene	1.46
	o-Dichlorobenzene	1.61
Bruceton	Carbon Disulfide	1.44±0.07
	1,4-Dioxane	2.07
	Pyridine	2.32
	Methanol	1.32±0.08
	Ethanol	1.44±0.01
	2-Propanol	1.40±0.04
	Methylene Chloride	1.65
	Chloroform	2.00
	Formamide	1.16
	Aniline	2.22±0.02
	2,6-Lutidene	1.90
	Chlorobenzene	1.81±0.06
	o-Dichlorobenzene	1.72
	Tetrahydrofuran	2.07
	Ethylenediamine	2.03
	N,N-Dimethylaniline	1.82±0.06
	Diethylether	1.54
	Dimethyl Sulfoxide	2.21

Table III (continued)

Illinois No.6(1)	Carbon Disulfide	1.46
	1,4-Dioxane	1.73
	Pyridine	1.94
	Methanol	1.28±0.04
	Ethanol	1.44
	Acetone	1.54
	Methylene Chloride	1.48
	2-Propanol	1.45±0.05
	Chloroform	2.21
	Formamide	1.26
	2,6-Lutidene	1.79
	Chlorobenzene	1.52
	o-Dichlorobenzene	1.60±0.02
	Aniline	1.82
Illinois No.6(2)	1,4-Doxane	2.00
	Pyridine	2.42±0.17
	Carbon Disulfide	1.41±0.01
	Diethylether	1.60
	Dimethyl Sulfoxide	2.15±0.08
Ashibetsu Bituminous	Chloroform	1.71
	1,4-Dioxane	1.66
	Pyridine	1.95
	Ethylenediamine	1.81
	Ethanol	1.51
	Dimethyl Sulfoxide	1.62
Odaira Lignite	Acetophenone	1.02
	1,4-Dioxane	1.15
	Pyridine	1.92
	Dimethyl Formamide	1.50
Nakago	Pyridine	1.66
Takamatsu	Pyridine	1.94
Bibai	Pyridine	1.82
Yubari I	Pyridine	1.89
Yubari II	Pyridine	1.80
Hashima	Pyridine	1.97
Yatake	Pyridine	1.02
Hongei	Pyridine	1.02

Table III (continued)

Bituminous Coal hvB Coal	2-Propanol	1.14
	Diethylether	1.15
	Dioxane	1.16
	Methanol	1.19
	n-Propanol	1.23
	Ethanol	1.25
	Ethyl Acetate	1.26
	Acetone	1.30
	Methyl Acetate	1.32
	Methyl Ethyl Ketone	1.49
	Tetrahydrofuran	1.59
	1,2-Dimethoxyethane	1.60
	Dimethylformamide	1.69
	Dimethyl Sulfoxide	2.04
	Pyridine	2.08
	Ethylenediamine	2.08
	1-Methyl-2-Pyrolidone	2.38
Furst Leopold	Pyridine	1.28
Hohenzollern	Pyridine	1.17
Heinitz	Pyridine	1.11
Heinitz vitrain	Pyridine	1.11
Heinitz durain	Pyridine	1.10
Heinitz fusain	Pyridine	1.05
Emscher Emil	Pyridine	1.05
Eschweiler	Pyridine	1.03
Greta (N.S.W.)	Chloroform	2.13
	Methyl Ethyl Ketone	2.44
	Dioxane	2.27
	Acetone	2.27
	Aniline	2.78
	Pyridine	2.00
	Quinoline	2.86
	Dimethylformamide	2.78
	n-Propanol	2.78
	Ethanol	1.82
	Methanol	1.61

Table III (continued)

Bulli (N.S.W.)	Chloroform	1.82
	Methyl Ethyl Ketone	1.82
	Dioxane	1.85
	Acetone	2.70
	Aniline	1.89
	Pyridine	1.92
	Quinoline	1.89
	Dimethylformamide	1.92
	Ethanol	1.75
	Methanol	1.72
PSOC Code No.418	Pyridine	2.13
791	Pyridine	2.10
414	Pyridine	2.15
211	Pyridine	2.22
207	Pyridine	2.07
402	Pyridine	2.10
341	Pyridine	1.59
1029	Pyridine	1.49
647	Pyridine	1.10
384	Pyridine	1.14
Bruceton	Pyridine	2.6
PSOC-219	Pyridine	2.3
PSOC-213	Pyridine	2.6
Illinois No. 6	Pyridine	2.5
Wyodak	Pyridine	2.3
Big Brown	Pyridine	2.4
SRS 800	Pyridine	2.2
SRS 600	Pyridine	1.9
SRS 400	Pyridine	0.85
SRS 300	Pyridine	0.45

non-polar solvents.

5. SWELLING IN HYDROGEN BONDING SOLVENTS

In the classification of coal solvents developed by Dryden, a good solvent for coal contains a nitrogen or oxygen atom possessing an unshared pair of electrons . Other things being equal, nitrogen compounds are better solvents than compounds containing oxygen. He inferred that the possibility of hydrogen bonding must play a part in these interactions[31].

From the data presented in Table III, the swelling ratios in hydrogen bonding solvents are much higher than those in the non-hydrogen bonding solvents particularly with solvents, containing nitrogen. Swelling ratios of different coal samples published by a variety of different researchers are shown in Figures 2 and 3. The swelling ratios seem to be constant from 60%C to 85%C (dmmf), then abruptly drop off above 85%C. This trend is observed in most of the coal samples.

The data from Bunte et. al.[41] are unusually low (Figs. 4-6). Similar to the other results, the swelling are constant from 60%C to about 88%C. A careful examination of the experimental details revealed no obvious sources of errors. Perhaps these coals are significantly weathered. These data are clearly inconsistent with the others and is not possible to identify the origin of the difference.

Figures 2 and 3 show correlations between swelling in pyridine and elemental composition. There is a correlation between swelling and the carbon contents of the coal samples which is most apparent in individual data sets. The swelling ratio is constant up to about 85%C then drops sharply. The Bunte et al. data are low but show the same trends as the other data. The data are consistent with the conclusion made by Honda and Sanada[40] and Peppas and Lucht[43] that swelling is constant from 65%C-87%C and then decreases abruptly above 87%C. Not unexpectedly, a correlation is also observed between swelling and the oxygen-carbon ratio. The swelling ratios are relatively constant between about 0.10 O/C and 0.35 O/C and drop off below 0.10 O/C. Again, the Bunte data are low. No correlations are observed between swelling and hydrogen content and H/C ratios.

Figures 4-6 present the plots of the swelling ratio versus carbon and hydrogen contents of some coal samples in hydrogen bonding solvents other than pyridine. No correlations are observed.

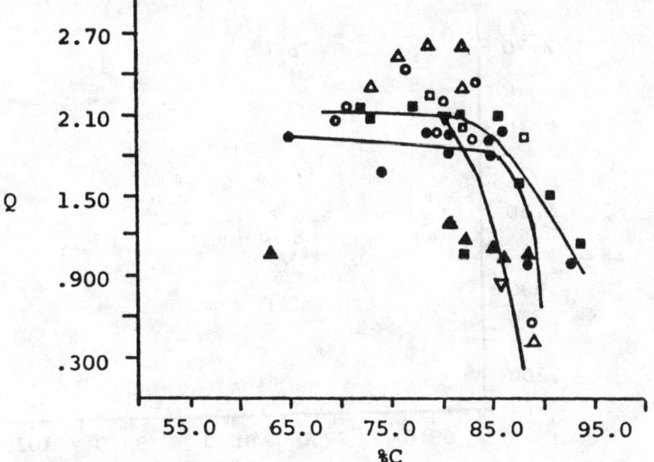

Fig. 2. Relation between equilibrium swelling and carbon content in pyridine at room temperature.

▼ Szeliga and Marzec
● Honda and Sanada
△ Larsen and Lee
▽ Van Bodegom et al.
□ Kirov et al.
○ Green
■ Peppas and Lucht
▲ Bunte et al.

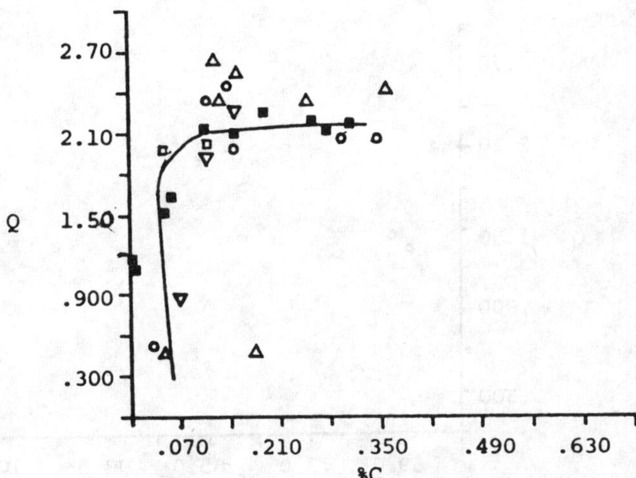

Fig. 3. Relation between equilibrium swelling and the oxygen-carbon ratio of coals in pyridine at room temperature.

○ Green
■ Peppas and Lucht
▽ Van Bodegom et al.
□ Kirov et al.
△ Larsen and Lee

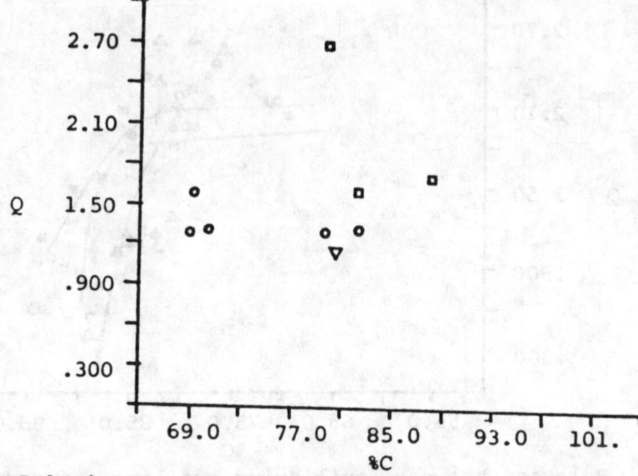

Fig. 4. Relation between equilibrium swelling and the carbon content of coals in methanol at room temperature.
o Green
□ Kirov et al.
▼ Szeliga and Marzec

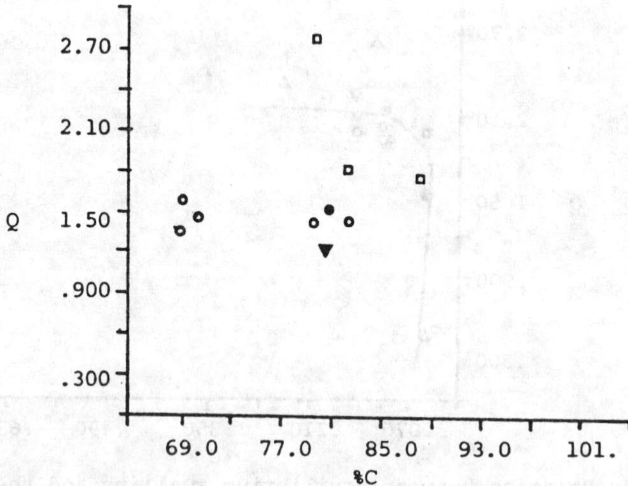

Fig. 5. Relation between equilibrium swelling and the carbon content of coals in ethanol at room temperature.
o Green ● Honda and Sanada
□ Kirov et al. ▼ Szeliga and Marzec

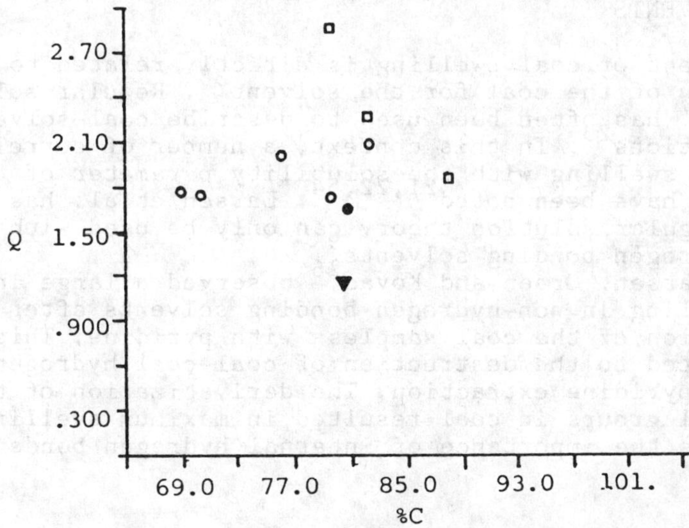

Fig. 6. Relation between equilibrium swelling and the carbon carbon content of coals in 1,4-dioxane at room temperature.

○ Green ● Honda and Sanada
▼ Szeliga and Marzec ◻ Kirov et al.

6. INTERACTIONS OF COALS WITH NON-HYDROGEN BONDING SOLVENTS

The extent of coal swelling is directly related to the affinity of the coal for the solvent[35]. Regular solution theory[57] has often been used to describe coal-solvent interactions[23]. In this context, a number of correlations of coal swelling with the solubility parameter of the solvent have been noted[21,22,54]. Larsen et al. has claimed that regular solution theory can only be used with non-hydrogen bonding solvents[9].

Larsen, Green and Kovac[26] observed a large increase in swelling in non-hydrogen bonding solvents after extraction of the coal samples with pyridine. This was attributed to the destruction of coal-coal hydrogen bonds during pyridine extraction. The derivatization of the hydroxyl groups in coal resulted in maximum swelling which supports the importance of internal hydrogen bonds in coals.

7. INTERACTIONS OF COALS WITH HYDROGEN BONDING SOLVENTS

Basic organic solvents tend to be good swelling solvents for coals. The high swelling caused by such solvents (hydrogen bond-accepting solvents in general) was attributed to the replacement of coal-coal hydrogen bonds with coal-solvent hydrogen bonds. This phenomenon causes the coal to swell more because the replacing of a coal-coal hydrogen bond by a coal-solvent hydrogen bond reduces the cross link density of the coal[26].

The most startling feature of these data is the rapid decrease in pyridine swelling above about 85%C. Two explanations are possible. The coals may become increasingly cross-linked at this high rank and this will decrease swelling. There are no obvious indications of increasing cross-link densities at carbon contents as low as 85%C, although most coal properties undergo sharp changes between 85%C and 89%C. Alternatively, the coal-pyridine interactions may decrease sharply. This might be due to the decrease in phenolic hydroxyl groups which can strongly interact with the basic pyridine. It is clear that an important structure transition of coals occurs here and an understanding of it should be assiduously sought.

8. PHYSICAL CONSEQUENCES OF COAL SWELLING

The swelling of coal refers to an increase in the volume of coal due to absorption of a solvent. It is not necessarily true that when coal swells, the weight uptake corresponds to the dimensional change in coal because coals have pores which are occupied by the solvent. Therefore, the necessity of correcting for the pore volume is critical in the gravimetric measurement of coal swelling. Methods for making this correction have been developed[34,35].

Dryden's studies of particulate coal swelling resulted in dramatic changes in the appearance of the coal[10d]. With good swelling agents, an appreciable fraction of the coal was extracted. The dried sample was highly distorted. The coals expanded greatly in size and cracked. These changes were interpreted as a consequence of the reorientation of the macromolecular chains, the driving force coming from the free energy of mixing of the liquid and the coal structure. This model assumes a gel structure of coal in which there are cross-links between the various macromolecular subunits of the coal. In the presence of selected liquids, the gel structure swells but the cross-linkages prevent dissolution of the coal structure. The high degree of swelling (about 100%) indicates that there are a number of flexible molecular subunits between cross-links in the average chain, however the high modulus of the unswelled coal suggests a relatively short chain length. This contradiction was finally resolved recently by the realization that native coals are in a glassy state[23].

An in-situ microscopic study of the swelling of polished coal surfaces and thin sections was done by Brenner[23]. A high degree of reversibility of the dimensional swelling was observed after the removal of the swelling solvent. This observation is typical of cross-linked networks. When the swelling was allowed to continue for a long period of time, it proceeded beyond the point where it can be reversed. This irreversibility was attributed[23b] to fractures or dislocations within the coal structure which occur when sufficiently high stresses are generated by uneven swelling of the coal structure. The non-uniform swelling of the coal was attributed to kinetic effects or differences in the swellability of the various microscopic subcomponents of the coal sample. The various inner regions of the coal particles are reached by the swelling agent at different times or the different subcomponents are swelled at different rates. Both of these occurrences will cause large mechanical stresses. Also, the different macerals are likely to have different swellabilities, hardnesses, and shapes and all of these differences can contribute to stresses as swelling occurs. Such stresses can result in cracking, dislocation, and fracturing of the structure which can not be reversed when the swelling agent is removed.

The studies involving the swelling of thin sections of coal in pyridine, ethylenediamine and n-propylamine were especially revealing[23b]. Swelling was reversible and the swollen samples were substantially more flexible than the original samples or the redried samples. The shapes of the thin dried coal samples were similar to the shapes of the samples before they were swollen. The most dramatic result observed in Brenner's study was the high degree of reversibility of the observed swelling. Samples which roughly doubled in volume, shrunk to the original size after drying. This could be repeated several times. The first time a thin section was swollen and dried, the resulting piece of coal was slightly smaller than the original, although retaining the same shape. This reduction in size was attributed to extraction of some of the coal material by the solvent and provides additional evidence for the mobility of the chains in the coal macromolecular structure. It may also be due to the change in packing efficiency of the coal macromolecular chain segments. The decrease in modulus which occurred upon swelling was attributed to the solvation of strong non-covalent interactions between chains, such as hydrogen bonds, which are present in the dry coal[15].

In a macromolecular structure, the amount of swelling is controlled by the cross-link density and by the magnitude of the interactions between the solvent and the macromolecules[1-4]. Thus, the measurement of swelling and the interactions between solvent and the macromolecules can be used to calculate the number average molecular weight between cross-links ($\bar{M}_c$) with the Flory-Rehner equation (equation 1). In this equation, $\bar{M}_c$ is the number average molecular weight between branch points, ρ is the coal density, V_s is the solvent molar volume, V is the volume fraction polymer at equilibrium swelling and the Flory χ parameter for coal-solvent pairs.

$$M_c = \rho_c V_s V^{1/3} / -[\ln(1-V) + V + \chi V^2] \quad (1)$$

Larsen, Kovac and Green pointed out that the Flory-Rehner equation is not applicable to coals because they are too highly cross-linked and their chains too stiff[26]. Kovac developed a simple non-Gaussian theory which describes the swelling of coal better and accounts for chain stiffness and the finite extensibility of the network in an approximate way[46]. The N in the Kovac equation describes the deviation from Gaussian behavior. In the limit of very low cross-link density, the Kovac equation (equation 2) reduces to the Flory-Rehner equation.

$$M_c = \frac{\rho_c \, Vs \, V^{1/3} + \rho_c \, Vs \, / \, N \, V}{-[\ \ln (1-V) + V + V + \chi V^2\ ]} \qquad (2)$$

The earlier work on the cross-link density of coal was done by Kirov et al.[21] and Sanada & Honda[22]. The measurement of the interaction parameter (χ) between solvent and the coal macromolecular structure is experimentally difficult. However, in spite of the difficulty, several researchers were able to measure indirectly and calculate the cross-link density of some coal samples[22,33]. The results showed a minimum in cross-link density at about 86%C which shows that as coalification process continues, the coal does not become increasingly cross-linked. Recently, researchers have published results which show no dependence of the interaction parameter (χ) on coal rank[47]. This raises the question of whether regular solution theory will lead to accurate determinations of the interaction parameter (χ) in coal systems[48].

9. DYNAMIC SWELLING

Dynamic swelling or sorption measurements of macromolecular systems yield information about the physical state (glassy or rubbery) of the network, the time scale of major relaxation processes, and the diffusion coefficient of solvent into the macromolecule. If the system is in the glassy state, it is possible to determine whether the sorption is due to Fickian diffusion and/or due to relaxations of the macromolecular chains. It is possible to estimate values of the diffusion coefficients and the relaxation constants of the macromolecular chains can be determined.

Sorption mechanisms in macromolecular systems may be defined in terms of two limiting cases. These are Fickian diffusion and Case II transport. Case II transport systems have been described in detail by Thomas and Windle[49], Peterlin, Astarita and Sarti[50] and Hopfenberg[51]. This system is characterized by a change of state from a glassy to rubbery state as it is penetrated by the solvent. In Case II, solvent uptake is directly proportional to time. On the other hand, Fickian diffusion usually occurs in rubbery systems and the solvent uptake is proportional to the square root of time.

The analysis of sorption data in macromolecular systems involves fitting the sorption data to the empirical equation below[28,43]. M_t is the mass of the solvent taken up at time t,

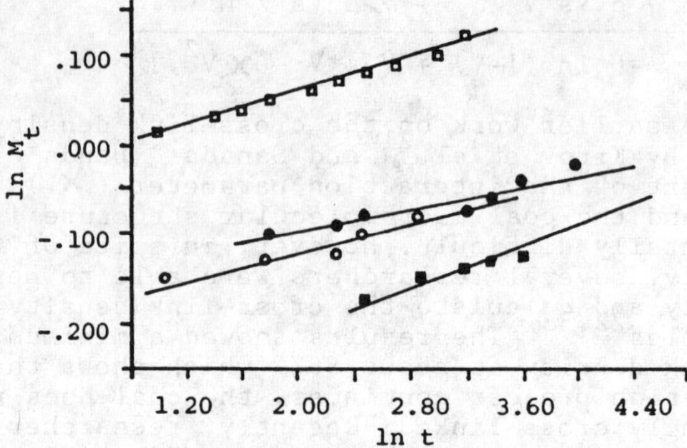

Fig. 7. Dynamic swelling by pyridine vapors.
□ Gasflammekohle; ■ Fettkohle; ● Chinolinrestkohle;
○ Gl

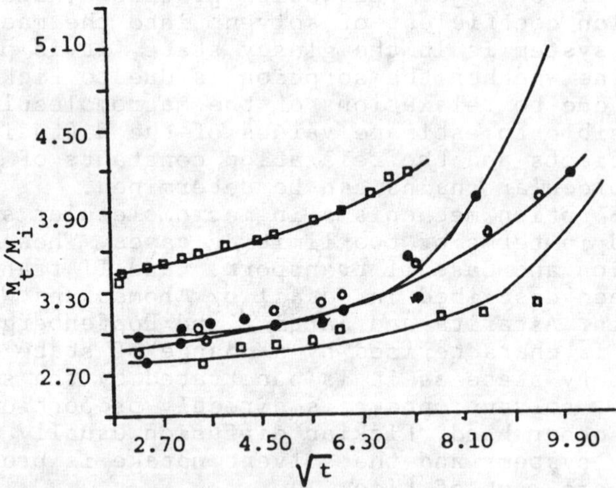

Fig. 8. Solvent swelling in pyridine.
□ Gasflammekohle; ■ Fettkohle; ● Chinolinrestkohl;
○ Gl

$$M_t / M = kt^n \quad (3)$$

M is the mass of solvent taken up at long times and k is a constant which depends on the structural characteristics of the material and the solvent-material interactions. The exponent n is used to indicate the type of diffusion and to infer state changes in the macromolecular systems. When n equals 0.5, the diffusion is Fickian; when n is 1.0, it is a Case II transport; when n is between 0.5 and 1.0, it is an anomalous transport and when $n>1.0$, the swelling material is likely to craze and fracture due to the tremendous osmotic pressure differences at the accelerating and advancing front. This type of transport mechanism is called Super Case II transport.

For analysis of dynamic swelling or sorption in coals, a slightly modified form of the empirical transport equation is used as shown below[27]. M_t is the mass of the solvent taken up

$$M_t = k't^n \quad (4)$$

at time t, and k' is a constant the same as k in equation 3. Equation 4 can only be used up to about 60% of the final volume uptake.

An analysis of the dynamic swelling data of four coals[52] using equation 4 is shown in Figures 7 and 8. These coals are (A) Fettkohle (bituminous) (B) Chinolinrestkohle (C) G1 and (D) Gasflammkohle. The elemental analyses of these coals are not available, however bituminous coal carbon content is between 87-89% or higher and gas flammable coal carbon content is between 80-85%[5]. The data are from reference 50. The plotted data in Figure 7 show a Fickian type of diffusion. The values of n as determined from the slopes of the straight lines in Fig. 7 are between 0.42-0.58. The plot of the solvent uptake versus the square root of time in Figure 8 shows deviation from straight lines after about 25-30 hours. This is attributed to the change in the physical state of coal from a glassy state to a rubbery state[27]. When a macromolecular structure is in its glassy state, large molecular motions are restricted, although segmental motion may still be exhibited. Increased concentration of a diluent in a macromolecular system like coal causes it to swell, decreasing its density and allowing increased bond rotation. In polymeric systems, this phenomenon effectively lowers the glass transition temperature of the polymer. In coals, however, no conclusive evidence yet has been found that a glass transition temperature exists, although several researchers have observed temperature-dependent transitions which are suggestive of glass transition temperatures[53,54].

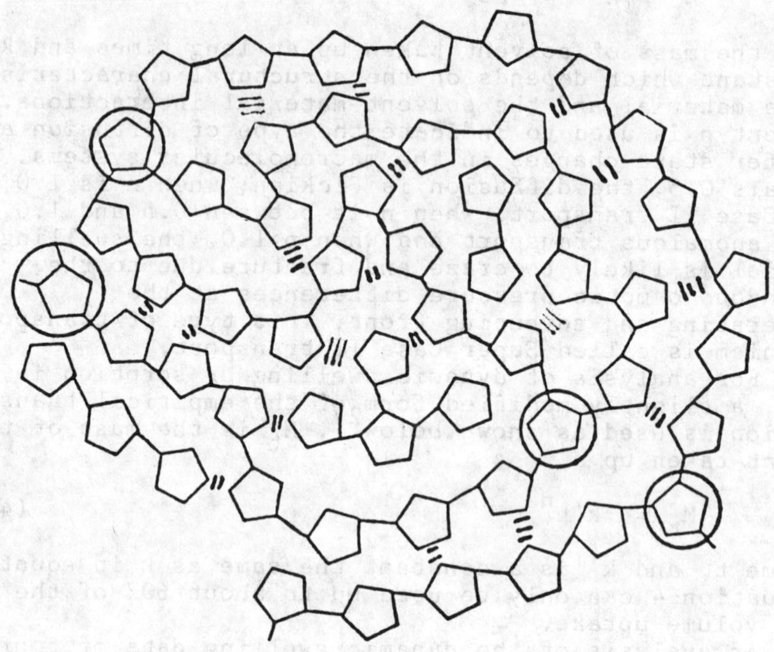

Fig. 9. Macromolecular structural model for coal. The pentagons are clusters, single lines represent rotationally free groups linking cluster, and the dash marks denote hydrogen bond cross links.

The solvent uptake plot, Figure 8, shows that Fettkohle (bituminous) absorbs more solvent than the rest of the coal samples and the Gasflammkohle absorbs the least solvent.There are not enough data to support any conclusions regarding the correlation of elemental composition with solvent uptake.

10. STRUCTURAL MODEL

The structural model of coal that is in best agreement with the swelling data is that of a covalent macromolecular network containing many hydrogen bonds which are cross links[21] (see Fig. 9). In a covalently bonded network, clusters are extensively hydrogen bonded to each other. The removal of the hydrogen bonds causes a large increase in $\bar{M}_c$ and the covalently cross-linked coal is somewhat flexible. In low rank bituminous coals the hydrogen bonds between clusters provide most of the cross-links. With the hydrogen bonds in place, $\bar{M}_c$ is low and this is responsible for the brittle character of bituminous coals. The destruction of the network of internal hydrogen bonds greatly increases $\bar{M}_c$ changing the character of bituminous coals and making them more flexible. The existence of extensive hydrogen bonding in the coal structure is a significant difference from classical coal structure models.

The observed high swelling in hydrogen bond-accepting solvents is explained by the existence of hydrogen bond cross-links in coals. The strong hydrogen bonding solvents replace the coal-coal hydrogen bonds with a new coal-solvent hydrogen bond. This increases $\bar{M}_c$ so swelling can increase. Non-hydrogen bonding solvents do not show enhanced swelling because the coal-coal hydrogen bonds remain intact.

11. SUMMARY OF OBSERVATIONS AND CONCLUSIONS

The data points for solvent swelling show a large degree of scattering and no visible trends are observed except for pyridine swelling. A relatively constant swelling ratio is observed in pyridine in coal samples from 65%C to 85%C, and a dramatic decrease in the swelling ratio is observed above 86% C. This observation could mean that swelling is a type property, that is, it is a function of individual coal rather than a function of coal rank. The absence of any trends could also mean poorly executed experiments or inconsistent experimental procedures among the several different researchers whose data are summarized. It is difficult to decide whether experiments were done properly. It is worthwhile mentioning that these experiments are easy.

The swelling data point to a covalent cross link structure of coal. This is supported by the high swelling observed in hydrogen bonding solvents. The strong hydrogen bonding solvents replace the coal-coal hydrogen bonds with new coal-solvent hydrogen bonds. If the coal-coal hydrogen bonds are active cross links, the replacement will result in a lower cross-link density which causes swelling to increase. The weak hydrogen bonding solvents result in less swelling since more coal-coal hydrogen bonds remain intact. The decrease in swelling in high carbon content coals could be attributed to possible physical stacking interactions between polynuclear aromatic (PNA's) in the coal structure although there is no supporting evidence in the literature. These interactions will not be easily altered by materials which are liquids and the high effective cross-link density will result in little swelling. Alternatively, the coals may be more covalently cross linked above 86% C.

The swelling and transport behavior of coals is similar to that of other cross-linked macromolecular networks. While diffusion rates can be enhanced and mechanical properties altered, there are no obvious, immediate, cost-effective applications of this knowledge.

ACKNOWLEDGEMENT. We thank PSE&G Research Corporation for support of this work.

REFERENCES

1. Flory, P. J. J. Chem. Phys., **1942** 10, 51.

2. Flory, P. J.; Rehner, J. Jr. J. Chem. Phys. **1943** 11(11), 521.

3. Flory, P. J.; Rehner, J. Jr. J. Chem. Phys., **1943** 11(11), 512.

4. Flory, P.J. Principles of Polymer Chemistry; Cornell University Press: Ithaca and London 1983.

5a. Vam Krevelen, D. W. Coal; Elsevier: Amsterdam, 1961.

b. Van Krevelen, D.W. Fuel **1966** 45, 99,229.

6. Vahrman, M. Fuel **1970** 49, 5.

7. Palmer, T. J. and Vahrman, M. Fuel **1972** 51, 22.

8. Shapiro, M. D.; Al'terman, L. S. Solid Fuel Chem. **1977** 11(3), 13.

9. Larsen, J. W.; Lee, D.; Shawver, S. E. Fuel Process. Techn. **1986** 12, 51.

10a. Dryden, I. G. C. Fuel **1950** 29, 197.

b. Dryden, I. G. C. Fuel **1951** 30, 39.

c. Dryden, I. G. C. Fuel **1951** 30, 145.

d. Dryden, I. G. C. Discuss. Far. Soc. **1951** 11 28.

e. Dryden, I. G. C. Fuel **1946** 25, 104.

11. Dreulen, D. J. W. Fuel **1949** 28, 231.

12. Bangham, D. H. Fuel **1949** 28, 231.

13a. Brown, H. R.; Waters, P. L. Fuel **1966** 45, 17.

b. Brown, H. R.; Waters, P. L. Fuel **1966** 45, 41.

14. Lazerov, L.; Angelova, G. Solid Fuel Chem. **1976** 10(3), 12.

15. Green, T.; Kovac, J.; Breener, D.; Larsen, J. W. in Coal Structure, Meyers, R. A., ed., Chapter 6, pp. 199-282, Academic Press, New YOrk, 1982.

16. Dryden, I. G. C. in Chemistry of Coal Utilization, Suplementary Volume, Lowry, H. H., ed., Wiley, New York, 1963.

17. Mukherjee, P. N.; Bhowmik, J. N.; Lahiri, A. Fuel **1957** 36, 417.

18a. Camier, R. J.; Siemon, S. R. Fuel **197** 57, 85.

b. Camier, R. J.; Siemon, S. R.; Stanmore, B. R. Fuel **1957** 36, 417.

19. Boas-Traube, S. G.; Dryden, I. G. C. Fuel **1950** 29, 260.

20. Larsen, J. W.; Choudhury, P. J. Org. Chem. **1979** 44, 415.

21. Kirov, N. Y.; Oshea, J. M.; Sergeant, G. D. Fuel **1968** 47, 415.

22. Sanada, Y; Honda, H. Fuel **1966** 45, 295.

23a. Brenner, D. Fuel **1983** 62, 1347.

b. Brenner, D. Fuel **1984** 63, 1324/

24. MacRae, J. C.; Mitchell, A. R. Fuel **1957** 36, 1553.

25. Morgans, W. T. A.; Terry, N. B. Fuel **1958** 37, 201.

26. Larsen, J. W.; Green, T. K.; Kovac, J. J. Org. Chem. **1985**

27. Peppas, N. A.; Larson, J. M.; Lucht, L. M>; Sinclair, G. W. Proceeding of the International Conference on Coal Science, August 15-19, 1983, Pittsburgh, PA.

28. Lucht, L. M.; Peppas, N.A. Prepr. Pap.-Am. Chem. Soc., Div. Fuel Chem. **1984** 29(1), 213.

29. Szeliga, J.; Marzec, A. Fuel **1983** 62, 1229.

30. Marzec, A. Chemia Stosowana **1981** XXV 3, 39.

31. Dryden, I. G. C. Fuel **1951** 30, 39.

32. Sanada, Y.; Honda, H. Bull. Chem. Soc. Japan **1967** 47, 831.

33. Kirov, N. Y.; O'Shea, J.; Sergeant, G. D. **Fuel 1967** 47, 831.

34. Nelson, J. R. Fuel **1983** 62, 112.

35. Nelson, J. R.; Mahajan, O. P.; Walker, P. L. Fuel **1981** 59, 831.

36. Peppas, N. A.; Lucht, L. M. in Investigation of the Cross-Linked Macromolecular Nature of Bituminous Coals, Final Report, DOE Contract No. DE-FG22-78E 13379, 1980.

37. Green, T. K.; Kovac, J.; Larsen, J. W. Fuel **1984** 63, 935.

38. Liotta, R.; Brown, G.; Issacs, J. Fuel **1983** 62, 781.

39. Green, T. K., Ph.D. Thesis **1984**, University of Tennessee.

40. Sanada, Y.; Honda, H. Fuel **1966** 45, 451.

41. Bunte, K.; Bruckner, H.; Simpson, H. G. Fuel **1933** 12, 268.

42a. Moore, B. Fuel **1931** 10, 436.

b. Moore, B. Fuel **1931** 10, 244.

43. Peppas, N. A.; Lucht, L. M.; Hill-Lievense, M. E.; Hooker, D. T. II in Macromolecular Structural Changes Changes in Bituminous Coals During Extraction and Solubilization, Final Technical Report, DOE Contract No. DE-FG22-80PC30222, 1983.

44. Larsen, J. W.; Lee, D. Fuel **1984** 64, 981.

45. Van Bodegom, B.; Van Veen, R. J. A.; Van Kessel, M. M.; Sinnige-Nijssen, M. W. A.; Stuiver, H. C. M. Fuel 64, 59.

46. Kovac, J. Macromolecules **1978** 11, 362.

47. Lucht, L. M.; Peppas, N. A. Conf. Phys. Chem. Coals **1980**.

48. Hsieh, S.-T; Duda, J. L. Proceedings of the ACS Division of Polymeric Materials: Science and Engineering, **1983** Vol. 51, 7-3.

49. Thomas, N. L.; Windle, A. H. Polymer **1983** 23, 529.

50. Astarita, G.; Sarti, G. C. Polymer and Eng. Sci. **1978** 18(5), 388.

51a. Hopfenberg, H. J. Membr. Sci. **1978** 3, 215.

b. Hopfenberg, H.; Frisch, H. L. J. Polym. Sci., Polym. Letters **1969** 405.

52. Agde, G.; Hubertus, R. Braunkohlenarch **1936** 46, 3-30.

53. Sanada, Y.; Honda, H. Fuel **1963** 42, 479.

54. Sanada, Y.; Mochida, N.; Honda, H. Bull. Chem. Soc. Japan 33, 1479.

55. Koenig, J. L. Chemical Microstructure of Polymer Chains **1980** John Wiley and Sons, New York.

56. Billmeyer, F. Jr. Textbook of Polymer Science **1962** John Wiley and Sons, New York.

57a. Hildebrand, J. H.; Scott, R. L. The Solubility of Non-Electrolytes **1964** Dover Publication Inc., New York N. Y.

b. Hildebrand, J. H.; Prausnitz, J. M.; Scott, R. L. Regular and Related Solutions Van Nostrand Reinhold Co., New York.

SOLID STATE ^{13}C NMR IN COAL RESEARCH: SELECTED TECHNIQUES AND APPLICATION

Francis P. Miknis
Western Research Institute
P.O. Box 3395, University Station
Laramie, WY 82071

ABSTRACT

Developments in solid-state ^{13}C NMR techniques and their applications in coal research are reviewed. These techniques include cross polarization and magic-angle spinning with high-power decoupling, dipolar dephasing, sideband suppression at high magnetic fields, dynamic nuclear polarization and combined rotation and multiple pulse spectroscopy. The quantitative reliability of ^{13}C NMR aromaticity measurements is also discussed. Results from recent studies indicate that not all the carbons in coal are observed in the solid-state ^{13}C NMR measurements. Thus, the question of the quantitative reliability of ^{13}C NMR in coal research remains unresolved. Selected applications of solid-state ^{13}C NMR for the study of pyrolysis and hydroliquefaction of coals are presented. Results of these studies indicate that solid-state ^{13}C NMR can provide valuable information about aromaticity changes that occur during coal processing.

1. INTRODUCTION

The application of nuclear magnetic resonance (NMR) to coal research began in 1955 when broadline ^{1}H NMR was applied to coals (1). However, it was not until the development of the solid-state ^{13}C NMR techniques of cross polarization (CP) (2) and cross polarization with magic-angle spinning (CP/MAS) (3) in the mid-1970s that coal researchers began to recognize the potential of NMR techniques for providing information about the basic carbon structure of coal. Of course, the great advantages of these techniques in coal research were that they could be applied to whole coals, little sample preparation was required other than grinding, and the techniques were nondestructive. Moreover, the techniques provided a direct measurement of the carbon aromaticity in coal, a much sought after parameter in coal research (4).

As often occurs when new advances are made in science, only a handful of laboratories were initially successful in developing or modifying NMR spectrometers to perform the CP/MAS NMR experiments. Yet,

Y. Yürüm (ed.), New Trends in Coal Science, 117–158.

it is interesting to note that some of the spectra first published to illustrate the techniques were of coals (5, 6) and oil shales (7, 8), materials that are largely insoluble and for which few good techniques existed to probe their molecular structure. Indeed it can be said that the solid-state ^{13}C NMR techniques of CP/MAS have caused a renaissance in studies of coal structure.

In this chapter, selected applications of solid-state NMR techniques to the study of coals and coal processing are discussed. This chapter does not include a comprehensive or critical review of the literature. Rather, the author's purpose is to use results presented in the literature to illustrate recent advances and applications of NMR techniques to the study of coals and to illustrate the state of the art about a decade after the emergence of these techniques in coal research. Omission of pertinent literature is an oversight of the author and is purely unintentional. Comprehensive reviews have been published (9-12), and recent books on solid-state ^{13}C NMR have been published (13, 14), as has the proceedings of a NATO Advanced Study Institute devoted to NMR in fossil fuel science (15). These should be consulted for more complete discussions of NMR techniques as applied to fossil fuels materials.

2. NMR IN SOLIDS

Before the development of CP/MAS ^{13}C NMR techniques, observations of NMR signals from coal provided little or no structural information other than confirming the high aromaticity of coal (16). This was due in part to problems associated with (1) strong ^{1}H - ^{13}C dipolar interactions in the solid-state that give rise to broad, featureless spectral bands; (2) chemical shift anisotropy, which also causes line broadening and unsymmetrical line shapes because of the many different orientations molecules in an amorphous state can assume in a magnetic field; and (3) long spin-lattice relaxation times, T_1's, for ^{13}C in solids (≈ minutes), which tend to negate signal averaging techniques. In addition, the ^{13}C nucleus is 1.1% naturally abundant, so that its sensitivity of detection is low, making signal averaging a must, provided that the problem of the long T_1's can be overcome. Items (1) and (2) are associated with resolution problems in solids, while item (3) is associated with the problem of sensitivity of detection of an NMR signal from a dilute system. It was the development of the techniques of cross polarization with high-power decoupling and magic-angle spinning that paved the way for the observation of "high-resolution" ^{13}C NMR spectra in solids. Virtually all new advances in solid-state ^{13}C NMR techniques first incorporate these techniques, followed by various forms of embellishments (e.g., higher magnetic field, different radio frequency (rf) pulse sequences, relaxation time behavior) (14).

High-Power Decoupling

For most solids the major line-broadening interaction for ^{13}C is the ^{1}H - ^{13}C dipolar interaction. The local field which the ^{1}H exerts on the ^{13}C magnetic moment is proportional to

$$H_{loc} \quad \alpha \quad \frac{\mu_H}{r^3}(1-3\cos^2\Theta) \qquad (1)$$

where μ_H is the ^1H magnetic moment, r is the internuclear distance between ^1H and ^{13}C, and Θ is the angle between r and the applied magnetic field, H_o. This interaction is largely eliminated by irradiating the ^1H spin system with a strong radio frequency field at its Larmor frequency. The effect of this is to induce rapid transitions between the ^1H energy levels that effectively make the ^1H's "transparent" to the ^{13}C nuclei and hence decouple them. Solids require much greater power levels than liquids because in liquids molecular tumbling motions tend to average the dipolar interactions to zero.

Magic-Angle Spinning

While high-power ^1H decoupling substantially reduces dipolar line broadening of ^{13}C spectral lines in solids, the resultant lines are still broadened by an anisotropic effect present in most samples, which cannot be removed by decoupling. This effect is referred to as the chemical shift anisotropy and without its removal, the spectral lines exhibit an anisotropic line shape that prevents resolution of the carbon types useful for structural studies.

The chemical shift anisotropy arises from the nonspherical electron density around the ^{13}C nuclei and is particularly prominent for aromatic and carbonyl (C = O) types of chemical bonds, that is, sp^2 hybridization (17). These types of carbons experience different shieldings of the magnetic field depending upon whether the bond axes are parallel or perpendicular to the applied field. In a polycrystalline or amorphous state all orientations between and including these extremes are possible.

The observed chemical shift in a solid, $\sigma_{obs} = \sigma_{zz}$, is given by Pinis et al. (2):

$$\sigma_{obs} = \sin^2\Theta\cos^2\phi\sigma_{11} + \sin^2\Theta\sin^2\phi\sigma_{22} + \cos^2\Theta\sigma_{33} \qquad (2)$$

where Θ and ϕ are their polar angles relating the principal axes (1, 2, 3) to the laboratory system (x, y, z) and σ_{ii} are the eigenvalues of the chemical shift tensor, σ. In liquids the isotropic chemical shift is given by

$$\sigma_{obs} = 1/3(\sigma_{11} + \sigma_{22} + \sigma_{33}) \qquad (3)$$

If $\Theta = 54°44'$, then $\cos^2\Theta = 1/3$, $\sin^2\Theta = 2/3$ and equation (2) reduces to the expression

$$\sigma_{obs} = 1/3\ (2\cos^2\phi\sigma_{11} + 2\sin^2\phi\ \sigma_{22} + \sigma_{33}) \qquad (4)$$

By physically placing the sample in the magnetic field, H_o, at the angle of 54°44', the "magic angle," and rotating the sample about this axis, $\sin^2\phi$ and $\cos^2\phi$ both average to one-half and σ_{obs} in equation (2)

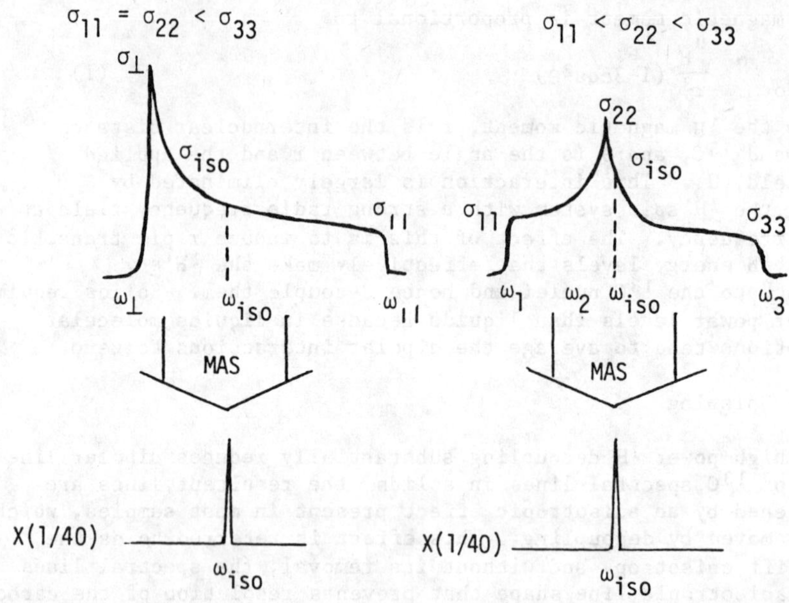

Figure 1. Theoretical effect of magic-angle spinning on the powder patterns due to chemical shift anisotropy. In the lower spectra the total intensities have been scaled down by a factor of about 40. (Reprinted from Maciel G. E., Magnetic Resonance Introduction, Advanced Topics and Applications to Fossil Energy, L. Petrakis, and J. P. Fraissard, eds, NATO ASI Series C, **124**, 98, (1984) by permission of the D. Reidel Pub. Co.©).

reduces to σ_{obs} in equation (3), provided the rotational rates are of sufficient speed. The theoretical effect of magic-angle spinning on the powder patterns because of a chemical shift anisotropy is shown in Figure 1 for the general case (right spectrum) and for the case of axial symmetry where $\sigma_{11} = \sigma_{22} < \sigma_{33}$ (18).

As can be seen, magic-angle spinning reduces the broadening caused by the chemical shift anisotropy and whatever residual broadening persists is due primarily to the many different chemical shifts of the various carbons in the sample. Line broadening caused by unpaired electrons, carbons bound to other magnetic nuclei, and incomplete decoupling may be still present. However, these broadening effects have not yet been investigated in detail to ascertain their importance in resolving the various carbon types. The spinning rate at the magic angle must be of the order of the chemical shift anisotropy broadening. Because the chemical shift anisotropy is field dependent, higher fields require higher spinning rates. For example, the chemical shift anisotropy for carbonyl carbons can cover a spectral range of 200 ppm (18), which is almost the entire chemical shift range for carbons under high resolution conditions. At a ^{13}C frequency of 50 MHz, this corresponds to a spinning rate of 10kHz. Generally, these high rates are not routinely achievable for any reasonably sized sample and lower spinning rates have been employed, even at the higher fields. This gives rise to problems associated with spinning sidebands. This aspect is discussed later.

Cross Polarization

The obstacle of sensitivity of detection of dilute spins (^{13}C) in solids has been overcome by the technique of cross polarization (2). Cross-polarization NMR relies on the presence of a system of abundant nuclear spins (^{1}H) in order to observe the NMR signal from a dilute spin (^{13}C). The main idea is to provide a mechanism for efficient transfer of polarization (hence signal intensity) from the abundant ^{1}H spin system to the dilute ^{13}C spin system.

When a sample containing ^{1}H and ^{13}C nuclei is placed in a magnetic field, H_o, the energy levels of these nuclei are split by an amount that depends upon the strength of the field, H_o, and a fundamental constant of the nucleus, the magnetogyric ratio, γ, (Figure 2, laboratory frame). The ratio, γ_H/γ_C, is nearly 4 so that for a given field, the ^{1}H energy levels are split four times greater than the ^{13}C energy levels. For a field strength of 1.4 Telsa, the splittings correspond to Larmor frequencies of 60.00 MHz and 15.08 MHz for the ^{1}H and ^{13}C nuclei, respectively. Because of this "mismatch" in energy levels, any energy

Laboratory Frame

$\omega_H = \gamma_H H_o$

$\omega_C = \gamma_C H_o$

^{13}C

^{1}H

Abundant Spins

Dilute Spins

Rotating Frame

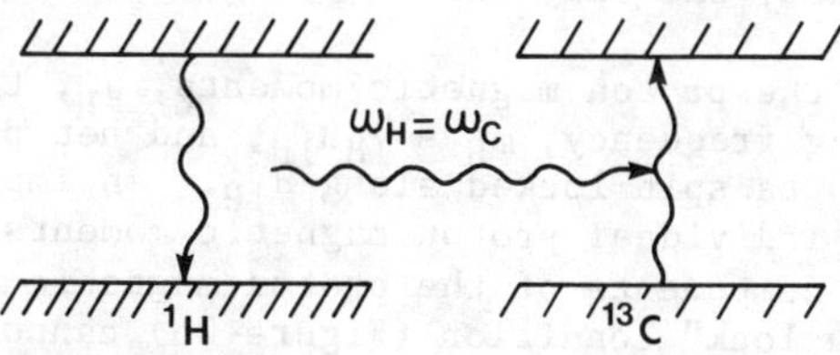

Figure 2. Schematic diagram of energy level matching in rotating frame under Hartmann-Hahn conditions.

given up by the ^1H spin system cannot be effectively utilized by the ^{13}C spin system. This mismatch is indicated schematically in the upper portion of Figure 2. However, it is possible to make a transformation to a rotating reference frame and "match" the ^{13}C and ^1H energy levels by satisfying the Hartmann-Hahn conditions, $\omega_H = \omega_C$ (19) where ω is the angular frequency. Under these conditions the energy levels of the two spin systems are equalized, and a ^1H spin flip can cause a ^{13}C spin to become aligned with the field, H_o, (i. e., populate the lower energy state). As a result energy is transferred efficiently between the abundant ^1H spin system and the dilute ^{13}C spin system. This process is called cross polarization and takes place much more rapidly than spin lattice relaxation processes. Thus, one doesn't have to wait the usual 3-5T_1 of ^{13}C in order to repeat the experiment for purposes of signal averaging. Also, because $\gamma_H/\gamma_C \approx 4$, a fourfold enhancement in ^{13}C signal intensity can be obtained from the cross-polarization process.

The CP experiment consists of four basic timed sequences of radio frequency (rf) pulses and is shown schematically in Figure 3. The left portions of Figure 3 represent the experimental timing sequence in the CP experiment, and the right portions depict what happens to the nuclei in the sample under the action of the rf pulses. The four-part procedure consists of (1) polarization of the ^1H spin system, (2) spin-locking in the rotating frame, (3) establishment of ^{13}C-^1H contact, and 4) observation of the ^{13}C free induction decay (FID).

Polarization of ^1H . The sample is prepared by applying a 90° rf pulse of intensity, H_{1H}, at the ^1H resonance frequency (60.00 MHz for a static magnetic field, H_o, of 1.4T, Figure 3a). The action of this pulse rotates the magnetization vector away from its equilibrium position along H_o (in the Z direction) and into the x'y' plane, along y'. It is assumed that the coordinate systems from which the ^1H and ^{13}C spins are viewed are rotating about H_o at their respective Larmor frequencies. That is, the ^1H coordinate system x'y' rotates at $\gamma_H H_o$, and the ^{13}C system (x''y'') at $\gamma_C H_o$.

Spin-locking. Immediately after the 90° pulse (within microseconds), a 90° phase shift is applied to the rf field, H_{1H}, to redirect it along y', colinear with the proton magnetization (Figure 3b). At resonance, and in the rotating frame, the only field acting on the ^1H spins is H_{1H}.

This field causes the proton magnetic moments, μ_H, to precess about H_{1H} at the corresponding frequency, $\omega_H = \gamma_H H_{1H}$, and net proton magnetization vector to be spin-locked along H_{1H}. An important fact is that the precession of individual proton magnetic moments about H_{1H} causes oscillating components of the proton magnetic moments along z. The so-called "spin-lock" condition (Figure 3b) cannot be maintained indefinitely because of the strong field H_o. As a result, the projection of proton magnetization along y' eventually will decay to zero, even in the presence of H_{1H}. The characteristic time for this to occur is called the spin-lattice relaxation time in the rotating frame and generally is denoted by $T_{1\rho}$.

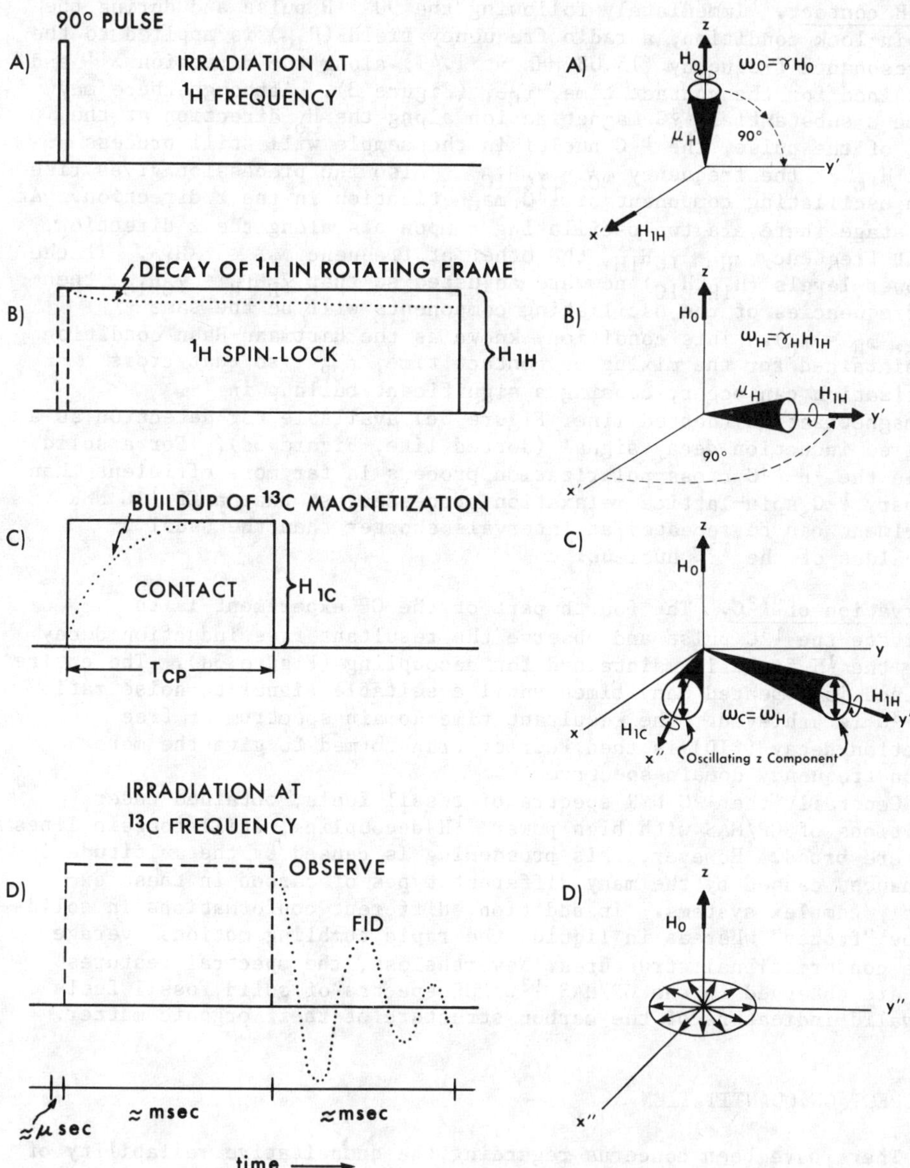

Figure 3. Timing sequence for cross polarization experiment a) polarization of ^{1}H, b) spin-locking of ^{1}H in rotating frame, c) ^{13}C-^{1}H contact under Hartmann-Hahn conditions, and d) observation of ^{13}C free induction decay.

^{13}C-^{1}H contact. Immediately following the 90° ^{1}H pulse and during the ^{1}H spin-lock condition, a radio frequency field (H_{1C}) is applied to the ^{13}C resonance frequency (15.08 MHz at 1.4T) along the direction X'' and maintained for the contact time, τ_{CP}, (Figure 3). Although there may not be a substantial ^{13}C magnetization along the H_o direction at the start of the pulse, the ^{13}C nuclei in the sample will still process about H_{1C} at the frequency $\omega_C = \gamma_C H_{1C}$. Also the precession gives rise to an oscillating component of ^{13}C magnetization in the z direction. At this stage there are two oscillating components along the z direction, one at frequency $\omega_H = \gamma_H H_{1H}$, the other at frequency $\omega_C = \gamma_C H_{1C}$. If the rf power levels (H_{1H},H_{1C}) now are adjusted so that $\gamma_H H_{1H} = \gamma_C H_{1C}$, then the frequencies of the oscillating components will be the same (i.e., $\omega_H = \omega_C$). This condition, known as the Hartmann-Hahn condition is maintained for the mixing or contact time, τ_{CP} , so that cross polarization can occur, causing a significant buildup in ^{13}C magnetization (dotted line, Figure 3c) available for detection as a ^{13}C free induction decay signal (dotted line, Figure 3d). For a solid sample the ^{1}H-^{13}C cross-polarization process is far more efficient than ordinary ^{13}C spin-lattice relaxation processes, so that a CP ^{13}C NMR experiment can be repeated at intervals shorter than the usual 3-5T_1 values of the ^{13}C nucleus.

Observation of ^{13}C. The fourth part of the CP experiment is to terminate the ^{13}C pulse and observe the resultant free induction decay while the ^{1}H field is maintained for decoupling (Figure 3d). The entire sequence is repeated many times until a suitable signal-to-noise ratio for ^{13}C is achieved. The resultant time domain spectrum or free induction decay (FID) is then Fourier transformed to give the more common frequency domain spectrum.

Generally the ^{13}C NMR spectra of fossil fuels, obtained under conditions of CP/MAS with high power ^{1}H decoupling, still contain lines that are broad. However, this broadening is caused by the multitude of resonances caused by the many different types of carbon in these extremely complex systems. In addition, different conformations in solids can be "frozen" whereas in liquids the rapid tumbling motions average these conformational structures. Nevertheless, the spectral features that are observed in the CP/MAS ^{13}C NMR spectra of solid fossil fuels are valid indicators of the carbon structure of their organic matter.

3. ASPECT OF QUANTITATION

There have been concerns regarding the quantitative reliability of CP/MAS ^{13}C NMR measurements of aromaticities in coals (10, 14, 22-28). The main concerns are (1) whether or not all the carbons are observed equally in a CP experiment, since different carbon types cross-polarize at different rates and, therefore, may not contribute equally to the signal intensity under the conditions of the experiment; and (2) whether the fraction of the observed carbons at true signal intensity is a high enough fraction to ignore the small loss of signal from unobserved carbons, or, whether the signal still represents the carbon functionality in coals if the fraction of observable carbons is too low.

Another concern arises from instrumental advances that allow CP/MAS measurements to be made at high magnetic field strengths. As mentioned in the previous section, spinning sidebands are generated by magic-angle spinning, and high magnetic field strengths require higher spinning speeds in order to keep sidebands from overlapping with signals in a spectral region of interest. This is particularly important for coals, because aromatic carbons constitute the major source of spinning sidebands in coal spectra. If the spinning speeds are not commensurate with the chemical shift anisotropy broadening associated with the high field, then aromatic sidebands will overlap the aliphatic spectral region and cause errors in the carbon aromaticity measurements. Techniques have been developed (29) for eliminating these contributions. Their applications to the NMR of coals are discussed later.

Quantitative Reliability of Cross Polarization

For CP/MAS ^{13}C NMR measurements at frequencies of 15-25 MHz, spinning speeds are high enough to eliminate sideband contributions by reducing them to the noise level, or by removing them from spectral regions of interest wherein they can be accounted for. However, the main problems regarding the quantitative reliability of the CP experiment still remain and have been discussed by a number of investigators (10, 14, 22-28).

As shown in Figure 3c, the buildup of ^{13}C magnetization depends upon a characteristic time constant, T_{CH}, and the signal intensity depends upon an instrumental contact time, τ_{CP}. In addition, the signal intensity is truncated by the proton rotating frame relaxation time, $T^H_{1\rho}$ (Figure 3b). The observed signal depends on the values of T_{CH}, $T^H_{1\rho}$ and the chosen instrumental contact time, τ_{CP}. Questions of quantitation arise because T_{CH} and $T^H_{1\rho}$ are different for different types of carbon, but only a single τ_{CP} is used for measuring the signal. Botto et al. (22) have recently measured T_{CH} and $T^H_{1\rho}$ for coals and coal macerals and have given a thorough discussion of how these parameters affect the reliability of solid-state ^{13}C NMR measurements.

An idealized representation of the carbon magnetization behavior versus contact time is shown in Figure 4 (22). The buildup of magnetization is governed by T_{CH} and the decay by $T^H_{1\rho}$. The analytic form of the magnetization is given by

$$M_t = Mo \exp(-\tau_{cp}/T^H_{1\rho})\ (1-\exp(b\tau_{cp}/T_{CH}) \qquad (5)$$

where $b = 1-T_{CH}/T^H_{1\rho}$. The condition, $T_{CH} \ll \tau_{CP} \ll T^H_{1\rho}$, would be the "ideal" situation. However, each carbon type has a characteristic T_{CH} and $T^H_{1\rho}$ that may not satisfy the ideal condition for a given τ_{CP}. T_{CH} depends on the number of protons in the vicinity of the carbons and is different, for example, for aromatic and aliphatic carbons. This is illustrated in Figure 5a-d for a series of maceral concentrates—

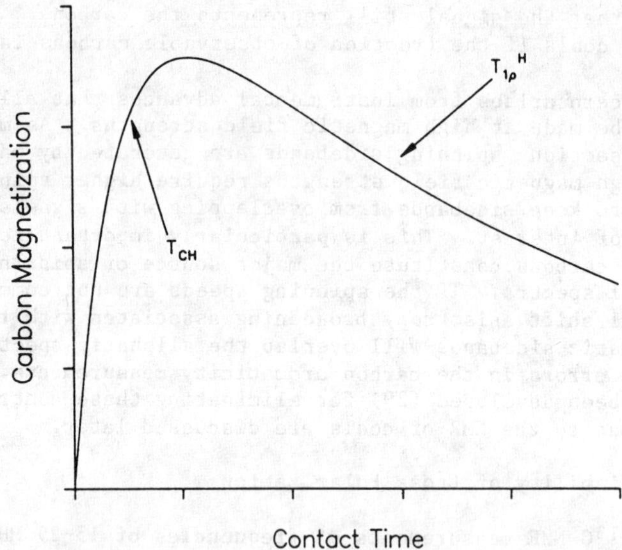

Figure 4. "Idealized" behavior of the carbon magnetization with the experimental contact time in the cross-polarization experiment (Reprinted from Botto, R. E., et al., Energy and Fuels, 1, 175 (1987) by permission of the American Chemical Society©).

resinite, sporinite, fusinite and vitrinite. Careful inspection of Figure 5 shows differences in the aliphatic and aromatic carbon signal intensities for each maceral. In addition, the aromatic carbons have maximum intensities at longer contact times than the aliphatic carbons. For the aromatic carbons, the contact times for maximum intensities range from 1.5 to 3.0 ms, while for the aliphatic carbons the contact times range from 0.5 to 0.8 ms for maximum intensity. These differences illustrate that a single contact time would not yield a spectrum of maximum polarization for both types of carbons. To overcome this problem, Botto et al. (22) fit their data (dotted lines, Figure 5) by a nonlinear least-squares method using equation 5 so that the individual curves could be extrapolated to zero contact time to obtain the full intensity spectrum independent of T_{CH} and $T_{1\rho}^{H}$ relaxation effects.

Despite the correction to zero contact time to obtain the full intensity spectrum, the question still remains as to whether CP/MAS results yield quantitative aromaticities. Carbon aromaticities obtained by other NMR measurements can be used to compare with the CP/MAS results. One method is to obtain the ^{13}C NMR spectrum under conditions of magic-angle spinning, without cross polarization, but with long delay times (i.e., a Bloch-decay or FID experiment). In this experiment the signal intensity is governed by the ^{13}C T_1's and is independent of T_{CH} and $T_{1\rho}^{H}$ effects. Since the ^{13}C T_1's are substantially longer than the $T_{1\rho}^{H}$ of the protons, these experiments require much longer times to elicit a ^{13}C spectrum. Nevertheless, Bloch-decay experiments have been employed by a number of investigators (22, 23, 30) to check the CP/MAS

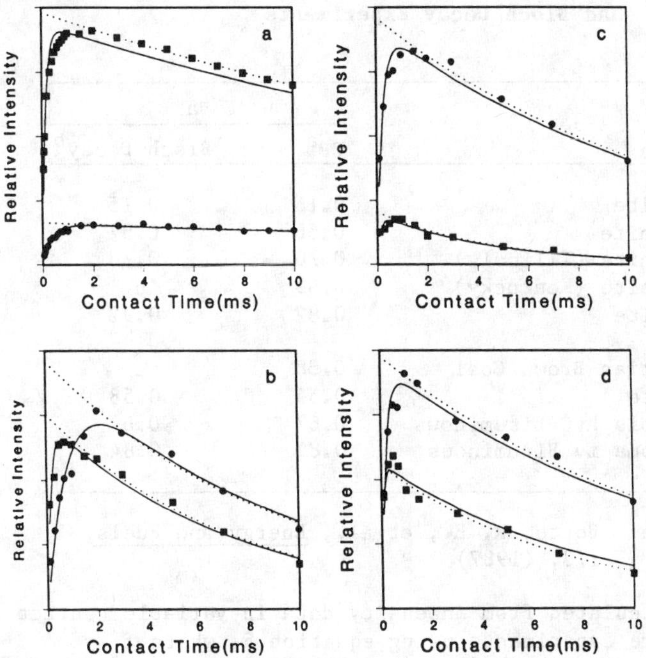

Figure 5. Variation in aliphatic (■) and aromatic (●) carbon signal intensity with contact time for maceral concentrates: (a) resinite; (b) sporinite; (c) fusinite; (d) Illinois vitrinite. The solid and dashed curves represent, respectively, the least-squares fits of the experimental data points to biexponential and exponential expressions as described in the text. (Reprinted from Botto, R. E., et al., Energy and Fuels, 1, 176 (1987) by permission of the American Chemical Society©).

results. In general, the CP/MAS values were found to be lower than those obtained from Bloch-decay experiments. The recent data (Table I) from Botto et al. (22) illustrate these differences. Except for the fusinite sample, there is a reasonably good agreement between the two methods, and the aromaticity values show the expected trend of increasing with increasing rank of coal.

The reasonable agreement for aromaticities obtained by two different NMR procedures still does not settle the matter of whether all carbons are being observed by NMR or, if not, whether the observable carbons are representative of the carbon distribution. Paramagnetic centers, for example, can cause rapid relaxation of carbons (31) and render them unobservable in both types of experiments. One method for testing this aspect of NMR is to use spin counting techniques (22, 32-35). In this procedure, an internal standard of known composition (and known ^{13}C spectrum) is homogenously mixed with the coal, and the total ^{13}C NMR spectrum is recorded, so the intensity, I_x , of the sample, and

Table I. Fraction of Carbon Aromaticity Estimated in CP and Bloch Decay Experiments

	f_a	
Sample	CP[a]	Bloch Decay[b]
Resinite	0.16	0.15
Sporinite	0.56	0.57
Vitrinite (Illinois)	0.70	0.73
Vitrinite (Kentucky)	0.69	c
Fusinite	0.82	0.92
Victorian Brown Coal	0.38	c
Lignite	0.55	0.58
Illinois hvC bituminous	0.67	0.67
Oklahoma mv Bituminous	0.81	0.84

Source: Botto, R. E., et al., Energy and Fuels, **1**, 173, (1987)

[a] Calculated from intensity data in variable contact time experiments using equation 5, where $f_a = M_o(\text{aromatic})/M_o(\text{aliphatic}) + M_o(\text{aromatic})$, estimated accuracy $\pm$ 0.01

[b] Estimated accuracy $\pm$ 0.02

[c] Not determined

the intensity, I_o, of the standard are measured. If the mass wt % carbon of the sample, m_x, and that of the standard, m_o, are known, then the percentage of carbon observed is obtained from

$$\%C_{obsd} = \frac{m_o\ (C_o\%)\ I_x}{m_x\ (C_x\%)\ I_o} \times 100 \qquad (6)$$

Botto et al. (22) have applied this technique on macerals and coals and found the results given in Table II. It is clear that the amount of observable carbon is different for the Bloch-decay and CP techniques, and in either method, the observable carbons decrease with increasing rank for the macerals. The loss of signal in the Bloch-decay experiments is attributed to the influence of paramagnetic centers, which cause severe line broadenings or shift the resonances outside the limits of detection. In the case of the CP spectra, a further loss in signal results from aromatic carbons being too remote from the protons to be adequately cross-polarized.

Table II. Percentage of Carbon Spins Observed in Bloch Decay and Cross-Polarization Experiments Using Hexamethylbenzene as an Internal Standard

Sample	% Carbon Observed[a]	
	Bloch Decay	CP
Resinite	100	105
Sporinite	90	90
Vitrinite (Illinois)	59	38
Fusinite	43	26
Lignite	65	55
Illinois hvC Bituminous	73	56
Oklahoma mv Bituminous	70	40

Source: Botto, R. E., et al., Energy and Fuels, **1**, 173 (1987)

[a] Estimate accuracy ± 10%

Thus, the problem of quantitation in CP/MAS NMR spectroscopy remains essentially unresolved. However, this technique is often the only way of obtaining structural information on solid materials such as coals. How confident the investigator is of his results depends on the nature of the investigation, and whether the CP/MAS results yield information that is supported by other available methods. For example, the yield of oil from oil shale pyrolysis can be nicely correlated with the aliphatic carbon content of oil shales determined by CP/MAS ^{13}C NMR and elemental analysis (36). Similarly, the residue carbon from oil shale pyrolysis and the fixed the carbon from coal proximate analysis correlate remarkably well with the aromatic carbon in the samples determined by NMR and elemental analysis (37). For these correlations it appears that the observable carbons are valid indicators of the process being studied.

Contributions from Spinning Sidebands

Earlier it was noted that magic-angle spinning produces spinning sidebands that can affect the quantitativeness of the aromaticity measurements. It was also noted that the sidebands change in relation to magnetic field strength so that the higher the field, the more severe the sideband problem becomes. Therefore, to overcome the sideband problem at high magnetic fields the sample must be rotated at a much faster rate (10 kHz at 50 MHz ^{13}C frequency versus 3 kHz at 15 MHz ^{13}C frequency) in order to eliminate the sideband problem. An example of the effect of sidebands on CP/MAS spectra for different field strengths is shown in Figure 6 (38). All spectra were recorded at a spinning rate of 4.2 kHz. The aromatic carbon sidebands for the 15 MHz spectrum occur at about 300 and -20 ppm and do not interfere with the

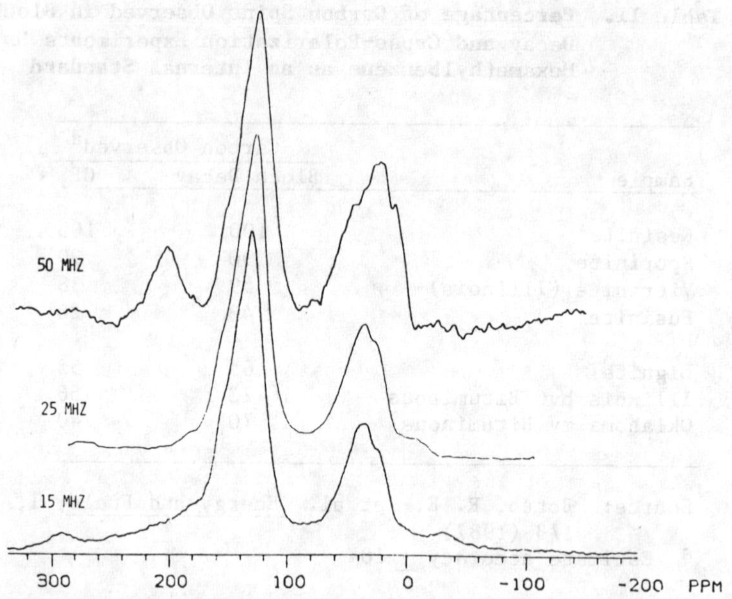

Figure 6. Comparison of ^{13}C CP/MAS spectra obtained at 15.1, 24.1, and 50.3 MHz. Powhatan No. 5 coal. (Reprinted from Sullivan, M. J., and G. E. Maciel, Anal. Chem., 54, 1613 (1982), by permission of the American Chemical Society©).

aliphatic carbon intensity. At 25 MHz the high field aromatic carbon sideband appears as a shoulder in the aliphatic carbon region at about 0 ppm. However, at 50 MHz the high field aromatic carbon sideband is under the aliphatic carbon envelope and thus provides a non-representative aliphatic carbon resonance.

While the sidebands are symmetrically located about the respective aromatic carbons giving rise to the signal, their intensities are not equivalent. This means that the contribution of the sideband that overlaps the aliphatic carbon signal cannot be ascertained by assuming it is equal to, or is some fraction of, the low field sideband intensity, and then subtracting this amount from the aliphatic carbon signal intensity. This method, however, has been applied to coals by assuming that the high field sideband has an intensity of 50% of the low field sideband (39). Reasonable agreement was obtained by this method when compared with aromaticities obtained at low field. However, in view of the complexity of the origins of sidebands (18), this method is an oversimplification.

Techniques for sideband removal. Ingenious pulsed NMR techniques have been devised to eliminate sidebands from high field spectra so that "normal" CP/MAS spectra are generated (29). These are referred to as PASS (phase alternated sideband suppression) and TOSS (total sideband suppression) techniques. A good account of the advantages and disadvantages of PASS and TOSS techniques has been provided by Axelson (14). The experimental aspects of these techniques will not be discussed. Instead, the salient features of these techniques, as applied to coals, is discussed.

The PASS and TOSS techniques remove sidebands through different approaches. In the PASS technique, spectra are recorded under differing experimental conditions (time delays) and the resultant spectra are linearly combined to remove the spinning sidebands. The TOSS technique removes all orders of spinning sidebands simultaneously in a single experiment. Therefore, there is a considerable time advantage using TOSS. However, the quantitativeness of the TOSS experiment has been questioned (14, 40). The main reason for this is that the TOSS experiment appears to overemphasize the aliphatic carbon portions of the signal, which would lead to an erroneous carbon aromaticity measurement. Nevertheless, the TOSS technique has recently been applied to a wide variety of fossil fuel materials that included coals of different rank, weathered and oxidized coals, asphaltenes, and stockpiled bituminous coals (41). In general, consistent aromaticities were obtained in most coals for normal CP/MAS and TOSS spectra. The more inconsistent results were for aromaticity measurements of subbituminous and oxidized coals. However, all the normal CP/MAS spectra were generated at 50 MHz by varying the spinning rate so that the high field sideband did not interfere with the aliphatic carbon signal. This approach is somewhat tenuous. A better method would be to compare the TOSS spectra with CP/MAS spectra recorded at low field where overlap of spinning sidebands is not a factor.

The PASS technique appears to be the more preferable of the two techniques for carbon aromaticity measurements in coal (14, 39, 42). An example of the application of PASS is shown in Figure 7 (42). The effectiveness of PASS for sideband removal is obvious. There are noticeable differences in the aliphatic portion of the signal that were attributable to insufficient suppression of background signals from the rotor material in the sample spinner. However, it is possible that the PASS spectrum is distorted because of the nature of the experiment. Newman et al. (43) found diminished relative methylene areas in the PASS spectra of New Zealand coals; Painter et al. (39) experienced difficulties in applying the PASS techniques in their coal studies and preferred the empirical approximation of estimating the high field sideband to be 50% of the low field sideband. Botto and Winans (42) obtained good agreement between their PASS results and aromaticities determined by other methods. However, only three coals of high volatile bituminous rank were studied. Thus, it appears that the quantitative reliability of both the PASS and TOSS techniques for sideband removal from ^{13}C NMR spectra of coals has not yet been adequately assessed.

The main reasons for using high-field NMR spectrometers in coal research are that, in principle, greater resolution and greater sensitivity should be achievable at higher fields. However, in practice, this need not be the case. For example, little resolution is gained at the higher fields (38, 44). This is illustrated in Figure 6 where it can be seen that the resolution in the 15 MHz and 50 MHz spectra are comparable. There is a gain in sensitivity in going to the

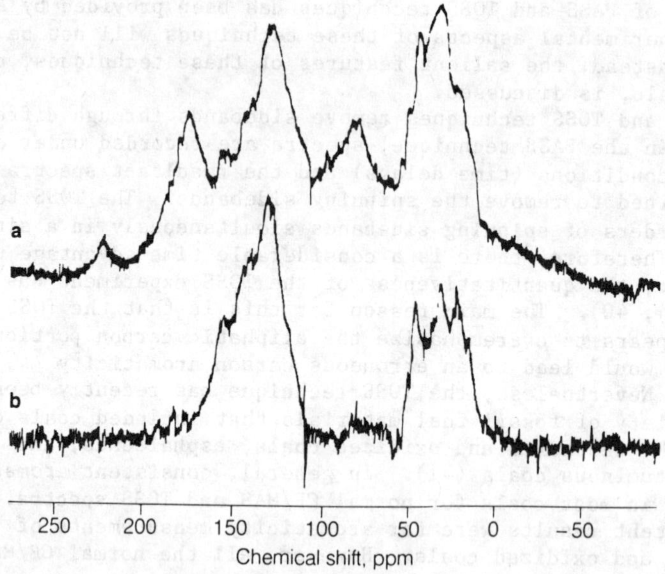

Figure 7. Solid-state 50 MHz ^{13}C spectra of high volatile bituminous coal using (a) conventional CP/MAS (b) the PASS method. (Reprinted from Botto, R. E., and R. E. Winans, Fuel, **62**, 382 (1983), by permission of Butterworth and Co., Ltd©).

higher field (38); however, this is partially offset by the fact that smaller sample sizes are used to achieve the faster spinning rates required to remove sidebands at the higher field. In view of the problems with quantitation using PASS and TOSS techniques, it can be expected that new spinners will be designed to achieve sufficiently high spinning rates (≈10 kHz) so that sideband removal at high fields will be obtained by spinning. However, the higher spinning rates may decouple weak ^{13}C-^{1}H interactions so that aromatic carbons remote from protons will not be observable. Thus, high spinning rates in conjunction with high-field spectrometers may have an adverse effect on quantitation. Based on the ratio of spinning speeds to magnetic field strength and sample size, the most desirable spectrometer frequency for ^{13}C measurements in coals appears to be about 25 MHz.

4. RESOLUTION ENHANCEMENT BY DIPOLAR DEPHASING

A variety of experimental approaches have been developed to improve the resolution of solid-state ^{13}C NMR spectra. Good discussions of these are found in the work of Sullivan and Maciel (25, 38), the review of Davidson (10), and the book by Axelson (14). Such techniques include variable temperature studies, variable frequency studies, mathematical enhancements and deconvolution techniques, and relaxation rate methods. One that is currently in vogue is a relaxation rate method called dipolar dephasing (DD) (45), also referred to as interrupted decoupling (38, 46). This experiment, a simple variation of the basic CP experiment (Figure 3a-d), is shown schematically in Figure 8.

The variation of the CP experiment is the switching off of the ^1H decoupler after the cross polarization contact time for a time, t_1, referred to as the dipolar dephasing time. During this time, t_1, the signals from different carbon types in the sample will decay exponentially at different rates depending upon the strength of the ^1H-^{13}C dipolar interactions in their local environments. The characteristic time for the decay is referred to as T_2, the spin-spin relaxation time. Carbons directly attached to hydrogens (primary, secondary, and tertiary carbons) experience strong ^1H - ^{13}C dipolar interactions and decay more rapidly than carbons not directly attached to hydrogen (quaternary carbons). When the ^1H decoupler is switched on again after a sufficiently long t_1, the resultant signal is due primarily to the quaternary carbons. However, because of their rapid rotation even in the solid-state, methyl (CH_3) carbon signals are not completely suppressed by the experiment and appear in the aliphatic portion of the signal (47). Generally, these contributions can be resolved from the quaternary aliphatic carbons on the basis of chemical shifts.

An idealized representation of a two-component decay (quaternary and tertiary aromatic carbons, for example) is shown in Figure 9 (14). Murphy et al. (45, 48) have found that the short decays are best fitted by a Gaussian or second-order exponential, while the longer decays are best approximated by a Lorentzian or first-order decay function. In an actual experiment dipolar dephasing (DD) spectra are recorded for a series of dephasing times, t_1, in order to establish the relaxation behavior of the system (Figure 9, circles). As might be expected, they

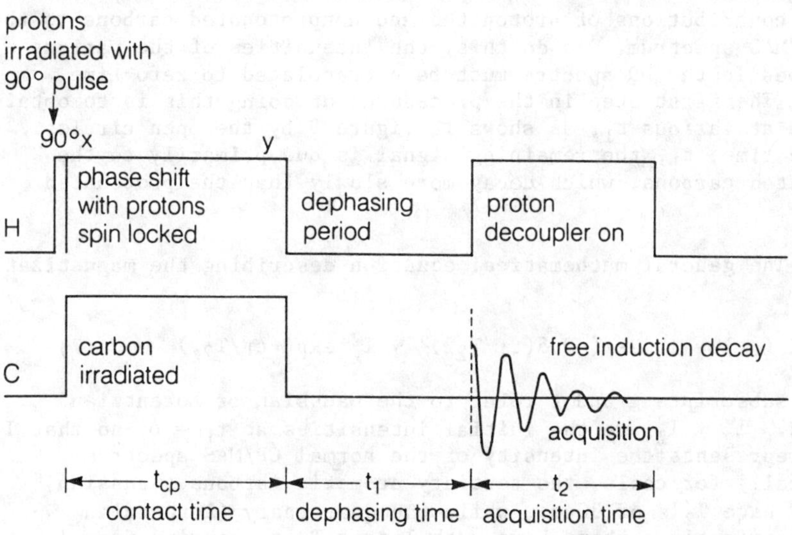

Figure 8. Timing sequence for dipolar dephasing.

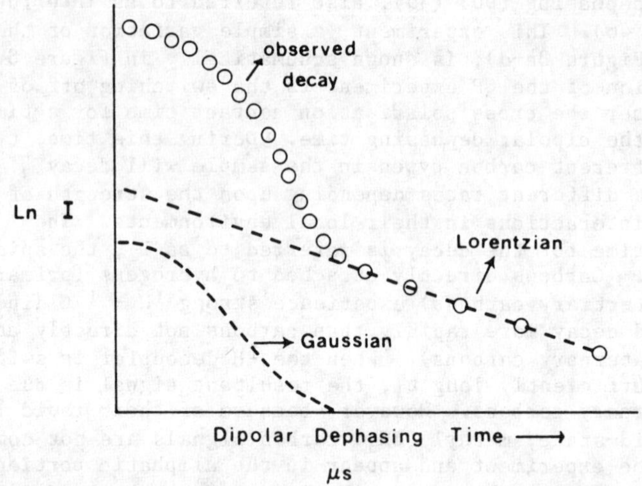

Figure 9. Schematic representation of the Gaussian-Lorentzian two-component decay in a dipolar dephasing experiment, (Reprinted from Axelson D. E., Solid State Nuclear Magnetic Resonance of Fossil Fuels, 160, (1985) Multiscience Publications Ltd, by permission of the Canadian Government Publishing Centre©).

are time-consuming experiments. Nevertheless, these experiments must be done to establish the characteristic decay constant, T_2, for the various protonated and nonprotonated carbons in the sample.

The general idea of the DD experiment is to "resolve" by relaxation times the contributions of protonated and nonprotonated carbons to the normal CP/MAS spectrum. To do this, the intensities of the various carbon types in the DD spectra must be extrapolated to zero time ($t_1 = 0$). The first step in the procedure for doing this is to obtain DD spectra at various t_1, as shown in Figure 9 by the open circles. After some time, t_1, the remaining signal is due primarily to the nonprotonated carbons, which decay more slowly than the protonated carbons. The general mathematical equation describing the magnetization is

$$I\ (\tau_1) = I_g^o \exp(-0.5(t_1/T_{2g})^2 + I_\ell^o \exp(-t_1/T_{2\ell}) \quad (7)$$

where the subscripts g and ℓ refer to the Gaussian or Lorentzian components. $I_g^o + I_\ell^o$ are the initial intensities at t_1 = o, so that I ($t_1 = 0$) represents the intensity of the normal CP/MAS spectrum.

Typically for coals, the tertiary aromatic carbons (Gaussian component) have T_2's of 30μs , while the quaternary (Lorentzian component) aromatic carbons have much longer T_2's, ranging from 150 to 300μs (49). Secondary and tertiary aliphatic carbons have T_2's in the range 20 to 50μs. In general the aliphatic carbon signals follow a single decay and the observed T_2 of this decay represents the lumped contributions from CH, CH_2 and CH_3 carbons (49).

For the aromatic carbons in coals, equation 7 reduces to

$$I\ (t_1 = 0) \simeq I_{\ell}^{o} \exp\ (-t_1 / T_{2\ell}) \qquad (8)$$

for t_1 >40μs. By fitting the data for times longer than this, $T_{2\ell}$ can be obtained and used to extrapolate the curve to t_1 = o, which yields I_{ℓ}^{o}, the contribution of quaternary aromatic carbon to the total aromatic carbon signal of the normal CP/MAS spectrum.

The contribution of tertiary (protonated) aromatic carbons can then be obtained by difference, that is,

$$I_g^o = I(t_1{=}0) - I_{\ell}^{o} \qquad (9)$$

The dipolar dephasing method has been applied by a number of investigators to obtain information about the amounts of protonated and nonprotonated carbons in coals (11, 45, 47-49). This information can be used to estimate the minimum number of condensed aromatic rings in coal (45) and to obtain other parameters (vida infra) of use for coal structure studies.

One disadvantage of the dipolar dephasing technique is that it is time-consuming because a number of DD spectra must be recorded to establish the relaxation behavior of the various carbon types. Wilson et al. (49) have made extensive measurements of this type and found that fractions of protonated, $f_a^{a,H}$, and nonprotonated, $f_a^{a,N}$, aromatic carbons could be approximated by recording spectra at only two times, t_1=0 (normal CP/MAS spectrum) and t_1=40μs. Correction factors are applied to the t_1=40μs spectrum to extrapolate the nonprotonated carbon intensities to t_1=40μs. Wilson et al. (49) were than able to derive a self-consistent set of various structure-related parameters using this method. Their basic equation for the aromaticity is

$$f_a = (1.21\ (0.88M) + P)\ /T \qquad (10)$$

where f_a is the carbon aromaticity, T is the total carbon signal at t_1=0, M is the measured aromatic carbon signal at t_1=40μs, and P is the protonated aromatic carbon signal at t_1=0. Equation 10 is used to obtain a value for P. From a knowledge of P the following can be derived:

$f_a^{a,H}$ = P/A, fraction of protonated (tertiary) aromatic carbon, where A is the total aromatic carbon signal intensity at t_1=0. Note that this fraction is identically equal to the hydrogen-to-carbon atomic ratio of the aromatic carbons, that is, $(H/C)_a$.

$f_a^{a,N}$ = $1-f_a^{a,H}$, fraction of nonprotonated (quaternary) carbon

f_a^H = $f_a^{a,H} \cdot$ fa, fraction of total carbon that is protonated aromatic carbon

Table III. Comparison of the $f_a^{a,H}$ Values Obtained from Multiple t_1 Studies and from a Series of Calculations Based on t_1 =0 and t_1=40μs

Sample	ASTM Rank	Multiple t_1 Studies	Calculations based on t_1=0 and t_1=40μs
PSOC 867	An	0.15 ± 0.02	0.12[a]
Braztah No. 9	hVa	0.39 ± 0.02	0.33
PSMC 53	mv	0.48 ± 0.02	0.55
PSMC 47	mv	0.51 ± 0.03	0.47
Blair Athol Vitrain (I)	--	0.29 ± 0.05	0.45
Beluga Lignite	lig	0.21 ± 0.09	0.18

Source: Wilson, M. A., et al., Anal. Chem., **56**, 933 (1984)

[a] Based on elemental analysis, not on t_1=0 and t_1=40μs spectra
H_a = (C/H) • f_a^H, fraction of aromatic hydrogen (hydrogen aromaticity) where C/H is the carbon/hydrogen atomic ratio obtained from elemental analysis

The limitations to this procedure are discussed by Wilson et al. (49). The obvious question is how well do the values for $f_a^{a,H}$, extrapolated from a single spectrum recorded at t_1=0, agree with those obtained from multiple t_1 measurements. These comparisons are shown in Table III for a series of coals of different rank. As seen, there is both agreement and disagreement between the two methods of measurement. Thus, for studies in which a large number of coals of varying rank are used to develop rank-dependent correlations, considerable scatter can be expected using the two-point method (11). However, the method may be useful for assessing structural changes that occur in a given coal during conversion (pyrolysis, donor solvent liquefaction, etc).

5. SENSITIVITY ENHANCEMENT BY DYNAMIC POLARIZATION

Mention has been made that there are advantages to be gained in the sensitivity of detection of ^{13}C by recording spectra at high magnetic field strengths. However, attendant with these fields are problems associated with spinning sidebands which tend to detract from the use of high-field NMR spectrometers for coal research. An intriguing approach, which shows enormous potential for overcoming the sensitivity problem, is the technique of dynamic nuclear polarization (DNP) being developed by Wind and co-workers (50-55).

The DNP effect uses free electrons to enhance either the proton signal in a CP experiment, or the ^{13}C signal directly in a free induction decay (FID) or Bloch-decay experiment. Details of how this is accomplished experimentally are given by Wind et al. (54, 55). Because coals contain a large number of free electrons ($\approx 10^{19}$ free electrons/g) a sizeable DNP enhancement effect is observable as illustrated by the times required to record the spectra in Figure 10a-d (54). Figure 10a is a spectrum from a free induction (Bloch-decay) experiment. The cross polarization (CP) spectrum (Figure 10c) was obtained via the procedures illustrated in Figure 3a-d. Figure 10b is a spectrum recorded by utilizing the free electrons to enhance the ^{13}C signal directly, and Figure 10d is recorded by first enhancing the ^{1}H signal by the DNP effect and then transferring the enhanced polarization to ^{13}C by cross polarization (indirect enhancement). For both DNP experiments the signal enhancements are obvious.

Equally obvious is the fact that the spectra are not all alike. There is a considerable difference in the aliphatic carbon signal intensity between the DNP-FID spectrum (Figure 10b) and the DNP-CP spectrum (Figure 10d). This can be explained in two ways: (1) the DNP enhancement is not uniform for all the carbons so free radicals within

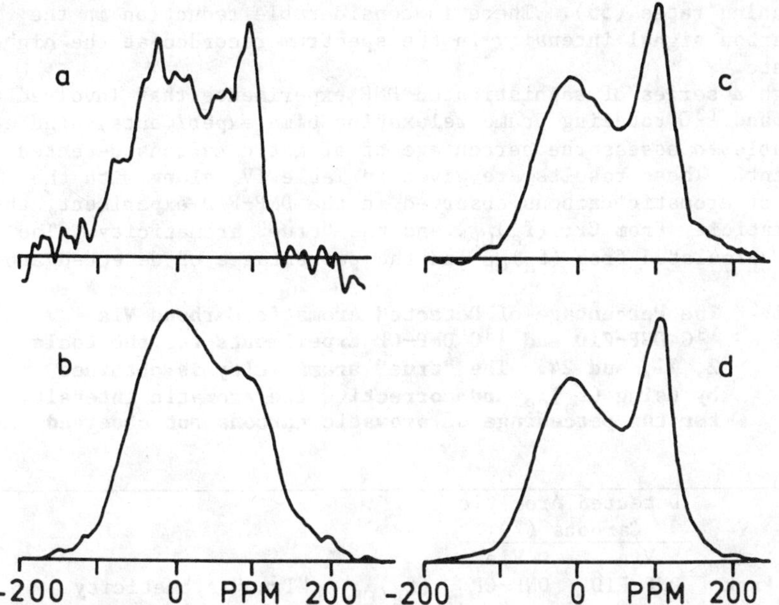

Figure 10. Solid-state ^{13}C NMR spectra of a high volatile bituminous coal measured with different techniques a) the FID experiment, 100 hrs measuring time, b) the DNP-FID experiment, 17 min measuring time, c) the CP experiment, 6.7 hrs measuring time, d) the DNP-CP experiment, 40 min measuring time (Reprinted from Wind, R. E., et al., Magnetic Resonance Introduction, Advanced Topics and Applications to Fossil Energy, NATO ASI Series C, **Vol 24**, 480, (1984), by permission of the D. Reidel Pub. Co.©).

aromatic clusters preferentially enhance the aromatic carbon signal over that of the aliphatic carbons, or (2) not all the aromatic clusters are observed in the DNP-CP (and CP spectra) because of weak ^{1}H-^{13}C dipolar couplings for aromatic carbons not attached to hydrogens. In a recent paper, Wind et al. (55) describes the same DNP and CP experiments with and without magic-angle spinning (MAS). Loss of aliphatic carbon signal was observed in both the DNP-FID and DNP-FID(MAS) spectra in contrast to the corresponding CP spectra. They concluded that the loss of aliphatic carbon signal in the DNP-FID experiments results from localization of the radicals in the aromatic region of the coal, which preferentially enhances the aromatic carbon signal.

DNP techniques have also been used to assess the quantitative reliability of CP and CP/MAS carbon aromaticity measurements. Wind et al. (55) studied 60 coals of varying rank and geographic locations throughout the world. They observed that the carbon aromaticities obtained by CP/MAS were systematically lower than those obtained by CP. The lower values obtained by CP/MAS were attributed to the MAS technique itself. Magic-angle spinning can reduce the intensity of carbons remote from protons so that some fraction of aromatic carbons is not observed, thereby lowering the measured aromaticity value. This is illustrated in Figure 11 where DNP-CP/MAS spectra are shown for a low volatile bituminous coal obtained at two spinning rates (55). There is considerable reduction in the aromatic carbon signal intensity in the spectrum recorded at the higher spinning rate.

Through a series of sophisticated DNP experiments that involved CP match time and ^{13}C rotating frame relaxation time experiments, Wind et al. (55) were able to assess the percentage of aromatic carbons detected in a CP experiment. These results are given in Table IV, along with the percentage of aromatic carbons observed in the DNP-FID experiment, the carbon aromaticity from CP, $(f_a)_{CP}$, and the "true" aromaticity. The latter was calculated from $(f_a)_{CP}$ and the percentages of detected aromatic

Table IV. The Percentage of Detected Aromatic Carbons Via ^{13}C DNP-FID and ^{13}C DNP-CP Experiments for the Coals 2, 11, and 24. The "true" aromaticity is obtained by using $(f_a)_{CP}$ and correcting the aromatic intensity for the percentage of aromatic carbons not observed

		Detected Aromatic Carbons (%)			
Coal No.	Rank	Via DNP-FID	Via DNP-CP	$(f_a)_{CP}$	"True" Aromaticity
2	An	80	46	0.98	0.99
11	lvb	90	56	0.89	0.94
24	mvb	90	65	0.85	0.90

Source: Wind, R. A., et al., <u>Fuel</u>, **66**, 876 (1987)

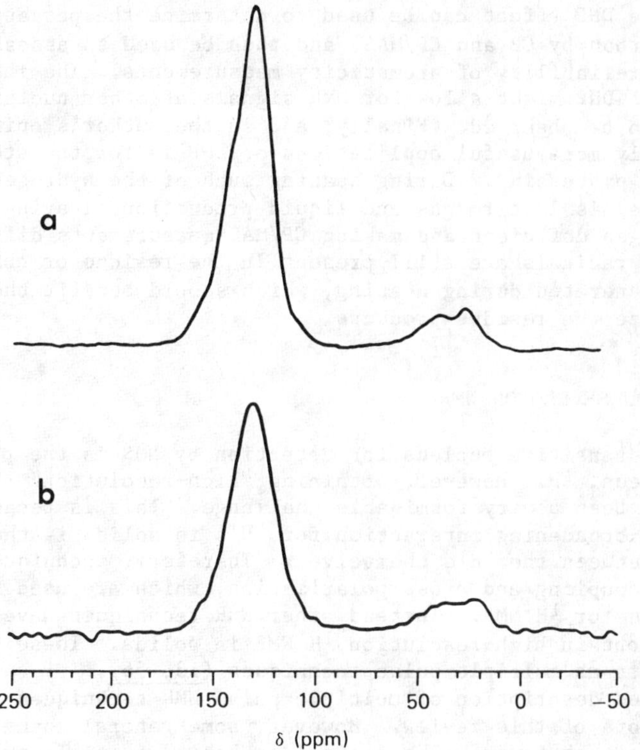

Figure 11. The ^{13}C DNP-CP (MAS) spectrum of a low volatile bituminous coal obtained at two spinning frequencies: a) f_r = 3.1 kHz, $(fa)_{CP\ (MAS)}$ = 0.87; b) f_r = 5.2 kHz, $(fa)_{CP\ (MAS)}$ = 0.82. From the non-spinning DNP-CP experiment a value $(fa)_{CP}$ = 0.89[5] was obtained (Reprinted from Wind, R. A., et al., Fuel, **66**, 882 (1987) by permission of Butterworth and Co. Ltd©).

carbons in the DNP-CP experiment, assuming 100% detection of the aliphatic carbons. The percentage of aromatic carbons detected by DNP-CP decreases with rank (coal 2 being the highest rank). However, the three coals ranged in rank from medium volatile to anthracite. Whether the trend continues to the lower-rank coals was not established. Nevertheless, the fact that only 46 to 65% of the carbons are observed by DNP-CP, and presumably lesser percentages would be obtained with DNP-CP/MAS, keeps the annoying problem of quantitation of solid-state ^{13}C NMR measurements still at the surface in NMR research on coals.

In summary, the sensitivity enhancements of DNP offer advantages for coal studies. The ^{1}H and ^{13}C enhancement factors obtained by the DNP effect might be used as a parameter to characterize coals (i.e., the larger the enhancement, the higher the rank). Information about the nature of paramagnetic centers in coals might be obtained since the DNP effect uses the free electrons which are localized around aromatic clusters. The DNP effect can be used to determine the percentages of observable carbon by CP and CP/MAS, and thus be used to assess the quantitative reliability of aromaticity measurements. The increased sensitivity of DNP might allow for NMR signals of other nuclei such as ^{17}O and ^{33}S to be observed. Finally, and in the author's opinion, one of the potentially more useful applications of DNP is for the study of coal and oil shale processing. During heating much of the hydrogen in coal (and oil shale) is lost to gas and liquid production, leaving the char or residue hydrogen deficient and making CP/MAS measurements difficult. However, free radicals are still present in the residue or char, some having been generated during heating, which should benefit the use of DNP to characterize the residue products.

6. MULTIPLE PULSE PROTON NMR

The most sensitive nucleus for detection by NMR is the proton, or hydrogen nucleus, ^{1}H. However, obtaining "high-resolution" ^{1}H NMR spectra in solids has been a very formidable challenge. This is because the dominant line-broadening interaction for ^{1}H's in solids is the dipolar interaction between the ^{1}H's themselves. Therefore, techniques such as high-power decoupling and cross polarization, which are used for ^{13}C, are not applicable for ^{1}H NMR. Instead other NMR techniques have had to be developed to obtain high-resolution ^{1}H NMR in solids. These techniques are referred to as multiple-pulse techniques (13, 56).

A detailed description of multiple-pulse NMR techniques is well beyond the scope of this review. However, some general comments can be made about the techniques. The basic idea behind multiple-pulse NMR techniques is to achieve line narrowing (i.e., eliminate the ^{1}H-^{1}H dipolar interaction) by manipulating the ^{1}H spin system with rf pulses, rather than by motion of the whole system as is done by MAS. This manipulation is done by applying a series of well-timed rf pulses in such a manner that on the average, the ^{1}H spin system is oriented along the magic angle in the rotating frame. The ^{1}H NMR signal is sampled in "windows" between each cycle of pulses, and Fourier transformed to yield the frequency spectrum. Because of the strict requirements on rf pulse widths, shapes, phasing and timing, the multiple-pulse techniques are probably the most difficult of the high-resolution solid-state NMR techniques to implement on a routine basis.

Even though the multiple-pulse techniques eliminate the major ^{1}H line-broadening interaction, the ^{1}H-^{1}H dipolar interaction, the ^{1}H spectral lines still exhibit a residual broadening because of the proton chemical shift anisotropy, which for coals is of the order of 16 ppm (57). This interaction can be eliminated by magic-angle spinning and the combination of multiple-pulse techniques with magic-angle spinning referred to as CRAMPS (combined rotation and multiple-pulse spectroscopy) (58).

For coals, ^{1}H NMR spectra obtained by CRAMPS do not exhibit a high degree of resolution. This is because of overlap of the multitude of resonances from the different ^{1}H types in coals analogous to that observed in ^{13}C CP/MAS spectra. Gerstein and coworkers (58, 59-62) were very

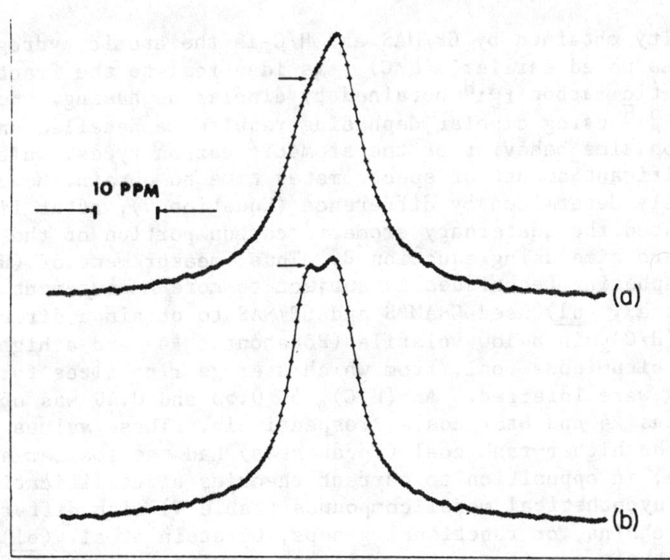

Figure 12. High-resolution solid state ^{1}H NMR spectra of (a) PSOC 501 and (b) Lovilia vitrains. (Reprinted from Gerstein et al., in Coal Structure, R. A. Meyers, ed, Academic Press Inc. New York, NY, 17, (1982), by permission of Academic Press Inc.©).

instrumental in developing the multiple-pulse and CRAMPS techniques for the study of coals. Their ^{1}H NMR spectra were recorded on a home-built spectrometer operating at a ^{1}H resonance frequency of 55 MHz. Two examples of CRAMPS spectra of coals are shown in Figure 12 (62). These spectra illustrate the state of the art circa 1980. As can be seen, resolution of the aromatic and aliphatic hydrogens would have require curve-resolving techniques in order to extract a proton aromaticity, $f_{a,H}$, of coal.

With the development of high-field spectrometers and different multiple-pulse sequences a moderate gain in resolution of CRAMPS spectra of coals has been obtained (63, 64). However, obtaining a proton aromaticity, assuming all protons are observed, will still require curve resolving. Despite this nuisance, the fact that the hydrogen distribution in coals can be measured directly using the CRAMPS technique should make this technique a useful addition to the arsenal of techniques for the study of coal structure.

Measurement of the hydrogen distribution in coal using CRAMPS and the carbon distribution using CP/MAS can provide valuable information about the structure of coal. In particular, these combined measurements provide a direct measurement of the aromatic hydrogen to aromatic carbon ratio, $(H/C)_a$ in coals since

$$(H/C)_a = (fa_{,H}/fa_{,C})\ H/C \qquad (11)$$

where $f_{a,H}$ is the hydrogen aromaticity obtained by CRAMPS, $f_{a,C}$ is the carbon aromaticity obtained by CP/MAS and H/C is the atomic hydrogen-to-carbon ratio. As noted earlier, $(H/C)_a$ is identical to the fraction of protonated aromatic carbon $f_a^{a,H}$ obtained by dipolar dephasing. However, measurement of $f_a^{a,H}$ using dipolar dephasing requires a detailed knowledge of the relaxation time behavior of the aromatic carbon types, which requires a significant amount of spectrometer time to obtain. Moveover, $f_a^{a,H}$ is ultimately determined by difference (Equation 9), after first having extrapolated the quaternary aromatic carbon portion of the relaxation to zero time using equation 8. Thus, measurement of $(H/C)_a$ using dipolar dephasing techniques is subject to more measurement errors.

Gerstein et al. (61) used CRAMPS and CP/MAS to obtain a direct measurement of $(H/C)_a$ in a low volatile (Pocahontas #4) and a high volatile (Star) bituminous coal, from which average ring sizes for the aromatic carbons were inferred. An $(H/C)_a$ of 0.53 and 0.40 was obtained for the Pocahontas #4 and Star coals, respectively. These values suggested that the higher-rank coal (Pocahontas) had the smaller aromatic carbon ring size, in opposition to current theories of coalification. With the aid of hypothetical model compounds (Table V) with differing numbers of side chains, or functional groups, Gerstein et al. (61) were able to explain their apparently inconsistent results. A value of $(H/C)_a$ of 0.53 is not inconsistent with an average ring size of

Table V. Average Aromatic Ring Size as a Function of $(H/C)_a$ and Connectivity

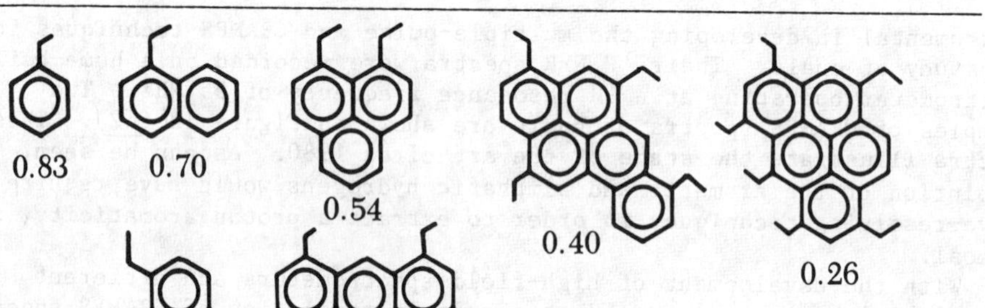

0.83 0.70 0.54 0.40 0.26

0.67 0.50

0.60 0.46 0.40 0.30

Source: Gerstein, B. C., et al. in ACS Adv in Chem. Series No. 192 Am. Chem. Soc., Washington, DC. 1981, Chapter 2

three, with two substituents on the polyaromatic ring (Table V). On the other hand, an $(H/C)_a$ of 0.40 is not inconsistent with an average ring size of two, with four substituents on the polyaromatic ring system, and is also consistent with an average ring size of four to six, with varying numbers of substituents (fourth column, Table V). Under the assumption that the lower-rank coal should have the smaller ring size Gerstein et al. (61) inferred that the Star coal had an average polyaromatic ring size no greater than two, and the Pocahontas coal had a polyaromatic ring size not greater than three. These conclusions appear to be justified and are supported by other studies (23, 62). Miknis et al. (23) noted that the width of the aromatic carbon band in CP/MAS spectra of coals of varying rank decreased with increasing rank, indicating that with increasing rank there is less substitution on the aromatic rings, and the aromatic ring sizes are larger in the higher-rank coals.

In summary, multiple-pulse ^1H NMR techniques are being developed that provide a direct measurement of the hydrogen type distribution in coals. This information can be used in conjunction with ^{13}C NMR techniques to obtain parameters that can be related to coal structure. However, as described above, some parameters, such as the $(H/C)_a$, do not provide unequivocal results about the aromatic ring size in coals and other data may be required to resolve ambiguities.

7. SELECTED APPLICATIONS

During the past decade there has been considerable activity in the application of NMR, both solid and liquid state, to the study of coal and coal products. Solid-state NMR has been applied to obtain information about the basic carbon structure of coals and how it changes with rank (23, 62, 65-71), the origin and evolution of coal (72, 73), and the behavior of coal under various heating and processing conditions (74-79). Davidson (10) has written an excellent review that summarizes many of these applications of NMR to coal research.

While it is important to understand the structure and origins of coal, it is equally important to know how coal behaves under various conditions that transform it into other products than can be effectively utilized. A large number of NMR studies have focused on before and after studies of the original coal and its pyrolysis products (10). However, to obtain detailed information about the chemical behavior of coal, it is necessary to obtain information about how this behavior changes with time and temperature under process conditions. In this regard NMR has not been extensively used, and in particular, the behavior of the solid coal and residue products under processing conditions has received little attention. This aspect of NMR applications to coal research is discussed in this section.

Pyrolysis

Most coal conversion processes involve heating; therefore pyrolysis is an important initial step in coal conversion processes. Although coal pyrolysis has been studied for decades, it still continues to be an active area of coal research. In part, this is the result of coal

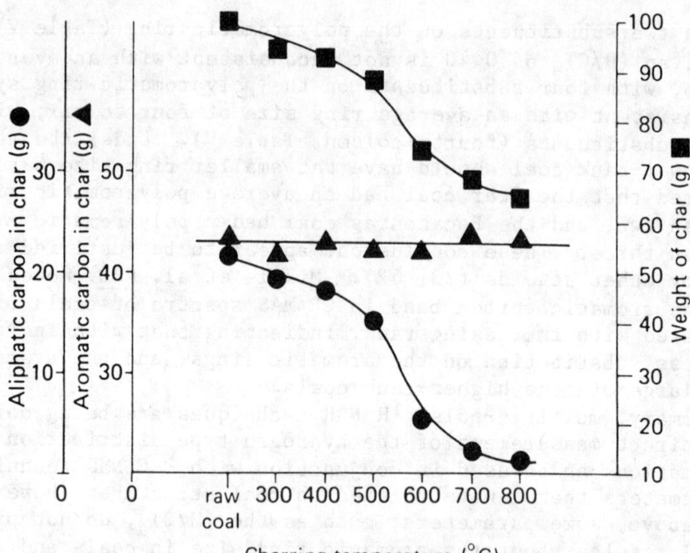

Figure 13. The calculated amounts of aliphatic and aromatic carbon in the char in relation to the temperature and the weight of char based on 100 g of raw coal used. (Reprinted from Chou M. M., et al., Liq. Fuels Technol., 2, 382 (1984) by permission of Marcel Dekker Inc.©).

researchers' awareness of the need to understand pyrolysis in relationship to the basic structure of coal. Solid-state ^{13}C NMR techniques offer one method for developing these relationships because they provide structural information about the parent coal and its residue products formed during pyrolysis. In conjunction with elemental analyses, conversion data, and liquid-state NMR, a detailed accounting can be made of changes in the mass distributions of aliphatic and aromatic carbons during pyrolysis (75, 76, 79). Such information may then provide clues to the chemistry and mechanisms of pyrolysis reactions.

Chou et al. (77) used CP/MAS ^{13}C NMR to study the chars produced in the temperature range of 300 to 800°C (Figure 13) from flash pyrolysis of Illinois No. 5 hvC bituminous coal. Elemental analyses and conversion data were obtained so that changes in the mass of aliphatic and aromatic carbon could be monitored throughout the temperature range studied. The results in Figure 13 show that the mass of aromatic carbon remains constant, even though the relative carbon aromaticities of the char increased with increasing temperature. The aliphatic carbon weight loss curve is similar to the total carbon weight curve loss, illustrating that aliphatic carbons are responsible for a large portion of the volatile carbon produced during pyrolysis. The aromatic carbons, being hydrogen deficient, prefer to remain in the char. However, the constancy by weight of aromatic carbon does not mean that the aromatic carbons have not undergone changes in their carbon structure. Tars produced from flash pyrolysis contain aromatic carbons. Some aliphatic carbon types must have "aromatized" for the aromatic carbon in the char to remain constant.

Vassallo et al. (78) used liquid- and solid-state ^{13}C NMR to determine the aromatic carbon balance from the flash pyrolysis of five Australian coals. They made before and after measurements on the coals and their products generated at 600°C in a fluidized-bed reactor; therefore the temperature dependence of the aliphatic and aromatic carbons was not determined. Their results (Table VI), however, were in disagreement with those of Chou et al. (77), in that the residue contained less aromatic carbon on a mass basis than the starting coal for each of the five coals studied. This could be the result of rapid removal of volatile matter from the coal particle, which would prevent coking of the volatile matter to produce aromatic carbons in the residue, and would also allow aromatic carbons in the coal to escape as volatile matter. In addition, differences in the experimental arrangement, heating rates, product collection, sample size, as well as differences in the basic coal structure and mineral matter between the two sets of coals could account

Table VI. Aromatic Carbon Balance on Coal Flash Pyrolysis Products from Australian Coals

	Liddell	Piercefield	Millmeran	Acland	Macalister
Coal					
Carbon Aromaticity	0.68	0.68	0.55	0.53	0.59
Aromatic C (wt %)	48.7	40.3	34.3	30.9	35.6
Char					
Carbon Aromaticity	0.79	0.88	0.82	0.83	0.80
Aromatic C (wt %)	37.4	35.6	27.2	21.1	28.1
Total Products					
Aromatic C (wt %)	47.4	46.3	41.6	34.5	39.6
% Change in Aromatic C	-2	15	21	12	11
% Conversion, Aliphatic C to Aromatic C	-5	32	26	13	16

Source: Modified from Vassallo et al., Fuel, **65**, 622 (1987)

for the discrepancy between the results of the two studies.

During pyrolysis of fossil fuels, a net increase occurs in the amount of aromatic carbon in the products (tar plus residue) over that in the starting material. This increase results from aromatization reactions of aliphatic moieties and the associated release of light, high hydrogen-content aliphatic species. Aromatization of hydroaromatic structures, such as tetralin, is a likely mechanism for contributing to the increase in aromatic carbon. By combining liquid- and solid-state ^{13}C NMR results with elemental analyses and conversion data, Vassallo et al. (78) were also able to quantify the amount of aromatic carbon in the products in excess of the aromatic carbon in the starting coal. Four of the five coals studied produced 11 to 21% more aromatic carbon than that present in the starting material. One coal (Liddell) did not produce more aromatic carbon in the products.

The increase in aromatic carbon in the products over that of the starting coal results from aromatization of the aliphatic carbon structures in the coal. This can occur either by dehydrogenation of hydroaromatic structures, or by ring closure of alkyl sidechains followed by subsequent dehydrogenation. Vassallo et al. (78) noted that the amount of aliphatic carbon aromatized might reflect differences in the amounts of straight chain versus hydroaromatic structures in coal since coals of comparable aromaticity aromatized to different extents.

A different approach to studying the fate of aliphatic and aromatic carbon during pyrolysis has been taken by Miknis et al. (79). They employed an isothermal method to study the pyrolysis of Illinois No. 6 and Wyodak coal in the temperature range of 375 to 425°C. The mass distributions of aliphatic and aromatic carbons are shown in Figure (14 a,b) for an isothermal temperature of 425°C. The carbon distribution data obtained as a function of time shows behavior similar to that shown in Chou et al. (77), which were obtained as a function of temperature (Figure 13). That is, for both the Illinois No. 6 and Wyodak coals, the aromatic carbon mass distribution remains constant, whereas the aliphatic carbon mass distribution shows a behavior with time similar to the total carbon. The isothermal experiments were conducted in a fixed bed of coal (10 to 20 grams, 20/45 mesh particle size) using a helium sweep gas to remove volatile products. The constant aromatic carbon may be the result of trapping and coking of some of the volatile matter before it can escape the coal (78, 80). Nevertheless, the behavior of the aromatic carbon mass distribution in the studies of Chou et al. (77) and Miknis et al. (79) provide some basis for correlations that have been developed between aromatic carbon in coal and the fixed carbon from proximate analysis (37, 80).

The extent of aromatization during isothermal pyrolysis also has been determined by combining solid and liquid ^{13}C NMR measurements and the carbon mass balance data (Figures 15 and 16) (79). For some experiments NMR carbon aromaticity and/or total organic carbon measurements were not obtained because insufficient quantities of tars were produced. In these cases the amount of aromatic carbon in the tar was estimated using the weight percentage of produced tar and the average values of organic carbon and/or carbon aromaticities from other experiments at the same temperature. These data are denoted by the symbol Θ (Figures 15 and 16). The increases in aromatic carbon content for the Illinois No. 6 and

Wyodak coals are about 18% and 10%. These results are in the same range of values as the results of Vassallo et al. (78) for Australian coals, which were heated to much greater temperatures.

An interesting feature of the data is that the net production of aromatic carbon approaches its limiting value during the early stages of pyrolysis (Figures 15 and 16). For example, at 425°C the net production

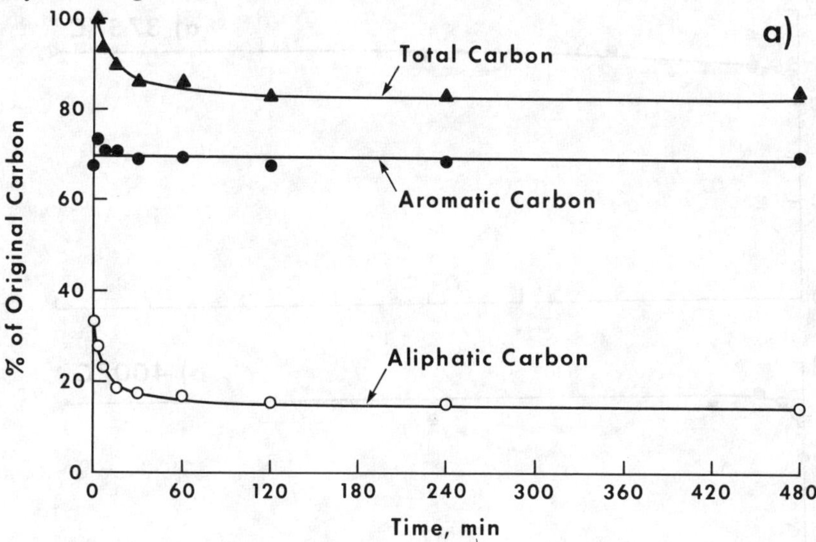

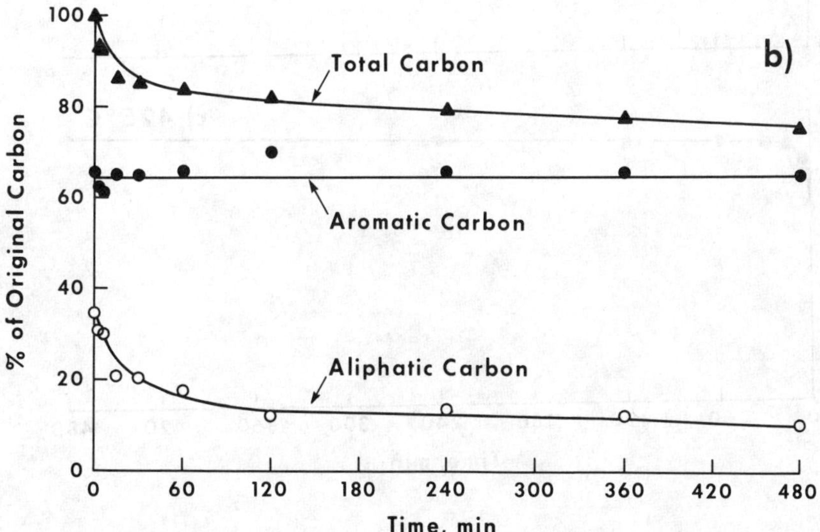

Figure 14. Distribution of carbon types vs. time at 425°C for a) Illinois No. 6 and b) Wyodak coal.

of aromatic carbon for the Illinois No. 6 coal has reached 95% of the limiting value within 2 minutes (Figure 15c). Similar behavior is noted for the Wyodak coal. Thus, aromatization of the aliphatic moieties appears to be a very facile chemical reaction; however, it is not possible from these data to determine to what extent these reactions occur directly in the solid coal or in the produced tars.

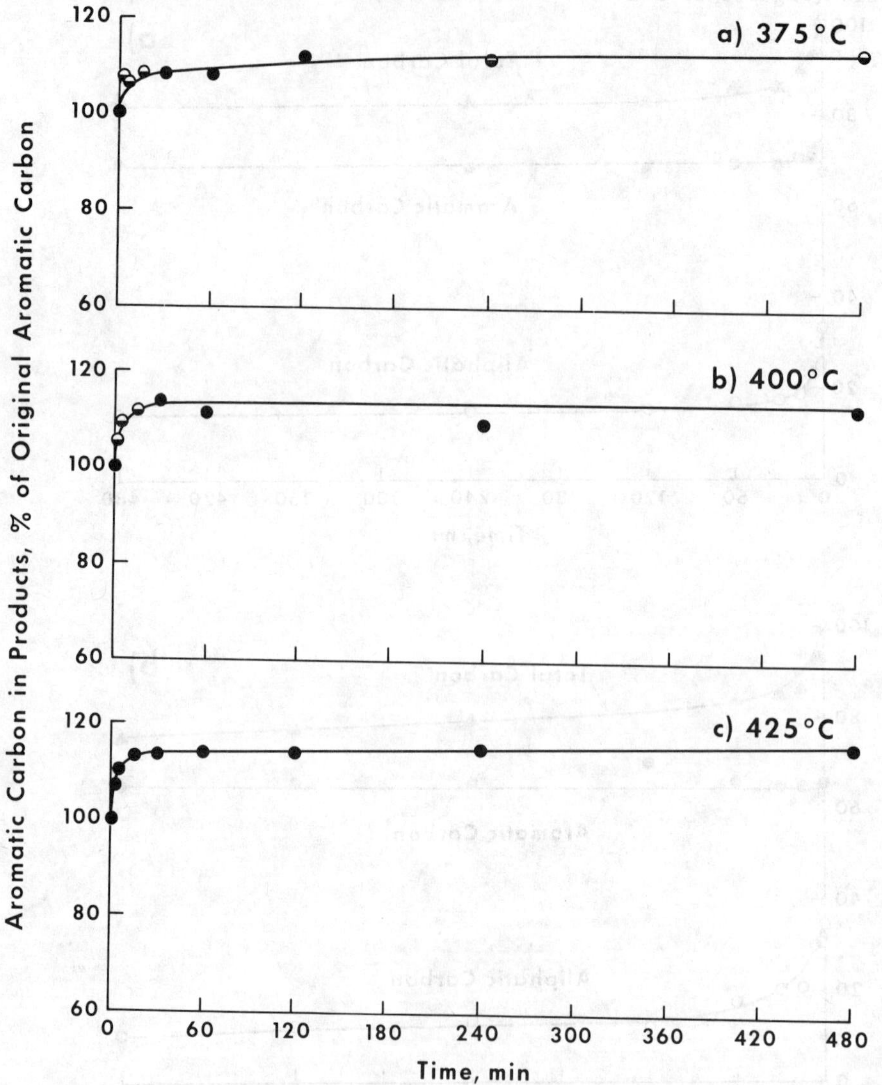

Figure 15. Total aromatic carbon in products vs. time for Illinois No. 6 coal at a) 375°C, b) 400°C, and c) 425°C.

Hydroliquefaction

Because coals are highly aromatic and therefore hydrogen deficient, they do not readily convert to liquids during pyrolysis. This is illustrated in Figure 17 where the Fischer assay pyrolysis oil yields are given for a Colorado oil shale and a Wyoming bituminous coal, along with

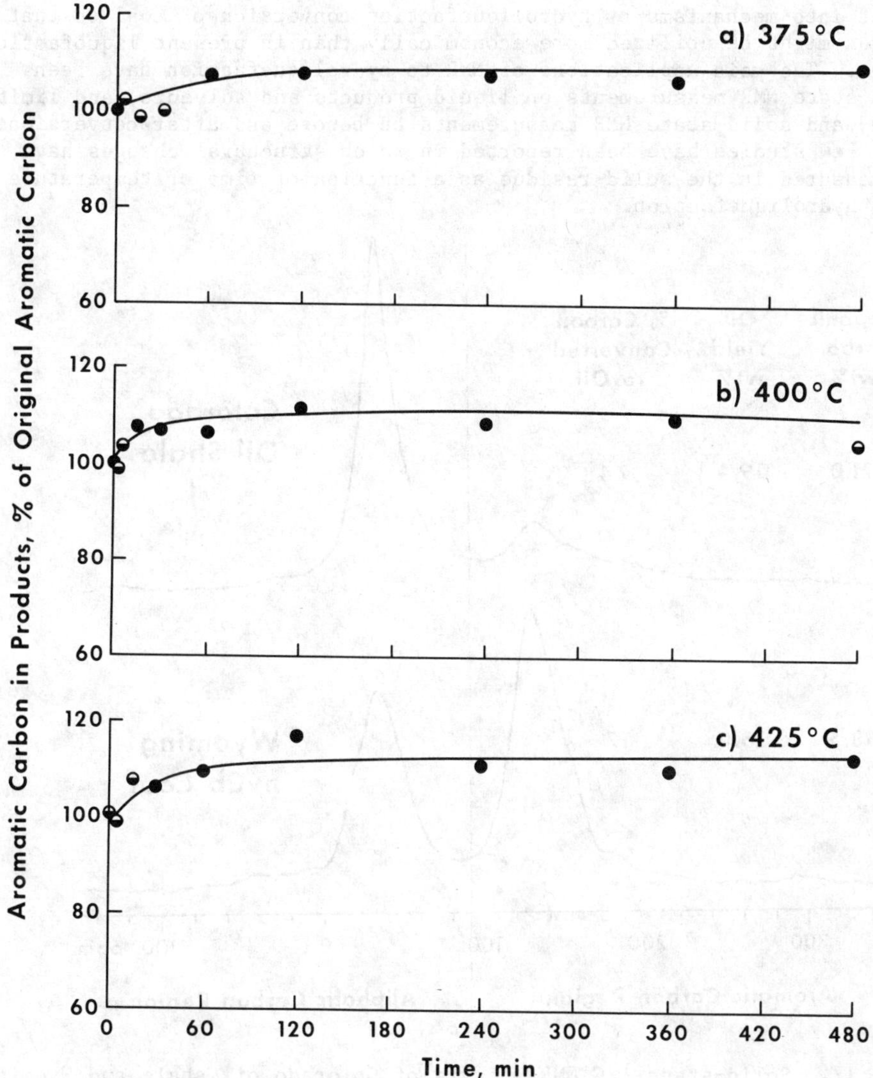

Figure 16. Total aromatic carbon in products vs. time for Wyodak coal at a) 375°C, b) 400°C, and c) 425°C.

the CP/MAS ^{13}C spectra of the two materials. The lower conversion to oil of the bituminous coal is largely a result of its high carbon aromaticity. Thus, in order to increase the yields of liquid products Numerous papers have been published on the subject of coal liquefaction, and two books (81, 82) discuss much of the relevant work done in this area to about 1980.

Because NMR techniques can be used to examine the full range of products produced during hydroliquefaction, they can used to provide insight into mechanisms of hydroliquefaction conversion of coal so that hydrogen might be utilized more economically than in present liquefaction schemes. The main applications of NMR to hydroliquefaction have been liquid-state NMR measurements on liquid products and solvents, and limited liquid- and solid-state NMR measurements on before and after conversions (10). Few studies have been reported in which structural changes have been measured in the solid residue as a function of time or temperature during hydroliquefaction.

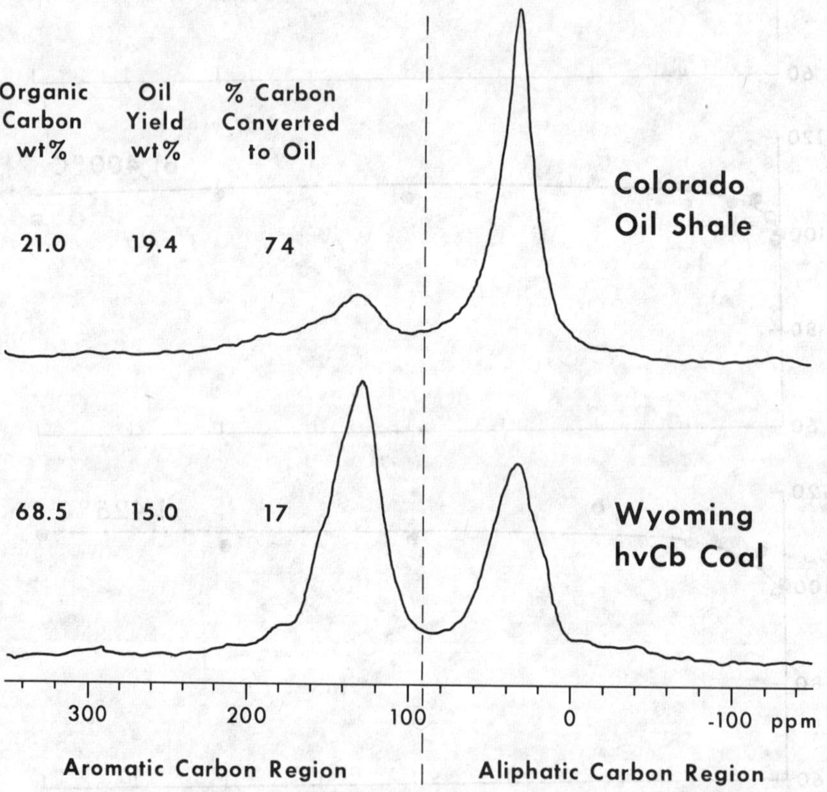

Figure 17. Solid-state ^{13}C NMR spectra of Colorado oil shale and Wyoming coal and percentage of conversion of organic carbon to oil during Fischer assay.

Wilson et al. (75) were the first to report the use of solid and liquid NMR to study changes in carbon aromaticity during liquefaction of an Australian bituminous coal (Liddell) in tetralin at 400 and 425°C. The aromaticities of the residues were found to increase with time and temperature, while that of the liquid products increased only slightly. On a mass basis, the aromatic and aliphatic carbons in the residue both decreased with reaction time, which is unlike what occurs in pyrolysis when the mass of aromatic carbon remains relatively constant (Figure 14). This result is not unexpected since conversion to liquids during hydroliquefaction is much greater than during pyrolysis.

Wilson et al. (75) were able to determine the importance of aryl ring hydrogenation, dealkylation, and hydrogenolysis for producing liquids during donor solvent liquefaction. They defined an aromaticity of the whole product, f_a^T, based on the mass balance of aromatic carbon in the residue and liquid product as

$$f_a^T = (Y_a^L + Y_a^R) / Y^C \quad (12)$$

where Y_a^L and Y_a^R are the yields of aromatic carbons in the liquid and residue product, and Y^C is the weight of carbon (m.a.f.) in the coal. If hydrogenation of aryl rings occurs, f_a^T will decrease with respect to the aromaticity in the starting coal. If hydrogenolysis or dealkylation occurs, then f_a^T should remain about the same as that of the starting coal. By plotting the change in the whole product aromaticity, f_a^T, versus the amount of hydrogen consumed from the tetralin (Figure 18), insight can be obtained as to the relative importance of hydrogenation, dealkylation and hydrogenolysis reactions (75). During the early stages of reaction in which hydrogen consumption is small, the total aromaticity decreases slightly because of aryl ring hydrogenation. Upon further reaction and hydrogen consumption, f_a^T increases slightly, rather than decreases. From this behavior, Wilson et al. (75) concluded that the major consumption of the hydrogen in tetralin is due to alkyl bond fission and hydrogenolysis reactions. However, because these reactions are temperature dependent, Wilson et al. (75) also concluded that temperature is the most important variable in determining the aromaticity of the products from coal hydroliquefaction.

Foster et al. (76) extended the work of Wilson et al. (75) by studying the liquefaction of Liddell coal under more representative isothermal conditions. Their experiments were done at 400 and 450°C for reaction times up to 120 minutes. They emphasized product yields at short reaction times (0-20 minutes), which were not studied by Wilson et al. (75). At 400°C, the total aromaticity of the products was similar to that of the starting coal (Table VII), indicating that hydrogenation of aromatic rings was negligible. At 450°C and longer reaction times, the product aromaticity was less than that of the starting coal. In addition, there was significant gas production (34% of product) at 120 minutes at 450°C. This observation, in conjunction with the decrease in aromaticity of the products, led Foster et al. (76) to conclude that aromatic rings were hydrogenated to hydroaromatic intermediates, which further reacted to form gases.

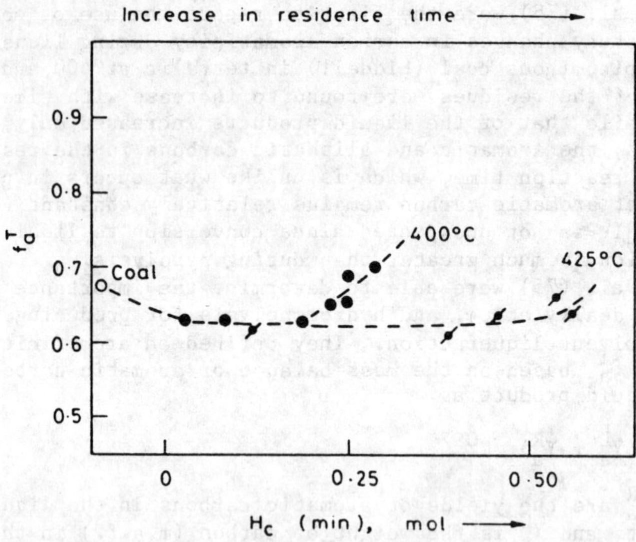

Figure 18. Change in aromaticity of the whole product f_a^T with minimum hydrogen consumption: (ɸ) 425°C; (●) 400°C. (Reproduced from Wilson, M. A., Ind. Eng. Prod. Res. Dev., **21**. 482 (1982) by permission of the American Chemical Society©).

Table VII. Total Aromaticities of Products of Isothermal Coal Liquefaction

Temperature, °C	Reaction Time, minutes	f_a^T
	coal	0.60
400	1.0	0.60
	5.0	0.58
	10.0	0.63
	20.0	0.62
	60.0	0.61
	120.0	0.61
450	0.5	0.63
	1.0	0.59
	2.5	0.61
	20.0	0.58
	60.0	0.55
	120.0	0.55

Source: Foster, N. R. et al., Ind. Eng. Prod. Res. Dev., **22**, 428 (1983)

As mentioned at the beginning of this section, NMR has been extensively applied to a variety of studies on coal. With regard to coal processing, most of the NMR studies have dealt with before and after characterization of the products. There have been few studies in which NMR measurements have been made to gain information about the behavior of coals during processing. However, because NMR measurements can be made on the full range of products produced during processing, they can provide valuable information about the chemistry of coal processing. With an understanding of the chemistry of coal conversion, improvements and novel approaches to coal conversion can be made. It can be expected that an increasing number of studies will appear in the literature with these objectives in mind.

ACKNOWLEDGMENT

The Colorado State University Regional NMR Center in Fort Collins, Colorado, USA, funded by the National Science Foundation Grant No. CHE 820-8821, is acknowledged for providing some of the NMR data used in this review. This material was prepared with the support of the U.S. Department of Energy, Grant No. DE-FG22-85PC80531. However, any opinions, findings, conclusions, or recommendations expressed herein are those of the author and do not necessarily reflect the views of the U.S. Department of Energy.

REFERENCES

1. Newman, P. C., L. Pratt, and R. E. Richard, Nature (London), **175**, 645 (1955).

2. Pines, A., M. G. Gibby, and J. S. Waugh, J. Chem. Phys., **54**, 569 (1973).

3. Schaefer, J., and E. O. Stejskal, J. Am. Chem. Soc., **98**, 1031 (1976).

4. Retcofsky, H. L., and T. A. Link in Analytical Methods for Coal and Coal Products, **vol. II**, C. Karr, ed., Academic Press Inc., NY, 1978, Chapter 24.

5. VanderHart, D. L., and H. L. Retcofsky, Fuel, **55**, 202 (1976).

6. Bartuska, V. J., G. E. Maciel, J. Schaefer, and E. O. Stejskal, Fuel, **56**, 354 (1977).

7. Miknis, F. P., and D. A. Netzel in Magnetic Resonance in Colloid and Interface Science, H. A. Resing and C. G. Wade, eds., ACS Symp. Ser. 34, Am. Chem. Soc., Washington, DC, 1976, Chapter 16.

8. Resing, H. A., A. N. Garroway, and R. N. Hazlett, Fuel, **57**, 450 (1978).

9. Miknis, F. P., Magn. Reson. Rev., **7**, 87 (1982).

10. Davidson, R. M., Nuclear Magnetic Resonance Studies of Coal, IEA Coal Research, London, 1986, Report No. ICTIS/TR32.

11. Wilson, M. A., and A. M. Vassallo, Org. Geochem., **8**, 299 (1985).

12. Snape, C. E., J. Anal. Chem., **324**, 781 (1986).

13. Fyfe, C. A., Solid State NMR for Chemists, CFC Press, Guelph, Ontario, Canada, 1983.

14. Axelson, D. E., Solid State Nuclear Magnetic Resonance of Fossil Fuels, Multiscience Publications Ltd., CANMET and Canadian Government Publishing Centre, 1985.

15. Petrakis, L., and J. P. Fraissard, eds., Magnetic Resonance Introduction, Advanced Topics and Applications to Fossil Energy, NATO ASI Series C, **vol 124**, D. Reidel Pub. Co., Dordrecht, Netherlands, 1984.

16. Retcofsky, H. L., and R. A. Friedel, Anal. Chem., **43**, 485 (1971).

17. Pines, A., M. G. Gibby, and J. S. Waugh, Chem. Phys. Let., **15**, 354 (1977).

18. Maciel, G. E., in Magnetic Resonance Introduction, Advanced Topics and Applications to Fossil Energy, L. Petrakis, and J. P. Fraissard, eds, NATO ASI Series C, **124**, D. Reidel Pub. Co., Dordrecht, Netherlands 1984, pp 71-110.

19. Hartmann, S. R. and E. O. Hahn, Phys. Rev., **128**, 3042 (1952).

20. Miknis, F. P., G. E. Maciel, and V. J. Bartuska, Org. Geochem., 1, 169 (1979).

21. Miknis, F. P., J. V. Bartuska, and G. E. Maciel, Am. Lab., **11**, 19 (1979).

22. Botto, R. E., R. Wilson, and R. E. Winans, Energy & Fuels, **1**, 173 (1987).

23. Miknis, F. P., M. J. Sullivan, V. J. Bartuska, and G. E. Maciel, Org. Geochem., **3**, 19 (1981).

24. Dudley, R. L., and C. A. Fyfe, Fuel, **61**, 651 (1982).

25. Sullivan, M. J., and G. E. Maciel, Anal. Chem., **54**, 1615 (1982).

26. Packer, K. J., R. K. Harris, A. M. Kenwright, and C. E. Snape, Fuel, **62**, 999 (1983).

27. Wilson, M. A., P. J. Collin, R. J. Pugmire, and D. M. Grant, Fuel, **61**, 959 (1982).

28. Wilson, M. A., A. M. Vassallo, P. J. Collin, and H. Rottendorf, Anal. Chem., **56**, 433 (1984).

29. Dixon, W. T., J. Magn Reson., **44**, 220 (1981).

30. Zilm, K. W., R. J. Pugmire, S. R. Larter, J. Allan, and D. M. Grant, Fuel, **60**, 717 (1981).

31. Pugmire, R. J., K. W. Zilm, D. M. Grant, S. R. Larter, J. Allan, J. T. Senftle, A. Davis, and W. Spackman, in New Approaches in Coal Chemistry, B. D. Blaustein, B. C. Bockrath, and S. Friedman, eds., ACS Symp. Series 169, Am. Chem. Soc., Washington, DC, 1981, pp 23-42.

32. Botto, R. E., R. Wilson, R. Hayatsu, R. L. McBeth, R. G. Scott, and R. E. Winans, ACS Div. of Fuel Chem. Preprints, **30**(4), 187 (1985).

33. Hagaman, E. W., R. R. Chambers, and M. C. Woody, Anal. Chem., **58**, 387 (1986).

34. Hagaman, E. W., and M. C. Woody, in Proc. Int. Conf. on Coal Science, Duesseldorf FRG, Sept. 7-9, 1981, Essen, FRG, Verlag Glueckauf, 807 (1981).

35. Kalman, J. R., in Magnetic Resonance Introduction, Advanced Topics and Applications to Fossil Energy, L. Petrakis, and J. P. Fraissard, eds, NATO ASI Series C, **124**, D. Reidel Pub. Co., Dordrecht, Netherlands 1984, pp 557-567.

36. Miknis, F. P., and G. E. Maciel, Magnetic Resonance Introduction, Advanced Topics and Applications to Fossil Energy, L. Petrakis, and J. P. Fraissard, eds, NATO ASI Series C, **124**, D. Reidel Pub. Co., Dordrecht, Netherlands 1984, pp 545-555.

37. Miknis, F. P., and P. J. Conn, Fuel, **65**, 248 (1986).

38. Sullivan, M. J., and G. E. Maciel, Anal. Chem., **54**, 1606 (1982).

39. Painter, P. C., D. W. Kuehn, M. Starsinic, A. Davis, J. R. Havens, and J. L. Koenig, Fuel, **62**, 103 (1983).

40. Sullivan, M. J., Magnetic Resonance Introduction, Advanced Topics and Applications to Fossil Energy, L. Petrakis, and J. P. Fraissard, eds, NATO ASI Series C **124**, D. Reidel Pub. Co., Dordrecht, Netherlands 1984, pp 525-533.

41. Axelson, D. E., Fuel, **6**, 196 (1987).

42. Botto, R. E., and R. E. Winans, Fuel, **62**, 271 (1983).

43. Newman, R. H., S. J. Davenport, and R. H. Meinhold, An Assessment of Carbon-13 Solid State NMR Spectroscopy for Characterization of New Zealand Coals, Lower Hutt, New Zealand, New Zealand Department of Scientific and Industrial Research Report No. CD-2346, May 1984.

44. VanderHart, D. L., W. L. Earl, and A. N. Garroway, J. Magn. Reson., **44**, 361 (1981).

45. Murphy, P. D., T. J. Cassady, and B. C. Gerstein, Fuel, **61**, 1233 (1982).

46. Opella, S. J., and M. H. Frey, J. Amer. Chem. Soc., **101**, 5854 (1979).

47. Schmitt, K. D., and E. W. Sheppard, Fuel, **63**, 1241 (1984).

48. Murphy, P. D., B. C. Gerstein, V. L. Weinberg, and T. F. Yen, Anal. Chem., **54**, 522 (1982).

49. Wilson, M. A., R. J. Pugmire, J. Karas, L. B. Alemany, W. R. Woolfenden, D. M. Grant, and P. H. Given, Anal. Chem., **56**, 933 (1984).

50. Wind, R. A., J. Trommel, and J. Smidt, Fuel, **58**, 900 (1979).

51. Wind, R. A., M. Duijvestijn, J. Smidt and J. Trommel, in Proc. Int. Conf. on Coal Science, Duesseldorf, FRG, September 7-9, 1981, Essen FRG, Verlag Glueckauf, pp. 812-815 (1981).

52. Wind, R. A., J. Trommel, and J. Smidt, Fuel, **61**, 398 (1982).

53. Wind, R. A., F. E. Anthonia, M. J. Duijvestijn, J. Smidt, J. Trommel, and G. McDevette, J. Magn. Reson., **52**, 424 (1983).

54. Wind, R. A., M. J. Duijvestijn, C. V. D. Lugt, J. Smidt, and J. Vriend Magnetic Resonance Introduction, Advanced Topics and Applications to Fossil Energy, L. Petrakis, and J. P. Fraissard, eds, NATO ASI Series C, **124**, D. Reidel Pub. Co., Dordrecht, Netherlands 1984, pp. 461-468.

55. Wind, R. A., M. J. Duijvestijn, C. V. D. Lugt, J. Schmidt, and H. Vriend, Fuel, **66**, 876 (1987).

56. Gerstein, B. C. Phil. Trans. Roy. Soc. Land., **A299**, 521 (1981).

57. Gerstein, B. C., in Analytical Methods for Coal and Coal Products, **Vol III** C. Karr, ed., Academic Press Inc., New York, 1979, Chapter 51.

58. Gerstein, B. C., R. G. Pembleton, R. C. Wilson, and L. M. Ryan, J. Chem. Phys. **81**, 561 (1980).

59. Gerstein, B. C., C. Chow, R. G. Pembleton, and R. C. Wilson, J. Phys. Chem., **81**, 565 (1977).

60. Ryan, L. M., R. E. Taylor, A. J. Paff, and B. C. Gerstein, J. Chem. Phys. **22**, 508 (1977).

61. Gerstein, B. C., L. M. Ryan, and P. D. Murphy, in Coal Structure, M. Gorbaty and K. Ouchi, eds, ACS Adv. in Chem Series, No. 192, Am. Chem. Soc., Washington D.C., 1981, Chapter 2.

62. Gerstein, B. C., P. D. Murphy, and L. M. Ryan, in Coal Structure, R. A. Meyers, ed., Academic Press, Inc., New York, NY, 1982, Chapter 4.

63. Rosenberger, H. G. Scheler, and K. H. Rentrop. Z. Chem., **23**, 34 (1983).

64. Rosenberger, H., G. Scheler, and E. Kunstner, Fuel (accepted)

65. Gerstein, B. C., in Magnetic Resonance Introduction, Advanced Topics and Applications to Fossil Energy, L. Petrakis, and J. P. Fraissard, eds, NATO ASI Series C, **124**, D. Reidel Pub. Co., Dordrecht, Netherlands 1984, pp. 409-439.

66. Dereppe, J. M., J. P. Boudou, C. Moreaux, and B. Durand, Fuel, **62**, 575 (1983).

67. Furimsky, E., L. Vancea, and R. Belanger, Ind. Eng. Chem. Prod. Res. Dev., **23**, 134 (1984).

68. Sfihi, H., M. F. Quinton, A. Legrand, S. Pregermain, D. Carson, and P. Chiche, Fuel, **65**, 1006 (1986).

69. Axelson, D. E., Fuel Proc. Technol., **16**, 257 (1987).

70. Neill, P. H., P. H. Given, and D. Weldon, Fuel, **66**, 92 (1987).

71. Russell, N. J., M. A. Wilson, R. J. Pugmire, and D. M. Grant, Fuel, **62**, 601 (1983).

72. Hatcher, P. G., I. A. Breger, and W. L. Earl, Org. Geochem, **3**, 49 (1981).

73. Hatcher, P. G., I. A. Breger, N. M. Szeverenyi, and G. E. Maciel, Org. Geochem., **4**, (1982).

74. Collin, P. J., R. J. Tyler, and M. A. Wilson, in Coal Liquefaction Products, **vol 1**, Schultz, H. D., ed., J. Wiley and Sons, New York, NY, 1983, 85-124.

75. Wilson, M. A., R. J. Pugmire, A. M. Vassallo, D. M. Grant, P. J. Collin, and K. W. Zilm, Ind. Eng. Prod. Res. Dev., **21**, 477 (1982).

76. Foster, N. R., M. A. Wilson, R. G. Weiss, and K. N. Clark, Ind. Eng. Chem. Prod. Res. Dev., **22**, 478 (1983).

77. Chou, M. M., D. R. Dickerson, D. R. McKay, and J. S. Frye, Liq. Fuels Technol., **2**, 375 (1984).

78. Vassallo, A. M., M. A. Wilson, and J. Edwards, Fuel, **65**, 622 (1987).

79. Miknis, F. P., T. F. Turner, and L. W. Ennen, ACS Div. Fuel Chem. Preprints, **32**(3), 148 (1987).

80. Solomon, P. R., Fuel, **60**, 3 (1981).

81. Whitehurst, D. D. (ed). Coal Liquefaction Fundamentals, Symp. Series 139, Amer. Chem. Soc., Washington, DC, 1980.

82. Whitehurst, D. D., T. O. Mitchell, and M. Farcasiu. Coal Liquefaction: The Chemistry and Technology of Thermal Processes, Academic Press, New York, NY, 1980.

A NEW APPROACH FOR QUANTITATIVE ^{13}C NMR SPECTROSCOPY OF COAL

Leon M. Stock and John V. Muntean
Department of Chemistry
University of Chicago
5735 S. Ellis Avenue
Chicago, Illinois 60637
USA

Chemistry Division
Argonne National Laboratory
9700 S. Cass Avenue
Argonne, Illinois 60439
USA

Robert E. Botto
Chemistry Division
Argonne National Laboratory
9700 S. Cass Avenue
Argonne, Illinois 60439
USA

ABSTRACT. The use of tetrakis(trimethylsilyl)silane as a qualitative chemical shift standard and as a quantitative internal standard for carbon-13 NMR spectroscopy of coals and the use of samarium(II) iodide as a reagent for the realization of more quantitative spectroscopic results through the selective removal of organic free radicals from coal are discussed.

1. INTRODUCTION

Miknis reviewed the scope and limitations of solid-state carbon-13 NMR spectroscopy for the analysis of fossil fuels in the preceding chapter (1). He pointed out that several research groups have established that only 50-70% of the carbon-13 nuclei in coals are observed in Bloch decay or CP MAS solid-state experiments (2,3,4). This chapter concerns recent work from our laboratories directed toward the achievement of more quantitative spectroscopic results (5,6). One phase of our work centers on the use of tetrakis(trimethylsilyl)-silane as a qualitative chemical shift standard and as a quantitative internal standard for the measurement of the observable quantity of carbon-13 nuclei, %C(obs). The other phase of this study focuses on the application of samarium(II) iodide for the selective removal

Y. Yürüm (ed.), New Trends in Coal Science, 159–167.

of organic free radicals from coal. Reduction of the radical content is essential for the realization of quantitative spectroscopic results.

2. MATERIALS AND METHODS

2.1 Materials

The subbituminous Wyodak coal that was used in this study was obtained from the premium sample program at the Argonne Laboratory. This coal contains 48.7%C, 3.3%H, 0.61%N, 0.64%S, 12.5%O (by difference), 27.8%water, 6.37%ash.

The methods for the preparation of tetrakis(trimethylsilyl)silane and samarium(II) iodide were presented in prior reports (7,8).

2.2 Procedures

The Wyodak subbituminous coal was dried for 18 hours at 65°C under vacuum. The dry coal contains 63.8%C, 4.83%H, 0.89%N, and 8.3% ash. Wyodak coal (1.0g) was suspended in tetrahydrofuran (100 ml) and stirred for 2 hours prior to the addition of samarium(II) iodide in tetrahydrofuran (100 ml). This mixture was stirred for 12 hours, blue samarium (II) iodide remained. Water (10 ml) was added to quench the reaction, the solvent was evaporated and the coal was collected and washed with dilute hydrochloric acid to remove lanthanide ions. The dried product contained 66.6%C, 4.89%H, 0.94%N, and 4.3%ash.

Electron spin resonance spectra were recorded on a Varian E-9 EPR spectrometer operating at 9 GHz with 100 kHz field modulation. Electron spin counting was performed by preparing standards of polystyrene doped with varying amounts of a stable organic free radical, 1,3-bis-diphenylene-2-phenylallyl radical, BDPA. Benzene was evaporated from a solution containing polystyrene and BDPA to provide a uniform, intimate solid mixture. All the samples were placed in 5 mm tubes to uniform sample depth (2 cm). Electron spin concentration in the coals was determined by double integration of the single broad resonance. The calibration curve generated by the three standards containing 10exp18, 10exp19, and 10exp20 spins/g encompassed all coal ESR spectra intensities tested. Solid state NMR spectra were recorded at 2.3T (25.18 MHz carbon-13) on a Bruker Instruments Model CXP 100 Spectrometer operated in the pulse Fourier transform mode with quadrature phase detection. The operating parameters in a typical cross-polarization experiment include a spectral line width of 10 kHz, a 2 ms contact time, a 2 s pulse repetition rate with a 56 kHz proton decoupling field.

Carbon spin counting experiments were performed by placing an external standard (TKS) in the lower portion of the rotor. A teflon spacer separated the TKS from the coal sample. Although rf field inhomogeity can cause intensity distortions for physically separated materials not placed coaxially to the NMR coil, experiments with other standards (hexamethylbenzene, adamantane, and glycine) were performed to define intensity corrections that had to be applied for the achievement of quantitative results. The realiability of this procedure, which advantageously isolates the diamagnetic reference compound from the paramagnetic coal, was tested in a variety of configurations.

3. RESULTS AND DISCUSSION

3.1 Tetrakis(trimethylsilyl)silane as a Chemical Shift Standard.

Axelson has discussed the chemical and spectroscopic properties of the materials that have been used as secondary chemical shift standards in carbon-13 NMR spectroscopy (9). He pointed out that the resonances of the commonly used secondary standards such as hexamethylbenzene, adamantane, and glycine overlap the resonances of many compounds. Earl and Vanderhart reported that potentially useful substances such as polydimethylsilane, polydimethylsiloxane and dodecamethylhexasilane have other important deficiencies (10). For example, polydimethylsilane, the most promising candidate in this group, exhibited an undesirably broad line (10). A suitable standard for solid-state NMR spectroscopy must meet several important criteria. It should exhibit a resonance that is quantitative, narrow, and distinct. It should also be grindable, nonvolatile, chemically inert, with a high solubility in some solvent so it can be easily removed from the sample under investigation. In addition, it should display a chemical shift that is independent of magnetic field strength (11) and have favorable relaxation properties and cross-polarization dynamics for use in Bloch decay and CP analyses.

We found that tetrakis(trimethylsilyl)silane (TKS) can conveniently be employed as a chemical shift reference standard for carbon-13 solid-state NMR spectroscopy. TKS can be readily synthesized and purified (7). The crystals (mp 319-321°C) are clear, white octahedrons that do not sublime at ambient conditions. The carbon-13 and proton resonances of TKS in solution appear at 2.67 ppm and 0.20 ppm, respectively.

The solid-state NMR spectrum of TKS has a single sharp resonance 3.50 ppm downfield from tetramethylsilane. This chemical shift was determined using 1,4-dioxane in a coaxial capillary tube. The chemical shift is independent of the magnetic field strength. The carbon spin lattice relaxation time, 5.3 seconds, is short enough to allow for convenient use in 90° pulse experiments.

The ^{13}C cross-polarization spectrum of cholesteryl acetate and TKS is shown in Figure 1

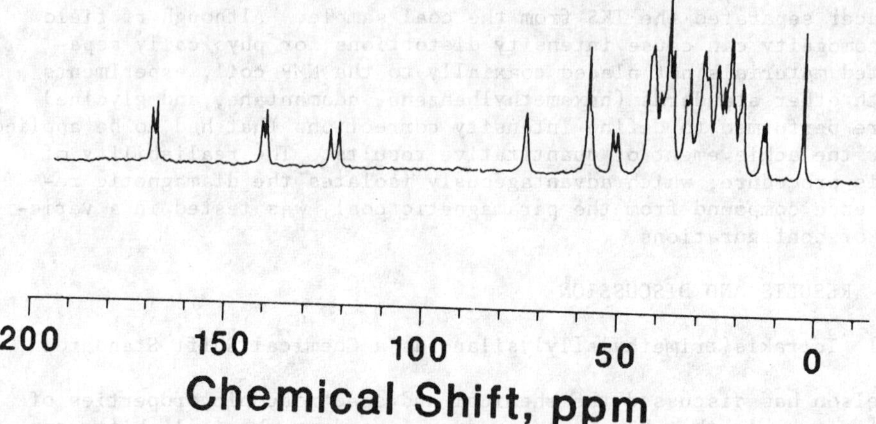

Figure 1. The solid-state CP/MAS spectrum of cholesteryl acetate (250 mg) and TKS (25 mg) with a 10 ms contact time and 10 Hz applied line broadening.

This representative result illustrates that TKS exhibits a distinct, narrow resonance separate from the resonances of most organic molecules. Thus, tetrakis(trimethylsilyl)silane has the chemical and spectroscopic properties necessary for use in solid-state NMR spectroscopy.

3.2 Tetrakis(trimethylsilyl)silane as a Quantitative Standard

Weighed mixtures of TKS and glycine, hexamethylbenzene, and adamantane were placed in a rotor. Spectra were recorded in Bloch decay experiments. Integrated NMR resonance intensities corresponded within 1% of the predicted values based on known composition of the samples.

3.3 The Selective Reduction of Organic Radicals in Coal

Miknis pointed out that a significant fraction of the carbon atoms in coals and coal macerals are invisible in solid-state carbon-13 NMR analyses (1). We recently devised a simple chemical approach to circumvent this problem exploiting samarium(II) iodide for the reduction of the radical content (6).

Organic free radicals and paramagnetic minerals both contribute to the loss in NMR signal strength, but the organic radicals have a particularly damaging effect. The inverse relationship between the observable carbon content, %C(obs), of a fossil fuel measured by reliable spin-counting techniques and the free-radical concentration measured in ESR experiments is well known. For example, when a Utah resinite that had a very low free radical content was doped with a stable organic radical, 1,3-bis-diphenylene-2-phenylallyl radical (BDPA), the %C(obs) determined by the addition of solid tetrakis(trimethylsilyl)silane to the solid resinite-BDPA mixtures was greatly reduced. In a90°-pulse experiment, the incorporation of BDPA at a concentration equivalent to 6.6 x 10exp19 spins/gram decreased %C(obs) to 63.

Very little information is available concerning the structures of the organic radicals in coal (12). However, it seems reasonable to postulate that these paramagnetic substances can be reduced to diamagnetic compounds as illustrated in the equations.

$$\text{CoalRad} + e \rightarrow \text{CoalAnion}^- \quad (1)$$

$$\text{CoalAnion}^- + \text{Proton Donor} \rightarrow \text{CoalH} + \text{Donor}^- \quad (2)$$

Preliminary experiments in which coal was treated with potassium in ammonia indicated that the organic free radical concentration could be reduced. Specifically, a demineralized sample of an Upper Freeport mv bituminous coal was reacted with potassium in liquid ammonia. The reduced coal was isolated, and its NMR and ESR spectra were recorded. Although the reduction reaction successfully decreased the ESR signal intensity by an order of magnitude and increased %C(obs) from 55 to 85, the aromatic compounds in the coal were also, as expected, reduced. The f_a value decreased from 0.82 to 0.62. This experiment verified the concept that the radicals in coal could be reduced, but the reaction with potassium was not adequately selective and the structure of the coal was significantly altered.

Elementary considerations suggested that the reduction potentials for the charged and uncharged organic radicals in coal would be lower than the potentials for the hydrocarbons and heterocycles in the coal macromolecule. This concept prompted the consideration of weaker reducing agents that might exhibit the desired selectivitiy.

Samarium(II) iodide was one of the first substances that was considered for selective reduction of the coal. This substance, which has an intermediate reduction potential, Sm(+2)/Sm(+3) =

-1.55V (13), has recently been used for targeted reduction reactions (14). It is also relevant that samarium(II) iodide can conveniently be prepared in tetrahydrofuran, an excellent coal swelling solvent (8).

The treatment of a suspension of Wyodak subbituminous coal in tetrahydrofuran with samarium(II) iodide proceeded smoothly as described in the Materials and Methods Section. The chemical and spectroscopic data for the dried coal and its reduction product are summarized in Table I.

TABLE I. Chemical and Spectroscopic Properties of Wyodak Subbituminous Coal and Samarium(II) Iodide Treated Wyodak Coal.

	Original Coal (Dried)	SmI_2 Treated Coal
%C (daf)	69.6	69.6
H/C	0.91	0.89
% Ash	8.3	4.3
R^a(spins/g)	4×10^{18}	1×10^{18}
%C(obs)	58	85
f_a (FID)	0.66	0.72
f_a (CP)	0.64	0.68

[a] Free radical concentration.

The original coal has 4 x 10exp18 spins per gram (or about 1 spin per 8750 carbon atoms). Spin-counting experiments reveal that only 58% of the carbon atoms are observed in conventional solid-state NMR experiments. The reduced sample has a greatly diminished ESR signal with about 1 spin per 35,000 carbon atoms. Spin-counting experiments indicate that 85% of the carbon atoms in the reduced sample are observed in the NMR experiment. The analytical data establish that this gain has been realized without a change in the carbon content, or in the hydrogen to carbon ratio of the coal.

The CP/MAS spectra of Wyodak subbituminous coal and the coal treated with samarium(II) iodide are shown in Figure 2.

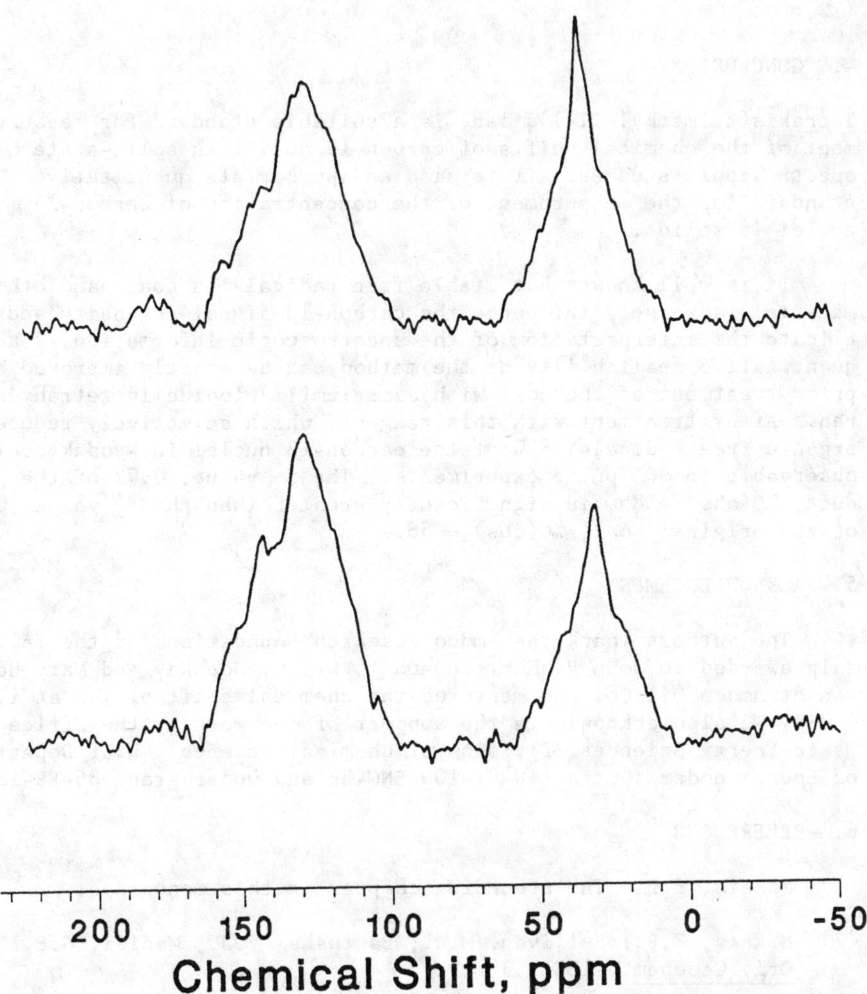

Figure 2. The solid state ^{13}C CP/MAS spectra of Wyodak subbituminous coal (top) and Wyodak coal treated with SmI_2 (bottom).

The spectrum of the treated coal exhibits markedly higher aromatic carbon content and shows more intense absorptions for nonprotonated carbons in the region 138-155 ppm. The f_a value for the reduced sample obtained in a 90° pulse experiment is also significantly higher, 0.72, than the value obtained for the original coal, 0.66. The observations indicate, not unexpectedly, that the aromatic carbons are most affected by the paramagnetic centers (12,15).

4. CONCLUSION

Tetrakis(trimethylsilyl)silane is a suitable standard for measurement of the chemical shifts of carbon-13 nuclei in solid-state spectroscopic studies. It is also an appropriate quantitative standard for the measurement of the concentration of carbon-13 nuclei in solids.

It is well known that stable free radicals in coals and other materials adversely influence the carbon-13 signal intensity and complicate the interpretation of the spectroscopic information. The quantitative realiability of the method can be greatly improved by the prior treatment of the coal with samarium(II) iodide in tetrahydrofuran. After treatment with this reagent, which selectively reduces the organic free radicals, 85% of the carbon-13 nuclei in Wyodak coal are observable in 90° pulse experiments. The f_a value, 0.72 of the product, %C(obs) = 85, is significantly greater than the f_a value, 0.66, of the original coal, %C(obs) = 58.

5. ACKNOWLEDGEMENT

The authors thank the Amoco Research Foundation for the fellowship awarded to John V. Muntean and to Dr. G. Joe Ray and Mary Henniman of Amoco Oil Co. who measured the chemical shift of TKS at 1.4 and 7.0T. We also acknowledge the support of the work by the Office of Basic Energy Sciences, Division of Chemical Sciences, U.S. Department of Energy under contract W-31-109-ENG-38 and under grant 86-ER-13573.

6. REFERENCES

(1) Miknis, F.P. The preceding chapter in this monograph.

(2) Miknis, F.P.; Sullivan, M.J.; Bartuska, V.J.; Maciel, G.E. Org. Geochem., 1981, 3, 19.

(3) Hagamann, E.W.; Chambers, R.R.; Woody, M.C. Anal. Chem., 1986, 58, 387.

(4) Botto, R.G.; Wilson, R.; Winans, R.E. Energy and Fuels, 1987, 1, 173.

(5) Muntean, J.V.; Stock, L.M.; Botto, R.E. J. Mag. Res., 76 1988, 000. In press.

(6) Muntean, J.V.; Stock, L.M.; Botto, R.E. Energy and Fuels, 1988, 2, 000. In press.

(7) Gilman, H.; Smith, L.L. J. Am. Chem. Soc., 1964, 86, 1454.

(8) Girard, P.; Namy, J.L.; Kagan, H.B. J. Am. Chem. Soc., 1980, 102, 2693.

(9) Axelson, D.E. "Solid State Nuclear Magnetic Resonance of Fossil Fuels: An Experiment Approach", Multiscience: Canada, 1985.

(10) Earl, W.L.; Vanderhart, D.L. J. Magn. Reson., 1982, 48, 35.

(11) Vanderhart, D.L. J. Chem. Phys., 1986, 84, 1196.

(12) Retcofsky, H.L. Coal Science, 1982, 1, 43.

(13) Morss, L.R. "Standard Potentials in Aqueous Solution", Chapter 20; Bard, A.J.; Parsons, R. and Jordon, J., Eds., Marcel Dekker: New York and Basel, 1985.

(14) Molander, G.A.; Etter, J.B. J. Org. Chem., 1986, 51, 1778.

(15) Wind, R.; Duijvestijn, M.J.; van der Lugt, C.; Schmidt, J.; Vriend, H. Fuel, 1987, 66, 876.

STRUCTURAL ANALYSIS OF COAL DERIVATIVES

K.D.Bartle
Department of Physical Chemistry
University of Leeds
Leeds LS2 9JT
UK

ABSTRACT. Methods for the structural analysis of coal extracts and liquefaction products are described. The required analytical data includes elemental molecular mass, functional group and ^{1}H and ^{13}C nmr analyses. If ^{13}C nmr is not used, assumptions must be made concerning aliphatic H/C ratios which may lead to errors. Additional analyses may be incorporated in least-squares minimization schemes. The structural scheme is useful for highlighting differences between coal extracts, but gives no information on the distribution of structures about the average; complementary analysis by voltammetric methods allows aromatic groups to be identified, while XPS gives information on nitrogen functionality. Variations in average structure with molecular mass must also be taken into account.

1. INTRODUCTION

Extracts and liquids derived from coal are multicomponent mixtures containing probably at least 10^4 different compounds. Because of this complexity, two parallel analytical approaches are necessary. Either: methods need to be devised which provide very high resolution-usually chromatographic; or average structural properties are derived, usually for the higher molecular mass constituents where the very large number of possible compounds makes even the highest resolution inappropriate.

Average structural properties for coal liquids were first determined from physical properties such as molar volume or refractive index in a general approach referred to as statistical structural analysis (1). This makes use of the fact that some molecular properties are additive, so that the value of a particular property can be calculated by summation of the contributions of the individual atoms or functional groups.

Y. Yürüm (ed.), New Trends in Coal Science, 169–186.

2. NMR BASED STRUCTURAL ANALYSIS

2.1. ^{1}H nmr

In the early 1960s statistical structural analysis was superceded by structural analysis schemes based on ^{1}H nuclear magnetic resonance (1). ^{1}H nmr differentiates a number of hydrogen types and provides, quantitatively, the distribution between them. An example of the ^{1}H nmr spectrum of the asphaltenes from a coal extract (2) with chloroform-d as solvent is shown in Figure 1. The assignments of the bands are listed in Table I, together with the typical proportions of the various forms of hydrogen.

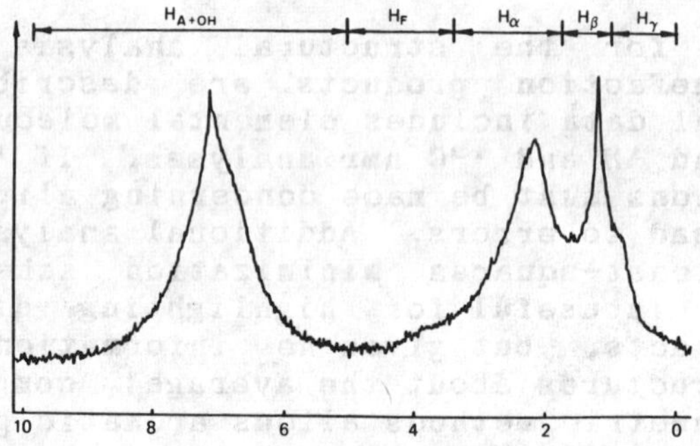

Figure 1. ^{1}H nmr spectrum of the asphaltenes of a coal extract (2).

The hydrogen distributions are combined with number average molecular mass and various elemental and functional group analyses so as to arrive at the properties of the average molecule. The often used Brown and Ladner method (3) gives the following parameters:

f_A ratio of aromatic carbon (C_{Ar}) to total carbon (C);
or,
the fraction of the total available outer edge positions which is occupied by constituents;
and
H_{Aru}/C_{Ar}, the atomic H/C ratio that the average aromatic skeleton would have if each substituent were replaced by a hydrogen atom.
These are derived from equations 1 to 3 where C/H and O/H are atomic ratios.

TABLE I

Assignments of Bands in ^{1}H nmr Spectra of Coal Asphaltenes

Hydrogen Type	Symbol	Chemical Shift, ppm	Hydrogen Distribution
Aromatic	H_A	6.0-9.0	
Phenolic	H_{OH}	5.0-9.0	42.5 (ar. + OH)
Ring joining methylene	H_F	3.4-5.0	4.8
Methyl, methylene and methine α to an aromatic ring	H(α)	1.9-3.4	31.2
β-methyl, methylene and methine β or further from an aromatic ring + paraffinic methylene and methine	H(β)	1.0-1.9	15.8
Methyl τ or further from an aromatic ring + paraffinic methyl	H(τ)	0.5-1.0	5.7

$$f_A = \{(C/H)-[H*(\alpha)/x]-[H*(\beta)/y]-[H(\tau)/3]\}(C/H)^{-1} \quad .....(1)$$

$$\sigma = \{[H*(\alpha)/x]+(O/H)\}\{[H*(\alpha)/x]+(O/H)+H*(Ar)\} \quad ......(2)$$

$$H_{Aru}/C_{Ar} = \{[H*(\alpha)/x]+(O/H)+H*_{Ar}\}\{(C/H)-[H*(\alpha)/x]-[H*(\beta)/y]-[H(\tau)/3]\}^{-1} \quad ....(3)$$

H*(α) = H(α)/H, ratio of α type to total hydrogen, determined by nmr,

H*(β) = H(β)/H, ratio of β type hydrogen to total hydrogen, again from nmr,

x and y are the atomic ratios of hydrogen to carbon in, respectively, α and β structures, both of which are assumed; and

$H*_{Ar}$ = H_{Ar}/H is the ratio of aromatic to total hydrogen determined by nmr with the assumption that 60% of the total oxygen is phenolic.

The main problems in applying equations 1-3 are that H/C ratios of aliphatic groups have to be assumed, while the incorporation of heterocyclic structures is not straightforward. Nonetheless, this approach has proved to be one of the most important concepts in coal chemistry and applications are legion (4). The sensitivity of values of structural parameters to x, y etc. has been assessed (5) and a number of schemes in which the Brown and Ladner method is combined with data derived by other spectroscopic methods (ultraviolet, infrared etc.) have also been proposed (1).

2.2. ^{13}C nmr

The development of ^{13}C nmr in the 1970s-principally the use of broad band decoupled pulse Fourier transform techniques-some of the chief stumbling blocks in the application of nmr based structural analysis (6). As long as measurements are made quantitatively (7), ^{13}C nmr gives carbon aromaticities directly. Moreover, the chemical shift range is much wider than in ^{1}H nmr and much more detailed information on the nature of aliphatic groups can be derived and quaternary and carbonyl carbon may be detected (8). To obtain reliable quantitative results, chromium acetylacetonate is added to shorten carbon relaxation times and reduce nuclear Overhauser enhancements and the proton decoupling is gated to suppress the remaining n.O.e. This procedure has been shown to yield accurate aromaticity values for both mixtures of model compounds and coal extracts (7). A typical ^{13}C nmr spectrum (2) is shown in Figure 2. The assignment of bands in the ^{13}C nmr spectra (Table II) is based on a detailed consideration of the chemical shifts of model compounds (9).

TABLE II

Assignments of Bands in ^{13}C nmr Spectra of Coal Asphaltenes

Carbon Type	Chemical Shift ppm	Carbon Distribution %
Carbonyl	170-210	-
Aromatic C-O	148-168	7.6
Aromatic C-C, C-H, C-N	100-148	68.6
Methylene, methine	22.5-60	16.0
Methyl	11-22.5	7.8

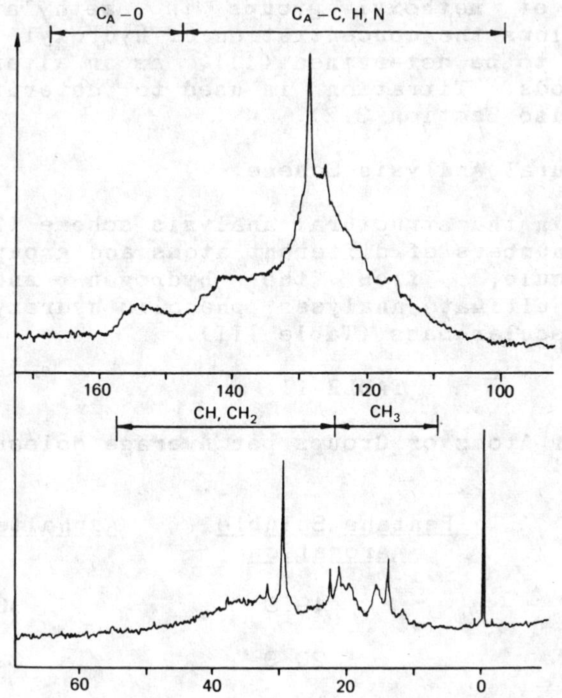

Figure 2. ^{13}C nmr spectrum of the asphaltenes of a coal extract (2).

Spin echo ^{13}C nmr is used (10) to obtain quantitative information on aliphatic CH, CH_2 and CH_3 groups from the CH_2 and CH + CH_3 subspectra (Figure 3). Quaternary aliphatic C is not usually observed.

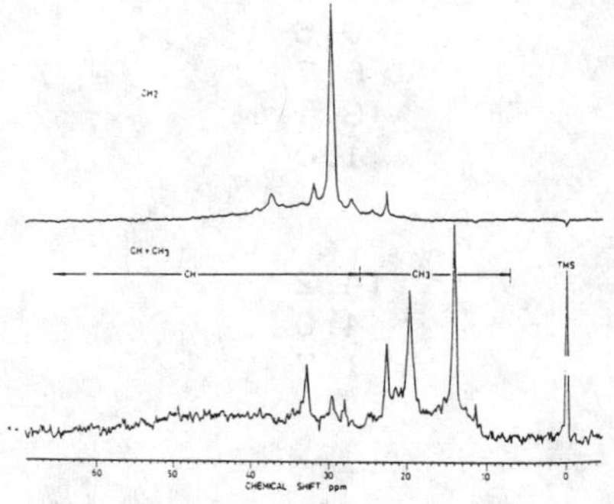

Figure 3. Aliphatic carbon ^{13}C nmr spectra of coal asphaltenes.

^{13}C nmr of methoxyl groups in methylated coal derivatives allows the concentration of hydroxyl (phenolic and carboxylic) to be determined (11), as an alternative to titration methods. Titration is used to determine basic nitrogen (see also Section 3.2).

2.3. The Structural Analysis Scheme

The first step in the structural analysis scheme (2,6) is to calculate the numbers of different atoms and groups in the average molecule, from the hydrogen and carbon distributions, ultimate analyses, phenolic hydroxyl content and average molecular mass (Table III).

TABLE III

Numbers of Atoms or Groups per Average Molecule

Atom or Group	n-Pentane Soluble Aromatics	Asphaltenes
Carbon	19.9	36.4
Hydrogen	20.9	32.6
Phenolic oxygen (pO)	0.2	2.0
Nonphenolic O (npO)	0.2	0.9
Nitrogen	0.06	0.5
Sulfur	0.08	0.2
Hydrogen atoms		
$H_{A,OH}$	9.6	13.8
H_F	0.5	1.6
$H(\alpha)$	6.7	10.1
$H(\beta)$	3.1	5.2
$H(\gamma)$	1.0	1.9
Carbon atoms		
C_A	14.2	27.7
CH_2,CH	4.0	5.9
CH_3	1.7	2.8

Average structural parameters are then calculated as follows and are summarised in Table IV.

TABLE IV

Definition of Average Structural Parameters

Carbon aromaticity	$f_A = C_A/C$
Number of alkyl groups per molecule	$AG = H(\alpha)/a$ (constant 'a' determined from spin echo nmr)
Alkyl chain length	$CL = [(C_{Ali}-H_{F/2})/AG$
Ring-joining groups per molucule	$RJG = H_{F/2} + npO + N + S$
Number of peripheral aromatic carbon atoms per molecule	$C_P = H_A + AG + 2\ RJG$
Number of internal aromatic carbon atoms per molecule	$C_{INT} = C_A - C_P$
Fraction of sites on aromatic skeleton occupied by alkyl and phenolic groups	$\sigma' = (AG + pO)/(AG + H_A)$
Degree of condensation of aromatic nucleus	$DC = (C_P-6)/(C_A+RJG-6)$

f_A is obtained directly from the ^{13}C spectrum. The number of alkyl or hydroaromatic groups per molecule AG is obtained with the aid of a value of x (see equation 1) determined with the aid of the numbers of CH_3, CH_2 and CH groups from the PCSE spectra and detailed chemical shift assignments. The number of ring joining groups per molecule, RJG, is calculated with the assumption that nonphenolic oxygen, nonbasic nitrogen and sulphur are found, respectively in aromatic ether, carbazole and dibenzothiophene structures. The assumption for nitrogen has been confirmed by XPS spectroscopy (see Section 3.2), while ^{13}C nmr can be used to determine the content of carbonyl groups.

From these values, the other structural parameters listed in Table IV follow: the number of peripheral and internal aromatic carbon atoms, the fraction of sites in the aromatic skeleton occupied by alkyl, hydroaromatic and

TABLE V

Structural Parameters

Parameter	Symbol	n-Pentane Soluble Aromatics	Asphaltenes
Aromaticity	f_A	0.71	0.76
Alkyl and hydroaromatic groups	AG	2.9	4.4
Average chain length	CL	1.8	1.8
Ring joining groups	RJG	0.6	2.4
Peripheral carbon	C_P	13.6	23.1
Internal carbon	C_{INT}	0.6	4.6
Degree of substitution	σ'	0.25	0.35
Degree of condensation of aromatic nuclei	DC	0.66	0.71

phenolic groups and a parameter indicationg the degree of condensation of the aromatic nuclei. Molecular structures may then be constructed to fit the numerical values (e.g. Table V) of the above parameters. An example of one such structure representing the asphaltene fraction from a coal extract is shown in Figure 4. Small aromatic groups are substituted with fairly short chain alkyl groups and linked together by methylene and heteroatomic groups. More than one structure fits the calculated parameter values. It must be emphasised that these structures are only the statistical averages of the very large number of individual components present, but the results do indicate that the asphaltenes are highly aromatic; the aromatic nuclei are small and that both alkyl and naphthenic groups are required to describe the aliphatic substituents (6).

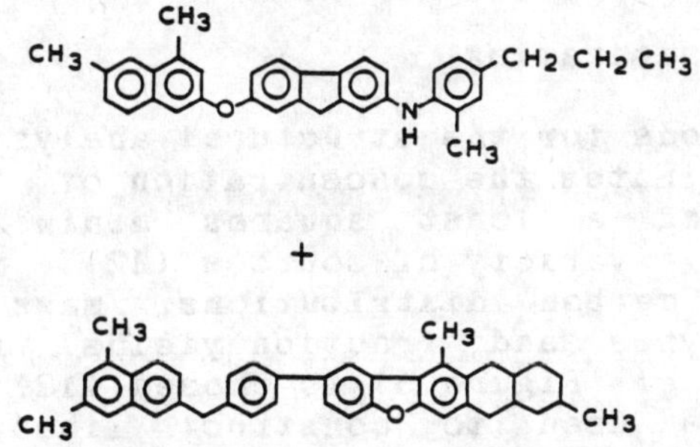

Figure 4. Average structure of the asphaltenes of a coal extract (2).

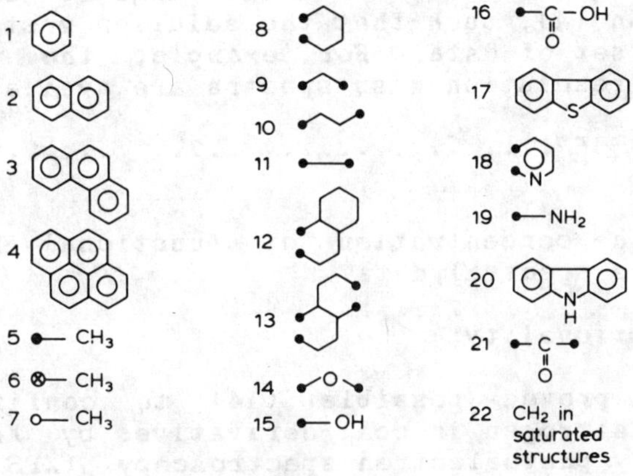

Figure 5. Coal derived liquid functional groups (13).

3. CURRENT DEVELOPMENTS IN STRUCTURAL ANALYSIS

3.1. Least Squares Minimization

The most recent methods for the structural analysis of coal derived materials estimates the concentration of functional groups by performing a least squares minimization on analytical data from a variety of sources (12); these may include hydrogen and carbon distributions, mass spectral data, ultimate analyses and fraction yields. A set of functional groups (e.g. Figure 5) is chosen (13) and the analytical data are used to construct linear balance equations in which the concentration of each atomic species b(i) is expressed as a sum of the contribution from the functional groups y(i), each with a suitable stoichiometric coefficient, A(ij).

$$\sum_{j=1}^{n} A(ij)\, y(i) = b(i), \quad (i = 1, \ldots, m) \quad \ldots\ldots\ldots\ldots\ldots\ldots (4)$$

The number of equations is less than the number of unknowns and there are constraints from nonnegative solutions.

The structure of the coal liquid is characterised (13) by selecting a single solution from the range of solutions defined by equation (4), such that the solution most closely matches a given set of data. For example, the function minimized if high resolution mass spectra are available is

$$P = \sum_{i=1}^{n} [y(i) - f(i)^2]^2 \quad \ldots\ldots\ldots\ldots\ldots\ldots\ldots\ldots\ldots\ldots (5)$$

where f(i) is the concentration of functional group i predicted from mass spectral data.

3.2. Nitrogen Functionality

It has recently proved possible (14) to confirm the functionality of nitrogen in coal derivatives by XPS. The technique of x-ray photoelectron spectroscopy (XPS), also known as electron spectroscopy for chemical analysis (ESCA), offers the possibility of a direct method for the determination of nitrogen containing functional groups in coal and low volatile coal products (15). XPS determines the binding energies of both core and valence electrons by measurement of the kinetic energies of electrons photoemitted from a solid when irradiated in vacuo with near monochromatic x-rays. Since the core electron binding energies are characteristic of the element, XPS provide elemental analysis for all elements except hydrogen and

TABLE VI

Binding energies relative to C (1s) at 285 eV for a number of model compounds

Compound	N (1s) Binding Energy ± 0.1 eV	Nitrogen Type
Phenanthridine	398.7	Tertiary
Phenazine	398.9	(Pyridine type)
1,3 H Dibenzo[a,i] carbazole	400.4	Secondary
Carbazole	400.3	(Pyrrole type)
1-Aminopyrene	399.4	Primary
1-Aminofluorene	399.4	Amines
1-Nitropyrene	406.2	Nitro
Quaternarised phenanthridine	401.4	Quaternary

helium. Small shifts in the characteristic binding energy can be measured and it is therefore possible to use XPS to identify the functional group in which the elemnt is present (Table VI). XPS has the advantage of constant sensitivity to an element irrespective of the functional group to which it may contribute. Functional group analysis obtained is therefore quantitative. Figure 6 shows how different pyrrole and pyridine type nitrogen groups contribute to a number of coal liquids and derived fractions (14).

3.3. Voltammetric Identification of Aromatic Groups

As was emphasised in Section 2.3 nmr signals only give information on the average behaviour of each atom in the different clusters of aromatic rings which are linked together in coal derived molecules. A range of unknown structural types with widely varying degrees of condensation can all contribute to the average. Recent work has shown, however, that electrochemical methods can be used to deduce the distribution of structures about the nmr determined average (16).

Thus the characteristic reduction potentials of many PAC have been determined and provide a basis for the detection of various functional groups in solvent extracts of coals. Given and Peover reported (17) the polarographic

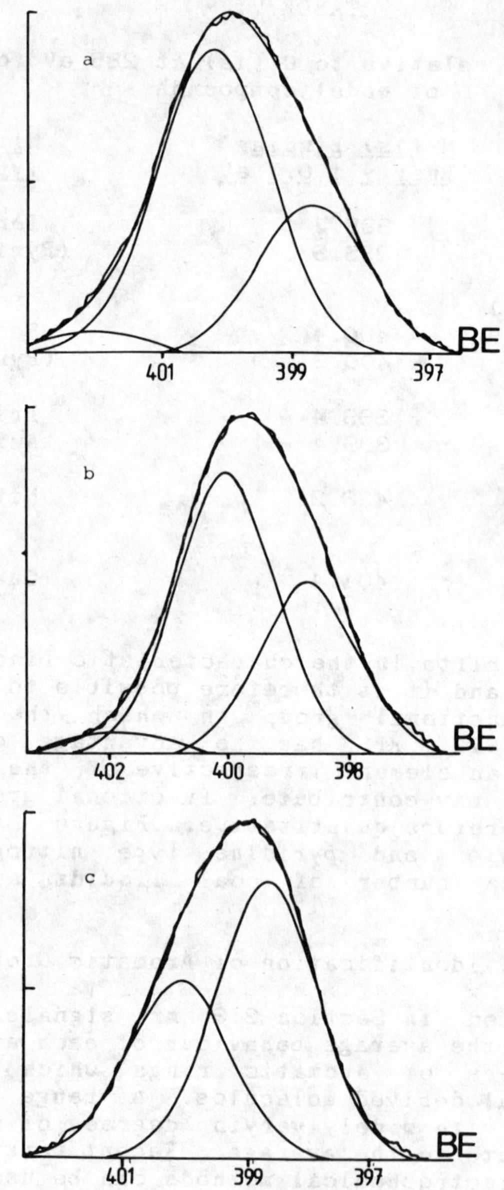

Figure 6. N (1s) XPS spectra for a) flash pyrolysis tar, b) hydrogenated flash pyrolysis asphaltene and c) hydrogenated flash pyrolysis bases showing synthesized components.

behaviour of coal extracts in dimethyl formamide (DMF) over twenty years ago, the polarograms were typically superpositions of sigmoid curves, with inflexions at negative potentials where several PAH and their quinone derivatives have half wave potentials. It was concluded that an appreciable fraction of the carbon in coal was present in polycyclic aromatic and quinone structures. Modern microprocessor based polarographic systems, coupled with dispense type mercury drop electrodes afford many advantages over those used in the early work. In particular, the contribution of charging current of the electrical double layer at current sampling is reduced, while a wide range of voltage wave forms in a variety of pulse modes can be employed; three element polarographic cells also reduce potential drop effects in the electrolyte and improve the precision of the measured reduction potentials. The hanging mercury drop electrode (HMDE) allows fuller use of the potential of the technique than the dropping mercury electrode which shows mechanical instability in DMF solutions (16).

Reduction potentials for a wide range of PAC have been determined by differential pulse voltammetry at the HMDE (16,18). Graphs of pulse current against concentration were linear. The presence and concentration of different PAC structures, e.g. anthracenes, fluoranthenes, pyrenes and chrysenes etc. could therefore be determined for coal derived oils and compared with results from analysis by open tubular column gas chromatography. Voltammetric curves for asphaltenes (A) and preasphaltenes (PA) from a variety of coal derivatives show (19) (Figures 7 and 8) considerable electrochemical activity, with voltammetric peaks at similar potentials to those of the corresponding lower MM material (see for example Figure 8). DPV of specially synthesised model compounds shows (19) how linkages through alkyl groups have only a small affect on E[½] values of common polycyclic aromatic groups so that the peaks in voltammograms of A and PA can be assigned to the reduction of specific aromatic species even though these are linked together.

On this basis, the A and PA of a supercritical gas (SCG) extract of lignite show little electrochemical activity (Figure 7) other than the reduction of aromatic clusters containing two aromatic rings (naphthalene, biphenyl, fluorene etc.). The A and PA from an SCG extract of bituminous coal show similar voltammograms to those from the lignite extract, except that small signals from more condensed structures such as phenanthrenes, fluoranthenes and pyrenes are also present (Figure 7). For A and PA from hydrogendonor solvent (HDS) or anthracene oil extracts of bituminous coals the reduction of the three and four ring aromatic structures are much more prominent (Figure 7). This evidence correlates well with the structural parameters determined by nmr. In particular, DPV confirms that the polycyclic aromatic structures suggested as being consistent with nmr derived average structures are actually present in the A and PA obtained from bituminous coals under relatively severe conditions of extraction and carbonization (20).

On a weight for weight basis, however, the electrochemical activity of A is considerably less than that of the corresponding (lower MM) pentane soluble (PS) fraction and the effect is even more marked for the PA (Figure 8). It could be argued that, since there is considerable overlap in the MM ranges of the solvent fractions of coal derivatives, the observed DPV peaks for A and PA are attributed to residual low MM material. Repeated re-precipitation of a number of samples of A was therefore carried out to remove as much low MM pentane soluble material as possible. Further traces were removed by Soxhlet extraction of reprecipitated asphaltene with n-pentane.

DPV of the resulting samples show that A retain their electrochemical activity, although at a lower level, gas chromatography of the A from an HDS extract after reprecipitation and Soxhlet extraction showed that fluoranthene and pyrene were present only at a concentration of less than 1% of that required to produce the electrochemical activity observed for the A near -2.20 and -2.50 V. Similarly, a fraction mainly containing molecules with MM between 250 and 600 obtained by preparative SEC for HDS extract asphaltenes gave a voltammogram corresponding to that of the parent A. These experiments confirm that signals in the voltammograms of asphaltenes from which low molecular mass material has been removed originate from molecules containing linked aromatic structures rather than independent aromatic molecules (19).

4. VARIATION OF THE STRUCTURE OF COAL DERIVED MATERIALS WITH MOLECULAR MASS

While electrochemical measurements afford evidence that the distribution of aromatic structures about the nmr determined average is probably quite small, it is also noteworthy that a variation of coal extract with molecular mass may also occur. Structural analysis is usually carried out on either the soluble portion of the extract or on gross fractions separated (6) by solvent extraction e.g. pentane soluble (oil), asphaltene and pre-asphaltene fractions; size exclusion chromatography (SEC) of coal extracts reveals (21) that the constituent molecules have MM ranging from as little as 200 to over 3000. The proposal structures are of course averaged over all of these molecules.

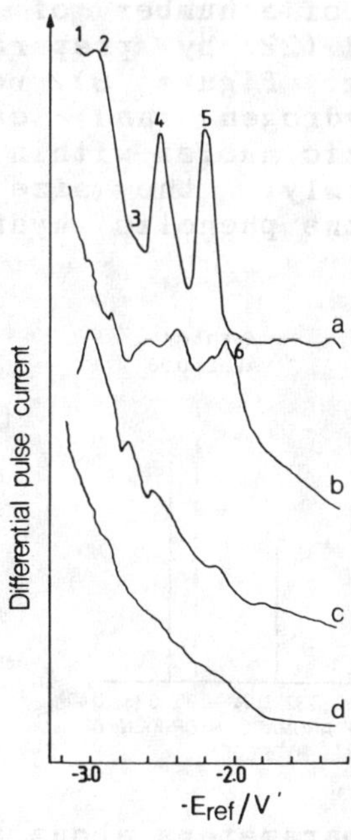

Figure 7. Differential pulse voltammograms (19) of asphaltene fractions from, a) hydrogen donor solvent extract of UK coal, b) hydrocracking residue from hydrogen donor solvent extract of a UK coal, c) supercritical gas extract of UK coal and d) supercritical gas extract of US lignite. Peak assignments, 1, two ring aromatics (naphthalene, biphenyl etc.), 2, phenanthrene structures, 3, chrysenes, 4, pyrenes, 5, fluoranthenes and 6, more condensed aromatic structures.

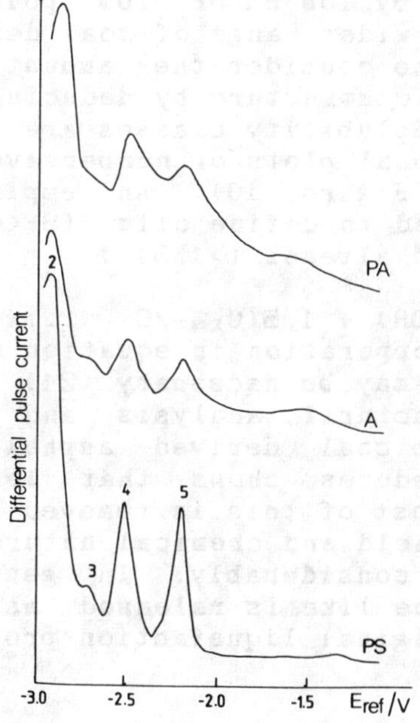

Figure 8. Differential pulse voltammograms (19) of pentane soluble (PS), asphaltene (A) and preasphaltenes (PA) of hydrogen donor solvent extract of a UK coal. Peak assignments as in Figure 7.

Asphaltenes and pre-asphaltenes of a number of high yield extracts of coal were separated (22) by preparative SEC. Analysis of these fractions (e.g. Figure 9) reveal that, with increasing MM the hydrogen and carbon aromaticities decrease and the aromatic nuclei within the structures are distributed nonuniformly; the size of aliphatic substituents increases and the phenolic hydroxyl content also varies.

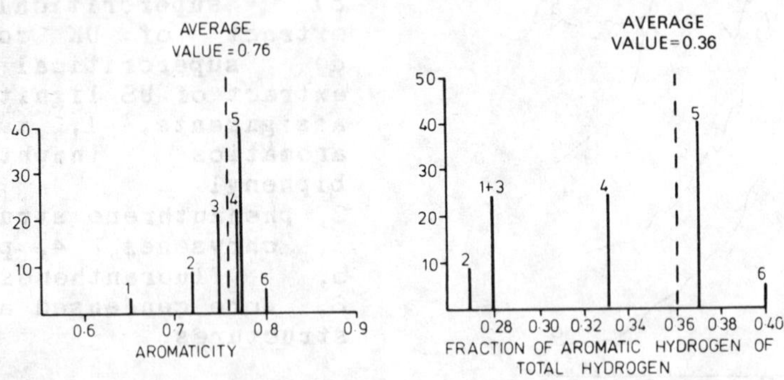

Figure 9. Distribution of structural parameters about their average values for coal asphaltenes (22).

The MM ranges for A and PA from the same extract overlap (21) and analyses of narrow MM subfractions show how benzene solubility is conferred by low MM or low polarity for high MM material (22). For a wider range of coal derived A and PA ot is also necessary to consider the amount and degree of condensation of aromatic structure by deducing the structural parameter C_{INT}/C. Solubility classes are then defined (23,24) by three dimensional plots of number average MM, % acidic OH and C_{INT}/C (Figure 10). An empirical solubility factor SP is then used to define oils (SP<0.4), asphaltenes (0.9-1.2) and pre-asphaltenes (>1.3) :

$$SP = 0.75 \log_{10} (M_n/200) + 0.1(\%OH) + 1.5(C_{INT}/C) \quad ....(6)$$

For more polar materials, incorporation in equation 6 of terms involving nitrogen content may be necessary (24).

Application of SEC, structural analysis and the solubility parameter concept to coal derived asphaltenes prepared by recommended procedures shows that low MM material is present, but that most of this is removed by a single reprecipitation. The yield and chemical nature of material thus released varies considerably. In general, material which is more asphaltene like is released as the concentration of oils in the original liquefaction products decreases (25).

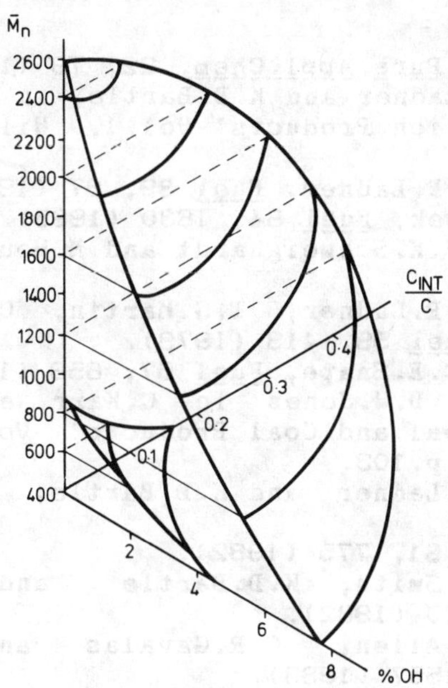

Figure 10. Possible volume defining solubility in n-pentane (inner volume) and benzene (outer volume) (23). Asphaltenes are defined by the volume between the boundary surfaces.

These studies which show how asphaltenes are a solubility class and not a chemically homogeneous group of compounds imply caution in proposing coal conversion mechanism from variations of yields of oils, asphaltenes and pre-asphaltenes. Since these groups are defined by at least three properties, asphaltenes from one conversion run may be quite chemically different from those from another e.g. higher MM, but lower in either or both of polarity or degree of aromatic condensation or lower in MM but higher in polarity etc. It is therefore dangerous to interpret the chemical reactions occurring in coal conversion simply from yields of asphaltenes etc.

REFERENCES

1. K.D.Bartle, Rev.Pure Appl.Chem. 22, 79 (1972).
2. C.E.Snape, W.R.Ladner and K.D.Bartle, in H.D.Schultz ed. 'Coal Liquefaction Products' Vol.1, Wiley, New York, 1983, p.70.
3. J.K.Brown and W.R.Ladner, Fuel 39, 87 (1960).
4. I.Lang and M.Hajek, Fuel 64, 1630 (1985).
5. H.L.Retcofsky, F.K.Schweighardt and M.Hough, Anal.Chem. 49, 585 (1977).
6. K.D.Bartle, W.R.Ladner, T.G.Martin, C.E.Snape and D.F.Williams, Fuel 58, 413 (1979).
7. W.R.Ladner and C.E.Snape, Fuel 57, 658 (1978).
8. K.D.Bartle and D.W.Jones in C.Karr ed. 'Analytical Methods for Coal and Coal Products' Vol.2, Academic, New York, 1978, p.103.
9. C.E.Snape, W.R.Ladner and K.D.Bartle, Anal.Chem. 51, 2189 (1979).
10. C.E.Snape, Fuel 61, 775 (1982).
11. C.E.Snape, C.A.Smith, K.D.Bartle and R.S.Matthews, Anal.Chem. 54, 20 (1982).
12. L.Petrakis, D.Allen, G.R.Gavalas and B.C.Gates, Anal.Chem. 55, 1557 (1983).
13. D.T.Allen, L.Petrakis, D.W.Grandy, G.R.Gavalas and B.C.Gates, Fuel 63, 803 (1984).
14. K.D.Bartle, D.L.Perry and S.Wallace, in J.A.Moulijn and F.Kapteijn eds.'Coal Characterisation for Conversion Processes 1986', Elsevier, Amsterdam, 1987, p.351.
15. D.L.Perry and A.Grint, Fuel 62, 1024 91983).
16. K.D.Bartle, C.Gibson, D.Mills, M.J.Mulligan, N.Taylor, T.G.Martin and C.E.Snape, Anal.Chem. 54, 1730 (1982).
17. P.H.Given and M.E.Peover, Fuel 39, 463 (1960).
18. A.J.Pappin, A.P.Tytko, K.D.Bartle, N.Taylor and D.G.Mills, Fuel in press
19. A.P.Tytko, K.D.Bartle, N.Taylor, I.O.Amaechina and A.Pomfret, Fuel in press.
20. A.P.Tytko, K.D.Bartle, N.Taylor, M.A.Thomson, W.Kemp and W.Steedman, Fuel 64, 1024 (1985).
21. K.D.Bartle, D.G.Mills, M.J.Mulligan, I.O.Amaechina and N.Taylor, Anal.Chem. 58, 2403 (1986).
22. D.G.Richards, C.E.Snape, K.D.Bartle, C.Gibson, M.J.Mulligan and N.Taylor, Fuel 62, 724 (1983).
23. C.E.Snape and K.D.Bartle, Fuel 63, 883 (1984).
24. C.E.Snape and K.D.Bartle, Fuel 64, 427 (1985).
25. C.E.Snape, K.D.Bartle, I.O.Amaechina and D.G.Mills, Fuel Proc. Technol. 15, 89 (1987).

ESTIMATING THERMOPHYSICAL PROPERTIES OF COAL LIQUIDS USING NMR SPECTRA

David T. Allen
Department of Chemical Engineering
University of California
Los Angeles, CA 90024

Leonidas Petrakis
Chevron Research Company
576 Standard Avenue
Richmond, CA 94802

ABSTRACT. The data provided by H1 and C13 NMR spectra have frequently been used to characterize the structure of coal derived liquids. Average structural parameters, average molecular structures and functional group concentrations can all be calculated based on NMR data. The NMR data can find much broader application, however, than merely characterizing structure. This paper reviews some of the basic methods for applying NMR to the problem of property estimation for coal derived systems. The NMR based estimates can be much more accurate than conventional property correlations.

1. INTRODUCTION

Coal derived liquids are complex mixtures containing thousands of components, many of which participate in hydrogen bonds, π - π bonds and acid-base interactions. Clearly the nature and extent of these intermolecular interactions profoundly affect the properties of the fuels. Yet, most property estimation methods for coal liquids rely on bulk characterization variables such as average molecular weight, average boiling point and specific gravity. To obtain the best possible accuracy and flexibility, property estimation methods for non ideal mixtures such as coal liquids must include some information about the concentration of the species participating in intermolecular interactions. That information can be derived from spectroscopic methods, such as Nuclear Magnetic Resonance (NMR). The purpose of this paper is to describe methods for interfacing NMR data with property estimation techniques. This process involves two steps. First, the NMR data must be

Y. Yürüm (ed.), New Trends in Coal Science, 187–196.

used to characterize the structures present in the coal liquids. Second, the structural characterization is utilized in property estimation methods.

Typical NMR spectra for coal liquids are shown in Figures 1 and 2. The spectra are divided into bands and the integrated intensities of the bands are the data used by various structural characterization methods. A detailed review of methods generally used to characterize coal liquids based on NMR spectra is given by Petrakis and Allen (1). Of the many approaches possible, three general methods have been employed most successfully to characterize the structures present in coal derived fluids. These approaches are characterization parameters, average molecule construction and functional group analysis. Each of the three methods will be described below and each will be applied to the same coal derived liquid.

1.1. Average Structural Parameters

The first approach to characterizing coal liquids involves calculating average structural parameters (2). The parameters describe structural features, such as the fraction of carbon that is aromatic, the number and length of alkyl substituents in a molecule, the percentage of aromatic carbons that are substituted and the number of aromatic rings per molecule. An example of this type of characterization is shown in Table I. Given sufficient data, structural parameters can provide a useful characterization of hydrocarbon mixture, however, the parameters approach present two difficulties. The first difficulty is assessing the validity of the assumptions made in calculating the parameters. In most cases it is difficult, if not impossible, to check the assumptions. The second difficulty with the parameters approach is that parameters represent average values and may not provide information about the actual components of the mixture. For example, a value of 3 for the average number of carbons in an alkyl chain could indicate either a uniform distribution of chain length about 3 or a bimodal distribution with high concentrations of very short and much longer aliphatic chains.

1.2. Average Molecules

A second approach to structural analysis utilizes NMR, elemental compositon and average molecular weight data to construct average or representative molecular structures. The structural formula of an average molecule can be determined in a straightforward manner from elemental analysis, NMR and average molecular weight data. After the structural formula is rounded to the nearest whole integer, the algorithm of Oka (3) can be used to find all possible structures that are consistent with the analytical data.

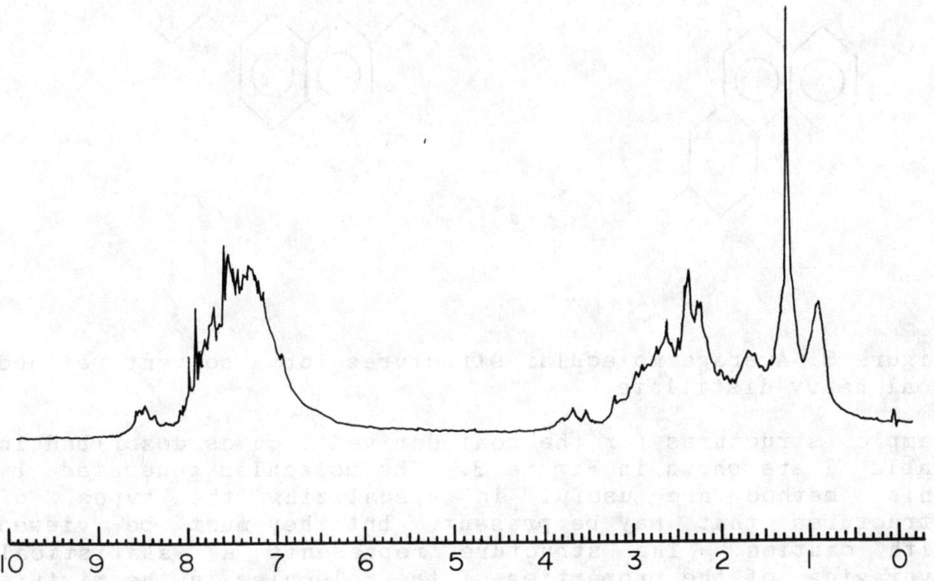

Figure 1. H1 NMR spectrum of a solvent refined coal heavy distillate.

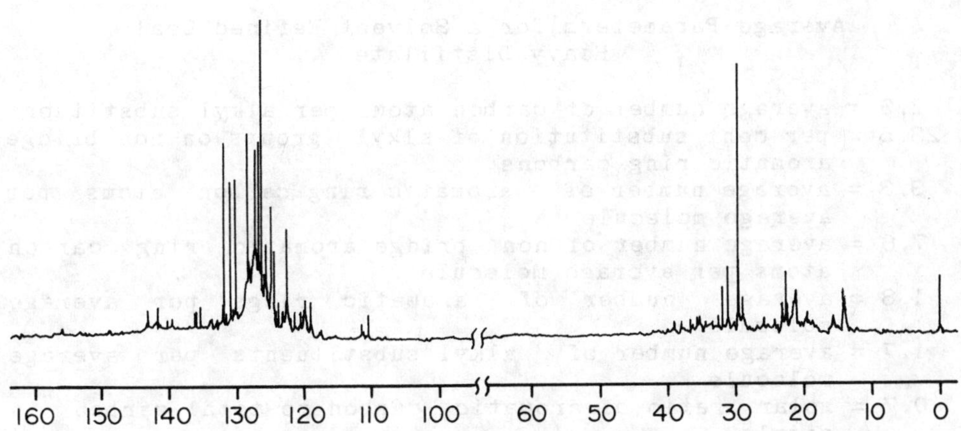

Figure 2. C13 NMR spectrum of a solvent refined coal heavy distillate.

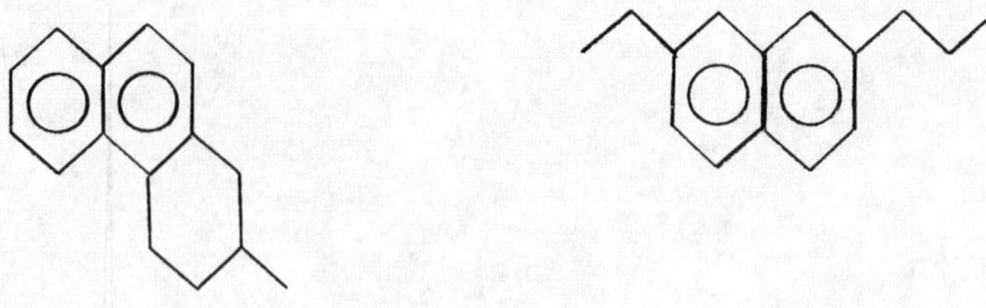

Figure 3. Average molecular structures for a solvent refined coal heavy distillate.

Sample structures for the coal derived liquids described in Table I are shown in Figure 3. The molecules generated by this method are useful in visualizing the types of structures that may be present, but they must be viewed with caution. The structure represents a statistical averaging of the properties of the molecules in the mixture and may or may not actually exist in the liquids. In addition, small variations in the average molecular formula can produce very different structures. As shown in Figure 3, even the same data can result in structures with significantly different properties, so the validity of the structures is questionable and comparisons between similarly treated samples are difficult to make.

TABLE I

Average Parameters for a Solvent Refined Coal Heavy Distillate

2.3	=	average number of carbon atoms per alkyl substituent
23.5	=	per cent substitution of alkyl groups on non bridge aromatic ring carbons
9.3	=	average number of aromatic ring carbon atoms per average molecule
7.6	=	average number of non bridge aromatic ring carbon atoms per average molecule
1.8	=	average number of aromatic rings per average molecule
1.7	=	average number of alkyl substituents per average molecule
0.7	=	molar ratio of aromatic carbon to total carbon in sample
174	=	average molecular weight (estimated from the H1 NMR spectrum)

Table II
Analytical Data for Solvent Refined Coal Heavy Distillate

Elemental Composition		^{1}H NMR Data	
C	89.5%	aromatic H (5.0-9.0 ppm)	42.7%
H	7.5%	alpha H (2.2-5.0 ppm)	27.7%
O	1.7%	beta H (1.0-2.2 ppm)	20.7%
N	0.7%	gamma H (0.5-1.0 ppm)	8.9%
S	0.6%		

^{13}C NMR Data	
aromatic carbon (60-160 ppm)	65.2%
carbon as CH (37-60 ppm)	3.1%
carbon as CH_2 (22.5-37 ppm)	19.3%
carbon as CH_3 alpha to an aromatic ring (20-22.5 ppm)	3.0%
carbon as CH_3 not alpha to an aromatic ring (0-20 ppm)	9.4%

Functional Group Distribution for an SRC-II Heavy Distillate Estimated from the Analytical Data

	Concentration (moles/g distillate)
Monoaromatics	1.89×10^{-3}
Diaromatics	3.28×10^{-3}
Hydroaromatic rings	1.72×10^{-3}
Beta CH	2.14×10^{-3}
Beta CH_2	3.26×10^{-3}
Beta(+) CH_3	2.23×10^{-3}
Alpha CH	2.23×10^{-3}
Alpha CH_2	4.60×10^{-3}
Alpha CH_2	8.3×10^{-4}
Phenols	1.06×10^{-3}
Carbazoles	5.0×10^{-4}
Dibenzothiophenes	1.9×10^{-4}

1.3. Functional Group Distributions

A third approach to the structural characterization of mixtures, termed functional group analysis, has been presented by Petrakis, Allen and coworkers (1, 4-6). The premise of this approach is that while the number of individual molecules in a mixture may be large, all these molecules are composed of a relatively small number of functional groups. This premise is supported for fuel mixtures by a large body of experimental evidence, including results from mass spectrometry (7), IR (8), NMR and chromatography (9). Since functional groups rather than individual molecules can largely determine chemical behaviour in hydrocarbon mixtures, a useful characterization of the mixture consists of listing the concentrations of the constituent functional groups. A characterization obtained using this method is given in Table II. The same coal liquid as in the previous two approaches has been used. This method is particularly useful in estimating thermophysical properties of the mixtures since it allows group contribution methods to be used directly.

2. PROPERTY ESTIMATION

The remainder of this paper will describe methods for interfacing the structural characterization data described above with property estimation methods. For the purposes of this work, the approaches for estimating thermophysical properties of coal liquids will be grouped under the general headings of group contribution theories, empirical correlations and corresponding states theories. The extent to which NMR data, through structural characterization, can be incorporated into these property estimation methods is described below.

2.1. Group Contribution Methods

Group contribution or group additivity methods make the assumption that each chemical functional group contributes a definite value to the property of the mixture, regardless of how the groups associate themselves into molecules. Thus, in the simplest cases, the property of a mixture can be estimated by multiplying the concentration of each group by the group's contribution. So, this method of estimating properties is easily interfaced with the functional group characterization of Table II. In the simplest group contribution cases, shown in Table III, the group concentration is multiplied by the group contribution and the summation is taken over all groups. The group contribution approach to property estimation can be more accurate than more traditional methods based on bulk properties, particularly in cases where concentrations of

Table III
Group Contribution Values for Thermodynamic Properties

Functional Group		Group Contributions				
		Heat Capacity		Critical Properties		
Number	Name	A_i	B_i	Δ_{Ti}	Δ_{Pi}	Δ_{Vi}
1	Monoaromatics	0.300	0.324	0.066	0.924	222
2	Diaromatics	0.444	0.461	0.110	1.54	368
3	Hydroaromatic rings	0.163	0.258	0.080	0.908	218
4	Beta CH	0.0101	0.100	0.012	0.210	51
5	Beta CH_2	0.070	0.046	0.020	0.227	53
6	Beta CH_3	0.0805	0.086	0.020	0.227	55
7	Alpha CH	0.0101	0.096	0.012	0.210	50
8	Alpha CH_2	0.0212	0.083	0.020	0.227	54
9	Alpha CH_3	0.050	0.069	0.020	0.227	54
10	Phenols	0.336	0.811	0.097	0.904	224
11	Carbazoles	0.644	0.465	0.156	1.938	467
12	Dibenzothiophenes	0.574	0.559	0.140	2.088	485

Liquid heat capacity for the coal liquid is given by
$C_p = \Sigma A_i X_i + \Sigma B_i X_i (T/1000)$
where A_i and B_i are the contributions to heat capacity of group i, the X_i are the group concentrations per gram of coal liquid from Table II and T is the temperature in °F. C_p is given in cal/ (g °C) or Btu/(lb_m°F).

Critical properties for the coal liquid are given by
$T_c(K) = T_b(K)[0.567 + \Sigma\Delta_{Ti}Yi - (\Sigma\Delta_{Ti}Yi)^2$
$P_c\ (atm) = MW(0.34 + \Sigma\Delta_{Pi}Yi)^{-2}$
$V_c\ (cm^3/gmol) = 40 + \Sigma\Delta_{Vi}Yi$
where MW is the average molecular weight of the coal liquid and Yi is the concentration of functional group i in an average molecule of mass MW.

Using these methods and the group concentrations of Table II:

Solvent Refined Coal Heavy Distillate Properties

$C_{p(liquid)} = 0.37 + 0.50\ (T\ (°F)/1000)$

$T_c = T_b/0.722$

$P_c = 22.4$ atm

$V_c = 626\ cm^3/gmol$

Table IV
Equation of State with Parameters Based on NMR Data (Alexander, 1985a,b)

Equation of State

$$P = \frac{RT}{V-b} - \frac{a(T)}{V(V+b)}$$

Parameters based on NMR data

$$a = a_c[1+m(1-T/T^*)^{1/2}]^2$$

$$T^* = 0.2027\ a_c/Rb$$

$$b = -0.1707 + 0.3929/C^{1/2} + 0.033131H_{aro} + 0.020902H_\alpha + 0.015881H_\beta$$
$$+ 0.012906H_\gamma - (0.10759H_{aro} + 0.06224H_\alpha + 0.3939H_\beta + 0.02355H_\gamma)/C$$

where b is in L/mol

$$a_c^{1/2} = (-3.437 + 3.990/C + 2.5930C^{1/2} + 0.20437H_{aro} + 0.04809H_\alpha + 0.10936H_\beta + 0.09500H_\gamma$$
$$- 0.2534H_{aro}^{1/2} + 0.2176H_\alpha^{1/2})/(1.000000 - 0.004973H_{aro} + 0.0009645H_\beta + 0.002973H_\gamma)$$

where a_c is in bar l^2/mol^2

$$m \equiv 3.010 + 0.6621C^{1/2} - 0.1256H_\beta^{1/2} - 0.0425H_\gamma^{1/2}$$
$$+ (2.223H_{aro} + 1.388H_\alpha + 1.316H_\beta + 1.086H_\gamma)/C$$

For those heavy fossil-fuel fractions containing heteroatoms, corrections are required; see Alexander (1985b) for details.

Characterization data are the following: C = number of carbon atoms per number-average molecule (PAM); H_{aro} = number of hydrogen atoms bonded to aromatic carbon atoms PAM; H_α = number of hydrogen atoms bonded to aliphatic carbon atoms which are in turn bonded to aromatic carbon atoms PAM; H_β = number of hydrogen atoms bonded to aliphatic carbon atoms which are not α and do not terminate a chain PAM; and H_γ = number of hydrogen atoms bonded to aliphatic carbon atoms which are not α and terminate a chain PAM.

Using this method, the characterization parameters and equation of state for the Solvent Refined Coal Heavy Distillate are:

$C = 13.4,\ \ H_{aro} = 5.6,\ \ H_\alpha = 3.7,\ \ H_\beta = 2.6,\ \ H_\gamma = 1.2$

$$P = \frac{RT}{V - 0.12\ \frac{l}{mol}} - \frac{62\ \frac{bar}{l^2 mol^2}\left[1 + 0.83\left[1 - \frac{T}{1300\ K}\right]\right]}{V\left[V + 0.12\ \frac{l}{mol}\right]}$$

non ideal species are high (10). The accuracy of these methods is obtained at the price of more detailed characterization data. The challenge associated with applying these methods is in identifying which sources of analytical data are most responsible for the improved accuracy (11). Once the information value of the analytical data has been assessed, optimized correlations can be developed.

2.2. Correlations

The simplest method for interfacing NMR data with property estimation techniques is through correlations. Alexander et al. (12.13) have used NMR data along with elemental composition data to generate structural parameters that were then used as correlating variables for equation of state parameters. NMR data have also been used as correlating variables for bulk thermodynamic properties. Examples are shown in Table IV.

2.3. Corresponding States Methods

Corresponding states methods make the assumption that fluids will have the same properties at the same reduced temperature, pressure and volume. The reduced temperature, pressure and volume are the ratio of the actual temperature, pressure and volume to the fluid's critical T, P and V. Corresponding states methods are useful for developing self consistent thermophysical properties for coal derived liquids. Interfacing data with equations of state and in particular, corresponding states equations is possible. The previous section described how equation of state parameters can be derived from correlations based on NMR data. NMR data can be incorporated into corresponding state equations of state by using the spectral information to estimate the critical properties. The critical properties (T_c,P_c,V_c) and the acentric factor can be estimated using group contribution methods, as shown in Table III. These properties can then be employed in the equations of state described by Brule (14).

REFERENCES

1. L.Petrakis and D.T.Allen, Nuclear Magnetic Resonance of Liquid Fuels, Elsevier, 1987, p.242.
2. D.R.Clutter, L.Petrakis, R.L.Stenger and R.K.Jensen, Anal.Chem., 44, 1395 (1972).
3. M.Oka, H.C.Chang and G.R.Gavalas, Fuel, 56, 3 (1977).
4. L.Petrakis, D.T.Allen, G.R.Gavalas and B.C.Gates, Anal.Chem.,55, 1557 (1983).
5. D.T.Allen, L.Petrakis, D.W.Grandy, G.R.Gavalas and B.C.Gates, Fuel, 63, 803 (1984).

6. D.T.Allen, D.W.Grant, K.M.Jeong and L.Petrakis, Ind. Eng. Chem. Process Des. Dev., **24**, 737 (1985).
7. J.T.Swansiger, F.E.Dickson and H.T.Best, Anal.Chem., **46**, 730 (1974).
8. P.R.Solomon, D.G.Hamblen and R.M.Carangelo, In Coal and Coal Products: Analytical Characterization Techniques, ACS Symposium Series No.205, 1982.
9. R.G.Ruberto, D.M.Jewell, R.K.Jensen and D.C.Cronauer, Adv. Chem. Ser. No.151, Chapter 3.
10. T.T.Le and D.T.Allen, Fuel, **64**, 1754 (1985).
11. D.T.Allen, Fluid Phase Equilibria, 30, 353 (1986).
12. G.L.Alexander, A.L.Creagh and J.M.Prausnitz, Ind. Eng. Chem. Fundam., 24, 301 (1985).
13. G.L.Alexander, B.J.Schawarz and J.M.Prausnitz, Ind. Eng. Chem. Fundam., **24**, 311 (1985).
14. M.R.Brule, C.T.Lin, L.L.Le and K.E.Starling, AIChE J., **18**, 616 (1982).

THE INFRARED SPECTRA OF COALS, 1980-1986

A.F.Gaines
Birkbeck College, University of London
Malet Street, London WC1E 7HX
England

ABSTRACT. Work between 1980 and 1986 has provided a routine infrared spectroscopy which gives quantitative information about the average structures present in a coal with a minimum of sample preparation. Techniques are available for the study of coal surfaces -all coal reactions are surface reactions- and for small areas of coal surface. The relative simplicity of infrared spectrometry and its ability to give information and composition and temperatures without causing any disturbance of the system being monitored renders the technique particularly useful for on line analysis. Its utility for the on line analysis of combustion, gasification and pyrolysis has already been demonstrated. There would be no difficulty in the continous monitoring of coal liquefaction and more work may be expected to permit on line analysis of coal feedstocks.

1. INFRARED TECHNIQUES

The years 1980-1986 saw a revolution in the infrared spectroscopy of coal and during this period over 200 papers on the subject were collated by Chemical Abstracts. The dawn of the revolution has already been reviewed (1). Infrared spectroscopy has always provided a rapid, relatively cheap, straightforward and versatile method for summarising the chemistry of a sample of coal. In previous decades the success of the technique has been limited by weak absorption by the sample, poor resolution of overlapping peaks, a pronounced and curved baseline caused by the scattering of radiation and by a difficulty in obtaining quantitative results inherent in the methods of preparing samples and in the uncertainity in the position of the baseline.

The problem of weak absorption can be overcome by repetition of carefully acquired spectra. Fourier transform infrared spectrometry (FTIR) achieves this happily: ten to a hundred scans being sufficient for a routine spectrum.

Y. Yürüm (ed.), New Trends in Coal Science, 197–218.

FTIR spectrometers became practical with the advent of microcomputers, these also encouraged the interfacing of 'data stations' with software for operations such as the scaling up of weak spectra, the resolution of overlapping peaks and the subtraction and straightening out of baselines.

Besides the routine simulation of spectra (1,2), (based, for example, on components indicated by the second derivative of the spectra) much software has been developed prompted by the recognition that an infrared spectrum contains a wealth of detailed chemical information even though peaks may overlap. There is increasing use of factor analysis, first, to show how a spectrum of a mixture is composed of a limited number of components (3-11) and secondly, to relate the chemical composition of a coal, as it is represented by its infrared spectrum, to those physico chemical properties of the coal which are important in its utilisation (12-15). The goal of the latter approach is to select a limited number of parameters from the infrared spectrum of a coal sample and from them to predict the technological behaviour of the coal.

As applied to coal, the use of data station to subtract and straighten baselines is alarming and referees and editors of journals should insist that, when appropriate, authors should mention the use of the practise so that readers may not be misled. The practise is alarming because many programmes draw a quadratic baseline of the form:

$$I(\bar{\nu}) = a\,\bar{\nu} + b\,\bar{\nu} + c\,,$$

where $I(\bar{\nu})$ is the intensity of the baseline at wavenumber $\bar{\nu}$ and a, b and c are constants determined by fitting the baseline to three points selected by the investigator. There is little evidence that the 'baseline equation' is of the form given and it is rarely possible to select points in different regions of the spectrum where the baseline is clearly defined. Infinitely to be preferred, is the use of photoacoustic detector (1, 16-20) or of diffuse reflectance infrared spectrometry (1, 21-33). Both techniques observe coal powders with the minimum of sample preparation and both give clearly defined, essentially horizontal baselines.

Photoacoustic detection has been used in many infrared investigations of coal surfaces (1, 16-20) and the physics of the detector has been elucidated in terms of the thermal and elastic constants of coal (34). When a coal surface absorbs radiation, localised generation of heat yields acoustic energy which disturbs the gaseous atmosphere in contact with the surface. The photoacoustic signals depend on the thermal expansion, A, of the sample, its bulk modulus, B, its density ρ, its specific heat, C, and its thermal conductivity, K. Two types of detector may be distinguished; a piezo electric transducer measuring changes in the solid and a microphone monitoring change in the gas. The intensity of the acoustic signal in the solid is proportional to the Gruneisen constant and AB/ ρC whilst the signal in the gas is proportional to $(K \rho C)^{-\frac{1}{2}}$. The behaviour of the two signals has been measured as a function of the rank of the coal (34). Better understanding has led to refinement of photoacoustic detectors (35,36).

Thanks largely to the work of P.R.Griffiths and his collaborators, the parameters defining diffuse reflectance infrared spectrometry are well established (1, 21-33). The method is very versatile and small samples, such as spots obtained in thin layer chromatography, may be examined readily. When compared with infrared spectra from potassium bromide discs, the relative intensity of peaks observed by diffuse reflectance infrared spectroscopy is often increased (31). Indeed, when infrared spectra of single crystals of minerals such as calcite and aragonite are compared with infrared spectra obtained by diffuse reflectance and from potassium bromide discs, it is the diffuse reflectance spectra which are more akin to the single crystal spectra and can be used to distinguish calcite and aragonite, the potassium bromide disc spectra apparently 'missing' some of the absorptions (37).

Other improvements in infrared technique have been reported. A diamond anvil has been used for preparing coal samples (38). Transmission spectra may be obtained from 10-30 μm films prepared from powdered coal (39). Thin coal sections, prepared by novel methods, have been probed by an infrared microspectrophotometer. The surface of the section is scanned systematically and spectra have been obtained from 25 μm diameter areas (40-42). The technique is capable of providing infrared spectra of maceral groups and, in favourable circumstances, of macerals, without the need to separate them from the parent coal. The interest in the surface chemistry of coals during the past decade has prompted the development of a variety of surface probes and attenuated total reflectance, well known as a method of obtaining infrared spectra of surfaces and adsorbed layers, has also been used to study coal surfaces (43).

Figure 1 shows diffuse reflectance FTIR spectra of three bituminous coals. These spectra are typical of those which may now be obtained routinely and quickly. It is arguable that such spectra provide more chemical information that can be obtained by any other single technique. One expects that, just as one obtains routine elemental analyses, so, in future, all coal laboratories will, as a matter of course, record well resolved infrared spectra of each coal they investigate. May this increase in the use of infrared spectrometry be accompanied by a decrease in the price of the instruments and an increase in the ability of laboratories to design their own software.

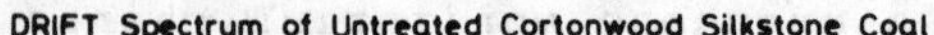

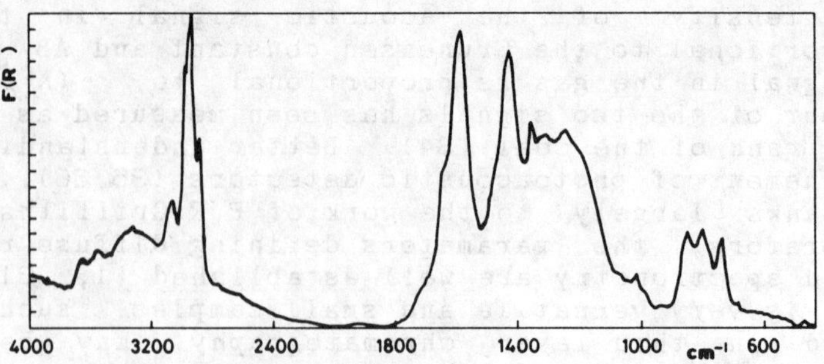

DRIFT Spectrum of Untreated Cresswell Coal

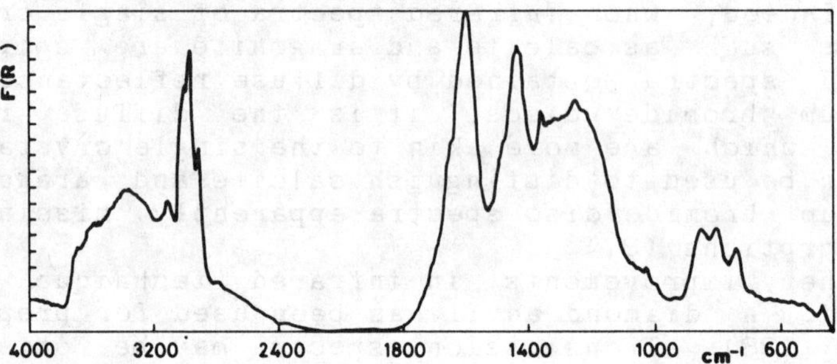

DRIFT Spectrum of Untreated Gedling Coal

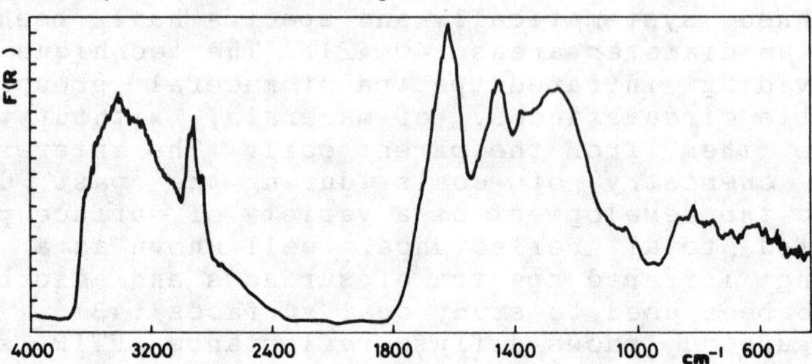

Figure 1. Diffuse reflectance spectra of coals.
y axis: Kubelka-Munk intensity function, x axis: cm^{-1}
Spectra from 1% of powdered coal in powdered potassium bromide.
Cortonwood coal: 87.2% C, 5.6% H, 3.9% O (dmmf)
Cresswell coal : 84.5% C, 5.5% H, 5.9% O (dmmf)
Gedling coal : 81.6% C, 5.2% H, 9.4% O (dmmf)

2. INFRARED ANALYSIS AND COAL STRUCTURE

2.1. Inorganics

Mineral matter being of consequence in most coal utilisation it is natural to employ infrared spectroscopy for its analysis. The technique has also been used for the analysis of fly ash (44) and of mine dust (45). Detailed studies of coal minerals are readily made after the organic matter has been removed by low temperature plasma ashing (46). The method is not entirely successful in leaving the minerals unchanged; thus, most of the nitrate present in the ash is formed from organic nitrogen in coal(47).

Clay minerals are particularly readily detected in coal (sharp peaks in the OH stretching region, large Si-O stretching vibration near 1000 cm^{-1}, peak at 901 cm^{-1} for example) and kaolinite has been analysed in coal in the presence of illite and montmorillonite (48). Kaolinite, itself, may vary in crystallinity and, and to a limited extent, in composition and this complicates its analysis (49). Interestingly, it has been observed using infrared spectroscopy that in coal liquefaction some hydrogen is transferred to the organic material via the OH groups of the associated kaolinite (50).

2.2. Organic Structure

Much of the work reported in this section is due to Professor P.C.Painter and his colleagues at Pennsylvania State University who have systematically applied FTIR to most aspects of coal structure.

2.2.1. C-H stretching vibrations. From coal spectra, similar to those shown in Figure 1, parameters describing an average structure can be determined from equations such as those of Ladner and Brown (51). Less assumptions have to be made if infrared spectroscopy be combined with a second technique such as C13 nmr (52-55). Thus, if the carbon aromaticity of coal is known from C13 nmr then the average aliphatic structure, CH_n, is deducible (55). It was found (55) that the proportion of CH_2 to CH groups increased with coalification whilst in agreement with previous work (56), the fraction of carbon atoms in methyl groups was independent of coalification. Surprisingly and confusingly, when absorption areas for aliphatic and aromatic CH groups were determined and compared with selected parameters of coalification for a set of vitrinites from the Lower Kittaning coal seam, the numbers of methyl and CH groups increased slightly with coalification whereas the number of CH_2 groups decreased (57-59). Much of the quantitative application of infrared spectroscopy to the structure of coal depends on the use of extinction coefficients which were originally chosen as averages of those from a wide variety of model compounds, relevant extinction coefficients are now known more precisely because the increased resolution of the C-H vibrations in coal spectra permits more precise assignments of the aliphatic linkages. Further knowledge has been obtained by comparing the infrared spectra of coals and coal extracts (60,61). Particularly valuable has been the use of the H1 nmr spectra of the extracts which, combined with the elemental analysis of extracted material, gives the number of each chemical type of proton present. From these numbers and the corresponding infrared spectra, the extinction coefficients of the absorptions by C-H vibrations may be deduced (62,63). The results go far toward resolving the inconsistency referred to previously (55,57). The intensities and wavelengths of C-H absorption correlated not only with rank but also with the technological properties of the coal (15,58,64).

2.2.2. Aromatic C-H deformation vibrations. Figure 1 shows overlapping, but resolvable, out of plane aromatic C-H deformation vibrations from which aromatic substitution patterns may be deduced (57). Unsurprisingly, coals contain a variety of substitution patterns, the mixture changing with rank.

2.2.3. Hydroxyl groups. In bituminous coals the broad absorption from 3600-2000 cm^{-1} which may obviously include absorption by adsorbed water if the coal has not been dried carefully, is mainly due to hydrogen bonded phenolic OH groups though a few alkyl OH and NH groups may also be present. These may be distinguished by acetylation (65). Spectra have been used for quantitative measurements of OH groups (65,67). These represent about 60 % of the oxygen in a bituminous coal (68).

2.2.4. Aromatic ring vibration. The major absorption of coals at 1600 cm^{-1} occurs in a region where both certain carbonyl groups and aromatic rings absorb. In pure compounds, aromatic ring vibrations give absorption not only at 1600 cm^{-1} but also at lower wavenumbers. The latter vibrations are difficult to observe in coals. Despite this, in bituminous coals the 1600 cm^{-1} absorption is usually ascribed almost entirely, to aromatic ring vibrations (69,70), their absorption intensity being enhanced by phenolic hydroxyl groups (71).

These subjects are occasionally reviewed (72,73). Table I summarizes much of the previous paragraphs.

2.2.5. Macerals and maceral groups. Although many of the spectra which have been discussed in previous paragraphs were obtained from vitrinite concentrates, most were obtained from samples of coal. The spectra therefore represent the average composition of a sample which is known to contain a variety of components having different chemical structures. There have been surprisingly few published spectra of macerals or maceral group concentrates other than vitrinite (74-77) though many more spectra are likely to be published in the future. Spectra from brown coals suggest the structure of the attrinite and densinite macerals to be similar (74). Textinite and textoulminite spectra appear to differ from gelinite and euoulminite spectra in that absorption in the 1500-1700 cm^{-1} region is mainly due to quinonoid carbonyl vibrations in the former pair of macerals and to carbonyl vibrations in carboxyl groups in the latter pair (75), aromatic ring vibrations also contributing. Spectra from bituminous coal samples indicate common structural patterns within maceral groups but there are also systematic differences within the groups (76,77). Absorption by C-O vibrations indicates that the C-O bonds in inertinites differ from those in vitrinites and exinites (77).

2.2.6. Changes with coalification. The study of coalified logs (78) by nmr and infrared spectrometry distinguishes at least three stages of coalification and helps to put infrared studies of rank changes into perspective. The first stage of coalification involved hydrolysis of the wood and loss of cellulose. The changes in spectra during the second stage were consistent with loss of methoxy groups, water and C3 side chains from lignite. Changes during the third stage, associated with the formation of the bituminous material, indicated the loss of oxygen rich humic acids. Several authors have studied the infrared spectra of humic acids (79-84). There has been growing realisation during recent years that humic acids from terrigenous and marine plants are different in structure and are formed by different mechanisms. Infrared spectra have been used to distinguish humic acids derived from lignites and from soil (79,80). The latter contained, almost exclusively, fractions that were aromatic whilst some high molecular weight fractions of lignite humic acid contained considerable amounts of aliphatic material. As lignites and coal become weathered or oxidised, more humic acids are formed, their yield being predictable from elemental analyses, and the structures of the humic acids gradually change (81). In fact the infrared spectra of aromatic humic acids from a lignite are similar to those of the parent fuel (82) and, accordingly, their chemical structures may be compared. Humic acids are well known to complex with cations through their carboxyl groups and this is readily demonstrated by changes in the absorption at 1700 and 1600 cm^{-1} (83). It has been shown (84) that much of the infrared absorption at 1050 cm^{-1} from lignite humic acids is due to the C-O stretching vibrations of polysaccharides.

TABLE I

Some Assignments in the Infrared Spectra of Coals

Wavenumber, cm^{-1}	Comment and Assignment
2956	asym. C-H stretching vibration in CH_3 and in hydroaromatic CH_2 (64)
2923	asym. C-H stretching vibration in CH_2 and in CH_3 α to an aromatic ring (64)
2891	revealed by curve fitting, due to C-H stretching vibration in methine and other groups (64)
2864	sym. C-H stretching vibration in CH_3 (64)
2850	sym. C-H stretching vibration in CH_2 (64), extinction coefficient 0.2 (62)
1770	C=O stretching vibration in acetylated aliphatic OH (65)
1656	conjugated C=O vibration, possibly quinone (70)
1614	aromatic ring vibration (70)
865	vibration of isolated aromatic C-H (57)
835	observed in extracts, revealed by curve fitting in coals, CH_2 rocking vibration + aromatic C-H out of plane vibration (57)
815	out of plane vibration of two and three adjacent aromatic C-H's (57)
785	as 835 cm^{-1}
750	vibration of four adjacent aromatic C-H's (57)

In general, lignites give little infrared information not observable in coals, save, of course, for a pronounced carbonyl-carboxyl peak at 1700 cm^{-1} (85,86) and, as with coals, the spectra may be correlated with technological properties (87). An interesting sequence of investigations of Australian coals (88-91) particularly of Victorian brown coals, included infrared spectrometry in showing that the concentration of phenolic components correlated well with the humotelinite maceral group which is composed of recognisable plant remains; the lighter colored lithotypes were rich in triterpenoids and the darker lithotypes were rich in phenolic components. Aromaticity and carbonyl content were the main gross chemical structural properties determining lithotype classification. The Australian work is a remainder that there has been insufficient investigation of the extent to which infrared spectra distinguishes solid fuels from different deposition.

2.2.7. Chemical structures present in coals. In order to obtain information about the actual structures present in coals as distinct from average structures it is generally necessary to degrade and solubilise the coals. Several such investigations featured infrared spectroscopy of extracts and of chemically treated coals (92-96). An Assam lignite (92) subjected to successive alkali degradation, depolymerisation and oxidation gave products whose spectra indicated the presence of two and three ring condensed aromatics possibly linked through quinonoid and furan bridges. Remnants of lignin and flavonoid structures were present. Extracts from 14 coals analysed by gas chromatography and infrared spectroscopy (93) showed a diverse distribution of alkanes, the ratio of cyclic and branched alkanes to normal alkanes being especially variable. The variability was attributed to differences in the lipid inputs to the deposition swamps rather than to differences in rank. Analysis of aliphatics is a theme of infrared characterisation of products from the reductive alkylation of coals (94,95). Particularly interesting was a study which attempted to distiguish the chemical structures in coking and non coking coals (95). While the aliphatic parts of the coking coals appeared to be comprised mainly of short and straight chain alkyl groups and alycyclic structures, there was much more branching in the aliphatic portion of the non coking coals. Aromatic-aromatic ethers appeared more prominent in the coking coals but aromatic-aliphatic and aliphatic-aliphatic ethers dominated the non coking coals. It would be interesting to know the petrography of these coals. The infrared spectra of depolymerised coals have been used to determine the probable effects of functionality on liquefaction behaviour (96).

3. INFRARED SPECTROSCOPY AND COAL UTILISATION

3.1. On Line Analysis

The rapidity of infrared analysis, the correlations between the infrared spectral and the technological properties of a coal, and the development of diffuse reflectance techniques suggest that it may be possible to develop infrared methods of on line analysis which will monitor coal feeds, the blending of coals, the beneficiation of coals, coals entering electricity generating stations and so forth. There has been little published work as yet, though a relatively low cost FTIR instrument has been developed for mobile analysis (97) and infrared spectrometry has been used for continuous monitoring of fluidised bed and entrained flow combustors (98-101). Observation of the emission from the hot combustor permits the deduction of the infrared absorption spectra of the particles suspended in the gas (100,101).

3.2. The Weathering, Aging and Oxidation of Coal

Coals are notoriously readiliy oxidised, even by air at ambient temperatures. The reaction reduces the calorific value of the fuel, renders the fuel crumbly and friable and can, in extreme cases, lead to spontaneous combustion; oxidation makes beneficiation more difficult and even small amounts of oxidation reduce the plasticity that coking coals attain during carbonisation and thereby prevent coke formation. Nevertheless, the reaction is still poorly understood and methods to suppress air oxidation are often primitive. Little wonder that infrared spectroscopy has often been used to study oxidation.

Like most reactions of coal, oxidation is a surface reaction and since the diffuse reflectance and photoacoustic techniques explore the coal near its external surface, several papers have been devoted to demonstrating the special utility of the techniques in monitoring oxidation (102-107). Obviously, changes in hydroxyl and aliphatic content and the formation of carbonyl groups are readily observed by the techniques, which can be applied in such a way as to monitor these changes as oxidation eats deeper into the coal structure (123). One result of these studies has been the development of sample containers permitting diffuse reflectance and photacoustic spectroscopy in controlled atmospheres. This is particularly helpful for determining the effects of increasing oxygen pressure and for distinguishing the effects of moisture on oxidation. Comparison (107) of diffuse reflectance infrared with Mossbauer spectroscopy and with Gieseler plasticity measurements showed that both the latter techniques were more sensitive to the onset of oxidation. Thus the plasticity of the coking coals was diminished before oxidation became observable by infrared spectroscopy and pyrites was oxidised more rapidly than coal substance. This is a significant observation since there is evidence, at least for some lignites, that the oxidation of pyrites is the match which causes the spontaneous combustion of moist lignite (131).

When one considers the organic structures in coals one would expect molecular oxygen to attack

1. the free radicals present (a reversible reaction, established several decades ago by electron spin resonance)
2. polyhydric phenol structures
3. aliphatic groups α to aromatic rings with the formation, first of hydroperoxides and, subsequently, of hydroxyl and carbonyl groups. This will be particularly true if, as frequently occurs in coals (132), there is branching at α the position. A significant proportion of the abundant literature on autoxidation should be applicable to coal.
4. cyclic hydroaromatics, which may decompose dehydrogenated.
5. mono hydric phenols at higher temperatures or over longer times.

Infrared evidence for the attack at free radical sites has been obtained recently (127), as has evidence for hydroperoxide intermediates (114,127) and the ubiquitous presence of surface peroxides (125). One notices no observations corresponding to the oxidation of polyhydric phenols though the presence of such structures in coal is established. At ambient temperatures the first infrared changes caused by the oxidation of coals are usually the loss of absorption by aliphatic C-H vibrations (67,115,118,122,126) and the growth of absorption by aryl-alkyl ketones (11,113); some authors detect carboxyl groups at this stage (110,114,115). At higher temperatures and longer times infrared spectral investigations indicate that oxidation results in cross linking of the coal, both ether (108,111,114,117) and ester (109,111,113) cross links being indicated.

The literature that has been summarised (108-127) is confusing and sometimes apparently contradictory. The attack by oxygen on a coal will depend, at least, on the proportions of reactive sites present in a coal (124,127), their chemical structure, the surface area of the coal, the distribution of macro and micropores (110), the presence of inorganic minerals, the amount of moisture present (122), the temperature, duration of oxidation and the pressure of oxygen. Few authors have characterised their system to this extent. If we are to unravel the mechanism of air oxidation, there appears to be a need for leading workers to select coal samples, agree on conditions of oxidation and exchange samples for analysis.

Two authors have used infrared techniques specifically to monitor the beneficiation of oxidised coals (128,130) and a study of oxidation during fluidised bed combustion (129) has been reported. Much work continues to relate oxidation and the inhibition of the development of plasticity (64,95,107-109,114,116). Most investigators consider that plasticity is inhibited by the formation of cross links during oxidation. This is undoubtedly true when oxidation is pronounced but, were one reviewing polymer chemistry, one would have expected to read of the effects of oxidation on initiation reactions. In coal terms, this probably means the attack of oxygen on the most reactive hydrogen present and the removal of this hydrogen so that, during pyrolysis, it is no longer available to attack bonds elsewhere in the system. This would imply the inhibition of plasticity by a decrease in bond scission during the early stages of pyrolysis. Such a hypothesis is consistent with the parallelism between the loss of absorption by aliphatic C-H groups and decrease in Gieseler plasticity and, in particular, it is consistent with the observation of decrease in Gieseler plasticity at levels of oxidation at which infrared spectral changes are insignificant (107).

3.3. The Combustion of Coal Particles

Combustion is obviously the overwhelmingly most important use of coal. The science of combustion has a long history and is a distinguished field of human endeavour. Much of this activity should be of interest to coal scientists who, nevertheless, are often divorced from it. Flames, for example, are rarely discussed at Coal Science conferences. The study of combustion has always been beset by the problem that any physical probe used for analysis invariably alters the flow of the combusting gases and may catalyse unwanted reactions. Remote sensing of optical properties is one of the few techniques which avoids the difficulty. The intensity of the emission from combustion has long been used to measure the temperature of the reactants. The monitoring of infrared emission permits measurement of relatively low temperatures (<900 K) (133), the method has been used to study temperatures in pulverised coal flames (134,136) and even to determine the temperatures of coal dust explosions (135). Of course, measurements in the infrared can also be used to give information about polyatomic species present during oxidisation. Reference has already been made (100) to the analysis of particulates in furnaces. A rather complete investigation of pulverised coal combustion, reviewed previously (1), uses FTIR to determine the temperature, composition and concentrations of the low molecular weight gases produced by the combustion (137-139). In a novel method of optimising the combustion of coal fired boilers, the ultra violet and infrared emission from the OH and CO_2 present in the flames have been used to control the behaviour of individual burners (140). The infrared optical properties of a power plant plume have been monitored (141).

3.4. Pyrolysis and Gasification of Coals

After their oxidation and combustion, the carbonisation (pyrolysis) and gasification of coals are their most valuable uses. Infrared spectroscopy is being used in two ways in pyrolysis.

First, the infrared spectra of coals have been used to predict -quantitatively- their pyrolysis behaviour (143,145). Of particular interest are the numbers and distribution of alkyl side chains which may be assumed to give gas on pyrolysis, the number of hydroxyl groups which give steam and the 'donatable hydrogen' which controls tar formation.

Second, infrared spectrometry has been used to analyse the pyrolysis products (142,144,146,151). Just as one of the most interesting developments in the infrared spectroscopy of combustion has been its use to monitor entrained beds so similar techniques are being developed to monitor continous rapid pyrolysis(142,146-148).

Gasification reactions may also be monitored by infrared spectroscopy (153,154,157). Infrared spectroscopy has been used to study the mechanism of heterogeneous catalysis of gasification by alkali metal salts mounted in carbonaceous lattices both by monitoring the products of reaction (155) and by studying the infrared spectrum of the catalyst before and after gasification (158-162). Since diffuse reflectance photo acoustic infrared spectroscopy can be used to give information about surfaces one expects the techniques to be increasingly employed in studies of surface catalysed gasification (156).

3.5. Coal Liquefaction Products

Numerous workers have characterised coal liquefaction products and tars by infrared spectrometry (163-208). Investigations have included the recognition of phenols as markers for coal derived material (163,164,174,185), the modification of the Ladner-Brown equation (167,169,177) when used to characterise liquefaction products and the identification of the different types of compound separable from liquefaction products (165,169,173,177,178,187). There have been notable developments in the microbore HPLC coupled to infrared detectors (191-193) and also in the identification of compounds eluted from a gas chromatograph by their matrix isolation followed by high resolution infrared spectroscopy (194-203). Matrix isolation is accomplished by the trapping of eluted compounds by freezing the carrier gas emerging from the gas chromatograph. Particular attention has been paid to the separation of nitrogen compounds and their identification by infrared spectroscopy (204-208).

3.6. Miscellaneous

The following investigations demonstrate the versatility of infrared spectroscopy. Infrared spectroscopy have been used to study the stability viz the change in coal concentration with time, of a coal/oil slurry (209). The change in surface groups when aqeous coal suspensions were electrolysed in the presence of iron or cerium oxidants has been monitored (210). The infrared absorption of sulphur dioxide has been used to develop elemental sulphur analysis via the combustion of coals and peats (211).

There have been studies of ash (212), the evolution of water from fly ash (213), the silica content of dusts (214) and the determination of trace quantities of vanadium iron and manganese in tar sands (215).

Acknowledgements

I am grateful to Dr.A.W.P.Jarvie of the University of Aston in Birmingham for the computerised literature survey on which this review is based and to the SERC (Great Britain) for financial support.

REFERENCES

1. New Developments in Coal Spectroscopy, NATO ASI, Nova Scotia, 1984.
2. R.F.Lacey, Anal. Chem.,58, 1404-1410 (1986).
3. M.A.Zilbergleit and V.M.Reznikov, Deposited Doc., SPSTL, 608 khp-D80 (1980).
4. M.McCue and E.R.Malinowski, Anal.Chim.Acta, 133, 125-136 (1981).
5. J.L.Koenig and M.J.M.Rodriquez, Appl.Spectrosc., 35, 543-548 (1981).
6. M.D.Erickson, Appl.Spectrosc., 35, 181-184 (1981).
7. P.C.Gillette, J.B.Lando and J.L.Koenig, Appl.Spectrosc., 36, 661-665 (1982).
8. P.C.Gillette and J.L.Koenig, Appl.Spectrosc., 36, 535-539 (1982).
9. P.M.Ovens, R.B.Lam and T.L.Isenhour, Anal.Chem., 54, 2344-2347 (1982).
10. L.T.Taylor, R.S.Brown, J.W.Hellgeth, C.C.Johnson and P.G.Amateis, Proc.Int.Conf.Coal Sci., Pittsburgh, 1983, p.639-642.
11. P.C.Gillette, J.B.Lando and J.L.Koenig, Anal.Chem., 55, 630-633 (1983).

12. P.M.Fredericks, P.R.Osborn and D.A.J.Swinkels, Fuel, 63, 139-141 (1984).
13. D.A.J.Swinkels, P.M.Fredericks and P.R.Osborn, PCT Int.Appl. W.O. 84/4594 (1984).
14. P.M./Fredericks, J.B.Lee, P.R.Osborn and D.A.J.Swinkels, Appl.Spectrosc., 39, 303-310 (1985) and 311-316 (1985).
15. D.W.Kuehn, A.Davies and P.C.Painter, Proc.Int.Conf.Coal Sci., Pittsburgh, (1983), p.304-306.
16. B.S.H.Royce, Y.C.Teng and J.Enns, Ultrason Symp.Proc., 2, 652-657 (1980).
17. D.J.Gerson, J.F.McClelland, S.Veysey and R.Markuszewski, Appl.Spectrosc., 38, 902-904 (1984).
18. T.Zerkia, Fuel, 64, 1310-1312 (1985).
19. W.L.Friesen and K.H.Michaelian, Infrared Phys., 26, 235-242 (1986).
20. J.C.Donini and K.H.Michaelian, Infrared Phys., 26, 135-140 (1986).
21. M.P.Fuller and P.R.Griffiths, Am.Lab., 10, 69-70, 72, 74-76, 78-80 (1978).
22. J.E.Carroll and W.M.Doyle, Proc.SPIE Int.Soc.Opt.Eng., 289, 111-113 (1981).
23. M.P.Fuller, I.M.Hamadeh, P.R.Griffiths and D.E.Lowenhaupt, Fuel, 61, 529-536 (1982).
24. D.E.Lowenhaupt, P.R.Griffiths, M.P.Fuller and I.M.Hamadeh, Proc.Iron.Making Conf., 41st, 39-45 (1982).
25. S.A.Yeboah, S.H.Wang and P.R.Griffiths, Appl.Spectrosc., 38, 259-264 (1984).
26. I.M.Hamadeh, S.A.Yehoah, K.A.Trumbull and P.R.Griffiths, Appl.Spectrosc., 38, 486-491 (1984).
27. S.H.Wang, Diss.Abstr.Int.B, 45, 2537 (1985).
28. E.L.Fuller, N.R.Smyrl, R.L.Howell and C.S.Daw, Prepr. Pap.Amer.Chem.Soc., Div.Fuel Chem., 29, 1-9 (1984).
29. S.A.Fysh, D.A.J.Swinkels and P.M.Fredericks, Appl.Spectrosc., 39, 354-357 (1985).
30. E.L.Fuller and N.R.Smyrl, Fuel, 64, 1143-1150 (1985).
31. S.H.Wang and P.R.Griffiths, Fuel, 64, 229-236 (1985).
32. P.J.Brimmer and P.R.Griffiths, Anal.Chem., 58, 2179-2184 (1986).
33. P.C.Painter, B.Bartges, D.Plasezynski, T.Plasezynsky, A.Lichtus and M.Coleman, Prepr.Pap.Am.Chem.Soc. Div.Fuel.Chem., 31, 65-69 (1986).
34. K.W.Johnson, K.L.Telschow and J.C.Crelling, Final Rept. to the Gas Research Instit., Project 82075 (1984).
35. N.M.Amer, US Pat.Appl. US 545338A0 (1984).
36. C.Morterra and M.J.D.Low, Mater.Chem.Phys., 12, 207-233 (1985).
37. E.R.Clark and A.F.Gaines, unpublished work, (1985).
38. J.R.Ferraro and L.J.Basile, Am.Lab., 11, 31-32, 34, 36, 38, 40-41 (1979).
39. W.A.Honeyball, J.W.Patrick and G.Sullivan, Appl.Spectrosc., 37, 209-210 (1983).
40. D.Brenner, Prepr. Pap. Am. Chem.Soc. Div.Fuel Chem.,

28,85-92 (1983).
41. D.Brenner, Proc. SPIE-Int. Soc. Opt.Eng., 411, 8-12 (1983).
42. D.Brenner, ACS Symp.Ser., 252, 47-64 (1984).
43. J.A.Mielczarski, A.Denca and J.W.Strojek, Appl.Spectrosc., 40, 998-1004 (1986).
44. R.M.Gendreau, R.J.Jakobsen, W.M.Henry and K.T.Knapp, Environ.Sci.Technol., 14, 990-995 (1980).
45. J.McCue, Brit.UK Pat. Appl., GB 2040035 (1980).
46. P.C.Painter, S.M.Rimmer, R.W.Snyder and A.Davies, Appl.Spectrosc., 35, 102-106 (1981).
47. P.C.Painter, J.Youtcheff and P.H.Given, Fuel, 59, 523-526 (1980).
48. P.C.Painter, R.W.Snyder, J.Youtcheff, P.H.Given, H.Gong and N.Suhr, Fuel, 59, 364-366 (1980).
49. R.B.Finkelman, F.L.Fiene and P.C.Painter, Fuel, 60, 643-644 (1981).
50. A.M.Vassalo, P.M.Fredericks and M.A.Wilson, Org.Geochem., 5, 75-85 (1983).
51. J.K.Brown and W.R.Ladner, Fuel, 39, 87-96 (1960).
52. P.C.Painter, A.Davis, M.E.Starsinic, J.L.Koenig, J.R.Havens, D.W.Kuehn, R.W.Snyder, J.T.Senftle, C.Rhoads and E.S.B.Riesser, Report DOE/PC/30013-F15 Order No. DE8400 3215 (1983).
53. M.E.Starsinic, Diss.Abstr.Int.B., 44, 314 (1984).
54. E.S.B.Riesser, Diss.Abstr,Int.B., 45, 1791 (1984).
55. P.C.Painter, M.E.Starsinic, E.S.B.Riesser, CRhoads and B.Bartges, Prepr.Pap.Am.Chem.Soc. Div.Fuel Chem., 29, 29-35 (1984).
56. A.K.Chatterjee and B.K.Mazumdar, Fuel, 47, 93-102 (1968).
57. D.W.Kuehn, A.Davis and P.C.Painter, ACS Symp.Ser., 252, 99-119 (1984).
58. D.W.Kuehn, A.Davies and P.C.Painter, Report DOE/PC/30013-4, Order No. DE84000539 (1983).
59. D.W.Kuehn, Diss.Abstr.Int.B, 44, 2361 (1984).
60. E.Benedetti, A.D'Alessio, P.Vergamini, A.Pennachi, P.Ghetti and G.Ghelardi, Riv.Combust., 34, 394-402 (1980).
61. N.E.Cooke, O.M.Fuller and R.P.Gaikwad, Fuel, 65, 1254-1260 (1986).
62. M.Sobkowiak, E.Reisser, P.H.Given and P.C.Painter, Fuel, 63, 1245-1252 (1984).
63. B.Riesser, M.Starsinic, E.Squires, A.Davies and P.C.Painter, Fuel, 63, 1253-1261 (1984).
64. J.T.Senftle, D.Kuehn, A.Davies, B.Brogoski, C.Rhoads and P.C.Painter, Fuel, 63, 245-250 (1984).
65. R.W.Snyder, P.C.Painter, J.R.Havens and J.L.Koenig, Appl.Spectrosc., 37, 497-502 (1983).
66. P.R.Solomon and R.M.Corongelo, Fuel, 61, 663-669 (1982).
67. P.B.Tooke and A.Grint, Fuel, 62, 1003 (1983).
68. J.S.Gethner, Fuel, 61, 689-697 (1982).

69. Y.Yurum, Thermochimica Acta, 113, 217-231 (1987).
70. P.C.Painter, M.Starsinic, E.Squires and A.A.Davies, Fuel, 62, 742-744 (1983).
71. P.C.Painter, R.W.Snyder, M.Starsinic, M.M.Coleman, D.Kuehn and A.Davis, Appl.Spectrosc., 35, 475-485 (1981).
72. P.C.Painter, R.W.Snyder, M.Starsinic, M.M.Coleman, D.Kuehn and A.Davies, ACS Symp.Ser., 205, 47-76 (1982).
73. P.C.Painter, M.Starsinic and M.Coleman, Fourier Transform Infrared Spectrosc., 4, 169-241 (1985).
74. A.F.Gaines, G.Ozyildiz and M.Wolf, Fuel, 60, 615-618 (1981).
75. M.Wagner, Bull.Acad.Pol.Sci.Ser.Sci.Terse., 29, 321-330 (1981).
76. R.J.Pugmire and D.M.Grant, Report, DOE/PC/30226-T3, Order No. DE 83013421 (1983).
77. G.R.Dyrkacz, C.A.A.Bloomquist and P.R.Solomon, Fuel, 63, 536-542 (1984).
78. P.G.Hatcher, I.A.Breger, N.Szeverenyi and G.E.Maciel, Org.Geochem., 4, 9-18 (1982).
79. S.Sipos and E.Sipos, Acta Phys.Chem., 25, 187-193 (1979).
80. S.Sipos, E.Sipos and A.Nagy, Acta Phys.Chem., 26, 103-110 (1980).
81. S.Raj, Prepr.Pap.Am.Chem.Soc.Div.Fuel Chem., 25, 58-66 (1980).
82. T.I.Balkas, O.Basturk, A.F.Gaines, I.Salihoglu and A.Yilmaz, Fuel, 62, 373-379 (1983).
83. T.Martinez, C.Romero and J.M.Gavilan, Fuel, 62, 869-870 (1983).
84. M.K.Barvah, Fuel, 65, 1756-1759 (1986).
85. S.Florea and G.Niac, An. Univ. Craiova [Ser] Mat. Fiz. Chem., 6, 71-75 (1978).
86. S.A.Benson, K.S.Groon, G.G.Montgomery and H.H.Schobert, Prepr.Pap.Am.Chem.Soc.Div.Fuel Chem., 29, 22-28 (1984).
87. P.R.Solomon and R.M.Carangelo, Report EPRI-AP-2115, Order No. DE 82901292 (1981).
88. T.V.Verheyen and R.B.Johns, Geochem.Cosmochim.Acta, 45, 1899-1908 (1981).
89. R.B.Johns, T.V.Verheyen and A.L.Chaffee, Proc. Int. Kohlenviss. Tag, 863-868 (1981).
90. T.V.Verheyen, R.B.Johns and R.J.Esdaille, Geochem.Cosmochim.Acta, 47, 1579-1587 (1983).
91. T.V.Verheyen, R.B.Johns, R.L.Bryson, G.E.Maciel and D.T.Blackburn, Fuel, 63, 1629-1635 (1984).
92. D.K.Sharma and M.K.Sarkar, Indian J.Technol., 21, 24-27 (1983).
93. S.Raj, Coal Technol., 2, 53-64 (1979).
94. H.Wachowska, A.Andrzejak and J.Thiel, Fuel, 64, 644-649 (1985).
95. G.Erbatur, O.Erbatur, A.Coban. M.F.Davis and G.E.Maciel, Fuel, 65, 1273-1280 (1986).

96. D.K.Sharma, Aufbereit-Tech., 26, 295-299 (1985).
97. D.A.C.Compton, M.Markelov, M.L.Mittlemass and J.G.Grasselli, Appl.Spectrosc., 39, 909-915 (1985).
98. M.O.Brewster and T.Kunitomo, Proc. ASME-JSME Therm. Eng. Jt. Conf., 4, 1357-1385 (1985).
99. D.E.Burrows, Adv.Instrum., 40, 1357-1385 (1985).
100. P.R.Solomon, R.M.Carangelo, P.E.Best, J.R.Markham and D.G.Hamblen, Prepr.Pap.Am.Chem.Soc.Div.Fuel Chem., 31, 141-147, 147a, 148-151 (1986).
101. P.R.Solomon, R.M.Carangelo, D.G.Hamblen and P.E.Best, Appl.Spectrosc., 40, 746-759 (1986).
102. M.G.Rockley and P.J.Devlin, Appl.Spectrosc., 34, 407-408 (1980).
103. N.R.Smyrl and E.L.Fuller, ACS Symp. Ser., 205, 133-145 (1982).
104. P.R.Griffiths, S.H.Wang, I.M.Hamadeh, P.W.Yang and D.E.Henry, Prepr.Pap.Am.Chem.Soc.Div.Fuel Chem., 28, 27-34 (1983).
105. P.R.Griffiths, Polym.Prepr.Am.Chem.Soc.Div.Polym.Chem., 25, 151-152 (1984).
106. P.L.Chien, R.Markuszewski and J.F.McClelland, Prepr.Am.Chem.Soc.Div.Fuel.Chem., 30, 13-20 (1985).
107. G.P.Huffman, F.E.Huggins, G.R.Danmyre, A.J.Pignocco and M.C.Lin, Fuel, 64, 849-856 (1985).
108. British Carbonisation Research Assoc., Carbonisation Res.Rep., 64, 23 (1979).
109. P.C.Painter, R.W.Snyder, D.E.Pearson and J.Kwong, Fuel, 59, 282-286 (1980).
110. R.Bouwman and I.L.C.Freriks, Fuel, 59, 315-322 (1980).
111. P.C.Painter, M.M.Coleman, R.W.Snyder, O.Mahajan, M.Komatsu and P.L.Walker Jr., Appl.Spectrosc., 35, 106-110 (1981).
112. P.C.Painter and C.Rhoads, Prepr. Pap. Am. Chem. Soc. Div. Fuel Chem., 26, 35-41 (1981).
113. J.C.Donini, S.A.LaCour, B.M.Lynch and A.Simon, Coal Phoenix '80's Proc.CIC Coal Symp., 64, 1, 132-134 (1981).
114. R.Liotta, G.Brons and J.Isaacs, Fuel, 62, 781-789 (1983).
115. P.M.Fredericks, P.Warbrooke and M.A.Wilson, Org.Geochem., 5, 89-97 (1983).
116. C.A.Rhoads, J.T.Senftle, M.M.Coleman, A.Davies and P.C.Painter, Fuel, 62, 1387-1392 (1983).
117. D.Joseph and A.Oberlin, Carbon, 21, 559-564 (1983).
118. J.R.Havers, J.L.Koenig, D.Kuehn, C.Rhoads, A.Davies and P.C.Painter, Fuel, 62, 936-941 (1983).
119. R.A.Johnson and R.P.Cooney, Aust.J.Chem., 36, 2549-2554 (1983).
120. J.S.Gethner, Fuel, 64, 1443-1446 (1985).
121. P.L.Chien, R.Markuszewski, H.G.Araghi and J.F.McClelland, Proc.Int.Conf.Coal Sci.Sydney, 818-821 (1985).

122. T.V.Verheyen, G.J.Perry, P.D.Nichols and D.C.White, Proc.Int.Conf.Coal Sci.Sydney, 822-825 (1985).
123. T.Zerlia, Appl.Spectrosc., 40, 214-217 (1986).
124. K.B.Anderson and R.B.Johns, Org.Geochem., 9, 219-224 (1986).

125. B.M.Lynch, L.I.Lancaster and J.T.Fahey, Prepr.Pap.Am. Chem.Soc.Div.Fuel Chem., 31, 43-48 (1986).
126. P.M.Fredericks and N.T.Moxon, Fuel, 65, 1531-1538 (1986).
127. J.S.Gethner, Appl.Spectrosc., 41, 50-63 (1987).
128. H.Hamya, K.H.Mickaelion and N.E.Anderson, Proc. Int. Conf. Coal Sci.Pittsburgh, 248-251 (1983).
129. E.L.Fuller, N.R.Smyrl, R.W.Smithwick and C.S.Daw, Prepr.Pap.Am.Chem.Soc.Div.Fuel.Chem., 28, 44-48 (1983).
130. P.R.Griffiths, Report DOE/PC/50797-T12, Order No. DE86011951 (1986).
131. B.Ratanasthien, Proc.Int.Conf.Coal Sci.Sydney, 463-466 (1985) and personal communication (1987).
132. R.Dogru, A.F.Gaines, A.Olcay and T.Tugrul, Fuel, 58, 823 (1979).
133. P.E.Best, R.M.Carangelo and P.R.Solomon, Prepr.Pap.Am. Chem.Soc.Div.Fuel.Chem., 29, 249-258 (1984).
134. R.A.Altenkirch, D.W.Mackowski, R.E.Peck and T.W.Tong, Chem.Phys.Processes Combust.Paper, 60 (1983).
135. K.L.Cashdollar and M.Hertzberg, Combust.Flame, 51, 23-35 (1983).
136. D.W.Mackowski, R.A.Altenkirch, R.E.Peck and T.W.Tong, Combust.Sci.Technol., 31, 139-153 (1983).
137. D.K.Ottesen and J.C.F.Wang, Prepr.Pap.Am.Chem.Soc. Div.Fuel Chem., 31, 136-141 (1986).
138. D.K.Ottesen and L.R.Thorne, Proc.Int.Conf.Coal Sci. Sydney, 351-354 (1985).
139. D.K.Ottesen and L.R.Thorne, Proc.Electrochem.Soc., 83-87, 425-433 (1983).
140. S.K.Batra, W.E.Cole and C.I.Metcalf, Coal Technol.8th, 177-196 (1985).
141. L.P.Stearns and R.F.Pueschel, Appl.Opt., 22, 1856-1860 (1983).
142. J.D.Freihaut, P.R.Solomon and D.J.Seery, Prepr.Pap.Am. Chem.Soc.Div.Fuel Chem., 25, 161-170 (1980).
143. P.R.Solomon and D.G.Hamblen, AIP Conf.Proc., 70, 121-140 (1981).
144. P.R.Solomon and D.G.Hamblen, Prepr.Pap.Am.Chem.Soc. Div.Fuel Chem., 27, 41-49 (1982).
145. J.D.Freihaut, D.J.Seery and M.F.Zabielski, Report DOE/MC/16221-1378, Order No. DE83008270 (1982).
146. P.R.Solomon and D.G.Hamblen, Report EPRI-AP-2603, Order No.DE83902950 (1983).
147. P.R.Solomon and D.G.Hamblen, R.M.Carangelo, J.R.Markham and M.B.DiTaranto, Prepr.Pap.Am.Chem.Soc.Div.Fuel Chem., 29, 83-93 (1984).

148. P.R.Solomon, D.G.Hamblen and R.M.Carangelo, Anal.Pyrolysis, published by Butterworth, London, p. 121-156 (1984).
149. J.S.Gethner, Appl.Spectrosc., 39, 765-777 (1985).
150. P.Chen, P.W.J.Yang and P.R.Griffiths, Fuel, 64, 307-312 (1985).
151. R.M.Carangelo, P.R.Solomon and D.J.Gerson, Prepr. Pap. Am.Chem.Soc.Div.Fuel Chem., 31, 152-161 (1986).
152. M.D.Erikson, S.D.Cooper, C.M.Sparacino and R.A.Zweidinger, Appl.Spectrosc., 33, 575-577 (1979).
153. P.R.Solomon and D.G.Hamblen, Report DOE/FE/05122-T1, Order No. DE82005658 (1981).
154. M.D.Erikson, S.E.Frazier and C.M.Sporacino, Fuel, 60, 263-266 (1981).
155. I.L.C.Freriks, H.M.H.van Weckem, C.M.Johannes and R.Bouwman, Fuel, 60, 463-470 (1981).
156. P.R.Griffiths, Report DOE/PC/30210-T4, Order No. DE83010618 (1982).
157. P.R.Solomon and D.G.Hamblen, Report DOE/METC 82-24, 275-278 (1982).
158. S.J.Yuh and E.E.Wolf, Fuel, 62, 252-255 (1983).
159. S.J.Yuh and E.E.Wolf, Prepr.Pap.Am.Chem.Soc.Div.Fuel Chem., 29, 206-211 (1984).
160. S.J.Yuh and E.E.Wolf, Fuel, 63, 1604-1609 (1984).
161. S.J.Yuh and E.E.Wolf, Ext. Abstr.Program Bienn.Conf. Carbon, 17th, 176-177 (1985).
162. M.B.Cenfontain and J.A.Moulijn, Fuel, 65, 1349-1355 (1986).
163. S.A.Farnum and E.A.Muth, Proc.M.D.Acad.Sci., 34, 11 (1980).
164. B.W.Farnum and S.A.Farnum, Proc.M.D. Acad.Sci., 34, 12 (1980).
165. E.A.Muth and S.A.Farnum, Proc.M.D. Acad.Sci., 34, 26 (1980).
166. D.G.Jones, H.Rottendorf, M.A.Wilson and P.J.Collin, Fuel, 59, 19-26 (1980).
167. J.M.Charlesworth, Fuel, 59, 865-870 (1980).
168. V.L.Weinberg, B.C.Gerstein, P.D.Murphy and T.F.Yen, Ext. Abstr.Program Bienn.Conf.Carbon 15th, 132-133 (1981).
169. S.E.Scheppele, P.A.Berson, G.J.Greenwood, O.G.Grindstaff, T.Aczel and B.F.Beier, Adv.Chem.Ser., 195, 53-82 (1981).
170. V.L.Weinberg, T.F.Yen, B.C.Gerstein and P.D.Murphy, Prepr.Pap.Am.Chem.Soc.Div.Pet.Chem., 26, 816-824 (1981).
171. V.A.Weinberg and T.F.Yen, Fuel, 61, 383-388 (1982).
172. L.J.S.Young, M.C.Li and D.Hardy, Prepr.Pap.Am.Chem. Soc.Div. Fuel Chem., 27, 139-144 (1982).
173. K.S.Seshadri and D.C.Cronauer, Prepr.Pap.Am.Chem.Soc. Div.Fuel Chem., 27, 64-75 (1982).
174. R.S.Brown and L.T.Taylor, Anal.Chem., 55, 723-730 (1983).

175. M.G.Strachan, A.M.Vassallo and R.B.Johns, Prepr.Pap.Am. Chem.Soc.Div.Fuel Chem., 28, 226-233 (1983).
176. Z.J.Stompel and K.D.Bartle, Fuel, 62, 900-904 (1983).
177. L.Petrakis, D.C.Young, R.G.Ruberto and B.C.Gates, Ind.Eng.Chem.Process Des.Dev., 22, 298-305 (1983).
178. K.S.Seshadri and D.C.Cronauer, Fuel, 62, 1436-1444 (1983).
179. L.Jones and N.C.Li, Fuel, 62, 1156-1160 (1983).
180. L.J.S.Young, N.C.Li and D.Hardy, Fuel, 62, 718-723 (1983).
181. P.G.Amateis and L.T.Taylor, Chromatographia, 18, 175-182 (1984).
182. T.F.Yen, W.H.Wu and G.V.Chilingar, Energy Sources, 7, 203-235 (1984).
183. P.G.Amateis, J.W.Hellgeth and L.T.Taylor, Prepr.Pap.Am. Chem.Soc.Div.Fuel Chem., 30, 225-231 (1985).
184. G.L.Alexander, A.L.Creagh and J.M.Prausnitz, Ind.Eng. Chem.Fundam., 24, 301-310 (1985).
185. J.B.Green, B.K.Stierwalt, J.A.Green and P.L.Grizzle, Fuel, 64, 1571-1580 (1985).
186. R.A.Winschel, G.A.Robbins and F.P.Burke, Conf.Coal Liquefaction 9th, 28/0-28/28 (1984).
187. K.S.Seshadri, D.C.Young and D.C.Cronauer, Fuel, 64, 22-28 (1985).
188. A.M.Mastral, V.L.Cebolla and J.M.Gavilan, Fuel, 64, 316-320 (1985).
189. L.Dixit, P.Kumar, H.C.Chandola and R.B.Gupta, Indian J.Technol, 24, 40-42 (1986).
190. J.Youtcheff, P.C.Painter and P.H.Given, Prepr.Pap. Am.Chem.Soc.Div.Fuel Chem., 31, 318-322 (1986).
191. C.C.Johnson and L.T.Taylor, Anal.Chem., 55, 436-441 (1983).
192. P.G.Amateis and L.T.Taylor, LC Mag., 2, 854-857 (1984).
193. L.T.Taylor, J.Chromatogr.Sci, 23, 265-272 (1985).
194. P.R.Solomon, D.G.Hamblen, R.M.Carangelo, J.R.Markham and M.R.Chaffee, Prepr.Pap.Am.Chem.Soc.Div.Fuel Chem., 30, 1-12 (1985).
195. A.A.Garrison, D.M.Hembree, R.A.Yoklev, R.A.Crocombe, G.Mamantov and E.L.Wehry, Proc. SPIE Int.Soc. Opt. Eng., 289, 150-153 (1981).
196. E.L.Wehry and G.Mamantov, ACS Symp. Ser., 169, 251-265 (1981).
197. D.M.Hembree, A.A.Garrison, R.A.Crocombe, R.A.Yokley, E.L.Wehry and G.Mamantov, Anal.Chem., 52, 1783-1788 (1981).
198. A.A.Garrison, G.Mamantov and E.L.Wehry, Appl. Spectrosc., 36, 348-352 (1982).
199. G.T.Reedy, D.G.Ettinger, J.F.Schneider and S.Bourne, Anal. Chem., 57, 1602-1609 (1985).
200. R.H.Hauge, L.Fredlin, J.Chu and J.L.Margrave, Prepr. Pap. Am.Chem. Soc. Div. Fuel Chem., 28, 35-43 (1983).

201. C.J.Chu, L.Fredin, R.H.Hauge and J.L.Margrave, Prepr. Pap. Am. Chem. Soc. Div. Fuel Chem., 29, 49-57 (1984).
202. C.J.Chu, L.Fredin, R.H.Hauge and J.L.Margrave, High Temp. Sci., 20, 51-73 (1985).
203. S.A.Cannon, C.J.Chu, R.H.Hauge and J.L.Margrave, Fuel, 66, 51-54 (1987).
204. L.G.Galya, D.C.Cronauer, P.C.Painter and N.C.Li, Prepr. Pap. Am.Chem. Soc. Div. Fuel Chem., 28, 137-143 (1983).
205. P.Burchill, A.A.Herod, J.P.Mahon and E.Pritchard, J. Chromatogr., 281, 109-124 (1983).
206. P.G.Amateis and L.T.Taylor, Anal.Chem., 56, 966-971 (1984).
207. V.P.Dmitrikov, Yu.V.Fedorov and V.M.Nabwach, Chromatographia, 18, 28-30 (1984).
208. L.T.Taylor, Report DOE/PC/40799-T11, Order No. DE85003753 (1984).
209. E.Benedetti, R.Santini, A.D.Alessio and P.Vergamini, Fuel, 61, 479-481 (1982).
210. P.M.Dhooge and S.M.Park, J. Electrochem. Soc., 130, 1539-1542 (1983).
211. E.S.Gladney, R.Raymond and N.W.Bauer, Am. Lab., 34, 36-38 (1985).
212. G.Platbroad and H.Barter, Anal. Chem., 57, 2404-2410 (1985).
213. L.M.Seaverson, J.F.McClelland, G.Burnet, J.W.Anderegg and M.K.Iles, Appl. Spectrosc., 39, 38-45 (1985).
214. T.L.Chien, Tung-Hai Hsueh Pao, 26, 699-712 (1985).
215. W.R.M.Graham, Prepr. Pap. Am. Chem. Soc. Div. Pet. Chem., 31, 608-612 (1986).

INTRODUCTION TO COAL PHOTOCHEMISTRY

J.R.Darwent* and A.F.Gaines
Birkbeck College
University of London
Malet Street
London WC1E 7HX
England

* Present Address:
Unilever Research
Port Sunlight Laboratories
Babington, Merseyside L63 3PWO

ABSTRACT. Invariably, coal extracts and solutions of coal liquids are not true solutions at all but contain, besides simple molecules, a range of much larger molecules and even small particles. Consequently, despite their aromaticity, such materials do not give ultraviolet absorption spectra, rather they scatter light. Information about the chemistry of the aromatic nuclei can be obtained, however, from the fluorescence of extracts, notably from their excitation spectra. Measurements of fluorescence commonly examine light emitted at right angles to the incident beam; the problems of scattering are thereby overcome. Even suspensions of 1 μm coal particles in water or chloroform emit fluorescence and the excitation spectra show mono, di and polyaromatic structures to be present.

The lifetimes of fluorescing states lie in the range 0.1-10 ns. This is shorter than the fluorescence lifetimes of simple aromatic molecules but akin to the fluorescence lifetimes of aromatic polymers. Thus the fluorescence behaviour of coal extracts and suspensions appear to be dominated by the ready transfer - both radiative and non radiative - of energy from one aromatic structure to the next. Energy is transferred ever more readily as the aromatic structures come closer together. It is an exothermic process in the sense that it is inevitably accompanied by loss of energy due to the vibration of molecules. Consequently energy emitted by monoaromatic nuclei becomes absorbed by diaromatics, reemitted at longer wavelengths and reabsorbed by polyaromatics and so on, until the energy becomes trapped in a structure from which further

Y. Yürüm (ed.), New Trends in Coal Science, 219–239.

energy transfer is impossible and fluorescence occurs. This is what happens when solid coals fluoresce and it is the reason why petrologists observe fluorescence from polished surfaces of coals only at long wavelengths not characteristic of the major aromatic species present. The highly aromatic inertinites and vitrinites transfer energy more efficiently and therefore fluoresce less than the less aromatic exinites.

The intensity and wavelengths of the molecular fluorescence from coal suspensions and coal extracts can be used to monitor the reactions of the aromatic nuclei which they contain. These nuclei exhibit the expected aromatic photochemistry and photooxidation, photohalogenation and photoreduction of coals have been studied. The reactions are limited by the extent to which the light and the reagents can penetrate the surface of the coal. There is evidence that particles of different coals have different morphologies.

The rates of surface reactions of coal systems will obviously depend on the surface areas present. Coal extracts contain species having a wide range of molecular sizes and usually species having a large surface area will react more rapidly than species having smaller surface areas. A good example of this is the reaction of photo excited coal extracts with acetone, the kinetics of which may be described by postulating that the species present in the extract possessed a Gaussian distribution of surface areas.

1. INTRODUCTION

Aromatic chemistry arose from the discovery of naphthalene and benzene in coal tar (1) and subsequently, it became clear that coals, themselves, were highly aromatic (2). Consequently, one may have expected coals to exhibit absorption of ultraviolet light characteristic of their aromatic structure and that this absorption would find ready utilisation in the technological analysis of fuels. How far this is from the truth.

Thin sections of coals, pyridine extracts of coals and even extracts of coal liquids such as solutions of asphaltenes, all give optical absorption spectra devoid of resolved peaks, the intensity of absorption increasing more or less exponentially as the wavelength of the light decreases. van Krevelen (3) supposed this behaviour to be a consequence of coals containing a great mixture of aromatic structures and he demonstrated that mixtures could be found that did, indeed, have very similar optical properties to coals, though the simulated absorption curve was not as smooth as that gives experimentally by coal systems.

Recently, Gethner (4) has demonstrated that absorption from thin sections of coal is dominated by scattering, much of the scaterring arising from the optical contrast between the coal and its pore structure. When differences in refractive index were diminished by filling the pores with the pores with solvent, scattering of light was greatly decreased, though there was still but little evidence of any absorption of light by the aromatic systems. Extracts of coals, such as pyridine extracts, will also scatter light. Hombach (5) showed that such extracts consist of small molecules, of molecules having molecular weights of 10^3 - 10^5 and of small particles (of diameter 1 μm). Both the larger molecules and the small particles will scatter light; indeed, the size distribution of the material in pyridine extracts - and of extracts of coal liquids in organic solvents - can be determined by light scattering, notably by photon correlation spectroscopy (Figure 1).

2. FLUORESCENCE OF PYRIDINE EXTRACTS

The problem of scattering is overcome in fluorescence measurements. Customarily, one monitors light at right angles to the incident beam and in particular, at a lower frequency than that of the incident beam. Consequently the results are free of scattering effects. The first measurements of the fluorescence of coal extracts were those of Hombach (6) and of Retcofsky, Friedel and Bredel (7) who showed that, when excited by near UV light, pyridine extracts gave maximum emission at about 440 nm. Many coal related materials such as humic acids and coal liquids emit strongly in this region. Rather more interesting are the excitation spectra. Excitation spectra of pyridine extracts of European carboniferous coals showed resolvable peaks in the 340, 390, 440 and 470 nm regions (8). Extracts of exinites and vitrinites fluoresced similarly. Typical spectra are shown in Figure 2. One cannot examine possible excitation maxima below 300 nm because of strong absorption in this region by pyridine itself. The observed excitation maxima are due to absorption by molecular species. That is, the excitation spectra are essentially absorption spectra and their detailed interpretation would furnish a description of the distribution of the aromatic structures present.

Although the fluorescence of pyridine extracts is due to the molecular species present, the quantum efficiencies for emission are of the order of 0.05 (8), significantly lower than efficiencies of solutions of pure organic compounds and similar to the quantum efficiencies of aromatic polymers. again, the lifetimes of the fluorescing states in pyridine extracts of coal range from 0.1 ns to

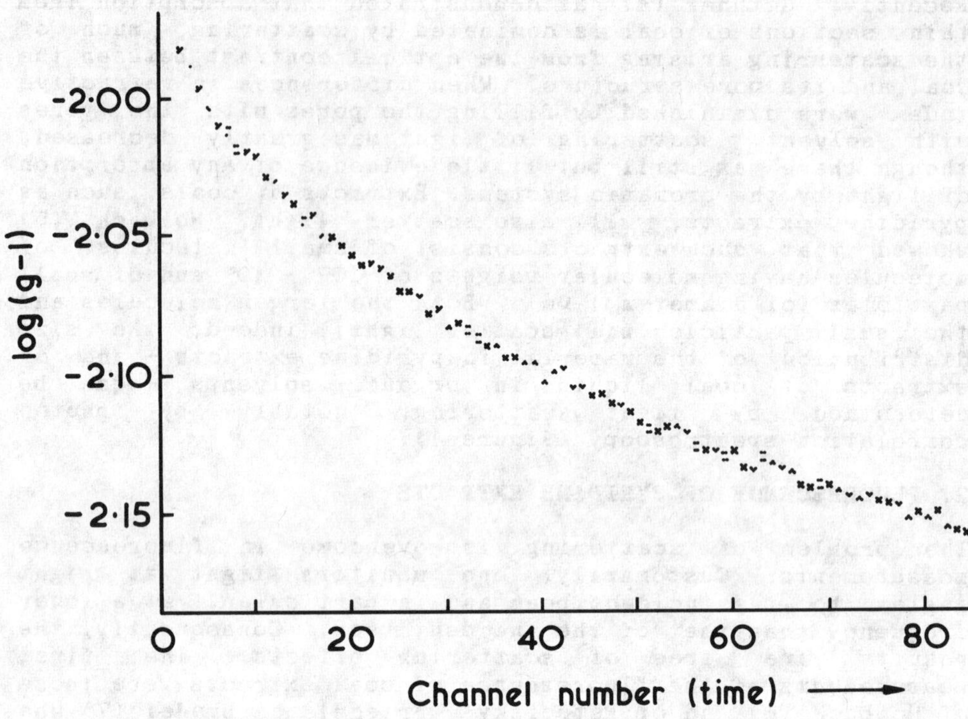

Figure 1. Typical photon correlation spectrum of a pyridine extract of a bituminous coal. The y axis is log(g-1), g being the auto correlation function and the x axis is 'channel number' i.e. time. A monodisperse solution gives a straight line plot whose slope is directly proportional to the rotational diffusion constant of the solute. The graph obtained from the pyridine extract shows a system with a wide range of rotational diffusion constants corresponding to small molecules (left), large molecules (centre) and small particles (right).

10 ns (8). Aromatic compounds in dilute solution have fluorescence lifetimes of between 5 and 500 ns (9) but when they are incorporated into aromatic polymers their lifetimes are as short as 1 ns or even less. Thus, not surprisingly, pyridine extracts of coals have fluorescence properties similar to those of aromatic polymers. It is worth considering this in some detail. Even in dilute solution, the aromatic nuclei within a polymer molecule are likely to lie relatively close together. In such circumstances absorbed light energy is transferred from one aromatic structure to another and relatively little is emitted; in fact, the aromatic nuclei can be as much as 100 nm apart and yet energy transfer remains significant (10). In other words, the quantum fluorescence efficiency is less than that of a dilute solution of simple aromatic compounds.

Two types of energy transfer may be distinguished

a. radiative

b. non radiative.

In radiative transfer of energy a second molecule absorbs the energy emitted by the first. Radiative transfer is particularly appropriate to a system, such as pyridine extracts of coals, where there is a variety of aromatic structures and absorption and emission wavelengths overlap (Figure 2).

Non radiative energy transfer also requires the overlap of excitation and emission wavelengths (11) but it occurs by means of interactions between the aromatic nuclei without the emission of light. Thus, the interaction between the dipole moment of an excited molecule and the dipole moments of neighbouring aromatic structures results in efficient transfer of energy (11) and it is this type of energy transfer which reduces the lifetime of the excited states - the fluorescence lifetime - in aromatic polymers.

Energy transfer, by whatever mechanism, commonly results in the formation of aromatic structures in vibrationally excited states. Fluorescence lifetimes of the order of nano seconds are much longer than periods of vibration (10^{-13}s) and the excited molecules easily revert to their ground vibrational state with a corresponding loss of energy. Thus, as energy is transferred from one aromatic nucleus to another, a portion is lost as kinetic energy. Successive transfer of energy therefore continues to aromatic structures having lower and lower lying electronic states until no further aromatic structures remain which are accesible. The 'trapped' energy is then emitted as fluorescence.

Energy transfer is a dominant feature of the fluorescence of coal extracts. The fluorescence spectra shown in the figures appear to suggest that polynuclear compounds are the predominant aromatic structures in coals.

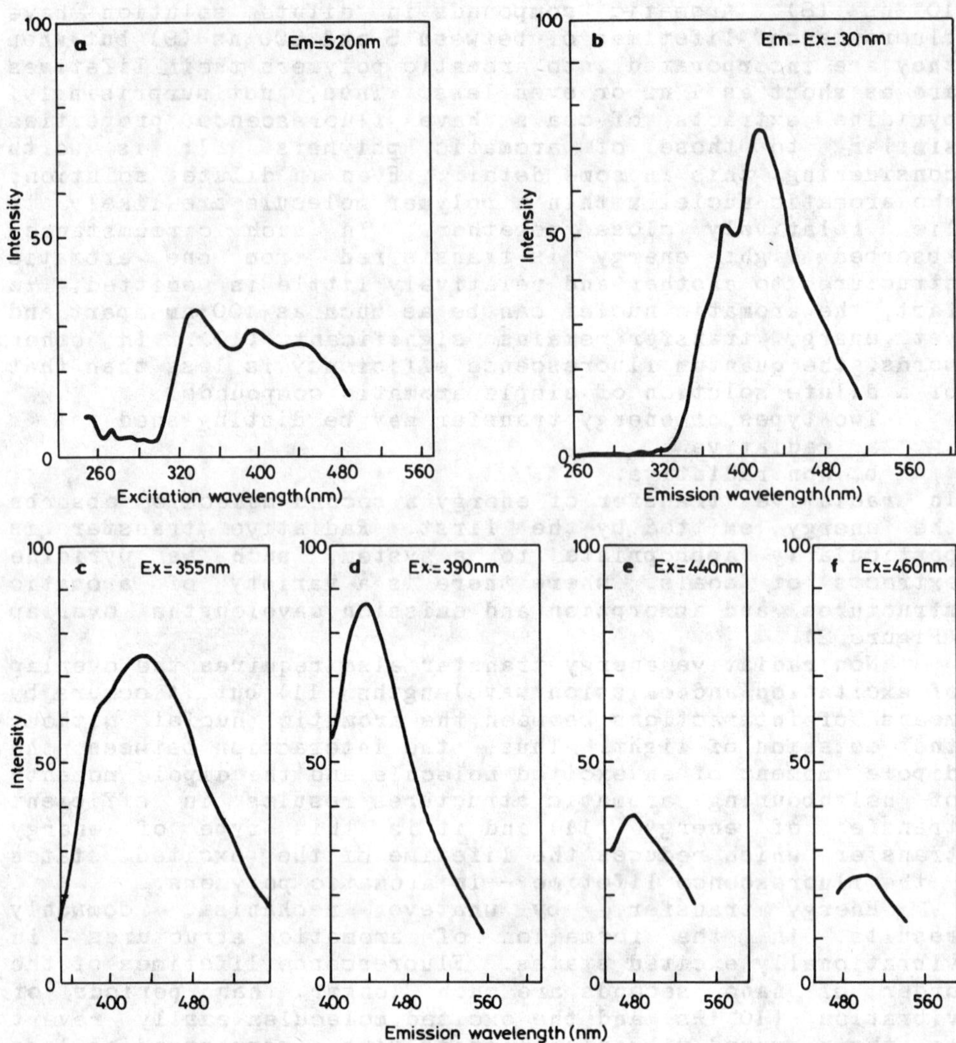

Figure 2. Fluorescence spectra of pyridine extract of a bituminous coal [81.6% C, 5.2% H, 9.4% O (Saf)].

a. corrected excitation spectrum emission 520 nm
b. synchronous spectrum, D = 30 nm
c. emission spectrum, excitation 335 nm
d. emission spectrum, excitation 390 nm
e. emission spectrum, excitation 440 nm
f. emission spectrum, excitation 460 nm

All spectra were obtained at the same concentration.

This is contrary to the accepted interpretation of X-ray diffraction studies (12) and of certain nmr studies (13) of coals. It seems likely, therefore, that the apparent predominance of polynuclear fluorescence is the result of energy transfer from smaller aromatics. Determination of the true distribution of aromatic structures will require the unravelling of the effects of energy transfer.

3. FLUORESCENCE OF COAL LIQUIDS

There is nothing special about pyridine extracts of coal; any 'coal solution' can be expected to fluoresce and the fluorescence of coal tars (14) and of carbon disulphide extracts of coals has been investigated (15).

Figure 3 shows the fluorescence of a supercritical toluene extract of a bituminous coal (16). The excitation spectra show peaks at about the same wavelengths as do excitation spectra of the pyridine extract of the same coal, though the relative intensities of the spectra are different. The study of the supercritical extract advanced the understanding of coal systems since the extract could be separated straightforwardly into pentane soluble, asphaltene and benzene insoluble fractions. The pentane

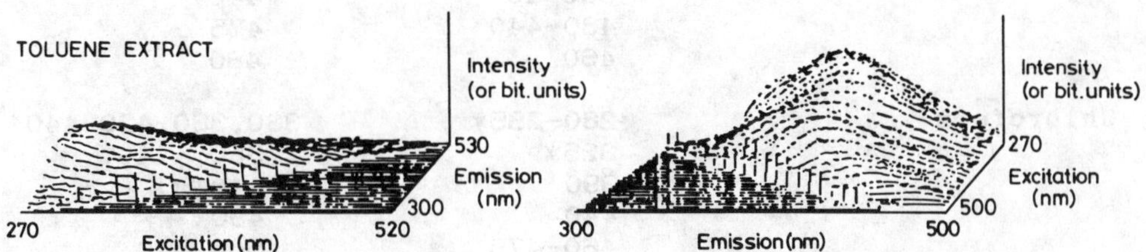

Figure 3. Fluorescence spectra of the supercritical toluene extract of Daw Mill (hvbA, H/C = 0.78) coal concentration: 0.2 ppm in chloroform.

solubles could be further separated into monoaromatic, diaromatic and polyaromatic fractions. Such detailed separation is not possible with pyridine extracts. By studying the fluorescence of its constituent fractions, it was possible to confirm that the fluorescence of the supercritical toluene extracts was indeed due to molecular species. Broad excitation maxima in the 280-290 nm region, the most intense of the extracts' excitation maxima, were ascribed to substituted monoaromatics and to diaromatics, whilst excitation maxima in the 330-360 nm region were attributed to such polyaromatic structures as anthracene and phenyl anthracenes. Excitation maxima around 390 nm suggested that some large polycyclic aromatic molecules were present (16).

4. FLUORESCENCE OF COAL SUSPENSIONS (8,17)

Both pyridine extracts and supercritical toluene extracts represent about 25% of a coal. It seems natural to explore whether one could determine the fluorescence of the whole of a coal. Table I summarizes fluorescence spectra of coal particles crushed to smaller than 1 μm diameter and suspended either in chloroform or in aqueous surfactant, the

TABLE I

Fluorescence of Suspensions of Bituminous Coal Powders (1 μm diameter).

Suspending Liquid	Excitation Maxima, nm	Corresponding Emission, nm
Water (& surfactant)	250	290, 370,440
	280-290	300-310
	330-350	380, 400-420
	390-400	440
	430-440	475
	460	480
Chloroform	280-285*	330,380,420-440
	325*	380, 440+
	390	440+
	440	490
	460-475	500-510

* Major maxima, of roughly equal intensity, forming a broad plateau of excitation

+ Major emission, wavelength varies with rank of coal (Figure 4).

coal being that from which the supercritical toluene extract was obtained. One had to work with very dilute suspensions in order to avoid concentration effects and in consequence, the spectra were noisy. The major excitation maximum of the aqueous suspension was below 270 nm and gave rise to intense emission at 370 nm which was obviously similar to that observed in the dilute solution of the supercritical toluene extract (and also seen in pyridine extracts; though in pyridine excitation cannot be observed below 300 nm) and assigned to polynuclear aromatic material. Excitation at 280-290 nm gave minor emission maxima at 300-310 nm probably from diaromatic material. Fluorescence of suspensions in chloroform could be excited by all wavelengths from 270 to 390 nm but emission increased in intensity as the excitation wavelength increased. Emission was a maximum at 440 nm but examination of a number of European carboniferous coals showed that this wavelength varied systematically with rank (Figure 4). Whereas methylation increased the intensity of emission from a chloroform suspension of coal, oxidation, whether by air or potassium permanganate, reduced the emission intensity.

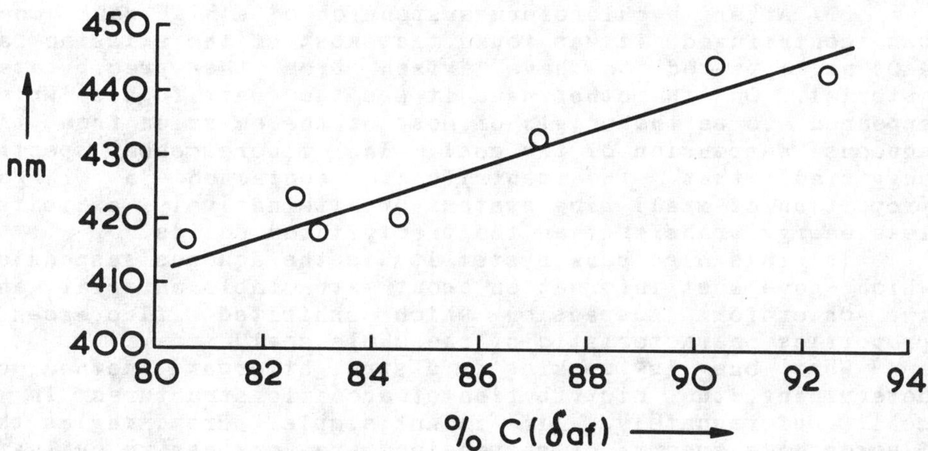

Figure 4. Wavelengths of the maximum emission of powdered bituminous coals suspended in chloroform plotted as a function of the carbon content of the coals.

The distinction between the fluorescence spectra of aqueous and chloroform suspensions of the same coal prompts the question as to whether the fluorescence arises from the particles or from extracted molecules. This can be answered by considering the effects of centrifuging the suspension. The result depend on the rank of the suspension:

a. Ultracentrifugation of a brown coal suspension in chloroform showed the excitation and emission spectra of the centrifugate to be nearly the same as those from the original suspension. The precipitate when resuspended in fresh chloroform gave fluorescence spectra which were less intense than those from the original chloroform suspension and similar to the spectra from the aqueous suspension.

After an aqueous suspension of the same brown coal was centrifuged, the precipitate, on resuspension in water, gave approximately two thirds of the emission intensity of the original suspension.

Thus the fluorescence of the chloroform suspension gave insight into the aromatic structures present in extractable material whilst the spectra of the aqueous suspension gave more information about the coal matrix. It is important to realise that not only is the previous sentence an oversimplification but that the chloroform centrifugate was not a true solution of simple molecules; it did not give a conventional absorption spectrum but, like most coal suspensions, it scattered light.

b. After a chloroform suspension of a high rank coal was centrifuged it was found that most of the emission at 440 nm appeared to have arisen from the precipitated material. On the other hand it was the centrifugate which appeared to be the origin of most of the emission from the aqueous suspension of the coal. The fluorescence spectra suggested that the centrifugate contained a larger proportion of small ring systems-or alternatively, exhibited less energy transfer-than the precipitated solids.

In this high rank system it was the aqueous suspension which gave most information about extractable material and the chloroform suspension which exhibited fluorescence properties characteristic of the whole coal.

What one is seeking is a straightforward method of determining the distribution of aromatic structures in a coal. Unfortunately, life is not simple. Promising as the fluorescence spectra of suspensions are, one has to evaluate each suspension carefully to determine the extent to which its spectra are representative of the coal material. Moreover, the effect of enhancement of the spectra at longer wavelengths caused by energy transfer has still to be resolved. On the other hand, the fluorescence spectra of coal suspensions could readily be used for the pragmatic characterisation of coals.

5. FLUORESCENCE OF COAL SURFACES

Petrologists routinely use the fluorescence emitted by polished surfaces of fossil fuels to delineate macroscopic fragments of the exinite-liptinite macerals (18). When illuminated by 365 nm light the exinite macerals usually emit yellowish light, though emission wavelengths can vary between 450 and 650 nm (18). Vitrinite fluoresces, more weakly than exinite, between about 600 and 700 nm (19). Weak red (650 nm) fluorescence is observed when polished surfaces of inertinite are illuminated with 450-490 nm light (19). Since the investigations of Jacob (18), coal petrologists have used the wavelengths of maximum emission, the intensity of the emission and the ratio of the intensities at 650 and 550 nm (the red-green quotient) to characterise fluorescence from polished surfaces and empirical correlations have been developed relating these parameters to the technological behaviour, the rank and the deposition of coals (19). Recently (20), the life times of fluorescence from polished surfaces have also been measured, the results have been interpreted as representing two characteristic lifetimes for each surface but it seems possible that further work will reveal distributions of lifetimes similar to those obtained from coal extracts.

The fluorescence studies by coal petrologists are in a sense a celebration of the investigations of the absorbance of light by aromatic crystals conducted by Davydov forty years ago (21). Compared to absorption spectra of dilute solutions, the crystal spectra were broadened and lines became split, maxima were shifted to longer wavelengths and their intensity was commonly found to be dependent on the polarisation of the incident light. Davydov deduced that the ready non radiative transfer of energy from an excited aromatic molecule to another ground state molecule generated bands of energy levels for the whole crystal - at least over patches of 1 μm diameter (22). Solid state physics formulates these phenomena as excitation theory (23) and they have been discussed already in the description of the flurescence of coal extracts and dispersions where the importance of energy transfer has been emphasised. In an aromatic crystal containing impurities and in a material such as coal, energy transfer is very efficient. Light is absorbed, continually transferred to structures having every lower excited states until eventually it becomes trapped and is then emitted as fluorescence of a relatively long wavelength. A 'trap' is a structure, such as the largest aromatic molecule present in the solid, though it may also be a defect structure in the lattice, which has no neighbours within a distance of about 100 nm which have lower energy excited states. Polynuclear compounds are astonishingly successful in scavenging energy transferred within aromatic solids. Thus, crystals of anthracene containing but one molecule of tetracene to every thousand molecules of anthracene emit light entirely characteristic of excited tetracene and containing no wavelengths characteristic of anthracene. This is the origin of the common green fluorescence of solid anthracene (24). Since such small quantities of impurity can be so important, it is not surprising that fluorescence properties vary from one sample of solid to the next: polished surfaces of coals give reproducable fluorescence only if one averages over many measurements and over many patches of surface. It is evident that emission from polished surfaces of coals tells one about the fluorescent traps which are present and is almost certainly unrelated to the chemistry of the major aromatic structures in the coal. Apparently, vitrinite and inertinite contain lower energy traps (larger polynuclear aromatic compounds) than exinites. Again, exinites fluoresce more intensely than vitrinites or inertinites because the latter maceral groups, probably because their higher aromaticity leads to great propiniquity of the aromatic structures, are much more efficient at transferring energy from one aromatic structure to another and losing vibrational energy as they do so. A key experiment here was that of Lin et al. (25) who showed that pyridine extracts which, as has been seen, exhibit molecular fluorescence when dilute, give similar (long wavelength) fluorescence to the parent coal when dried out. Further understanding of the fluorescence from surfaces of maceral groups might be obtained by studies of the polarisation of emitted light and especially by studies of excitation rather than of emission spectra.

6. REACTIONS OF IRRADIATED COAL EXTRACTS

Thirty years ago Bent and Brown (26) irradiated both coal suspensions and benzene and chloroform extracts of coal. When oxygen was bubled through irradiated (mercury lamp) suspensions of coal in water, esr spectra showed the free radical content of the coal to have increased but no change was observed in the ultimate analysis. However, when oxygen was bubled through irradiated benzene or chloroform extracts, the colour of the extract changed from dark red to pale yellow and a precipitate formed which contained abundant carbonyl groups. When coal was suspended in carbon tetrachloride and irradiated, not only did the free radical content increase, but infra red spectroscopy showed chlorine to have been incorporated into the coal structure.

Bent and Brown's work (26) demonstrated that, as one might expect, coals and coal liquids are markedly photoreactive, the extent of reaction being limited only by the ability of the light and the reagents to penetrate the coal. Of course, so far as reagents are concerned, this is true of most coal reactions.

7. REACTIONS OF IRRADIATED EXTRACTS WITH ACETONE AND CHLOROFORM

Since the molecular fluorescence of coal suspensions and extracts is a property of their aromatic nuclei, it may be used to monitor any chemical reactions in which the nuclei participate. We have used fluorescence to follow the chemical reactions of irradiated coal extracts. Figure 5 shows the results of irradiating supercritical toluene extracts of Daw Mill coal dissolved in oxygen free acetone. These solutions fluoresced as described previously and they were sufficiently dilute that their emission intensity was directly proportional to the concentration of the extract. It will be seen that irradiation produced a reduction in the subsequent fluorescence of the solutions. A 900 w 'Applied Photophysics' Xenon discharge lamp was used for irradiation, it produced little light below 330 nm and experiments showed the most effective irradiating wavelengths to be around 390 nm. Consequently, one expects irradiation to have promoted polyaromatic nuclei to excited singlet states. Acetone does not absorb light above 330 nm and irradiation must result in reaction of excited polyaromatics, presumably with the carbonyl group of the solvent molecules.

Not only the supercritical toluene extract but also its pentane soluble, benzene soluble (asphaltenes) and benzene insoluble fractions reacted with acetone or chloroform when irradiated. Reaction reduced the subsequent fluorescence of the solution throughout the normal range of excitation wavelengths (Figure 6). Since emission intensity was proportional to concentration, graphs of emission intensity as a function of duration of irradiation gave information about the kinetics of the irradiation reaction: when solutions of naphthalene, dimethyl quinoline, anthracene, chrysene and pyrene were irradiated in the same way as the coal liquids, all gave first order plots: the logarithm of the emission intensity was directly proportional to time. (It should be noted, however, that the fluorescence intensity of chrysene and pyrene, unlike the other aromatics tested and unlike all the coal liquids, increased on irradiation and it seems clear that neither of these nuclei can have been present in the Daw Mill toluene extract). In contrast to the behaviour of the pure aromatic compounds, the Daw Mill coal extracts gave curved semilogarithmic plots when tested for first order kinetics. However, the kinetics were not of higher order (e.g. second) because, when normalised by the initial fluorescence intensity, decay

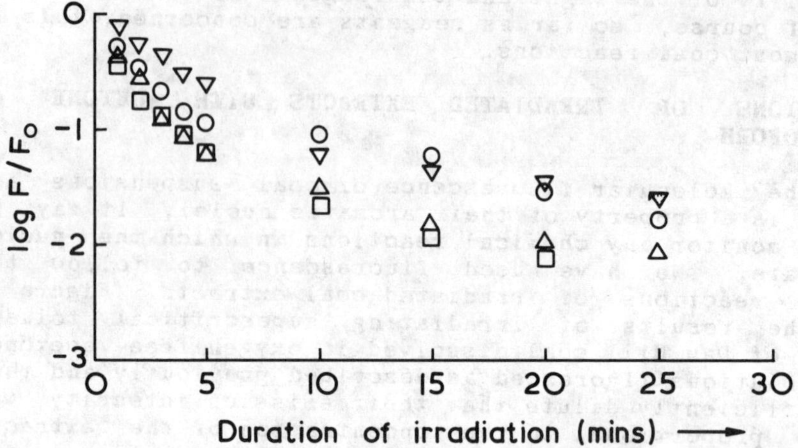

Figure 5. Typical graphs showing the decay in fluorescence intensity, F (excitation, 370 nm; emission, 420 nm) caused by irradiation of acetone solutions of Daw Mill extracts. A xenon lamp was used for irradiation.

△: material extractable by supercritical toluene extract
▪: pentane soluble fraction of supercritical toluene extract
o: asphaltenes from supercritical toluene extract
▽: benzene insoluble fraction of supercritical extract

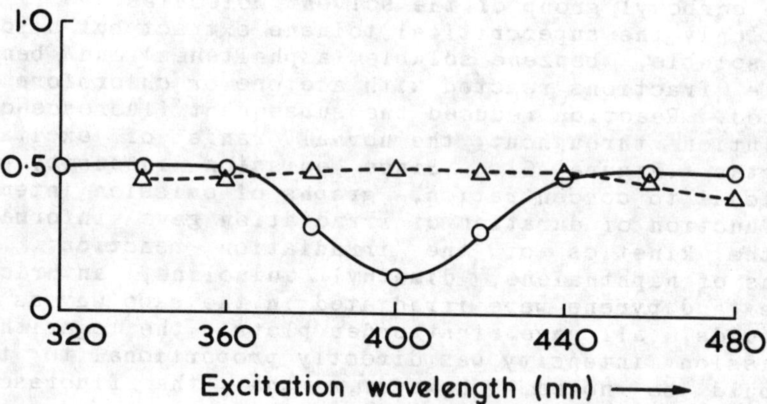

Figure 6. Typical graphs showing the effect of five minutes irradiation (Xenon lamp) on the excitation spectra of Daw Mill coal extracts.

△ : material extracted by supercritical toluene redissolved in acetone
o : pyridine extract diluted with an equal volume of chloroform

curves from any one extract all lay on a common curve irrespective of the initial concentration of extract. Superficially the curves (Figure 5) suggest a series of consecutive reactions, at first rapid and eventually slow. The slow reactions have similar rates to those observed for naphthalene, dimethyl quinoline and anthracene, it is the initial rapid reactions that require explanation. It is tempting to state that the results suggest the materials contain specific chemical structures which are highly photoreactive but, as we shall see, such an explanation is unnecessary. The rate of absorption of light by an extract during irradiation is given by the equation

$$d(S^*)/dt = -d(S)/dt = 4\pi r^{*2}(S) \qquad \text{(a)}$$

where

r* : mean radius of an 'extract molecule' or an extract particle in solution

(S*) : concentration of excited aromatics in the surface of the molecules or particles

(S) : concentration of ground state aromatics in the surface of the molecules or particles.

Products, P, may be supposed to form by reaction of excited aromatics at a rate

$$d(P)/dt = -d(S^*)/dt = k_2(S^*).$$

If one molecule of aromatic gives one molecule of product, and if one may neglect (S*) as being tiny, then one has $(P) = (S_o) - (S)$, (S_o) being the initial surface concentration of aromatics. Assume the intensities of fluorescence of the aromatics and their products are F(S) and αF(P) respectively, F and α being constants and $0 \leq \alpha \leq 1$. Then it is readily shown that

$$d\Phi/dt = -4\pi r^* I \epsilon \Phi \qquad \text{(b)}$$

where

I : rate of intensity of irradiation

ϵ : appropriate extinction coefficient

Φ=[(fluorescence at time t)-(final fluorescence of system)]/ [(intl. fluorscnce.of system)-(fnl. fluorscnce of system)]

Of the various assumptions made in deriving equation (b) the most pertinent is the use of the mean radius r*. As we have seen, that coal extracts give scattering rather than absorption spectra implies that these 'solutions' contain small particles and large molecules as well as simple molecules. Clearly there is a contribution of molecular radii, r*, and it is the molecules and particles with large values of r which react rapidly at the beginning of irradiation and the molecules with small values of r which react slowly at rates similar to those of naphthalene and anthracene. Equation (b) implies that extracts with similar distributions of r will give similar kinetic curves. Albery et al. (27) givethe mathematical analysis when the number of molecules, n, having a radius, r, obeys a Gaussian distribution

$$n = N \exp(-x^2) \quad \text{and} \quad \ln r = \ln r^* + px \qquad \text{(c)}$$

where, N is the number of molecules having a mean radius, r* and p governs the width of the distribution. Albery et al. (27) show how, with such a distribution of molecular radii, equation (b) may be generalised and integrated to give

$$\bar{\phi} = \pi^{-1/2} \int_{-\infty}^{\infty} \exp(-x^2) \exp[-\tau \exp(2px)]dx \qquad \text{(d)}$$

where $\tau = 4r^{*2}\pi I\epsilon t$.

The plots of fluorescence as a function of irradiation time in acetone did, in fact, satisfy equation (d) reasonably well, the fit generating two parameters, the average rate constant, $4\pi r^* I\epsilon$ and the 'spread parameter', p. Figure 7 illustrates the results which are summarised in Table II. The average rate constants were all of the same order of magnitude, as one would expect.

TABLE II

Summary of the Kinetics of the Decay in Fluorescence Caused by Irradiation of Acetone Solutions of Material Entracted from Daw Mill Coal

Material	$4\pi r^2 I\epsilon$, min^{-1}	p	r*, μm
Supercritical toluene extract	0.20	1.25	1.45 ± 0.12
Pentane solubles from toluene extract	0.15	1.00	1.63 ± 0.03
Asphaltenes from toluene extract	0.30	1.00	3.10 ± 0.60
Benzene insolubles from toluene extracts	0.35	1.00	2.00 ± 0.30

Symbols are defined in equations (a) and (c). r* was determined from light scattering experiments performed at 25°C.

The plots of fluorescence as a function of irradiation time in chloroform did not give reasonable agreement with equation (d). Moreover, reaction in chloroform was more rapid than in acetone though the value of the mean radius r* in chloroform was less than in acetone. It is clear that the model of irradiation we have described, though it describes many of the qualitative features of the kinetics, is too simple to describe the chloroform results quantitatively.

The most important result of these experiments has been the recognition that the kinetics of coal liquids are complicated in consequence of the fact that the liquids possess a range of molecular and particle sizes. This truth is not confined to photochemistry; it will be equally valid in such fields of coal utilisation as liquefaction.

The reaction of irradiated supercritical toluene extracts of Daw Mill coal with acetone and chloroform led one to expect that pyridine extracts would also react with chloroform when irradiated and that subsequent decay in their fluorescence could be used to monitor the reaction. This expectation was realised, reaction with chloroform was

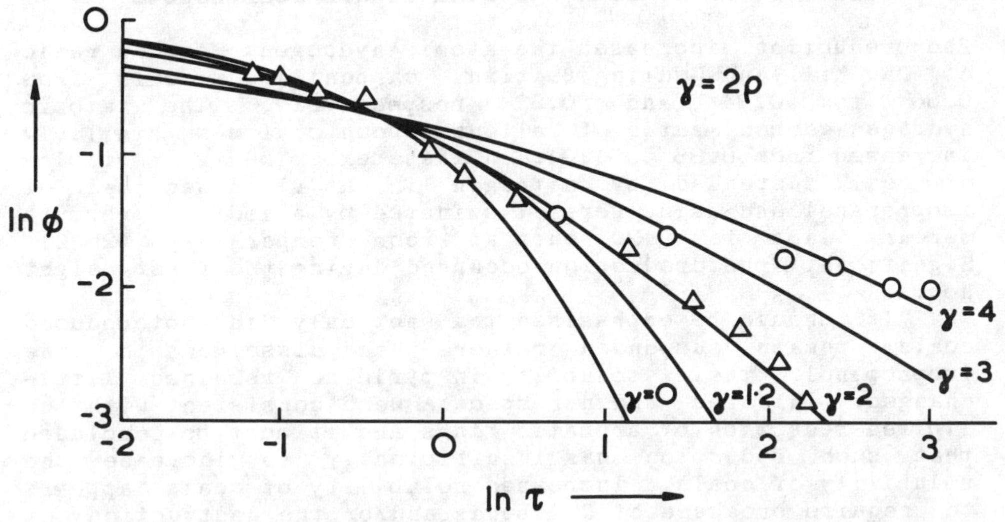

Figure 7. Kinetics of the decay in fluorescence intensity (excitation, 370 nm: emission 420 nm) caused by irradiation of asphaltenes extracted from Daw Mill coal. Symbols are defined in equation (d)
: typical results of irradiating on acetone solution
o : typical results irradiating a chloroform solution

rapid and the kinetics were again complicated. However, whereas irradiation of chloroform solutions of supercritical toluene extracts reduced subsequent fluorescence more or less uniformly at all excitation wavelengths, irradiation of pyridine extracts in the presence of chloroform caused the subsequent fluorescence from 400 nm excitation to be diminished much more dramatically than fluorescence generated by excitation at other wavelengths. This is illustrated in Figure 6. This phenomenon was observed with pyridine extracts of several bituminous coals and vitrinites and also with supercritical toluene extracts which were dissolved in equal quantities of pyridine and chloroform. The phenomenon thus required coal material, pyridine, chloroform and light. Acetone had a similar but much less marked effect to chloroform. One deduces that in the excitation of pyridine extracts of coals the fluorescence centred at excitation wavelengths of 400 nm was due to species solvated by pyridine. The structure of the solvated complex remains to be examined. One notes that fluorescence can be used as a probe of solvent-coal interaction.

8. PHOTOREDUCTION OF COAL AND COAL-LIQUID SUSPENSIONS

Photoreduction increased the atomic hydrogen: carbon ratio of Daw Mill and Gedling British, carboniferous coals from 0.80 to 0.84 and 0.92 respectively. The atomic hydrogen:carbon ratio of Yallourn brown coal was similarly increased from 0.86 to 1.01. In these experiments, particles of coal suspended by nitrogen in a fluidised bed of isopropanol and amine were illuminated by a medium pressure mercury lamp for 120 hours at room temperature (25°C). Significant photoreduction occurred during the first eight hours.

It should be emphasised that not only did photoreduced coals remain suspended rather than dissolved in the isopropanol, their solubilty in pyridine remained little changed. Infrared and nmr spectra were consistent with the limited reduction of aromatic rings and it must be concluded that such reduction was insufficiently to increase the solubility of coals. Increased solubility of coals appears to require breakage of C-C bonds and/or the destruction of tertiary structure. In contrast, similar photoreductions of the supercritical toluene extract of Daw Mill coal and of its pentane soluble, asphaltene and benzene insoluble fractions led to alightening of their colour and to greatly increased solubility. It is well established (28) that when aromatic hydrocarbons such as naphthalene are irradiated in the presence of amines, reaction proceeds through electron transfer to yield such products as:

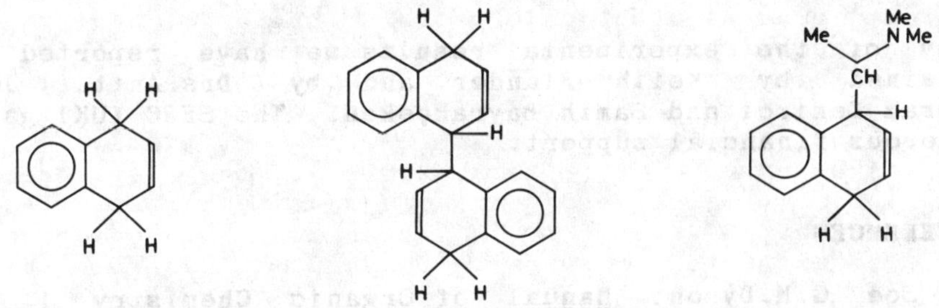

Photoreduction of the asphaltene extracted from Daw Mill coal was studied in detail. Consistent with the formation of compounds similar to those above, the recovered product had an atomic hydrogen:carbon ratio of 1.165 compared with 0.865 before photoreduction and nitrogen had become incorporated into the structure. Proton magnetic resonance spectra showed a substantial decrease in aromatic hydrogen, the formation of a small but significant portion of olefinic hydrogen and marked changes in the 2-3 ppm chemical shift region consistent with the creation of N-alkyl and ArCH groups. It appears that Daw Mill asphaltene reacted similarly to naphthalene. The limited solubility of coals restricted photochemistry to surface reactions in which only products analogous to dihydronaphthalene can have been formed.

Thus, the results of all the investigations indicated that the photoreduction of coals did not initiate free radical chains and that, consequently, quantum yields were low and generation of solubility was limited. Photoreduction does not appear a feasible method of coal conversion.

Not all coals were susceptible to photoreduction: thus Gedling coal could be photoreduced but Cresswell coal could not. These coals had similar chemistries, and there was no obvious reason why one should be photoreducible and the other not. Consequently, one postulates that Gedling and Cresswell must have different surface structures. Scanning electron micrographs showed particles of Gedling vitrinite to be composed of thin (0.2 μm) parallel sheets like the pages of a book whereas particles of Cresswell were wrapped in vitrinite sheets in a manner analogous to an onion wrapped in skins. The Cresswell 'onion structure' may be less penetrable by certain reagents and more work is needed to determine to what extent morpholigies of vitrinite particles differ and whether these differences correlate with the chemical reactivity of the particles.

ACKNOWLEDGEMENTS

Many of the experimental results we have reported were obtained by Keith Flunder and by Drs.Anita Jones, Bayram Demirci and Samih Bayrakçeken. The SERC (UK) gave us generous financial support.

REFERENCES

1. See G.M.Dyson, Manual of Organic Chemistry 1, 131, Longman, Green and Co., (1950).
2. D.W. van Krevelen, Coal, Elsevier, (1961).
3. ibid page 180.
4. J.S.Gethner, J. Chem. Soc., Faraday Trans. 1 81, 991, (1985).
5. H.P.Hombach, Fuel 61, 215, (1982).
6. H.P.Hombach, Thesis, University of Munster, (1977).
7. H.C.Retcofsky, T.J.Brendel and R.A.Friedel, Nature 240ps, 18, (1977).
8. E.R.Clark, J.R.Darwent, B.Demirci, K.Flunder, A.F.Gaines and A.C.Jones, Energy and Fuels 1, 392, (1975).
9. R.S.Becker, Theory and Interpretation of Fluorescence and Phosphorescence, Wiley, (1969), p.138.
10. R.S.Barltrop and J.D.Coyle, Excited States in Organic Chemistry, Wiley, (1975), p.118.
11. ibid page 121.
12. P.B.Hirsch, Proc. Roy. Soc. A226, 143, (1954).
13. D.E.Wemmer, A.Pines and D.D.Whitehurst, Phil. Trans. Roy. Soc. A300, 15, (1981).
14. J.R.Kershaw, Fuel 57, 299, (1978).
15. J.A.G.Drake, D.W.Jones, B.S.Causey and G.F.Kirkbright, Fuel 57, 663 (1978).
16. H.B.Aigbehinmua, J.R.Darwent and A.F.Gaines, Energy and Fuels 1, 386, (1987).
17. S.Bayrakçeken and A.F.Gaines, unpublished work, (1987).
18. Stach's Textbook of Coal Petrology, Gebrüder Borntraeger, 2nd Edition, (1975).
19. K.Brown, C.F.K.Diesel, M.Wolf and E.Wolff-Fischer, Proc. Int. Conf. on Coal Science, 649, Sydney, (1985).
20. C.R.Landis, G.W.Sullivan, M.W.Pleil, W.C.Borst and J.C.Crelling, Fuel 66, 984, (1987).
21. A.S.Davydov, J. Exptt. Theor. Physc. (USSR) 18, 210, (1948).
22. A.S.Davydov, Igvest. Akad. Nauk (SSSR) Ser. Fiz. 15, 605, (1951)
23. D.S.McClure, Solid State Phys. Adv. in Res. and Appl. 8, 1, (1959).
24. D.P.Craig and S.M.Walmsley, Excitons in molecular crystals: theory and applications, Benjamin, (1968).

25. R.Lin, A.Davis, D.F.Bensley and F.J.Derbyshire, Prepr. Pap. Am. Chem. Soc., Div. Fuel Chem. 30 315, (1985).
26. R.Bent and J.K.Brown, Residential Conf., Science in the use of coal, Instit. of Fuel, 48, (1958).
27. W.J.Alberry, P.N.Bartlett, C.P.Wilde and J.R.Darwent, JACS 107, 1854, (1985).
28. Reference 10 page 283.

COAL CHARACTERIZATION BY MEANS OF CURIE-POINT PYROLYSIS TECHNIQUES

Peter J.J.Tromp, Jacob A.Moulijn
Institute of Chemical Technology
University of Amsterdam
Nieuwe Achtergracht 166
1018 WV Amsterdam
The Netherlands

Jaap J.Boon
Mass Spectrometrometry of Macromolecular Systems
FOM Institute for Atomic and Molecular Physics
Kruislaan 407
1098 SJ Amsterdam
The Netherlands

ABSTRACT. Curie-point pyrolysis mass spectrometry (PyMS) was performed at 1043 K to characterize a set of 40 coal samples, ranging from anthracite to lignite. Six coal samples, representative for the entire range of coal ranks, were selected from the set of 40 coals to study the pyrolysis product mixtures as a function of the coal rank in more detail by Curie-point pyrolysis gas chromatographic techniques (PyGC and PyGCMS). PyMS revealed that the number and nature of the pyrolysis products, particularly the degree of oxygen functionality of the one ring aromatic compounds formed, are characteristics for the rank of a coal sample. Furthermore, the formation of hydrogen sulphide is highly characteristic for high sulfur coals. Classification of the coal samples according to their rank order is possible on the basis of their Curie-point pyrolysis mass spectra. However, additional characterization of the coals by the conventional analyses and tests is still necessary due to the qualitative character of the Curie-point pyrolysis technique. The Curie-point PyGC analyses give fundamental knowledge on the metamorphism of coal precursor material like lignins and on the development of the molecular structure composition of coals during maturation. Evidence of the so called mobile phase of coals and its composition are obtained by gas chromatographic analysis of the product mixtures evolved from the coals at a Curie-pont pyrolysis temperature of 631 K.

Y. Yürüm (ed.), New Trends in Coal Science, 241–269.

1. INTRODUCTION

Coal is a heterogenous and structurally complex material, consisting of an organic and an inorganic part (1). It is widely used in a large number of industrial processes. In general, coal utilization processes can be divided into two main classes:

1. processes in which coal is used as an energy resource (combustion, gasification), or as a raw material to produce liquid fuels (liquefaction) or chemicals.
2. processes in which coal is a raw material for the production of coke. Coke is a highly porous, carbon rich solid and acts as a reduction agent in blast furnaces in the steel industry.

For every kind of coal utilization process specifications have to be met with respect to the chemical and physical properties of the coal. Hence, a large number of standard analyses and tests have been developed to characterize coal samples both chemically and physically. Examples of conventional standard coal characterization methods are (2): proximate analysis (moisture, ash, volatile matter and fixed carbon content), elemental analysis (C, H, N, S and O content), petrographic analysis, calorific value, swelling index, dilatometry, plastometry, etc. On the basis of these characterization methods coals are classified according to a rank order. With increasing coal rank, coals are assigned to one of the following classes: peat, lignite, subbituminous coal, bituminous coal, anthracite (1). Furthermore, coals are divided in coking and non-coking coals.

Some of the conventional characterization methods are quite time consuming and the exact meaning of the experimental results is not always clearly understood. More recently, however, fast acting, computer assisted coal characterization techniques have become available, like Electron Spin Resonance (ESR) (3), solid state C13 Nuclear Magnetic Resonance (NMR) (4-6), Fourier Transform InfraRed spectroscopy (FTIR) (7,8) and analytical Curie-point pyrolysis techniques (9,10). Beside characterization for conversion processes the modern analytical techniques are also directed towards a fundamental investigation of the molecular structure of coal (11,12).

This paper deals with the characterization of coals by means of Curie-point pyrolysis techniques. In these techniques small coal samples are pyrolyzed and the volatile pyrolysis products are analyzed. Usefullness of the techniques for coal classification purposes and analysis with these techniques of the product distribution as a function of the coal rank will be evaluated. The product distribution is important for two reasons. First of all, every coal conversion process includes a pyrolysis step of the coal. Secondly, the product distribution gives fundamental information on the building blocks of the coal structure on a molecular scale.

1.1. Principles of the Curie-Point Techniques

Depending on the analysis mode of the volatile pyrolysis products distinction can be made between Curie-point pyrolysis mass spectrometry (PyMS), -gas chromatography (PyGC) and -gas chromatography in combination with mass spectrometry (PyGCMS).

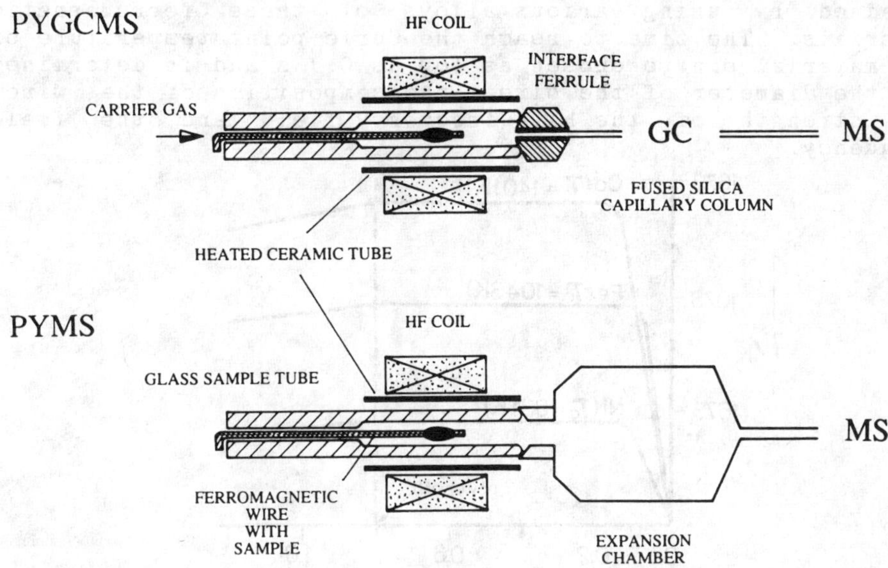

Figure 1. Schematic diagrams of the Curie-point pyrolysis unit on the Packard GC/JEOL GCMS system (PyGCMS) and the FOMautoPyMS (PyMS). The pyrolysis in the PyGCMS system takes place in the carrier gas flow and the pyrolysis product mixture is swept to the inlet of the capillary column for GC or GCMS analysis. In the PyMS system pyrolysis takes place in vacuo, the pyrolysis product mixture expands in the gold coated expansion chamber, which is heated and leaks to the ion source of the mass spectrometer (MS). The degree of condensation of heavy fractions in the pyrolysate on the walls of the glass sample tube can be regulated by the temperature of the heated ceramic tube.

Analytical Curie-point pyrolysis uses the metal surface of thin ferromagnetic wires (d <1 mm) as the sample (ug scale) carrier. A glass lined sample tube with a ferromagnetic wire is placed in an induction coil which is connected to a high frequency power supply (see Figure 1). When the high frequency field is switched on, the ferromagnetic wire is inductively heated to its Curie-point temperature, at which the wire loses its ferromagnetism and becomes paramagnetic. This causes a drastic loss in energy absorption from the high frequency field in the coil. When the dimensions of the ferromagnetic wire and the high frequency properties are properly matched, the equilibrium temperature of the wire stabilizes close to the Curie-point temperature. Typical temperature versus time profiles for wires of different ferromagnetic materials (pure Ni, Fe and Co) are shown in Figure 2. Intermediate curves can be obtained by using various alloys of these ferromagnetic materials. The time to reach the Curie-point temperature of the material can be chosen as low as 0.1 s and is determined by the diameter of the wire, the composition of the wire, the strength of the high frequency field and the field frequency.

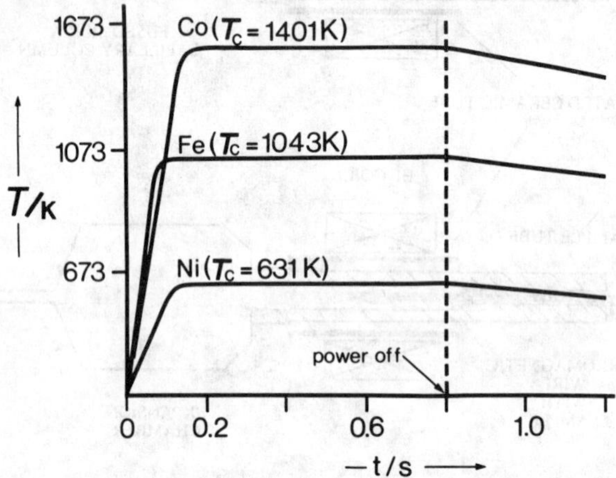

Figure 2. Temperature versus time profiles and Curie-point temperatures for pure Ni, Fe and Co wires. In PyMS the ferromagnetic wires are inductively heated in vacuo; in this case a heating time to the Curie-point temperature in the order of 0.1 s can be obtained. In PyGC and PyGCMS the wires are heated in the presence of a helium carrier gas flow. As a consequence the heating time in PyGC and PyGCMS is about 1 s.

Thus analytical Curie-point pyrolysis guarantees a well defined temperature-time history and fast and reproducible heating of a small amount of sample (µg scale). The advantage of a small amount of sample is that secondary condensation reactions are minimized and therefore, a simpler product distribution is obtained, which better reflects the chemical structure of the original sample. However, these extremely small sample sizes might cause irreproducible results. As Curie-point PyMS is a fast technique, the degree of irreproducibility is checked by this same technique performing the experiments in triplicate. For each analysis a fresh wire and glass sample tube (see Figure 1) is used to prevent contamination of the reaction zone, which would cause background and memory effects. Product residence time in the hot zone (only the ferromagnetic wire is heated) is designed to be short, again to prevent formation of secondary (recombination) pyrolysis products. Volatiles either expand in vacuo to the ion source region in the PyMS mode or when PyGC and PyGCMS are used, are swept away by the carrier gas and cooled to the temperature of the oven of the gas chromatograph.

After pyrolysis the non-volatile part of a sample, i.e. the char, remains on the ferromagnetic wire and is therefore not analyzed. Some pyrolysis products, volatile enough to escape from the sample, may condense on the walls of the glass sample tube or (in PyMS) in the expansion chamber of the mass spectrometer. To prevent excessive condensation of these products the glass liner and the expansion chamber have to be heated.

2. EXPERIMENTAL

2.1. Sample Preparation and Coating

A world-wide collection of 40 coal samples, ranging from anthracite to lignite, were obtained from the Dutch centre of coal specimens (SBN). The analytical data of the coals are given in Table I. The coal samples were ground in a nitrogen glove box to a particle size smaller than 100 µm. The samples for the pyrolysis experiments were taken from slurries in water, prepared by impregnation of the ground coal samples with deionized water until the pore volume was filled. Aliquots taken from these slurries were applied with a spatula to separate disposable ferromagnetic pyrolysis wires and subsequently, the samples were air dried. The samples were analyzed within 1 h of coating. The amount of sample on a wire is estimated to be in the order of 50 µg. For each wire a clean Pyrex glass sample tube was used.

2.2. Pyrolysis Mass Spectrometric Analysis and Data Processing

The instrumentation used for the Curie-point PyMS investigations -the FOMautoPyMS- is shown schematically in Figure 3. It is a fully automated system built at the FOM Institute for Atomic and Molecular Physics in Amsterdam, The

Table I. Analytical data of the 40 coal samples.

sample	SBN-code	country	ASTM-rank	ash	cal.value	C	H	S	O	Vol.Matter	$\%R_0$	Vitrinite	Exinite	Inertinite	Minerals
				(wt%,dry)	(MJ/kg,DAF)	←		wt%	(DAF)	→		←		%	→
#1	GB04	G-Britain	anthracite	2.1	34.9	94.1	2.81	0.61	1.66	4.6	3.72	76.9	0	21.8	1.1
#2	DE06	BRD	anthracite	5.4	34.7	90.1	3.16	1.92	3.45	7.7	2.98	71.9	0	22.3	5.8
#3	FR08	France	semi-anth.	12.7	35.4	90.0	3.67	0.85	4.45	8.8	2.29	72.1	0	14.9	13
#4	VC09	China	semi-anth.	8.4	35.1	92.3	2.97	0.21	3.76	10.1	3.02	39	0	56	5
#5	DE10	BRD	semi-anth.	9.0	34.9	89.8	3.58	0.52	4.83	11.6	2.26	42.3	0	49.7	8
#6	DE12	BRD	lv bit.	27.7	36.0	89.0	4.24	0.60	4.74	15.5	2.04	61.9	0	11.5	26.6
#7	SA12	S-Africa	semi-anth.	14.8	35.0	89.3	3.66	1.25	3.56	11.7	2.08	30.6	0	55.9	13.5
#8	GB14	G-Britain	lv bit.	10.3	35.9	90.1	4.07	1.14	3.34	16.8	1.64	75.9	0	17.8	6.3
#9	DE16	BRD	lv bit.	7.1	34.7	89.2	4.40	0.83	4.12	17.1	1.65	70.7	0	24.2	5.1
#10	FR17	France	mv bit.	28.6	34.8	88.1	4.74	0.89	4.90	23.5	1.33	67.6	0.2	12.4	19.8
#11	US19	USA	lv bit.	5.1	36.4	90.1	4.59	0.84	2.84	19.9	1.52	82.7	0.4	13.1	3.8
#12	AU20	Australia	mv bit.	8.4	36.3	89.7	4.64	0.62	3.10	22.8	1.32	81.2	0	15.4	3.4
#13	DE20	BRD	lv bit.	3.3	36.3	89.5	4.55	0.77	3.75	20.2	1.42	76.2	0	21.2	2.6
#14	SA23	S-Africa	mv bit.	15.9	32.4	81.9	4.34	0.72	11.25	30.1	0.75	28.8	4.6	58.9	7.7
#15	DE24	BRD	mv bit.	7.5	35.3	88.2	4.62	1.05	4.66	22.9	1.45	80	0	13	7.0
#16	SA25	S-Africa	mv bit.	15.7	32.6	82.4	4.28	0.84	10.83	30.8	0.71	25.7	6.1	57	11.2
#17	DE26	BRD	mv bit.	3.3	36.1	88.0	4.91	0.83	4.53	26.4	1.17	68.7	4.9	23.6	2.8
#18	AU27	Australia	mv bit.	12.8	35.3	86.2	4.84	2.40	4.86	23.9	1.09	39.7	5.9	47	13.3
#19	GB28	G-Britain	hvB bit.	20.0	33.0	81.1	4.94	3.07	9.62	33.9	0.75	74.9	5.6	12	7.5
#20	SA29	S-Africa	hvC bit.	39.1	30.5	76.1	5.26	1.69	15.57	42.8	0.73	25.1	8.1	25.4	41.4
#21	AU29	Austalia	hvC bit.	15.8	29.9	79.1	2.93	0.19	16.74	32.0	0.58	15.6	3.7	62.5	18.2
#22	AU29	Australia	hvB bit.	18.9	33.3	82.0	4.93	0.43	10.93	37.8	0.72	55.6	4.2	30.1	10.1
#23	BE30	Belgium	hvA bit.	3.5	35.9	86.6	5.08	0.87	5.95	31.9	0.99	58.9	11.9	26.6	2.1
#24	DE30	BRD	hvA bit.	4.0	35.6	86.5	5.11	1.09	5.65	31.2	0.98	65.8	11.0	21.3	1.9
#25	US33	USA	hvA bit	5.3	35.8	86.5	5.33	1.03	5.33	34.5	0.95	81.1	4.7	12.1	2.1
#26	BE33	Belgium	hvB bit	14.3	33.5	82.3	5.13	0.68	10.28	35.3	0.79	61.1	6.8	20.1	12.0
#27	AU35	Australia	hvB bit.	12.2	33.1	81.5	4.80	1.05	10.72	36.0	0.74	60.3	4.0	27.3	8.4
#28	FR36	France	hvA bit.	5.3	34.2	82.2	5.16	0.90	10.45	37.9	0.82	86.1	5.5	3.6	4.8
#29	CO37	Columbia	hvA bit.	2.1	33.7	81.9	4.66	0.53	11.45	38.1	0.64	63.5	4.2	30.9	1.4
#30	DE38	BRD	hvB bit.	7.5	33.7	82.1	5.28	1.09	9.48	41.1	0.77	74.0	6.2	15.2	4.6
#31	US40	USA	hvB bit.	10.7	33.5	79.7	5.57	3.67	9.48	45.5	0.54	74.0	10.3	9.2	6.5
#32	SA40	S-Africa	hvB bit.	11.1	33.0	80.5	5.36	1.04	12.91	40.0	0.72	79.6	4.7	9.0	6.7
#33	US42	USA	hvA bit.	10.5	34.4	81.3	5.39	4.92	7.05	46.7	0.65	68.3	13.4	12.3	6.0
#34	BE42	Belgium	hvC bit.	46.9	31.3	78.4	5.12	1.45	13.04	44.1	0.81	45.6	6.4	7.5	40.5
#35	US43	USA	hvB bit.	10.4	33.0	79.5	5.03	4.48	9.48	47.0	0.47	73	7.2	15.2	4.3
#36	FR45	France	hvC bit.	9.4	32.1	78.4	5.07	4.33	10.82	46.0	0.40	69.4	8.6	14.9	7.1
#37	FR45	France	subbit.	18.9	30.0	74.1	5.00	6.22	12.87	55.5	0.48	63.2	14.6	15.7	6.5
#38	AU52	Australia	subbit.	18.3	31.6	75.0	5.07	0.25	18.71	53.3	0.43	58.8	22.7	4.1	14.7
#39	DE53	BRD	lignite	3.6	25.9	65.9	4.41	0.31	28.67	53.6	0.45	80	-	13	-
#40	AU54	Australia	subbit.	39.0	31.9	75.7	6.49	0.68	15.96	59.0	0.49	38.6	28.0	0	33.4

Netherlands (13). It consists of an automatic sample exchanging system which can contain up to 36 samples, an internally heated Curie-point pyrolysis reactor and a quadropole mass spectrometer with ion counting detector. The pyrolysis products are generated in vacuo (10 Pa) and reach the ion source via an expansion chamber, which is held at a temperature of 475 K to avoid excessive condensation of the pyrolysis products. The purpose of the expansion chamber is to broaden the pressure wave generated by the explosive evolution of the volatile pyrolysis products. A small inlet to the ion source guarantees enough time to obtain a sufficient number of mass scans, representative for the pyrolysate. About 200 mass scans are multichannel averaged to one final mass spectrum of the sample. The pyrolysis productas are ionised in the ion source by low voltage electron impact. The low energy ionisation process minimizes fragmentation of the pyrolysis products and promotes the formation of molecular ions. The ions are detected by a quadropole mass spectrometer. Whereas the pyrolysis reactor, the glass liner and the expansion chamber are heated, the ion source and the mass spectrometer housing are not. The liquid nitrogen cooled screen around the ion source acts as a cryopump for pyrolysis products which escape ionisation. Finally, recording of the pyrolysis mass spectra is performed by a computer.

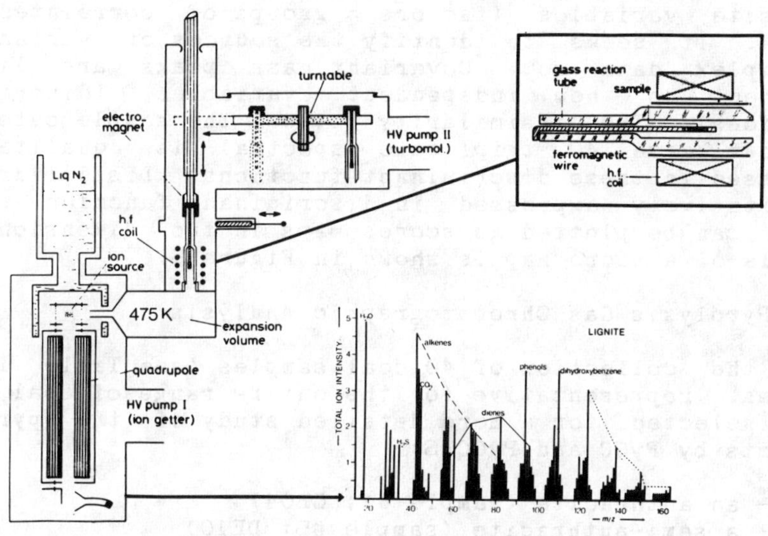

Figure 3. Schematic diagram of the automated Curie-point pyrolysis mass spectrometer system FOMautoPyMS.

For this study the mass spectrometer conditions were as follows: temperature of the expansion chamber, 475 K; ionization, 15 eV electron impact; mass ranged scanned, m/z: 25-225; scan speed, 10 scans/s; total number of averaged spectra, 200.

It was determined in a previous study (14) that a Curi-point temperature of 1043 K (pure iron wires) was optimal for the pyrolysis of the coal samples. Pyrolysis conditions were: heating time, 0.1 s; total heating time, 0.8 s. A heating time of 0.1 s means that the heating rate of the samples is in the order of 7000 K/s. All 40 coal samples were analyzed in triplicate in two consecutive days. For each run a single integrated spectrum was recorded and stored in a computer. In this paper pyrolysis mass spectra of the coal samples are presented as an average of the three separate analyses for visual evaluation.

The stored pyrolysis mass spectra (120 spectra, for every one of the 40 coal samples) were also studied by means of factor-discriminant analysis. Principles of this mathematical technique and the application of this procudure to pyrolysis mass spectra have been described (15,16). In brief, "Principal Components" and "Discriminant Analysis" are two mathematical techniques to reduce the size of a complex data set and to find relations and trends within the data set. Principal components is a type of factor analysis which transforms a given set of variables (in this case normalized mass peak intensities) into a new set of composite variables (factors = groups of correlated mass peaks). It seeks to identify the sources of variance in a complex data set. Covariant mass peaks are linearly combined to new independent variables (discriminant functions). The dissimilarity between the sample categories (in our case 40 triplicate spectra) is qualitatively expressed in these discriminant functions. Dissimilarity is quantitatively expressed in discriminant function scores, which can be plotted as scores maps in two dimensions. An example of a score map is shown in Figure 6.

2.3. Pyrolysis Gas Chromatographic Analysis

From the collection of 40 coal samples (see Table I) six samples, representative for the entire range of coal ranks, were selected for a more detailed study of the pyrolysis products by PyGC and PyGCMS:

- an anthracite (sample #1; GB04)
- a semi-anthracite (sample #5; DE10)
- a medium volatile bituminous coal (sample #12; AU20)
- a high volatile A bituminous coal (sample #23; BE30)
- a high volatile B bituminous coal (sample #30; DE38)
- a lignite (sample #39; DE53)

The pyrolysis reactor, especially designed for capillary gas chromatography, has been described in detail by Boon et al. (17). The glass sample tube was directly sealed to a Kalrez interphase with the inlet of a capillary column (Figure 1). Helium, used as carrier gas, swept the pyrolysis products from the glass sample tube into the column. The temperature of the glass sample tube was held at 450 K to prevent excessive condensation of the pyrolysis products. Gas chromatographic separation of the pyrolysis product mixture was performed on a 60 m fused silica capillary (ID 0.32 mm) column coated with DB-1 (film thickness 0.25 μm) using a Carlo Erba Model 4200 gas chromatograph equipped with a flame ionization detector (FID). The detector signal was both transferred to a recorder and to a Nelson Analytical 3000 data system. The column was first held isothermally at 323 K for 300 s and then temperature programmed to 598 K at 5 K/min. For the lignite, however, a 50 m fused silica capillary (ID 0.32 mm) column coated with CP-Sil5 was used. In this case the column was first held at 273 K for 300 s and then temperature programmed to 593 K at 5 K/min. The choice of the capillary columns, in combination with the temperature programme in the gas chromatograph precluded the separation of the pyrolysis products with a boiling point lower than n-hexane. Identification of the pyrolysis products was performed under identical conditions with Curie-point PyGCMS using a Packard 438S gas chromatograph combined with a JEOL DA - 5000 computer system. Mass spectra were obtained at 70 eV electron impact ionization.

Pure iron wires with a Curie-point temperature of 1043 K were used to pyrolyze the coal samples. For the bituminous coal samples (AU20, BE30 and DE38), however, also nickel wires with a Curie-point temperature of 631 K were used to study the thermal desorption of components adsorbed on or trapped within the macromolecular network of these coals. Evidence for the existence of these components in coal has been given in literature; they are termed the "mobile phase" of coals (18).

Pyrolysis conditions were as follows: heating time to final temperature (either 1043 or 631 K), 0.9 s; total pyrolysis time, 4 s; helium atmosphere.

3. RESULTS AND DISCUSSION

3.1. Pyrolysis Mass Spectrometry (PyMS) of the Worldwide Collection of 40 Coal Samples

In Figures 4 and 5 a selection of average mass spectra of 16 coal samples out of the collection of 40 samples is shown. The spectra are separately scaled upon their most abundant mass number. The mass spectra, ordered to decreasing coal rank, illustrate the trends in composition of the pyrolysis

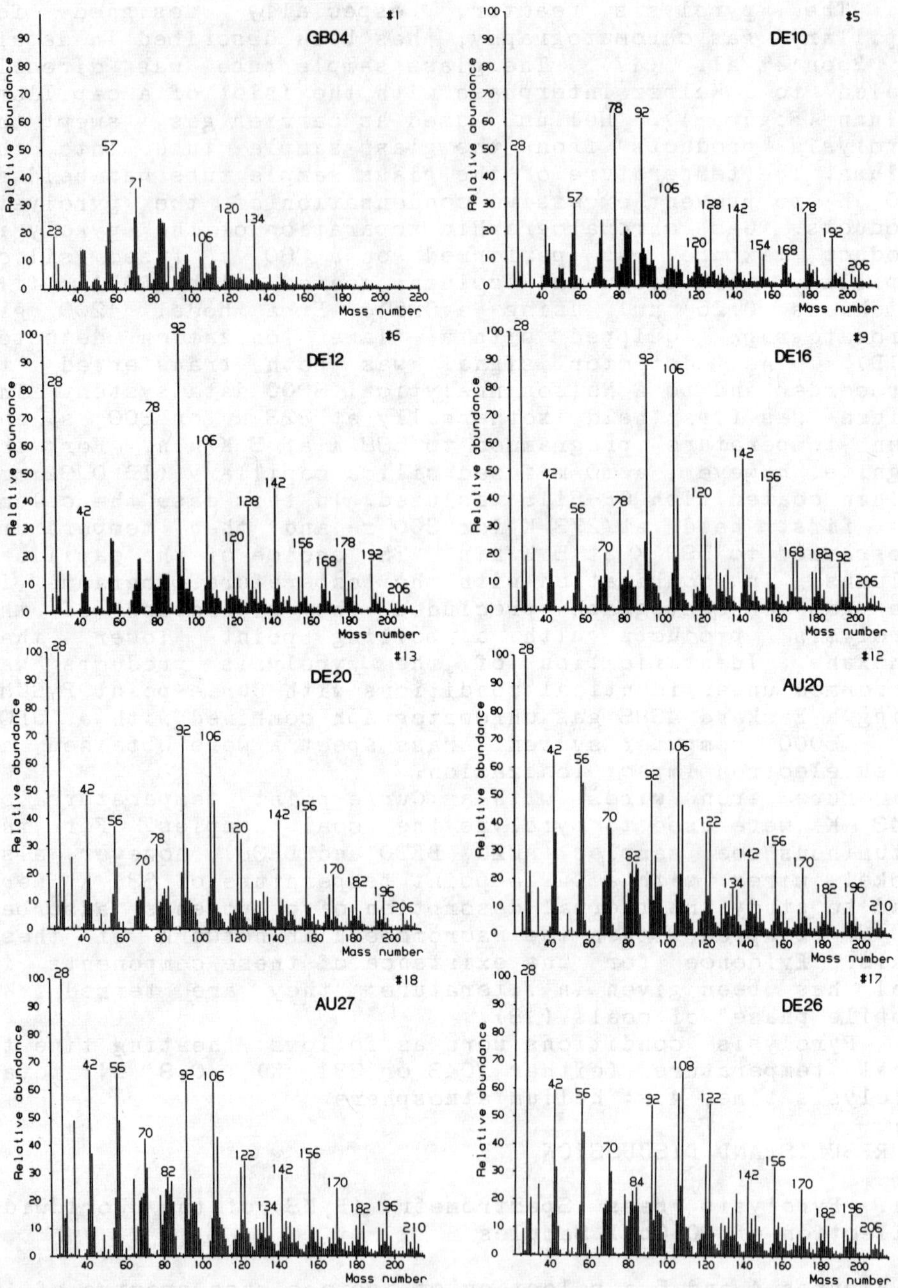

figure 4. Average mass spectra of selected coal samples, from anthracite (upper left) to medium−volatile bituminous coal (lower right). The numbers in the spectra correspond with the numbers in table I. The chemical structures and corresponding mass numbers of compounds, present in the mass spectra, are given in table II.

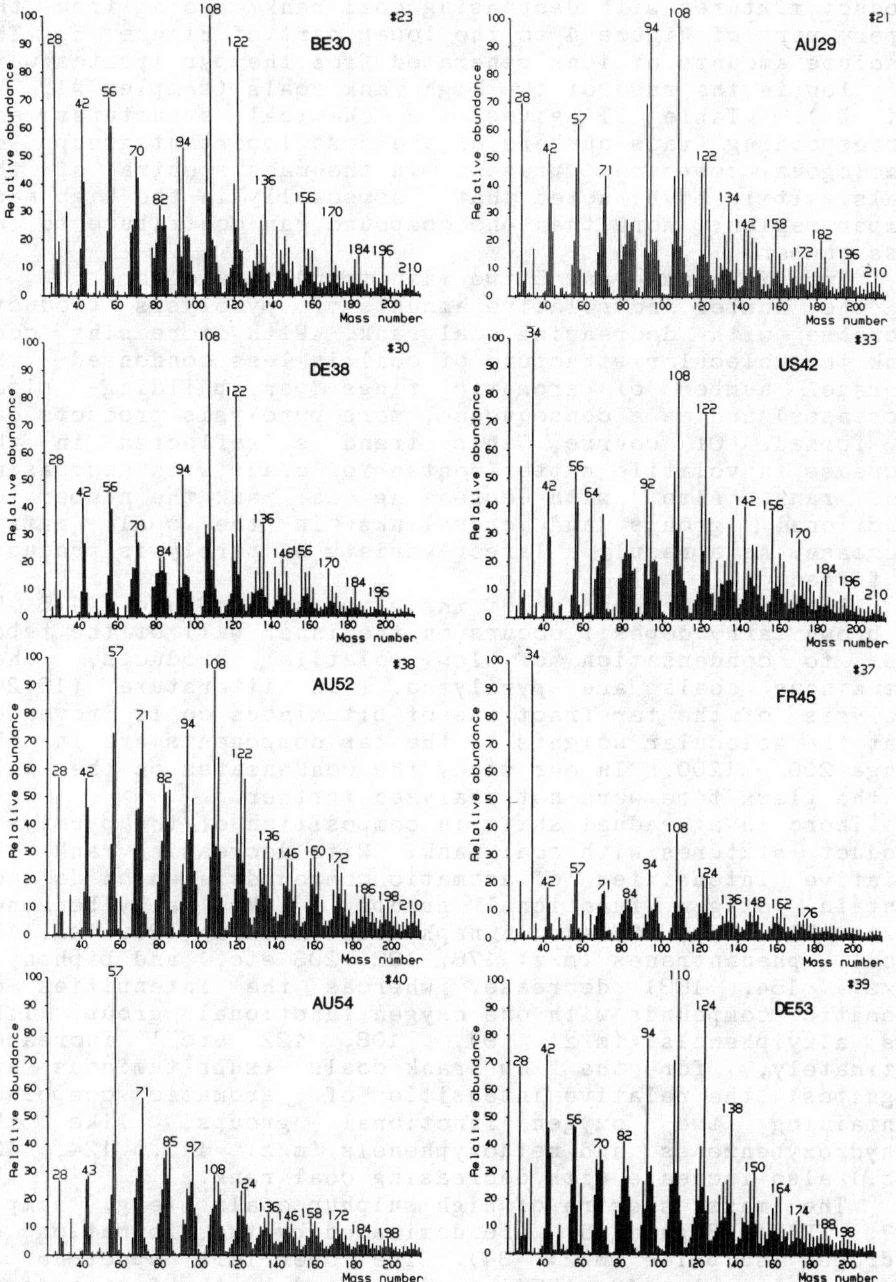

figure 5. Average mass spectra of selected coal samples, from high−volatile A bituminous coal (upper left) to lignite (lower right). The numbers in the spectra correspond with the numbers in table I. The chemical structures and corresponding mass numbers of compounds, present in the spectra, are given in table II.

product mixtures with decreasing coal rank, going from the upper part of Figure 4 to the lower part of Figure 5. The absolute amounts of ions generated from the pyrolysates were very low in the case of the high rank coals (samples #1, #2 and #4). Table II gives the chemical structures and corresponding mass numbers of the most important groups of homologous compounds present in the mass spectra of the coals. It is to be noted that, especially in the high mass number region, more than one compound can contribute to one mass number.

From Figures 4 and 5 two main trends are clear:

1. the number and relative amounts of pyrolysis products increase with decreasing coal rank. With decreasing coal rank the molecular structure of coal is less condensed (the average number of aromatic rings per building block decreases) and as a consequence, more pyrolysis products can be formed. Of course, this trend is reflected in the increase in volatile matter content of coals with decreasing coal rank. Also, with decreasing coal rank the number of functional groups and crosslinks in the coal matrix increase. As a result a larger variety of pyrolysis products is formed.

Despite the heating of the glass sample tube (475 K) a brown tarry deposit occurs on the inner wall of the tube, due to condensation of low volatile products, when bituminous coals are pyrolyzed. In literature (19,20) analysis of the tar fractions of bituminous coals revealed that the molecular weights of the tar components are in the range 200 - 1200. In our study the condensates on the wall of the glass tube were not analyzed further.

2. There is a gradual shift in composition of the pyrolysis product mixtures with coal rank. With decreasing rank the relative intensities of aromatic compounds, which do not contain oxygen functional groups, like alkylbenzenes (m/z: 78, 92, 106 etc.), naphthalenes (m/z: 128, 142, 156 etc.), phenantrenes (m/z: 178, 192, 206 etc.) and biphenyls (m/z: 154, 168) decrease, whereas the intensities of aromatic compounds with one oxygen functional group, like the alkylphenols (m/z: 94, 108, 122 etc.) increase. Ultimately, for the low rank coals (subbituminous and lignites) the relative intensities of aromatic compounds containing two oxygen functional groups, like the dihydroxybenzenes and methoxyphenols (m/z: 110, 124, 138 etc.) also increase with decreasing coal rank.

The mass spectra of high sulphur coals (e.g. sample #37, FR45; Figure 5) are dominated by the formation of hydrogen sulphide (m/z: 34). The lower left spectrum in Figure 5 (sample #40, AU54) is dominated by the formation of alkanes (m/z: 57, 71, 85 etc.), which can be explained by the high exinite content of this coal.

To examine the shift in composition of the pyrolysis products with coal rank more profoundly factor-discriminant

TABLE II. Chemical structures and corresponding mass numbers of the most important compounds and groups of homologous compounds present in the mass spectra of the coals as shown in Figures 4 and 5.

groups of homologous compounds	chemical structures with corresponding mass numbers
BENZENES	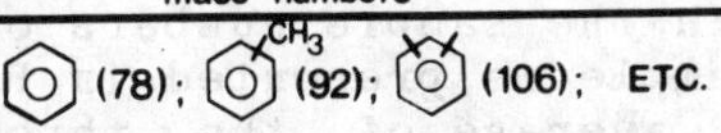 (78); CH_3 (92); (106); ETC.
NAPHTHALENES	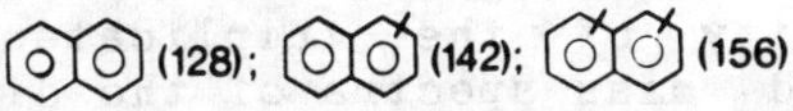 (128); (142); (156);
PHENANTHRENES	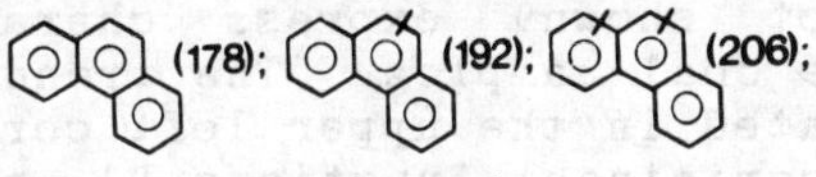 (178); (192); (206);
BIPHENYLS	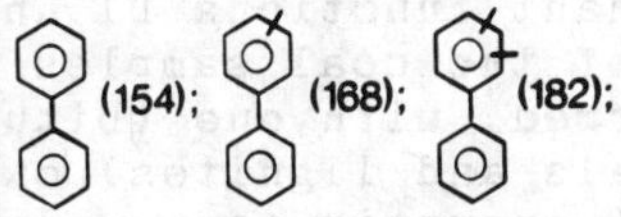 (154); (168); (182);
PHENOLS	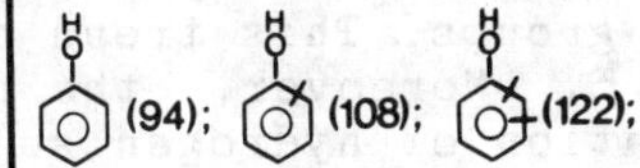 H O (94); H O (108); H O (122);
HYDROXYINDENES	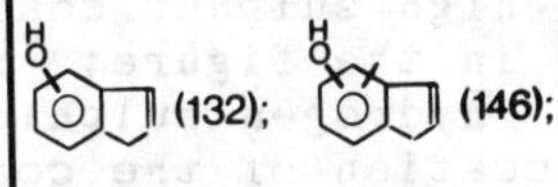 H O (132); H O (146);
NAPHTHOLS	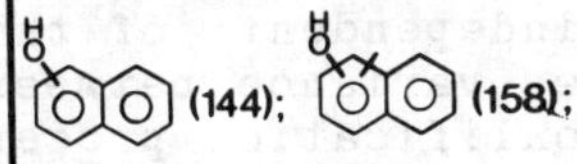 H O (144); H O (158);
DIHYDROXYBENZENE	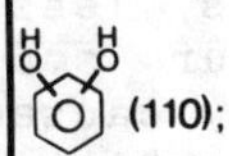 H O H O (110);
METHOXYPHENOLS	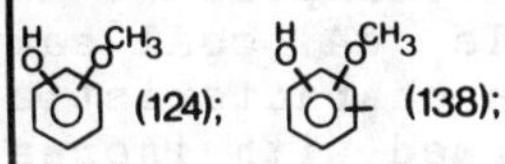 H O O CH_3 (124); H O O CH_3 (138);
SULFUR COMPOUNDS	H_2S (34); CH_3SH (48);  S (84);
ALKANES	$C_3H_7^+$ (43); $C_4H_9^+$ (57); $C_5H_{11}^+$ (71); ETC.
ALKENES	C_2H_4 (28); C_3H_6 (42); C_4H_8 (56); C_5H_{10} (70);
CARBON-OXIDES	CO (28); CO_2 (44);

<u>NOTE</u>: the analysis technique does not discriminate between isomers containing one ethyl or two methyl groups, etc. However, in general methyl compounds are more abundantly generated in coal pyrolysis than ethyl, propyl compounds.

analysis was applied to the 120 (40 coals in triplicate) mass spectra obtained. The result is shown in Figure 6 as a discriminant score map, which describes the difference between the spectra quantitatively. The discriminant functions D1 and D2 account for 50 % of the variance between the mass spectra. Higher discriminant functions are disregarded in this presentation. The numbers in the figure correspond with the sample numbers of the coals in Table I. Every coal sample is presented in Figure 6 by one point, which is the average of the three PyMS analyses. The reproducibility of the triplicates was very good. The reconstructed mass spectra of the discriminant functions D1 and D2 (not shown) express characteristic differences between the coal samples. The trends in these differences are illustrated in the upper left corner of the figure.

The discriminant functions D1 and D2 indicate that with decreasing rank of the coal samples more aromatic pyrolysis products are formed with one (bituminous coals) or two (subbituminous coals and lignites) oxygen functional groups at the expense of aromatic compounds, which do not contain oxygen functional groups. This trend is also clearly visible in Figures 4 and 5. Moreover, the discriminant functions express the formation of hydrogen sulphide as being highly characteristic. The high sulphur coals (samples #31, #33, #35, #36 and #37 in the figure; see also Table I) are projected along this "hydrogen sulphide-axis". The gradual shift in relative location of the coal samples in Figure 6 with coal rank, independent of the country of origin, indicates that the variance between the mass spectra is mainly due to the coalification process. The input variation due to paleo environment is less important, with the exception of the high sulphur coals. The high sulphur contents of coals are generally caused by marine influences.

The anthracites (samples #1 and #2) and one of the semianthracite (sample #4) coal samples are projected in Figure 6 outside the characteristic trend observed in the pyrolysis products formed with increasing coal rank. This is explained by the high structural stability of these anthracitic coal samples. Due to this high stability hardly any volatile products and therefore, no characteristic products as well, were formed upon Curie-point pyrolysis of these coals at 1043 K.

Other "outliers" in Figure 6 are the coal samples with the numbers 14, 16 and particularly 21. The coal samples with #21 is projected along the "hydrogen sulphide-axis", whereas it hardly contains any sulphur (0.19 %). However, the relative locations of the coal samples in Figure 6 is strongly similar to their relative locations on a H/C versus O/C coalification band (21), as shown in Figure 7. The origin of the H/C and O/C axes is chosen at the upper right corner of Figure 7 to illustrate the similarity between Figure 6 and Figure 7. The special character of the high

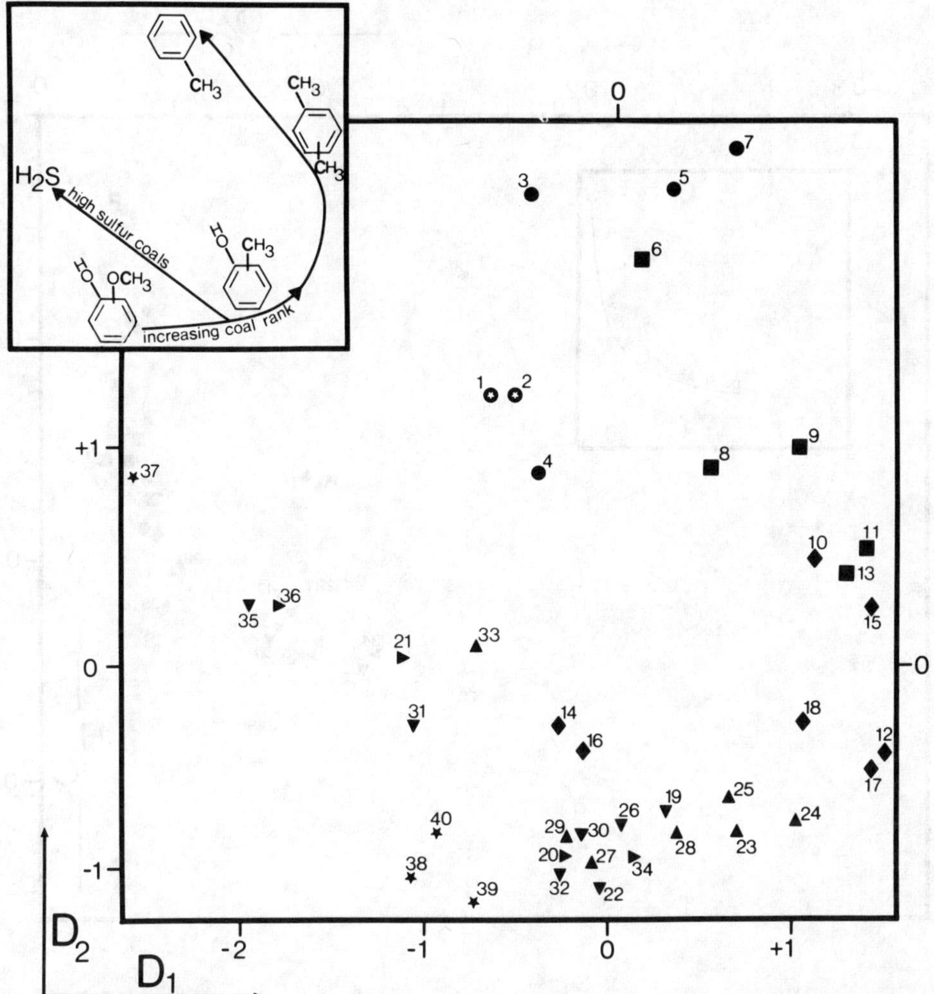

Figure 6. Discriminant function score plot of the mass spectra, obtained by Curie–point pyrolysis mass spectrometry at 1043 K, of the 40 coal samples (✪ :anthracite, ● :semi–anthracite, ■ :low–volatile bituminous coal, ◆ :medium–volatile bituminous coal, ▲ :high–volatile A bituminous coal, ▼ :high–volatile B bituminous coal, ▶ :high–volatile C bituminous coal, ★ :subbituminous coal and lignite). The numbers in the figure correspond with the sample numbers of the coal samples, as shown in table I. Every coal sample is represented by one point (centroid), which is the average of the triplicate PyMS analyses. The euclidic distances between the points is a measure for the difference in composition of the pyrolysis product mixtures, generated from the coal samples. The insert shows the most important trends in the composition of the pyrolysis product mixtures of the coals, deducted from the reconstructed mass spectra of the discriminant functions D_1 and D_2 (not shown).

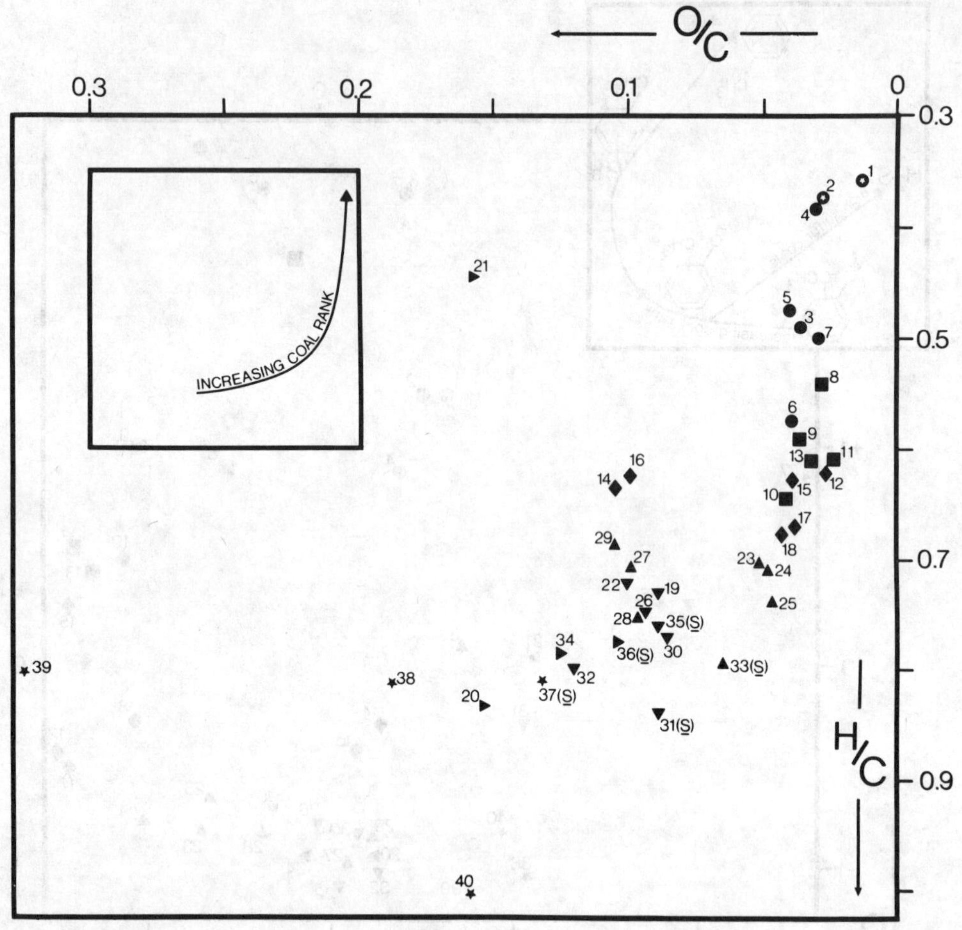

figure 7. Relative location of the 40 coal samples on a H/C versus O/C coalification band (✪ :anthracite, ● :semi–anthracite, ■ :low–volatile bituminous coal, ◆ :medium–volatile bituminous coal, ▲ :high–volatile A bituminous coal, ▼ :high–volatile B bituminous coal, ▶ :high–volatile C bituminous coal, ★ :subbituminous coal and lignite). The numbers in the figure correspond with the sample numbers of the coal samples, as shown in table I. The high–sulfur coals are indicated by (S̲). The origin of the H/C versus O/C axes is chosen in the upper right corner of the figure to illustrate that the relative location of the coal samples is similar to that in figure 6.

sulphur coals, which emerges from Figure 6, is not reflected in the H/C versus O/C coalification band, of course. The high sulphur coals in Figure 7 are indicated by (S̲).

The low H/C and O/C values of the anthracitic coal samples #1, #2 and #4 explain that hardly any products are formed upon Curie-point pyrolysis of these coal samples. The small amounts of pyrolysis products determine the locations of these coals in Figure 6. Furthermore, the high oxygen contents of the coal samples #14, #16 and #21, deviating from the coalification band, are also reflected in the relative locations of these coal samples in Figure 6. These three high oxygen containing coals also exhibit high inertinite contents.

Thus, visual evaluation of the separate Curie-point pyrolysis mass spectra, in combination with factor discriminant analysis of the mass spectra of a large collection of coal samples is a fast and reliable method to characterize the coal samples according to the conventional coal classification. The relative amounts of pyrolysis products and the degree of oxygen functionality of the one ring aromatic compounds, evolved upon Curie-point pyrolysis of a coal sample at 1043 K, indicate the rank of the coal sample. Furthermore, high sulphur coals are characterized by the formation of hydrogen sulphide. Strong correlations are found between the pyrolysis mass spectra and conventional coal parameters such as elemental analysis. Similar conclusions were drawn in a Curie-point PyMS study of 102 coal samples from the Rocky Mountain region (9,22). In this study the selected coal samples consisted of bituminous and subbituminous coals only. Furthermore, ferromagnetic wires with a Curie-point temperature of 883 K were used; the lower the Curie-point temperature of the wires, the more the mass spectra reflect compounds formed from thermal desorption instead of pyrolysis of the coal matrix.

However, due to the qualitative nature of Curie-point PyMS characterization of coals for conversion processes according to the conventional standard analyses and tests are still necessary. With respect to the coal conversion processes the advantage of the Curie-point pyrolysis technique is that it involves rapid pyrolysis of the coal sample (heating rate = 7000 K/s). Especially in coal combustion and in entrained flow or fluid bed gasification of coal, the coal particles are also rapidly heated to the reaction temperature. So, Curie-point PyMS may qualitatively predict the organic emissions to be expected from the use of a certain coal in different coal utilization processes. However, it is not yet clear how coking properties of coals are reflected in the mass spectra. In this respect microscopic observation of the coal samples after Curie-point pyrolysis is possible. The morphology of the coal particles after pyrolysis can give an indication on the thermoplasticity of a coal. Also, when pyrolyzing bituminous coals a tarry deposit is formed on the glass sample tube, which surrounds the ferromagnetic wire. As coking properties are only exhibited by bituminous coals, possible correlations can be found between the coking properties of a bituminous coal, as obtained with the conventional tests and the composition of the tarry deposit formed. However, due to the small sample size the tarry deposits were not further analyzed in our study. Analysis of this fraction of the coal would be possible by pyrolyzing the coal samples directly within the ion source of a mass spectrometer, for example with a direct electron impact probe.

Solomon et al. (20,23) analyzed the tar fractions of different coals directly within a field ionization mass spectrometer (FIMS). However, pyrolysis of the coal samples was not performed on line but in an entrained flow reactor. For bituminous coals complex mixtures containing compounds with a molecular weight up to 800 were analyzed; the average molecular weight of the compounds in the mixture was 300-400. Suuberg et al. (19,24) used liquid chromatography to analyze the tar fractions of coals; the molecular weights of the tar components of bituminous coals were reported to be in the range 200-1200. Schulten (25) performed pyrolysis field desorption mass spectrometry of a bituminous coal in the ion source of a mass spectrometer and observed organic building blocks of coal in the mass range up to >3000 mass units.

3.2. Pyrolysis Gas Chromatography (PyGC and PyGCMS)

Six coal samples, ranging from an anthracite to a lignite, were selected from the collection of 40 coals to study the composition of the volatile pyrolysis product mixtures as a function of the coal rank in more detail and to quantify the relative amounts of pyrolysis products or homologous groups of products. The three bituminous coals (#12, #23 and #30; coded: AU20, BE30 and DE38) were also analyzed, using pure nickel wires with a Curie-point temperature of 631 K to study the thermal desorption of components adsorbed on or trapped within the coal matrix. The results are shown with decreasing rank of the coal samples in Figures 8-12. The most important pyrolysis products or groups of products have been indicated in the figures. It must be emphasized again that the choice of the capillary column, in combination with the temperature programme in the gas chromatograph precludes the separation of the pyrolysis products with a boiling point lower than n-hexane.

The same trends observed with PyMS are clearly visible in the gas chromatograms. With decreasing coal rank the amount and number of pyrolysis products increase significantly and there is a gradual change in the composition of the pyrolysis product mixture. For the anthracite and semianthracite (Figure 8) hardly any aromatic

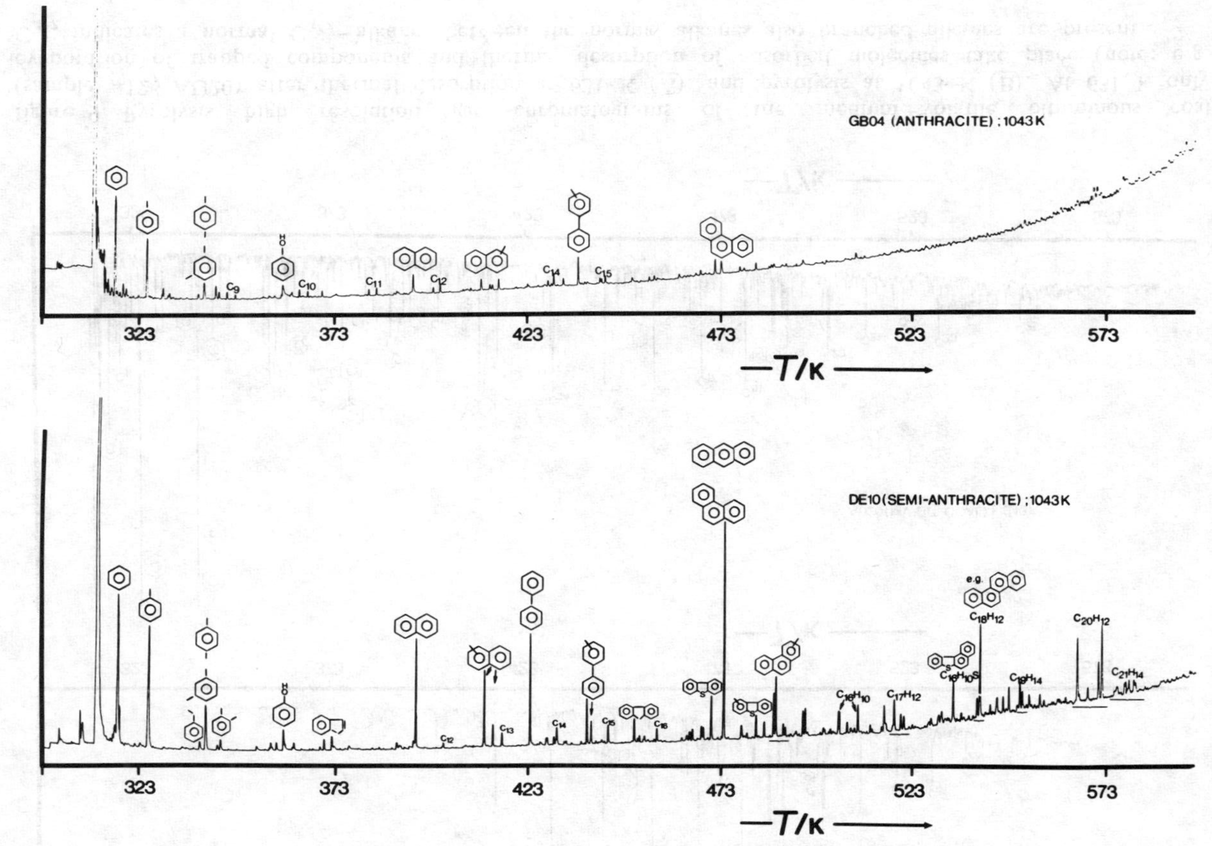

figure 8. Pyrolysis high resolution gas chromatograms of A: the anthracite, and B: the semi–anthracite after pyrolysis at 1043 K.

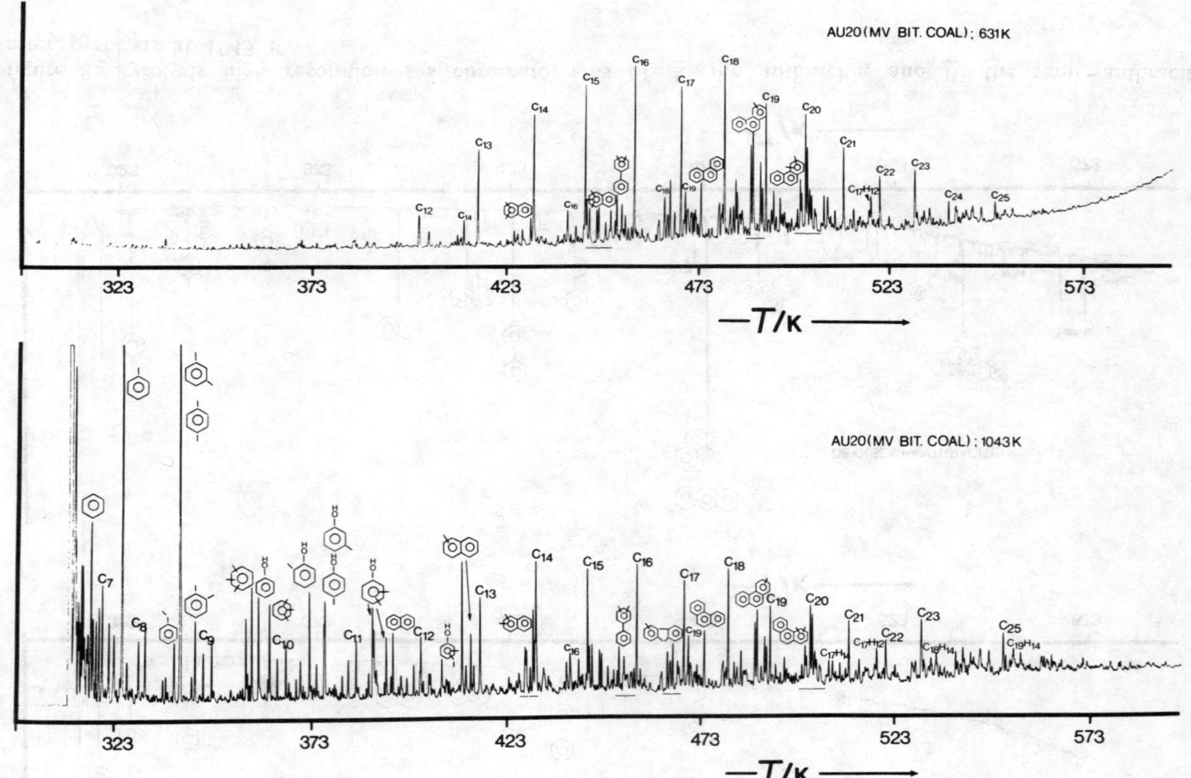

figure 9. Pyrolysis high resolution gas chromatograms of the medium−volatile bituminous coal (sample #12; AU20) after thermal desorption at 631 K (A), and pyrolysis at 1043 K (B). At 631 K only evaporation of trapped components and thermal desorption of adsorbed molecules take place (note: e.g. C_{17} indicates a normal C_{17}−alkane, between the normal alkanes also branched alkanes are present).

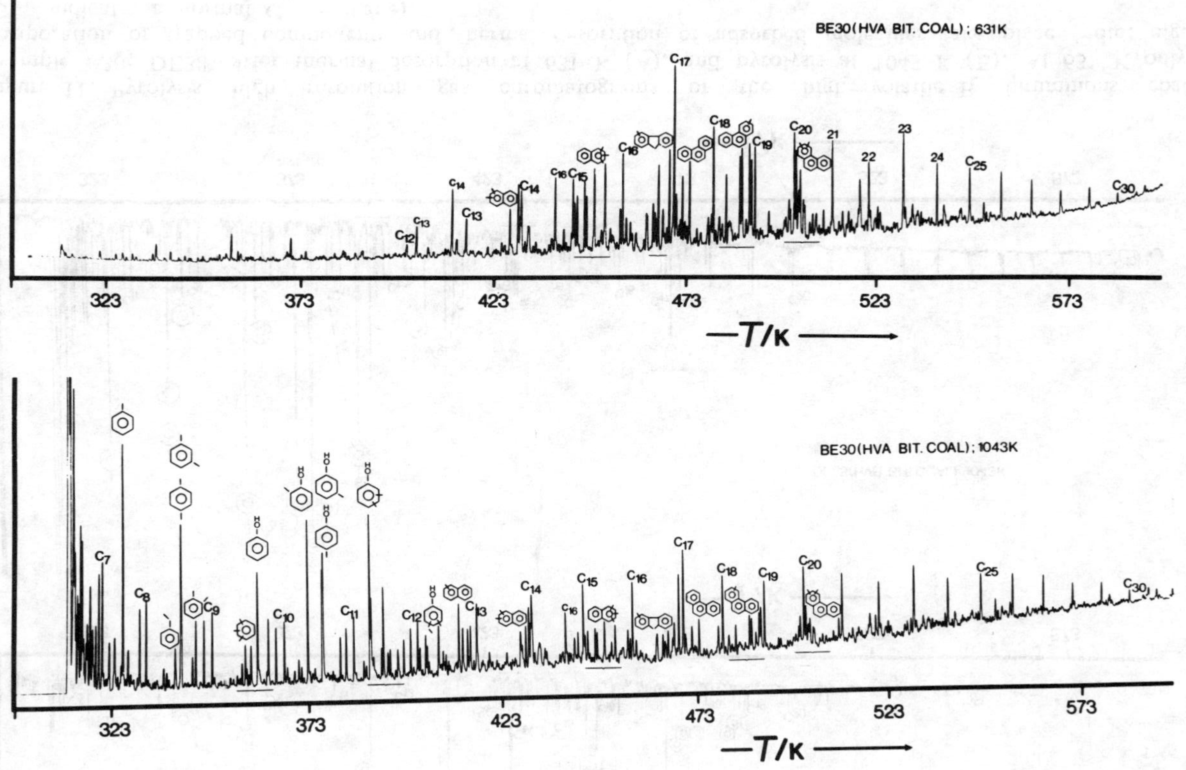

figure 10. Pyrolysis high resolution gas chromatograms of the high−volatile A bituminous coal (sample #23; BE30) after thermal desorption at 631 K (A), and pyrolysis at 1043 K (B). At 631 K only evaporation of trapped components and thermal desorption of adsorbed molecules take place (note: e.g. C_{17} indicates a normal C_{17}−alkane).

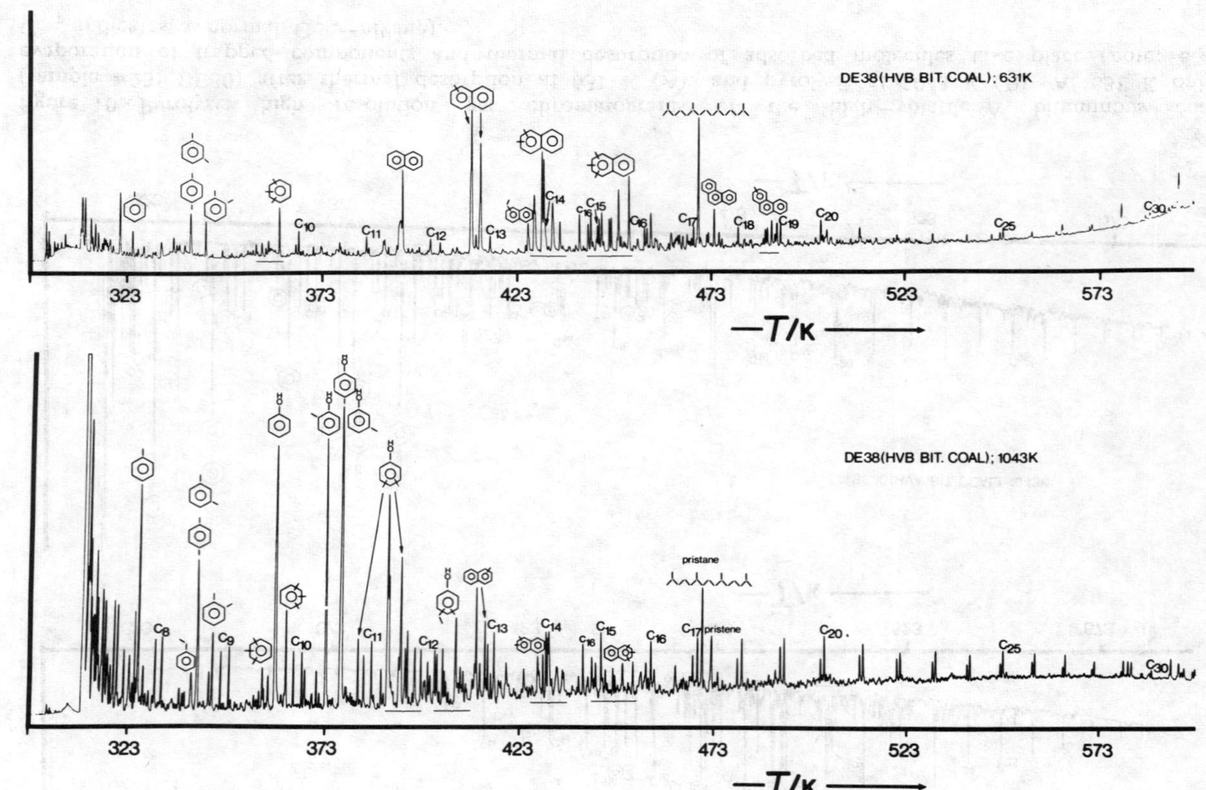

figure 11. Pyrolysis high resolution gas chromatograms of the high−volatile B bituminous coal (sample #30; DE38) after thermal desorption at 631 K (A), and pyrolysis at 1043 K (B). At 631 K only evaporation of trapped components and thermal desorption of adsorbed molecules take place (note: e.g. C_{20} indicates a normal C_{20}−alkane).

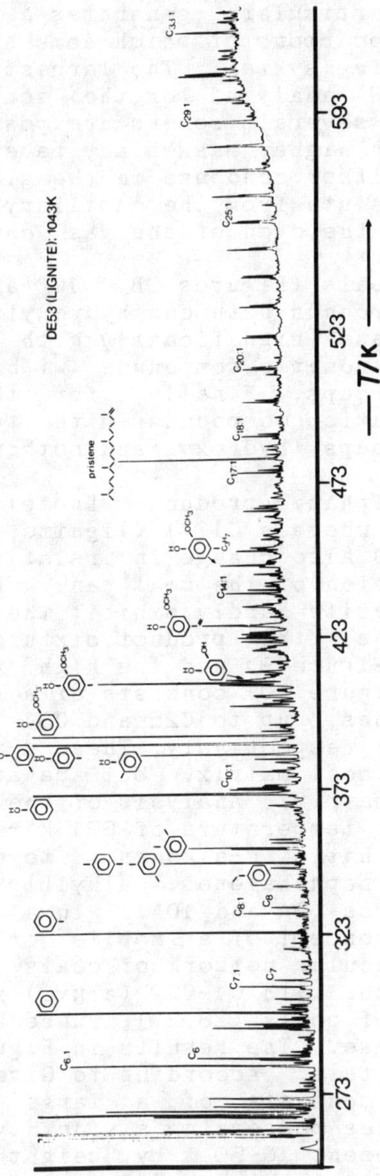

figure 12. Pyrolysis high resolution gas chromatogram of the lignite (sample #39; DE53) after pyrolysis at 1043 K (note: the gas chromatogram shows normal–alkene/alkane doublets; $C_{17:1}$ indicates a normal C_{17}–alkene. After thermal desorption at 631 K (not shown) only minor amounts of the odd C_{27}–C_{33} alkanes from residual plant waxes were detected.

pyrolysis products with oxygen functional groups are formed. The semianthracite, in particular, generates a variety of polycyclic aromatic hydrocarbons of which some also contain sulphur in the aromatic ring system. The largest polycyclic aromatic hydrocarbon (PAH) analyzed for the semianthracite is $C_{21}H_{14}$ (mass: 266); several isomers are possible with this mass. PAH's with even higher masses may be evolved from the coal sample. They either condense in the glass sample tube (450 K) or may not elute from the capillary column at the final temperature of the oven of the gas chromatograph (598 K).

For the bituminous coals (Figures 9B, 10B and 11B) the formation of aromatic compounds with one hydroxyl functional group ("phenols") increases significantly with decreasing rank, relative to the aromatic compounds without oxygen containing functional groups. Finally, for the lignite (Figure 12) also aromatic compounds with two oxygen containing functional groups (hydroxyl and methoxy groups) are formed.

Furthermore, the aliphatic products [note: e.g. C17 indicates a n-C17-alkane, whereas C17:1 (lignite, Figure 12) indicates a n-C17-alkene] also change in relative amounts and composition as a function of the coal rank. Whereas the anthracite and semianthracite hardly exhibit the formation of aliphatic hydrocarbons, the product mixtures of the medium volatile (AU20, Figure 9) and the high volatile A bituminous coal (BE30, Figure 10) consists of a significant part of aliphatic n-alkanes, up to C25 and C31 for the mv and hvA bituminous coals, respectively. These n-alkanes are no integral part of the coal matrix, but are adsorbed or trapped within the coal matrix. Analysis of the two coal samples at a Curie-point temperature of 631 K reveals the evolution of these aliphatic n-alkanes, together with a large variety of alkyl naphthalenes, alkylbiphenyls and alkylphenanthrenes (Figures 9A and 10A). In a debate in literature (18) on the concept of a mobile or molecular phase within the macromolecular network of coals, homologous series of aromatic compounds with C1-C22 (alkyl) side chains released on liquefaction of coals (26,27), were believed to be part of this mobile phase. The results in Figures 9A and 10A are in agreement with this. According to Given (28) the mobile phase of coals consists of a large number of relatively small molecules possessing a wide diversity of structures and may represent 10-50 % by weight of whole coal, except for coals of high rank. Unfortunately, due to the small sample size used in the Curie-point pyrolysis techniques no absolute amounts of products can be quantified in our study. Furthermore, a tarry deposit occurred on the glass sample tube upon Curie-point pyrolysis at 1043 K of the bituminous coals; part of this deposit might originate from the mobile phase.

With further decrease in coal rank Curie-point pyrolysis at 1043 K of the high volatile B bituminous coal and the lignite reveals an increase in the formation of n-alkenes at the cost of n-alkanes. For the lignite the peak between n-C17:1 and n-C19:1 is due to 1-pristene, which is thought to derive from tocopherols, also known as E-vitamins, reacted with the polymeric skeleton of the lignite (29). For the hvB bituminous coal also the saturated analogue of 1-pristene, pristane is found. The experiment at the Curie-point temperature of 631 K reveals that pristane is thermally desorbed from the coal. At the same time the abundance of n-alkanes decreases drastically in the gas chromatograms of the hvB bituminous coal (Figure 11A). The 631 K experiment with the lignite is not shown; only minor amounts of the odd C27-C33 alkanes from residual plant waxes were detected. These results indicate that with decreasing coal rank the n-alkanes are no longer part of the mobile phase but are increasingly formed, together with the n-alkenes, by pyrolysis of the coal matrix itself. It has been reported that these n-alkanes and n-alkenes originate from a highly aliphatic (polyethylene like) biopolymer, which was recognized as an important component of certain modern and fossil plant cuticles (30,31). This biopolymer is highyly resistant to biodegradation. Similar biopolymers, forming aliphatic hydrocarbon chains (alkene/alkane doublets) upon pyrolysis, have also been recognized in the outer walls of an alga (32) and in the rootlets of heathers and their peat (33).

The change in aliphatic products with increasing coal rank, as observed in our Curie-point PyGC and PyGCMS study, suggests that upon diagenesis of the aliphatic biopolymer in the earth crust, instead of the alkene/alkane doublets, only n-alkanes are formed which are trapped within the bituminous coal matrix (mobile phase). With further increase in coal rank (semianthracite and anthracite) the n-alkanes disappear from the coal layer by geothermal desorption and may lead to a separate oil formation.

It should be noted that the two coal samples (#12 and #23; coded AU20 and BE30), which show the evolution of the series of n-alkanes together with a large variety of alkylaromatics when a Curie-point temperature of 631 K is used, exhibit thermoplasticity (softening, swelling and resolidification) upon heating. This might indicate that the presence of a mobile phase, in combination with its composition, correlates with the thermoplastic properties of a bituminous coal sample. Further study will be directed towards the elucidation of the phenomenon of thermoplasticity in relation to the composition of coals.

Other important biopolymeric precursors of coal are lignins, which originate from cell walls of higher plants. It is important to realize that peatification of plant matter leads to extensive loss of polysaccharide and also to modifications in the lignin, e.g. loss of methoxy groups. A model structure of a guaiacyl (with o-methoxyphenoxy units) lignin is shown in Figure 13 (34) and can be used to explain the aromatic pyrolysis product formation as a function of the coal rank.

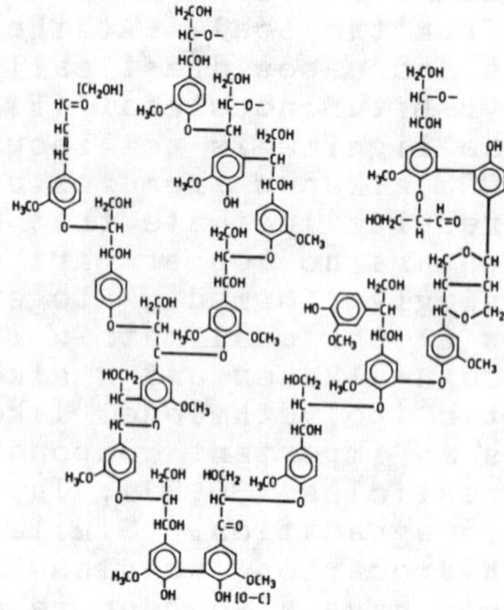

Figure 13. Structural scheme of a guaiacyl lignin [after Adler (34)]

The structure of the lignite (DE53), used in our study, albeit biodegraded, probably still exhibits some resemblance to the structure of a lignin; this explains the presence of the o-methoxyphenols in the gas chromatogram of the lignite (Figure 12). The high abundance of alkylphenols and alkylbenzenes indicates that conversion of methoxy groups to hydroxyl groups and dehydroxylation of aromatic ring components has occurred during the degradation of the lignins. During a period of million years the sedimentary material loses oxygen in the form of water (dehydroxylation) and carbon dioxide (decarboxylation) and becomes more condensed by the pressure of subsequent sediments and by the increasing temperatures with increasing depth in the earth crust. Due to these geological processes an increase in coal rank proceeds.

The loss of oxygen results in the shift with increasing coal rank from the formation of aromatic compounds with more than one oxygen functional groups to aromatic compunds with no oxygen at all, as observed in this study by the Curie-point pyrolysis techniques. Condensation of the aromatic coal structures causes a higher thermal stability of the coal matrix and results in a decrease in the amount and the number of pyrolysis products with increasing coal rank. From these condensation reactions also hydrogen is generated. The hydrogen can be used to hydrogenate the aliphatic biopolymer to form the alkane fraction in the coal samples.

4. CONCLUSIONS

Curie-point pyrolysis mass spectrometry (PyMS), in combination with factor discriminant analysis is a fast and reliable method to characterize and classify a large collection of coal samples, independent of their origin, according to their rank order. The number and nature of the pyrolysis products, particularly the degree of oxygen functionality of the one-ring aromatic compunds formed are characteristics for the rank of the coal sample. Furthermore, the formation of hydrogen sulphide is highly characteristic for high sulphur coals. However, characterization of coals for conversion processes according to the conventional analyses and tests is still necessary, due to the qualitative character of the Curie-point pyrolysis techniques. It is not yet clear how thermoplastic properties of coals are reflected in the mass spectra; more correlation studies are necessary to clarify this issue. The organic emissions of coal utilization processes like coal combustion and gasification can be qualitatively predicted as a function of the coal used.

More detailed information on the composition of the pyrolysis product mixtures as a function of the coal rank is obtained with Curie-point pyrolysis gas chromatography (PyGC and PyGCMS). The composition of the pyrolysis product mixtures as a function of the coal rank gives fundamental knowledge on the metamorphism of coal precursor material and on the development of the molecular structure composition of coals during maturation.

No absolute quantification of the product mixtures is possible with the Curie-point techniques due to the small sample size. After pyrolysis of the bituminous coals a tarry deposit occurs on the glass sample tube. Evidence of the so called mobile phase of bituminous coals and its composition is obtained by analysis of the product mixtures evolved from the coals at a Curie-point temperature, at which only evaporation and thermal desorption occur.

5. ACKNOWLEDGEMENT

The experimental work was performed at the FOM Institute for Atomic and Molecular Physics in Amsterdam. The technical assistance of B.Brandt-de Boer (PyMS), A.Tom (PyGC) and G.B.Eijkel (PyGCMS and data processing) is greatfully acknowledged. The group for mass spectrometry of macromolecular systems is financially supported by the Foundation for Fundamental Research on Matter (FOM), a subsidiary of the Netherlands Foundation for Pure Research (ZWO). The research reported here was also supported by the Netherlands Foundation of Chemical Research (S.O.N.) and by the Dutch Ministry of Economic Affairs, within the framework of the Dutch National Coal Research Programme, which is managed by the Project Office for Energy Research (PEO).

6. REFERENCES

1. P.J.J.Tromp and J.A.Moulijn, this book.
2. "Methods of Analyzing and Testing Coal and Coke", Bureau of Mines, Bulletin 638, Pittsburgh, Pa., 1967.
3. H.L.Retkofsky, in "Coal Science", volume 1 (Eds. M.L.Gorbaty, J.W.Larsen and I.Wender), Academic Press, New York, 1982, p.43.
4. K.D.Bartle and D.W.Jones, in "Analytical Methods for Coal and Coal Products", volume 2 (Ed. C.Karr Jr), Academic Press, New York, 1978, Chap. 23.
5. H.L.Retkofsky and T.A.Link, ibid, Chap. 24.
6. B.C.Gerstein, P.H.Murphy and L.C.Ryan, in "Coal Structure", (Ed. R.A.Meyers), Academic Press, New York, 1982, p.87.
7. P.C.Painter, M.M.Coleman, R.W.Snyder, O.Mahajan, M.Komatzu and P.L.Walker Jr., Appl. Spectrosc. 35(1), 106 (1981).
8. P.R.Solomon, Fuel 60, 3 (1981).
9. H.L.C.Meuzelaar, A.M.Harper, G.R.Hill and P.H.Given, Fuel 63, 640 (1984).
10. P.R.Philp, T.D.Gilbert and J.Friedrich, in "Chemical and Geochemical Aspects of Fossil Energy Extraction" (Eds. T.F.Yen, F.K.Kawahara and R.Hertzberg), Ann Arbor Science, Ann Arbor, 1984, p.63.
11. H.L.C.Meuzelaar, B.L.Hoesterey, W.Windig and G.R.Hill, Fuel Process.Technol. 15, 59 (1987).
12. K.M.Thomas, in "Carbon and Coal Gasification" (Eds. J.L.Figueiredo and J.A.Moulijn), NATO ASI Series No.105, Nijhoff Publishers, Dordrecht, The Netherlands, 1986, p.57.
13. H.L.C.Meuzelaar, J.Haverkamp and F.D.Hileman, in "Pyrolysis Mass Spectrometry of Recent and Fossil Biomaterials; compendium and atlas", Elsevier Publ.Co., Amsterdam, 1982.
14. P.J.J.Tromp, J.A.Moulijn and J.J.Boon, Fuel 65, 960 (1986).
15. W.Windig, P.G.Kistemaker and J.Haverkamp, Jour.of Anal.and Applied Pyrolysis 3, 199 (1981).
16. R.Hoogerbrugge, S.J.Willig and P.G.Kistemaker, Anal.Chem. 55, 1711 (1983).

17. J.J.Boon, A.D.Pouwels and G.B.Eijkel, Trans. Biochem. Soc.UK 15, 170 (1987).
18. P.H.Given, A.Marzec, W.A.Barton, L.J.Lynch and B.C.Gerstein, Fuel 65, 155 (1986).
19. P.E.Unger and E.M.Suuberg, Fuel 63, 606 (1984).
20. P.R.Solomon and D.G.Hamblen, "Chemistry of Coal Conversion" (Ed. R.H.Schlosberg), Plenum Press, New York, 1985, p.121.
21. D.W. van Krevelen, "Coal, Typology-Chemistry-Physics-Constitution", Elsevier, Amsterdam, 1981.
22. A.C.Harper, H.L.C.Meuzelaar and P.H.Given, Fuel 63, 793 (1984).
23. P.R.Solomon and H-H.King, Fuel 63, 1302 (1984).
24. E.M.Suuberg, P.E.Unger and W.D.Lilly, Fuel 64,956(1985).
25. H.R.Schulten, Fuel 61, 670 (1982).
26. Z.Mudamburi and P.H.Given, Org.Geochem. 8(3),221 (1985).
27. Z.Mudamburi and P.H.Given, Org.Geochem. 8(6),441 (1985).
28. P.H.Given, in "Coal Science", volume 3 (Eds. M.L.Gorbaty, J.W.Larsen and I.Wender), Academic Press, New York, 1984, p.63.
29. H.Goossens, J.W. de Leeuw, P.A.Schenck and S.C.Brasell, Nature 312 (5993), 440 (1984).
30. M.Nip, E.W.Tegelaar, J.W. de Leeuw, P.A.Schenck and P.J.Holloway, Naturwissenschaften 73, 579 (1986).
31. M.Nip, E.W.Tegelaar, H.Brinkhuis, J.W. de Leeuw, P.A.Schenck and P.J.Holloway, Org.Geochem. 10,769(1986).
32. C.Largeau, S.Derenne, E.Casadevall, A.Kadouri and N.Sellier, Org.Geochem. 10, 1023 (1986).
33. D.G. van Smeerdijk and J.J.Boon, submitted for publication in Jour. of Anal.and Applied Pyrolysis.
34. E.Adler, Wood Sci.Technol. 11, 169 (1977).

CHROMATOGRAPHIC TECHNIQUES IN COAL SCIENCE

K.D. Bartle
Department of Physical Chemistry
University of Leeds
Leeds LS2 9JT
U.K.

ABSTRACT. Chromatographic methods for the analysis of coal derivatives are reviewed. Column chromatography may be used to fractionate on the basis of polarity or functionality; size exclusion chromatography separates on the basis of molecular size and allows molecular mass distributions to be deduced. High resolution analytical separations of low MM compounds are readily achieved by capillary gas chromatography; high performance liquid chromatography, especially on microcolumns, and supercritical fluid chromatography are both applicable to higher MM compounds.

1. INTRODUCTION

The complexity of coal-derived materials generally makes necessary their separation into simpler fractions based on polarity, functionality, or molecular size before detailed characterization is attempted. Column chromatography is used extensively as a means of fractionating coal liquids; this may vary from a very simple chemical type separation into gross fractions to more complex separation using more than one method. Size-exclusion chromatography (SEC) is commonly used to separate coal extracts and liquids on the basis of molecular size.

For high-resolution chromatographic analysis, capillary column gas chromatography (GC) is the method of choice for relatively volatile (molecular mass less than 300) compounds but for non-volatile or thermally unstable molecules, high performance liquid chromatography (HPLC) or supercritical fluid chromatography (SFC) is usually employed.

2. LIQUID CHROMATOGRAPHY

2.1. OPEN-COLUMN LC.

The simplest and probably most commonly applied chromatographic separation of coal liquids is on silica gel[1] to yield aliphatic hydrocarbons (eluted by n-pentane), neutral aromatics (benzene) and a polar fraction (methanol or tetrahydrofuran). The method is applicable to fairly high molecular

Y. Yürüm (ed.), New Trends in Coal Science, 271–285.

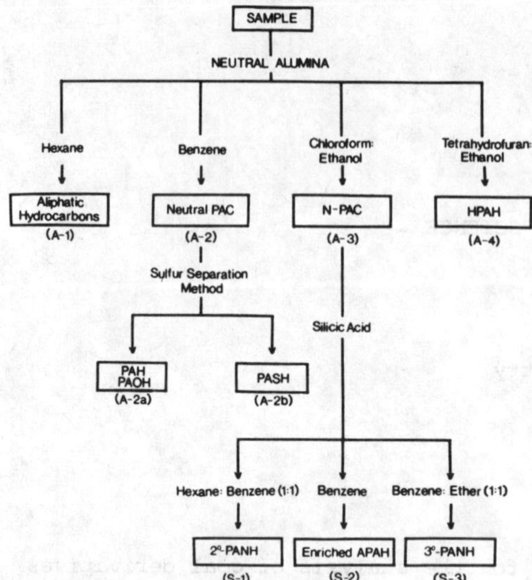

FIGURE 1

Chemical class separation scheme for synthetic fuel products

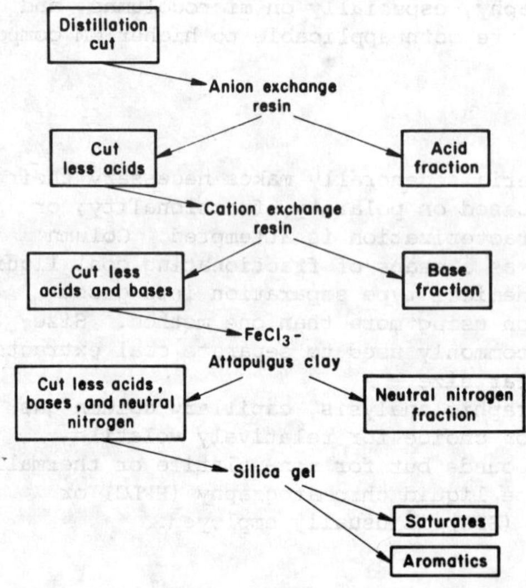

FIGURE 2

SARA separation scheme for coal-derived liquids

mass (MM) fractions such as asphaltenes. Sequential elution from silica gel with specific solvents (SESC) yields[2] further sub-fractions based on polarity (e.g. alkylphenols). A combined silica/ alumina column gives improved separation[3]. Aromatic fractions may be further fractionated into mono-(single ring), di-(two ring), and polyaromatics ($\geq$ 3 ring) on an alumina column by elution with benzene/pentane and dichloromethane/ pentane mixtures.

Very comprehensive open-column LC methods for the separation of coal liquids into discrete chemical classes have been developed[4] based on some of the above procedures. Fractionation on small quantities of neutral alumina and silicic acid allowed (Figure 1) separation into aliphatic hydrocarbons, polycyclic aromatic hydrocarbons and sulphur heterocycles, nitrogen polycylic aromatic compounds and hydroxylated polycyclic aromatic hydrocarbons. The nitrogen-containing aromatic compounds were further separated into secondary nitrogen polycyclic heterocycles, amino substituted aromatics, and tertiary nitrogen polycyclic aromatics (azaarenes).

In subsequent work the scheme has been modified to permit separation of hydroaromatic compounds, and of polycyclic aromatic nitriles.[5] The SARA separation (Figure 2), originally devised for the fractionation of petroleum products, has been applied to coal products.[6] Acidic and basic components are removed on anion and cation exchange resin columns, and neutral nitrogen-containing compounds are separated by complexation with $FeCl_3$ deposited on Attapulgus clay. Aliphatics and aromatics are then separated on silica gel and alumina.

2.2. HIGH-PERFORMANCE LIQUID CHROMATOGRAPHY

Since its inception in the early 1970's, HPLC has been widely used in the separation of polycyclic aromatic compounds.[7,8] It offers a variety of stationary phases capable of providing unique selectivity both for functional group types and for difficult-to-separate isomers, because of interactions of the solute with both stationary and mobile phases. HPLC thus provides both a useful fractionation technique and the means of high resolution analysis of compounds with MM up to 600.[9]

A silica or amino column with elution by an alkane gives first the saturated hydrocarbons followed by the aromatics. Polar compounds are eluted by backflushing. Class fractionation of coal liquids by functional group has also been investigated for a wide variety of other normal-phase columns (chemically bonded NO_2, CN, diol, sulphonic acid), and NH_2 and NO_2 phases found to be the most selective for fractions containing hetero-functions.[10,11] Excellent resolution of PAH from azaarenes has been reported on a nitrophenyl phase,[11] about 70% of a coal tar could be analysed with this phase. Normal phase HPLC is commonly used to separate PAH on the basis of ring number (Figure 3); with alkyl derivatives eluting with the parent. Such separations usually precede analysis by another chromatographic method.

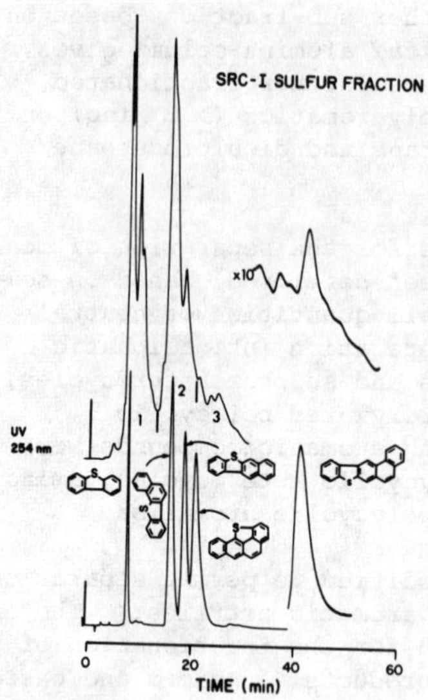

FIGURE 3

Normal-phase HPLC fractionation of polycyclic aromatic sulphur hydrocarbon fraction from SRC-1 coal liquid, and standard compounds

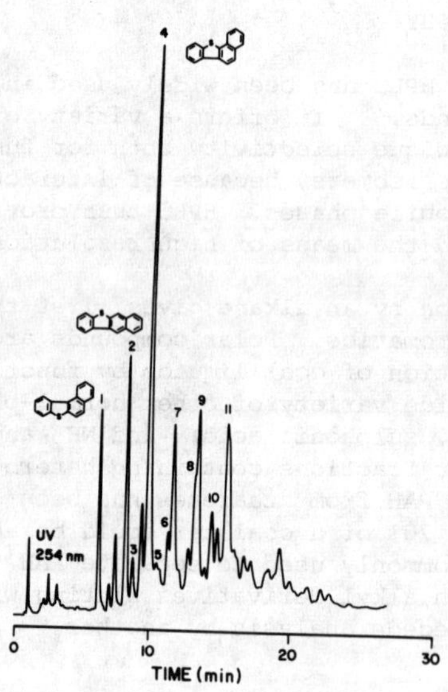

FIGURE 4

Reverse-phase HPLC analysis of fraction 2 (see Figure 3) of PASH fraction from SRC-1 coal liquid

High resolution HPLC analysis is generally achieved on reverse phase columns, such as octadecylsilane (ODS). Numerous separations of PAH and polycyclic aromatic sulphur heterocycle (PASH) fractions from coal have been reported,[7,8] which show the excellent selectivity for the separation of PAC isomers and alkyl derivatives. For example Figure 4 shows the separation of the four-ring PASH fraction separated from the SRC-1 coal liquid in Figure 3. The great advantage of reverse phase packings is their compatibility with a variety of mobile phases and with gradient elution. Selective separations are achieved on the basis of the length-to-breadth ratio of solutes: the more nearly linear molecules are retained longer.

Microcolumn HPLC has been proposed as a method of improving the efficiency of HPLC separations, and has proved capable[12] of resolving many constituents of coal derived materials containing between 5 and 9 rings (Figure 5). The columns are typically 200 µm internal diameter and are packed with 3 µm ODS particles, to generate over 200,000 theoretical plates, although very long analysis times are necessary. Fractions containing nitrogen compounds from coal liquids were also separated on similar columns[13]: over 170 peaks were resolved and 600 nitrogen compounds were characterized by mass spectrometric and fluorescence analysis of trapped peaks.

2.3. SIZE-EXCLUSION CHROMATOGRAPHY

The distribution of molecular mass (MM) is the factor of greatest interest (after chemical composition) in judging the suitability of coal extracts and derivatives as feedstocks for further processing to transport fuels, and raw materials for chemical industry etc.[11] The method of choice for determining MM distributions is high-performance size exclusion chromatography (SEC), in which molecules of different size are separated according to their degree of penetration into the pores of a gel packed into a column as small diameter spheres. The gel is typically a three-dimensional network of cross-linked polymeric chains of controlled porosity such as cross-linked polystyrene, and a number of mobile phases, most usually tetrahydrofuran (THF) have been investigated. Eluted components are detected by monitoring a physical property such as UV absorption or refractive index. The MM range of coal liquids is between approximately 200 and 5000. The recent advances in the technology of gas chromatography and supercritical fluid chromatography have extended the limits of these techniques, but of available chromatographic techniques only SEC is capable of providing information on MM distributions above 1000.

The choice of eluting solvent is vital in SEC. Unless a solvent minimizing solute self-association and adsorption on the column is chosen, large errors may result in MM distributions.[14] THF has been found generally preferable, but the polystyrene/THF combination is inappropriate for coal-derived materials containing alkanes (eluted early) or highly condensed aromatic molecules retained by an adsorption mechanism.[15] Low MM phenols,

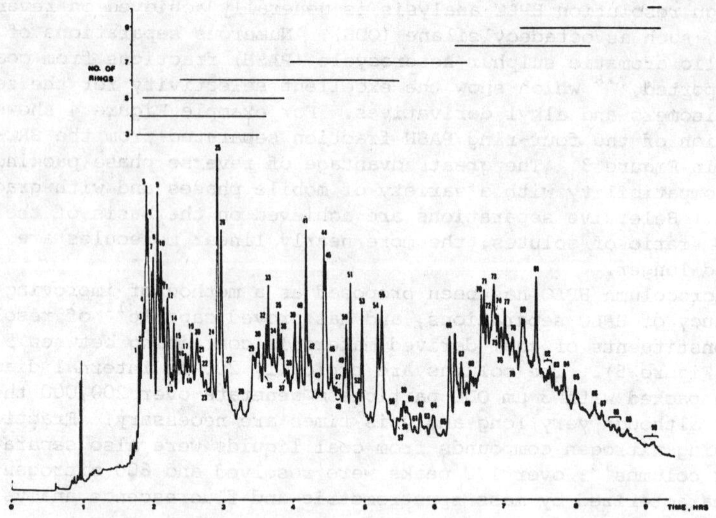

FIGURE 5. Microcolumn HPLC separation of SRC-II fuel oil nitrogen compound fraction.

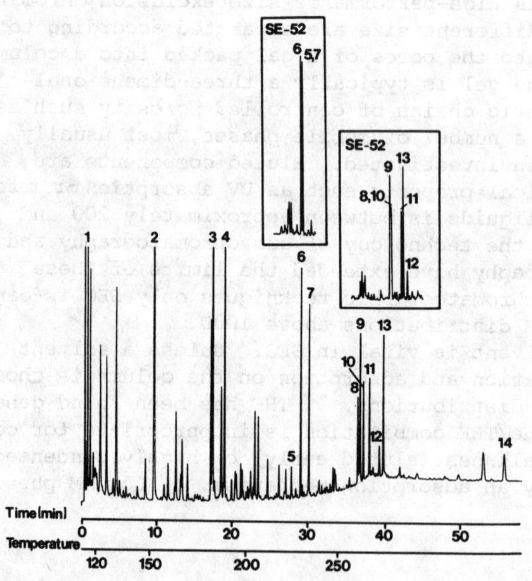

FIGURE 6.

Gas chromatogram of coal tar on mesogenic polysiloxane stationary phase. Inset, portions of chromatogram of the coal tar on SE-52 stationary phase.

some other oxygen-containing compounds and nitrogen compounds are also eluted early because solvation by THF increases the apparent molecular size.[16] This effect produces apparently bimodel distributions for e.g. pyrolysis tars but has little influence in the SEC of asphaltenes. For coal derivatives containing material insoluble in THF, quinoline and trichlorobenzene have been recommended as solvents.[17]

Reliable MM distributions for coal derivatives can only be derived if the detector responds equally to a given concentration of solute throughout the MM range.[14] For the differential refractive index detector, a reversal of polarity with increasing MM is observed. The distribution of groups which absorb in convenient regions of the infra red spectrum is not uniform. Calibration graphs of specific response of both UV absorption and evaporative analyser[18] detectors versus $\overline{M}_n$ have been constructed for narrow sub-fractions of coal derivatives. The evaporative analyser response for materials with MM <300, and which are lost before detection, is markedly reduced; but above MM $\simeq$300 the response is much more uniform than that of the UV detector.[18]

Narrow polystyrene fractions are often employed as calibration standards in SEC of coal derivatives via a graph of log MM against retention volume, V_R, the derived constants A and B in the equation

$$\log MM = A - BV_R$$

are used to calculate the distribution of MM. However, polystyrene is not a satisfactory model for the retention behaviour of coal derived molecules especially above 1000. A variety of alternative procedures both rigorous and empirical have been investigated, but calibrations should be made[15] with narrow preparative SEC sub-fractions of coal derivatives with $\overline{M}_n$ determined by vapour pressure osmometry. Alternatively, other, more suitable polymer standards such as polyacenaphthylene should be sought.

A comparison of errors in SEC determinations of average MM for coal derivatives shows indeed that errors originating in the size exclusion separation process are small; the dominant errors arise from calibration and detection procedures. Appropriate calibration standards and detector response factors must be employed.[14]

Molecular mass distributions of high MM coal derivatives are determined by high performance SEC by interfacing flow rate sensor and detector to a microcomputer. Detector response and flow rate are sampled at intervals and from stored calibration data a table of retention time, detector response and MM is produced. Number and weight average MM, and a normalised mass distribution curve are calculated with facility for taking account of the variation of detector response with MM.[14]

The use of a highly polar mobile phase dimethylformamide with a neutral macroporous gel, modifies the separation process considerably from one based mainly on molecular size, to include significant contributions from association, partition and adsorption effects.[19] The calibration of the column with model compounds (mainly constituents of tars), and previously characterised fractions of coal derived materials has shown that it is possible to define regions of the chromatogram which depend on both size and chemical type. This twin selectivity provides the basis for a rapid and unique characterisation of complex materials, such as tars, coal extracts, pitches etc., which contain numerous individual components varying considerably in molar mass and chemical type. The technique can detect small changes in the relative concentrations of groups of species and hence, monitor the effect of changes in process variables.

3. GAS CHROMATOGRAPHY

For the low MM constituents of coal derivatives, gas chromatography (GC) is the method of choice.[20] The extreme complexity of coal derivatives also demands the greatest possible resolution in their analysis and, in this respect, GC on packed columns even as long as 20m has fallen short of that available on capillary columns. Fused-silica columns are now used universally, and if surface activity is controlled by silanization treatments trace compounds are eluted as sharp peaks. For most analytical work, columns 10-25m in length are suitable, with internal diameter 0.2-0.3mm.[21]

A variety of stationary phases have been employed for hydrocarbon separations, but SE-52 and SE-54 (methylsilicone gums with 5% phenyl groups) have found most application,[22] especially after free-radical crosslinking to improve thermal stability. Silicone/liquid crystal blends have been observed to give remarkably selective separations, but use is limited by the low thermal stability.[23] Poly-(mesogenmethyl) siloxanes are, however, gum phases which show high column efficiencies and stabilities but retain high selectivity for PAC isomers of coal origin (Figure 6) and have wide nematic temperature range (70-300°C). More polar stationary phases such as polyethylene glycols, Superox 20M etc., are required for the capillary GC separation of coal-derived phenols.[21] Nitrogen containing compounds may be separated on SE-52 coated capillaries, but a promising phase for amino compounds is the polarisable biphenyl siloxane.[23]

The flame ionisation detector is most commonly used in the GC analysis of coal-derived oils, but the most versatile selective detectors here are the nitrogen and sulphur selective NPD and FPD. Figures 7 and 8 show dual trace chromatograms of sulphur- and nitrogen compound enriched fractions from solvent-refined coal. Complete identification of constituents of coal-derived mixtures separated by capillary GC is best achieved by combined GC/mass spectrometry.[21] Routine analyses may be made, however, with the aid of retention index systems, especially that of Lee.[24]

4. SUPERCRITICAL FLUID CHROMATOGRAPHY

Although chromatography with a supercritical fluid as mobile phase was reported more than twenty years ago, the advantages of supercritical fluid chromatography (SFC) have only recently been fully realised - in particular its rapidity, flexibility and ability to allow the analysis of substances which cannot be analysed by gas chromatography (GC). The separating power of open-tubular column GC is unparalleled, but its applicability is restricted by the limited volatility and thermal stability

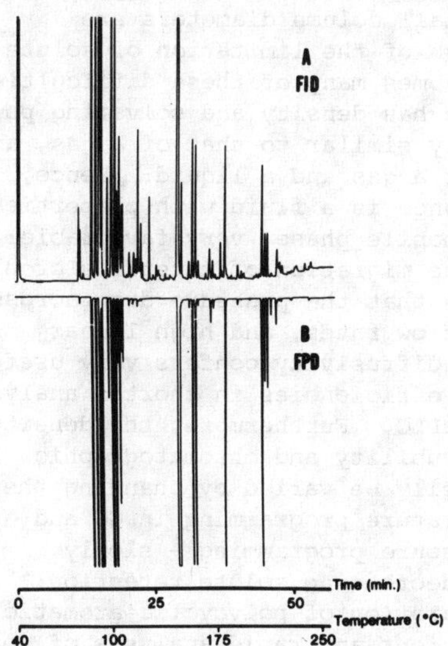

FIGURE 7

Flame-ionization (FID) - flame photometric detection (FPD) dual trace gas chromatograms of the sulphur heterocycle fraction of a coal liquid.

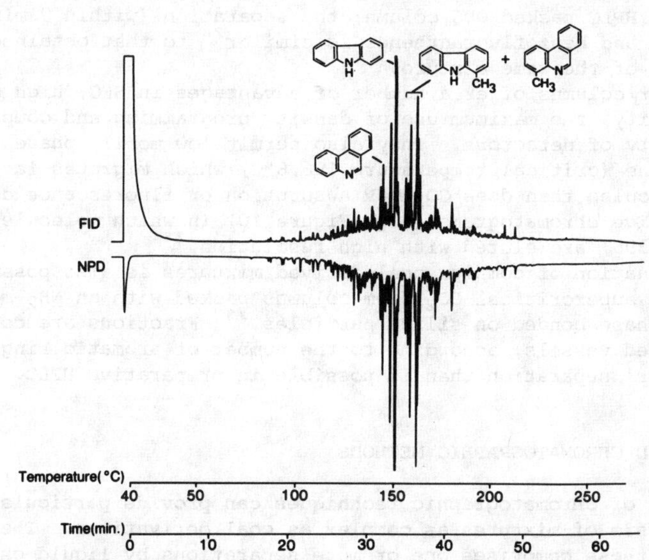

FIGURE 8. Flame-ionization (FID) - nitrogen selective detection (NPD) dual trace gas chromatogram of the nitrogen fraction of a coal liquid.

of many organic compounds; less volatile compounds can be analysed by HPLC, but long analysis times and very small column diameters are required for efficient separations because of the limitation of solute diffusion in the mobile phase. SFC overcomes many of these difficulties.[25]

Above its critical point a substance has density and solvating power approaching that of a liquid but viscosity similar to that of a gas, and diffusivity intermediate between those of a gas and a liquid. Hence, above the critical temperature the substance is a fluid with properties which make its use as chromatographic mobile phase very favourable: extraction and solvation effects allow the migration of materials of high molecular weight; the low viscosity means that the pressure drop across the column is greatly reduced for given flow rates, and high linear velocities can be achieved; and the high diffusivity confers very useful mass-transfer properties, so that higher efficiencies in shorter analysis times are possible than are achieved in HPLC. Furthermore, the density of the supercritical fluid and hence the solubility and chromatographic retention of different substances can easily be varied by changing the applied pressure. The analogue of temperature programming in GC and gradient elution in HPLC is SFC with pressure programming - slowly increasing the mobile phase density and decreasing solute retention.[26]

Many applications of SFC in the separation of polycyclic aromatic compounds have been reported,[27] and efficient and rapid analysis of coal-derived oils is possible with CO_2 as mobile phase at low temperatures. For example, Figure 9 is the SFC chromatogram of an anthracene oil on a conventional HPLC packed ODS column; the separation (within 9 minutes) of benzopyrenes and benzofluoranthenes is similar[26] to that obtained by capillary GC of the same mixture.

Capillary columns offer a number of advantages in SFC; high efficiency and sensitivity, the maximum use of density programming and compatibility with a variety of detectors. They also permit low mobile phase flow for example *n*-pentane (critical temperature 196.6°C) which migrates larger coal derived molecules than does CO_2. UV absorption or fluorescence detection is used to give chromatograms (e.g. Figure 10) in which molecules beyond ovalene (MM 398) are eluted with high resolution.[28]

Fractionation of complex coal-derived mixtures is also possible by elution with supercritical CO_2 from columns packed with an NH_2-modified stationary phase bonded on silica particles.[29] Fractions are collected in pressurised vessels, according to the number of aromatic rings, with much 'cleaner' separation than is possible in preparative HPLC.

5. COMBINED CHROMATOGRAPHIC METHODS

The coupling of chromatographic techniques can provide particular advantages in the analysis of mixtures as complex as coal derivatives. The most powerful of these comprises one or more separations by liquid chromatography into fractions based on chemical type or MM followed by a high resolution analysis of the constituents of the fraction by capillary GC.

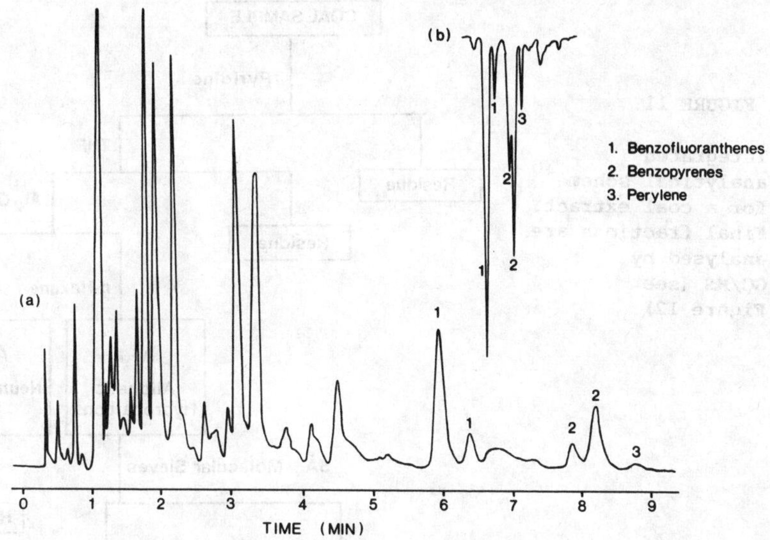

FIGURE 9. Chromatograms of a coal-derived oil: (a) packed column SFC with CO_2 mobile phase at 42°C; (b) 14m capillary gas chromatography column at 250°C

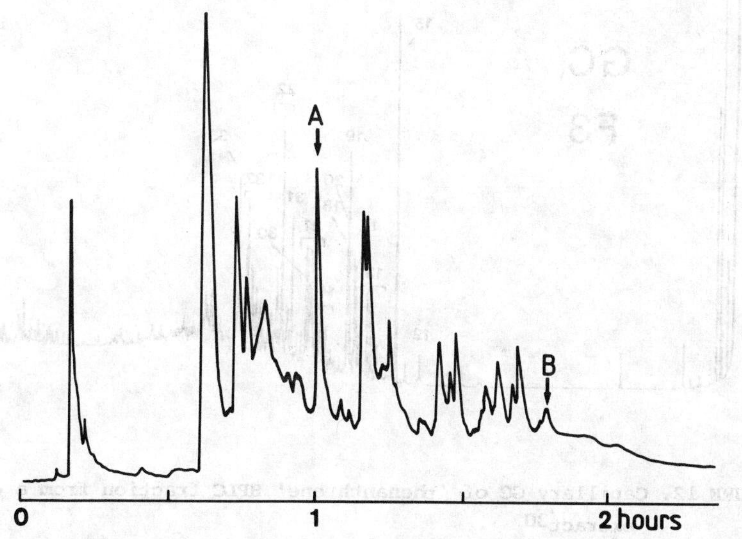

FIGURE 10. SFC chromatogram of coal-tar pitch on 10m capillary column; mobile phase, pentane at 220°C;coronene and ovalene elution marked A and B.

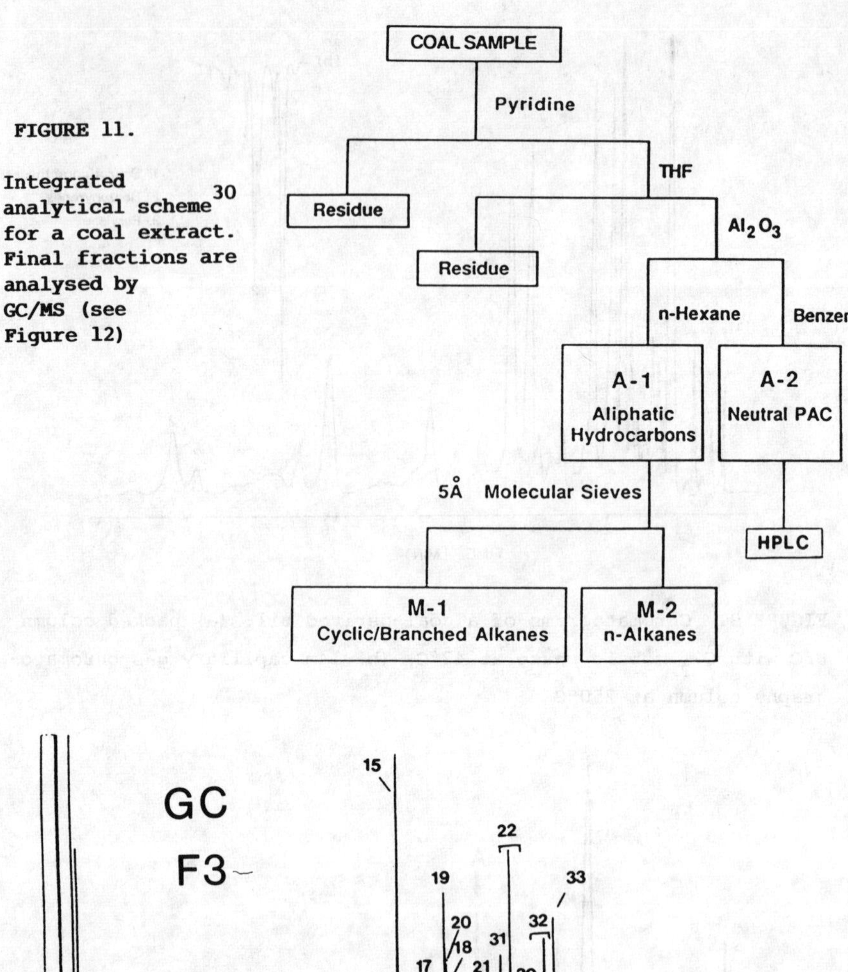

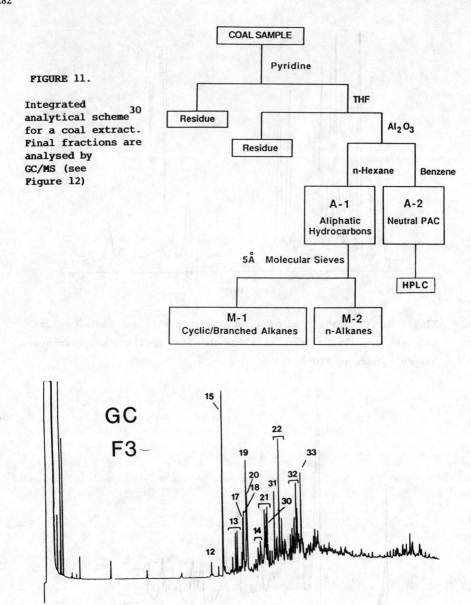

FIGURE 11.

Integrated analytical scheme[30] for a coal extract. Final fractions are analysed by GC/MS (see Figure 12)

FIGURE 12. Capillary GC of 'phenanthrene' HPLC fraction from a coal extract[30]

Fully integrated analytical methods based on this strategy have greatly increased knowledge concerning the composition of coal extracts and liquefaction products. For example, Figure 11 shows the fractionation[30] of a coal extract by open-column chromatography on alumina followed by normal-phase HPLC which gave neutral PAC sub-fractions based on ring number; the numerous constituents were identified by capillary column GC/MS (e.g. the phenanthrene fraction, Figure 12). Capillary GC/MS of fractions based on compound type has allowed the detailed analysis of polycyclic aromatic sulphur compound, carbazole, amine, azaarene, nitrile, hydroxyaromatic, hydroaromatic, n-alkane and branched/cyclic alkane fractions of coal derivatives.[4,5,30]

Most recently such techniques have been automated[31] by linking microbore HPLC columns to capillary GC columns through a retention gap. The system is totally automated by pneumatically activated valves and allows switching of HPLC fractions directly into the GC column.

REFERENCES

1. K.D. Bartle, W.R. Ladner, T.G. Martin, C.E. Snape and D.F. Williams, Fuel, 58, 413 (1979).

2. M. Farcasiu, Fuel, 56, 9 (1977).

3. M.K. Marsh, J. Chromatogr., 283, 173 (1984).

4. D.W. Later, M.L. Lee, K.D. Bartle, R.C. Kong and D.L. Vassilaros, Anal. Chem., 53, 1612 (1981).

5. R.B. Lucke, D.W. Later, C.W. Wright, E.K. Chess and W.C. Weimer, Anal. Chem., 57, 633 (1985).

6. P.H. Given, D.C. Cronauer, W. Spackman, H.L. Lovell, A. Davies and B. Biswas, Fuel, 54, 34,40 (1975).

7. S.A. Wise in A. Bjorseth ed. 'Handbook of Polycyclic Aromatic Compounds', Volume 1, Chapter 5 (Marcel Dekker, New York, 1983).

8. S.A. Wise in A. Bjorseth and T. Ramdahl eds. 'Handbook of Polycyclic Aromatic Compounds', Volume 2, Chapter 5 (Marcel Dekker, New York, 1985).

9. K.D. Bartle, M.L. Lee and S.A. Wise, Chem. Soc. Rev., 10, 113 (1981).

10. A. Matsunaga, Anal. Chem., 55, 1375 (1983).

11. K.D. Bartle, G. Collin, J.W. Stadelhofer and M. Zander, J. Chem. Tech. Biotech., 29, 531 (1979).

12. M. Novotny, A. Hirose and D. Wiesler, Anal. Chem., 56, 1243 (1984).

13. C. Borra, D. Wiesler and M. Novotny, Anal. Chem., 59, 339 (1987).

14. K.D. Bartle, D.G. Mills, M.J. Mulligan, I.O. Amaechina and N. Taylor, Anal. Chem., 58, 2403 (1986).

15. K.D. Bartle, M.J. Mulligan, N. Taylor, T.G. Martin and C.E. Snape, Fuel, 63, 1556 (1984).

16. N. Evans, T.M. Haley, M.J. Mulligan and K.M. Thomas, Fuel, 65, 694 (1986).

17. I.C. Lewis and B.A. Petro, J. Polym. Sci., 14, 1975 (1976).

18. K.D. Bartle, N. Taylor, M.J. Mulligan, D.G. Mills and C. Gibson, Fuel, 62, 1181 (1983).

19. M.J. Mulligan, K.M. Thomas and A.P. Tytko, Fuel in press.

20. C.W. White in A. Bjorseth ed. 'Handbook of Polycyclic Aromatic Hydrocarbons', Volume 1, Chapter 13 (Marcel Dekker, New York, 1983).

21. M.L. Lee, F.J. Yang and K.D. Bartle, 'Open Tubular Column Gas Chromatography' (Wiley Interscience, New York, 1984).

22. M.L. Lee, M. Novotny and K.D. Bartle, 'Analytical Chemistry of Polycyclic Aromatic Compounds', (Academic Press, New York, 1981).

23. K.D. Bartle in A. Bjorseth and T. Ramdahl eds. 'Handbook of Polycyclic Aromatic Hydrocarbons', Volume 2, Chapter 6 (Marcel Dekker, New York, 1985).

24. M,L. Lee, D.L. Vassilaros, C.M. White and M. Novotny, Anal. Chem., 51, 768 (1979).

25. P.A. Peaden and M.L. Lee, J. Liqu. Chromatogr., 5, (Supp. 2) 179 (1982).

26. K.D. Bartle, I.K. Barker, A.A. Clifford, J.P. Kithinji, M.W. Raynor and G.F. Shilstone, Analyt. Proc., in press,

27. W.P. Jackson, R.C. Kong and M.L. Lee in M. Cooke and A.J. Dennis, eds. 'PAH Mechanisms, Methods and Metabolism', p. 609 (Battelle Press, Columbus, Ohio, 1985).

28. K.D. Bartle, S. Wallace, N. Taylor, M. Raynor, A. Grint, G. Proud, D. Raynie and M.L. Lee, Fuel in press.

29. R.M. Campbell and M. Lee, *Anal. Chem.*, in press.

30. H.C.K. Chang, M. Nishioka, K.D. Bartle, S.A. Wise, J.M. Bayona, K.E. Markides and M.L. Lee, *Fuel*, in press.

31. I.D. Davies, M.W. Raynor, P.T. Williams, G.E. Andrews and K.D. Bartle, *Anal. Chem.*, in press.

THE OXYGEN FUNCTIONAL GROUPS IN BITUMINOUS COAL

Leon M. Stock
Department of Chemistry
University of Chicago
5735 South Ellis Avenue
Chicago, Illinois 60637
USA

ABSTRACT The recent contributions from our laboratory and from other laboratories concerning the distribution of oxygen functional groups in Illinois No. 6 bituminous coal and related materials are discussed.

1. INTRODUCTION

The design of selective methods for the disassembly of the complex macromolecules of bituminous coal would be greatly assisted by the acquisition of realiable data concerning the structural characteristics of these substances. Information about the structures of the oxygen-containing compounds is particularly important for two reasons. First, oxygen atoms are present in relatively high abundance in the bituminous coals and, second, the carbon-oxygen bonds in these coals should be susceptible to homolytic and heterolytic cleavage. As a consequence, these linkages offer an obvious point of attack for the disassembly of the macromolecule under relatively mild conditions. This situation notwithstanding, surprisingly little can be said with certainty about the actual environments of the oxygen functional groups that are present in the coal. Recent contributions directed toward the detection and determination of the relative abundances of the oxygen functional groups and their structural environments are discussed in this report.

2. RESULTS

Many early contributions centered on the determination of the relative abundances of hydroxyl, carboxyl, and etheral groups rather than on the structural environments of these groups. We undertook work on the latter issue through a study of the reductive alkylation of Illinois No. 6 coal (1).

Y. Yürüm (ed.), New Trends in Coal Science, 287–303.

The reductive alkylation experiments were performed with methyl-1-C-13 and butyl-1-C-13 methyl-d_3 and butyl-1,1-d_2 iodides to introduce nuclei that could readily be studied by magnetic resonance. The reactions, eq (1) and (2), were carried out in tetrahydrofuran at ambient temperature in the presence of naphthalene.

$$\text{Coal} + n\text{K} \rightarrow \text{Coal Polyanion}(\text{K}^+)_n \qquad (1)$$

$$\text{Coal Polyanion}(\text{K}^+)_n + n\text{RI} \rightarrow \text{Coal}(\text{R})_n + n\text{KI} \qquad (2)$$

Analytical work including studies with labeled tetrahydrofuran established that Illinois No. 6 coal consumed 22.4 ± 0.6 equivalents of potassium/100C. The number of alkyl groups added to the coal was determined in several different ways. The results obtained in our laboratory and by Larsen and Urban indicate that 12 methyl groups/100C are incorporated into this reductively methylated coal (1,2).

Inasmuch as only 12 methyl groups appeared in the product, it was first necessary to establish the fate of the remainder of the potassium. Accordingly, the reduction reaction was carried out in a mixture of tetrahydrofuran and tetrahydrofuran-d_8 and quenched in water. The reduced product contained 2.4 deuterons/100C. The reactions between the coal carbanions and tetrahydrofuran are influenced by a primary kinetic isotope effect. The literature suggests that k(H)/k(D) ranges from 3to 5 (3). Adoption of the intermediate value of 4 implies that 9.6 protons/100C are added to the coal from the solvent under the reaction conditions. Because the original coal contained mineral matter including 0.82% iron pyrite, we believe that another equivadent of potassium is consumed in side reactions with such materials. Thus, the three reactions, reduction and alkylation, reduction and hydrogen transfer from the solvent, and reduction of mineral matter account for about 21 equivalents of potassium/100C within 5% of the amount of potassium consumed in a typical reaction.

The reductively alkylated Illinois No. 6 coal obtained is quite soluble in organic solvents and the deuterium and carbon-13 NMR spectra can be readily recorded. A representative spectrum is shown in Figure 1. The chemical shift information available in the literature for methyl-C-13 and butyl-1-C-13 groups in a broad array of structural environments greatly limit the assignments of the structures of the compounds in regions 1, 2, and 3 of the NMR spectrum of the reductively methylated coal, Table I.

Table I. Structural Assignments Based on Chemical Shift Data

Region 1

Vinyl Ethers

Primary Alkyl Ethers, RCH_2OCH_3

Primary Benzylic Ethers, $ArCH_2OCH_3$

Hindered Aryl Ethers

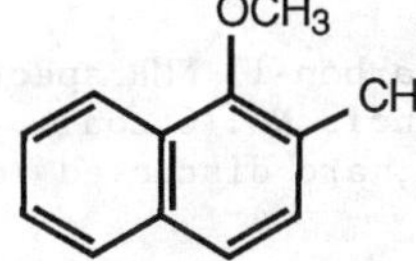

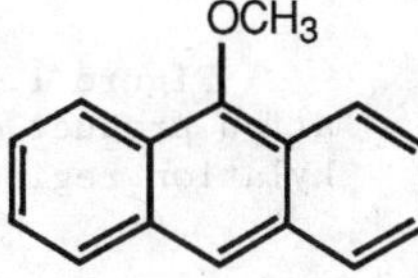

2 alkyl groups — 1 alkyl group, 2 peri interactions — 1 peri interaction

Region 2

Secondary Alkyl Ethers, R_2CHOCH_3

Secondary Benzylic Ethers, $ArCH(R)OCH_3$

Unhindered Aryl Ethers, PhenylOCH_3, NaphthylOCH_3, AnthrylOCH_3

Dihydroxyl Aryl Ethers

Region 3

Tertiary Alkyl Ethers, R_3COCH_3

Aryl and Alkyl Esters, $ArCO_2CH_3$ and RCO_2CH_3

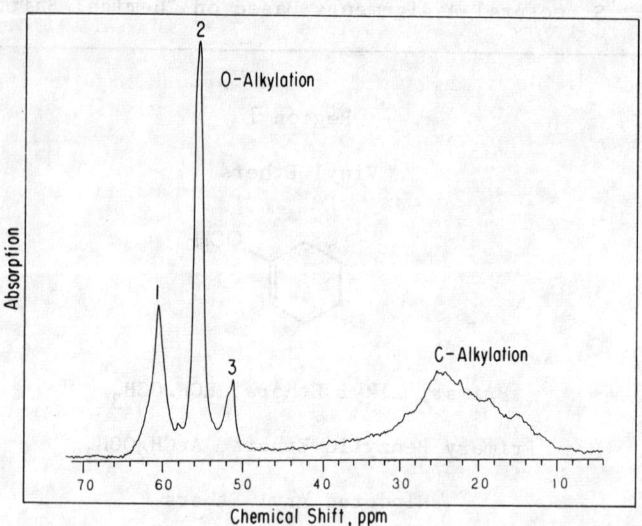

Figure 1. The carbon-13 NMR spectrum of the reductive methylation product of Illinois No. 6 coal. The resonances in the O-alkylation regions, 1-3, are discussed in the text.

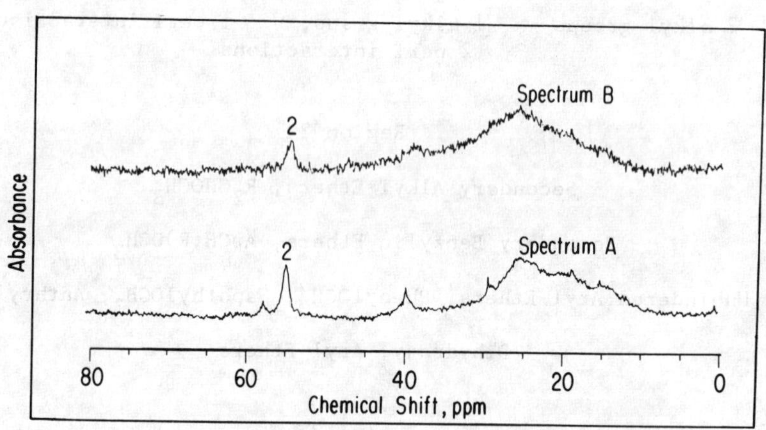

Figure 2. The carbon-13 NMR spectrum of the reductive methylation product shown in Figure 1 following (A) treatment with lithium iodide in 2,4,6-collidine and (B) treatment of the product with boron trifluoride etherate and 1,2-ethanedithiol.

Because the reductive alkylation products are soluble in conventional solvents, it is possible to transform them into other substances and to reexamine the NMR spectrum of the new products to define the nature of the functional groups undergoing reaction. We used mild basic hydrolysis, mild acidic hydrolysis, a selective demethylation reaction with lithium iodide in 2,4,6-collidine (4,5) and an equally selective alkyl methyl ether cleavage reaction with boron trifluoride etherate and a dithiol (6,7) in our work. Representative results are shown in Figure 2 for selective demethylation with lithium iodide and acid-catalyzed aliphatic ether cleavage.

The spectroscopic results obtained in chemical conversion reactions were, for the most part, quite readily interpreted. For example, the disappearance of the resonance in region 3 after basic hydrolysis eliminates tertiary methyl ethers from consideration and the absence of change in region 1 of the spectrum following mild acid hydrolysis negates the presence of methyl enol ethers. The reactions with lithium iodide proved equally straight forward with 96%, 89%, and 100% reductions in the intensities in regions 1,2, and 3, respectively, Figure 2A. The residual intensity in region 2 of the spectrum can best be rationalized on the basis of 1,2-dimethoxyaryl compounds. As shown in equation 3, these compounds un-

$$\text{C}_6\text{H}_4(\text{OCH}_3)_2 \xrightarrow[\text{170°C, 48 hrs, } N_2]{\text{LiI, 2,4,6-collidine}} \xrightarrow[H_2O]{H^+} \text{C}_6\text{H}_4(\text{OH})(\text{OCH}_3) \quad (3)$$

dergo monodemethylation with iodide ion, even when the reagent is in great excess (5a,5b) and also resist acid-catalyzed demethylation.

Chemical conversion of the reductive alkylation product obtained with methyl-d_3 iodide provided additional information. In particular, the spectrum of the reductively methylated coal in 2-bromopyridine at 145°C led to the appearance of a sharp resonance at 3.4 ppm in the deuterium NMR spectrum. Treatment of the coal with boron trifluoride and ethanedithiol eliminated this signal. Hence, the presence of alkyl methyl ethers was confirmed and, equally important, it was established that the deuterons in the alkyl methyl-d_3 ethers were in much more mobile environments than the aryl methyl-d_3 ethers.

All of this information was assembled to provide the oxygen functional group distribution summarized in Table II.

Table II. Normalized Relative Abundances of the Hydroxylic Groups in the Reductive A&kylation Products of Illinois No. 6 Coal as Determined by Chemical Conversions and Carbon and Deuterium NMR Spectroscopy.

Region 1
Primary Alkyl Ethers, 0.1 Methyl/100C Hindered Aryl Ethers, 2.2 Methyl/100C
Region 2
Dihydroxyl Aryl Ethers, 0.6 Methyl/100C Simple Aryl Ethers, 3.3 Methyl/100C
Region 3
Aryl and Alkyl Methyl Carboxylates, 0.8 Methyl/100C

3. DISCUSSION

Chemical and spectroscopic methods have been widely applied for the detection and quantitative determination of the abundances of hydroxyl, carbonyl and etheral groups in coals. The extensive work in this area was reviewed by van Krevelan (8), by Wender and his colleagues (9), by Given (10), and by Attar and Hendrickson (11). To narrow the scope of this discussion, we shall focus attention on studies of the oxygen functional groups in the often studied, high volatile C, bituminous coals from the Illinois No. 6 seam. Studies that have been carried out with the raw coal or its derivatives with selective reagents under mild reaction conditions will receive emphasis. Thus, the many excellent contributions concerning the oxygen functional groups in the products of high temperature liquefaction and gasification reactions will not be discussed because such reactions almost invariably alter the chemical structures drastically.

3.1 Chemical Analyses

It is well recognized that it is very difficult to measure many chemical properties of coal accurately and quantitatively. The determination of the total oxygen content of coal constitutes one of the more difficult challenges for the experimentalist. Recent work by the Kentucky group (12) on the organic oxygen content of a variety of North American coals led them to conclude that the conventional by-difference approach is clearly inadequate. However, they point out that the by-difference method developed by Given and his associates (13-15) yields results that are in reasonable accord with other more accurate measurements as illustrated by the results shown in Table III.

Table III. Organic Oxygen Content of Four Coals, Percentage of Coal.[a]

Sample[b]	ASTM/BD	GIVEN/BD	PC/DMC	FNAA/DMC
548	15.0	13.1	13.8	13.9
680	12.4	9.3	10.6	10.4
854	11.8	11.1	11.1	11.5
866	14.6	12.8	14.2	14.0

[a]Reference 12. The terminology used in the original publication has been adopted ASTM/BD and GIVEN/BD are by-difference methods, PC/DMC is pyrolysis coulometry with demineralized coal and FNAA/DMC is fast neutron activation analysis with demineralized coal.

[b]The samples were obtained from the Pennsylvania State University, Department of Energy Sample and Data Bank.

Clearly, the organic oxygen contents determined by pyrolysis-coulometry and fast-neutron activation analysis on demineralized coals are in good agreement with the values obtained by the approach used by Given and his associates (13-15). While it is well recognized that a fully unambiguous measurement of the organic oxygen content may not be realized, it is evident that suitable methods are available for the determination of organic oxygen and that the results can be discussed with some confidence.

The compositions of several samples of Illinois No. 6 coal including the pristine premium sample, which is available from the Argonne National Laboratory are summarized in Table IV.

Table IV. The Elemental Analyses of Several Illinois No. 6 Coals.

	Argonne National Laboratory	Monterey Mine	Peabody Mine	Burning Star Mine
Weight Percent, daf				
C	77.8	78	76.5	76
H	5.7	5.5	5.5	4.8
N	1.4	1.4	1.3	1.3
S	2.6	3.4	4.8	3.9
O	9.1[a]	11.5[b]	13.2[a]	13.5[c]
Atomic Ratios				
C	100	100	100	100
H	87.3	84.5	85.3	74.3
N	1.5	1.6	1.5	1.5
S	1.3	1.6	2.3	1.9
O	8.8	11.0	13.0	13.2

[a] The organic oxygen content was determined by-difference, ASTM/BD.

[b] The organic oxygen content has deduced using fast neutron activation analysis, FNAA/DMC.

[c] The organic oxygen content was determined by chemical analysis of the functional groups present in the sample.

A broad range of chemical methods have been used for the detection and determination of the oxygen functional group distributions in bituminous coals. Zhou and his coworkers at Oklahoma State University (16) have recently discussed this work and have described the appropriate empirical relationships that can be used to relate to the hydroxyl content of the coal to its rank. In brief, there is little doubt that between 50 and 60% of the organic oxygen atoms in Illinois No. 6 coal occur in hydroxyl groups. This feature is secured by titration (17) and methylation (18) data as well as conventional silylation and acetylation reactions (16,19). Chemical methods for the detection of carbonyl groups in neutral molecules suggest that these entities are much less abundant than the hydroxyl group and occur with a frequency no greater than

1/100C (16,19) in this Illinois coal. The carboxylic acid concentration determined by chemical analyses is also low, no greater than 1/100C (16,19).

3.2 Nuclear Magnetic Resonance Spectroscopy

The solid state carbon-13 NMR spectrum of many bituminous coals can readily be apportioned into several distinct regions of which the aromatic and aliphatic areas are most distinct. NMR spectroscopists often assign the signal intensity downfield of 162 ppm to carbonyl carbon atoms (20,21). A surprisingly intense signal, which constitutes 5% of the total signal strength, is observed in this region for the Illinois No. 6 coal. Thus, there appears to be a serious discrepancy between the results of the chemical analyses discussed in the previous section and the data obtained by NMR spectroscopy. However, the literal interpretation of the NMR spectrum cannot be accepted without reservation for three reasons. First, the signal intensity in conventional solid state carbon-13 NMR spectrum underestimates the actual carbon content by about 40%. In this situation, it is not prudent to rely upon NMR data to determine the relative abundance of a particular functional group. Second, as Maciel and his coworkers (21) have pointed out, the line widths of the carbon-13 resonances in the lower rank coals such as high volatile bituminous C coal are quite large. Consequently, the normal resonances of the ipso carbon atoms in hydroxy- and methoxyaromatic compounds may be broadened into the downfield region. Third, the family of model compounds used to define the spectral regions may be incomplete. In point of fact, compilations of data indicate that the ipso carbon atoms in some oxygen-rich ethers and phenols do appear below 160 ppm (22). These considerations suggest that the chemical analyses provide a more reliable estimate of the quantity of carbonyl groups in the Illinois No. 6 coal. It is evident, however, that forthcoming advances in the resolution and sensitivity of solid state NMR spectrometers will eventually provide considerable insight concerning the nature of the carbonyl groups.

Problems in the resolution of spectral transition in solid state carbon-13 NMR spectroscopy also limit the application of the method for the definition of the relative abundances of aliphatic carbon atoms bonded to etheral oxygen atoms. Careful examination of the NMR spectrum of Illinois No. 6 coal indicates that the abundance of such nuclei must be quite small (20). The contrast between the results for lignites and bituminous coals from Turkey and America is striking. The absorptions for carbon atoms in methoxy groups and alkoxy groups which can be readily observed in the lignites, are absent from the spectra of bituminous coals (23-25). Specifically, Erbatur and coworkers investigated a a series of Turkish coals and their extracts (24). They found that the resonances of the carbon atoms of methoxyl groups and the secondary carbon atoms of other etheral groups could be observed in certain low rank lignites with 59.4 or 71.0% C but that these were absent from bituminous coals with greater than 82% C (daf). These results provide good support for the view that methoxyl groups and other alkyl aryl ethers are present in very low abundance in the Illinois coals.

Erbatur and coworkers also measured the integrated intensity in the 165-210 ppm region of the C NMR spectra of the Turkish lignite and bituminous coals (24). They concluded that the carbonyl carbon atom content was below the detection limit in the bituminous coals. These results also are in accord with the data obtained for the Illinois coals.

3.3 Infrared Spectroscopy

Infrared spectroscopy has been used for many years for the study of coals and their derivatives, but the development of Fourier transform infrared spectroscopy coupled with the techniques of diffuse reflectance and photoacoustic spectroscopy and software tactics to deconvolute the spectra have considerably increased the scope of the method and, as a consequence, many new contributions have appeared during the past few years.

Solomon and Carangelo proposed that the intensity of the absorption at 3450 cm(-1) in the infrared spectra of coals could be used for the quantitative measurement of the hydroxyl group concentration (26). They based their conclusion on the relationship between the band intensity observed in the infrared spectrum and the hydroxyl group content measured by classical acetylation procedures. They, as all the other workers in this field, point out that the samples must be prepared with special care.

Painter and his group at Pennsylvania State University have also exploited Fourier transform infrared spectroscopy for the study of the oxygen functional groups in coals (27). They considered the use of the oxygen-hydrogen stretching frequencies for this purpose but concluded that coal derivatives provided more suitable substrates for study and, accordingly, recorded the difference spectra of acetylated and unacetylated coals. The absorptions in the carbonyl region consisted of a number of overlapping bands which could be resolved by deconvolution. The spectral lines were assigned to acetylated phenols, alcohols, and amines. Based on the average extinction coefficients observed for pure acetyl derivatives of simple hydroxy and amino compounds, Painter and his group concluded that the acetyl groups added to bituminous coals with 78-80% C(daf) were distributed among amine (0.4%), aliphatic hydroxyl (1.3%), and phenolic hydroxyl (4.0%) groups.

Chu and her coworkers prepared O-methyl-d_3 Illinois No. 6 coal and examined its infrared spectrum (27). Although the absorptions are broad and complex as had been predicted by Painter and his colleagues (27), the Rice group was able to analyze the spectra using the carbon-deuterium stretching absorptions in the 2100-2300 cm(-1) region as well as other deformation and stretching modes. It was found that there were three distinct frequencies at 2217, 2251, and 2188 cm(-1) and that the Illinois No. 6 coal exhibited its most intense absorption at 2217 cm(-1) with only very modest intensity at 2188 cm(-1). The authors concluded on the basis of the results for other coals, that the intensity of these bands depended upon the aromatic character of the coal and attributed the band at 2217 cm(-1) to aryl methyl-d_3 ether groups and the band at 2188 cm(-1) to alkyl methyl-d_3 groups. Data for simple methyl esters, methyl alkyl ethers and methyl aryl ethers support these assignments. In addition, this research group was able to provide further evidence for the assignment by correlating the changes in absorption intensity with chemical reactivity.

Rather sophisticated self deconvolution techniques are usually required to detect the carboxyl carbonyl absorptions in most whole bituminous coals (29). Stuart recently pointed out that the acid-washing of certain bituminous and other lower ranking coals increased the intensity of this absorption (30). Inasmuch as the increase in the carbonyl absorption correlated with an increase in the acid content of the sample, he concluded that the raw materials probably contained carboxylate salts which exhibited their principal infrared absorptions at much lower frequencies than the free carboxylic acids. Thus, measurement of the absorption intensity in the customary carboxylic acid region of the spectrum probably underestimates the actual concentration these substances in most coals.

New advances in instrumentation and in data analyses will certainly contribute more structural information. But, it must be recognized that coal is a very complex mixture of diverse materials and that the extinction coefficients of the infrared active absorptions, for example, the carbon-oxygen double bond stretching frequency of carboxylic acids, depend upon the structures of the aliphatic, aromatic, or heterocyclic units to which the functional group is bonded. As a consequence, infrared spectroscopy, although a very powerful qualitative method, has an inherent limitation that prevents its exploitation for highly accurate quantitative measurements of the composition of complex mixtures such as the ones that exist in bituminous coals, their extracts, and their reactive products.

3.4 Information from Reductive Alkylation

The chemical and spectroscopic data discussed in the previous paragraphs implies that the 12 oxygen atoms in a representative

Illinois No. 6 coal are distributed among 5 or 6 aryl hydroxyl groups, 1 carboxylic group, and 5 or 6 etheral groups with lower abundances for other carbonyl groups and aliphatic hydroxyl groups. The analyses of the reductive methylation products presented in Table II provide more information about the structures of these materials.

First, the reductively alkylated coal contains 0.8 carboxyl groups/100C. This result is comparable with previous estimates based upon chemical methods and upon determinations of the methyl esters in coal extracts. The facility with which these esters may be observed in the carbon NMR spectra of alkylated coals contrasts sharply with the complexities encountered in other chemical or spectroscopic methods. Two factors may contribute to the difference. The acids may be present as salts in the original coal (30), or alternatively, the acids may be entrapped and inaccessible deep within the crosslinked macromolecular coal structure (31-33).

Second, aliphatic methyl ethers are present in very small quantities, 0.1 primary alkyl ethers/100C, with no secondary or tertiary alkyl methyl ethers, or primary benzylic methyl ethers. The results suggest that there are very few aliphatic alcohols in this coal. Unfortunately, little can be said about the occurrance of dialkyl ethers which resist reduction or of aryl benzyl ethers which readily cleave to produce phenols.

Third, aryl methyl ethers are present in significant abundance. They are distributed among conventional aryl methyl ethers, 3.3/100C, hindered aryl methyl ethers, 2.2/100C, and dimethoxy aryl compounds, 0.6/100C. The reduction reaction of Illinois No. 6 coal produces 2 new hydroxyl groups/100C. A comparison of the carbon NMR spectra of soluble methylated extracts with the spectra of the reductive methylation products suggests that the major difference in the functional group distribution of the coal extract and the reduction product arises as a consequence of the cleavage of ethers to produce hindered phenols of the type shown in Table II.

Given and his associates proposed several years ago that benzenediols and triols were chemical fossils. The appearance of such compounds in this bituminous coal is compatible with observations made by Bimer and Hayatsu and their associates (34,35). They showed that mono-, di- and trihydroxybenzoic acids were produced in selective oxidation reactions of bituminous coals (34,35). Our results are in accord with their findings and support the view that lignin is the most probable precursor of the macromolecular component of the vitrinite of bituminous coal.

The distribution of hydroxyl and carboxyl groups in the coal and in the reductive alkylation product are summarized in Table IV.

Table IV. Distribution of Functional Groups in the Starting Material and in the Product.

Functional Group	Groups/100C Extract Original Coal	Groups/100C Reductive Alkylation Product
Carboxylic Acids	0.6	0.8
Hindered Phenols	0.7	2.2
Simple Phenols	2.8	3.3
Dihydroxy Compound	0.4	0.6
Ethers	6.0	3.0

Three etheral linkages did not cleave during the reduction reaction. These results are in accord with previous work by Wachowska and Pawlak (36) and Carson and Ignasiak (37). Wachowska and Pawlak (36) suggested that the unreacted ethers might be heterocyclic compounds and Carson and Ignasiak (37) emphasized the deactivating influence of oxyanions on the rates of ether cleavage reactions of diaryl ethers such as the anion formed from 2-hydroxydiphenyl ether.

Several of the ethers that are likely to be present in bituminous Illinois No. 6 coal will not undergo reductive alkylation reactions. Among this group are benzofurans, xanthenes, hydroxyaryl ethers.

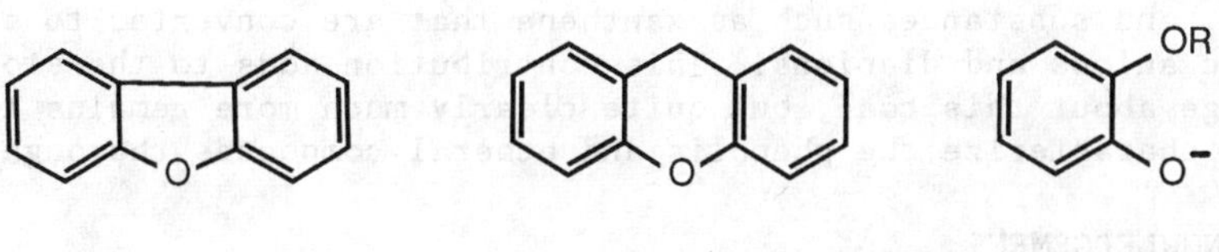

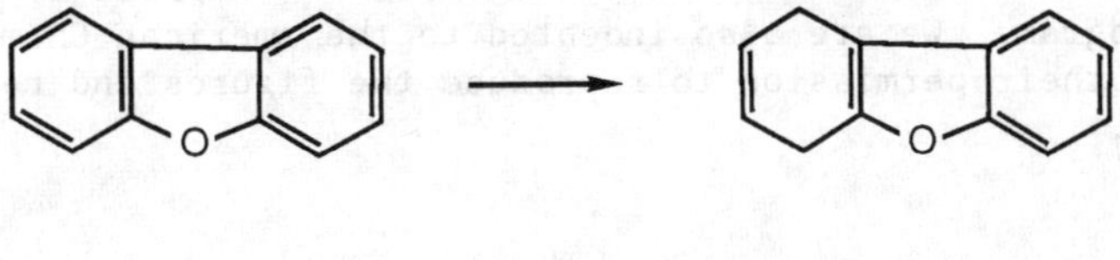

Dibenzofurans often undergo reduction rather than cleavage (38). Xanthanes and hydroxyaryl ethers form relatively stable anions, which

do not cleave, under the conditions used for the reduction of coal with potassium naphthylene(-1) (39). The reduction potentials for the dialkyl ethers are so great that these compounds do not cleave (40).

These suggestions find support in other results for Illinois No. 6 coals including detailed analyses of titration curves. Dutta and Holland point out that certain of the phenolic groups in this coal exhibit rather low pK values and note that such observations are compatible with the occurrence of 1,2-dihydroxy compounds (41). In addition, Winans and Hayatsu and their associates found that the vigorous oxidation of an Illinois bituminous coal provided xanthonecarboxylic acids (42). These concepts are also compatible with the presence of dibenzofurans and related oxygen heterocycles that selectively undergo reduction. Specifically, the Argonne group has demonstrated that the oxidation of bituminous coals from the Illinois Basin produce six different dibenzo- and naphthobenzofurancarboxylic acids in readily detectable amounts (42).

4. CONCLUSION

The solubilization of Illinois No. 6 coal by reductive alkylation under very mild reaction conditions has enabled us to record relatively high resolution carbon NMR spectra of the O-methylation products. The finding that this coal contains about 5 to 6 aryl hydroxyl groups, 5 or 6 ethers, and 1 carboxyl group is in accord with other qualitative information obtained by other independent procedures. In addition, however, more thorough chemical and spectroscopic studies have established that this coal contains readily detectable amounts of hindered phenolic compounds and dihydroxy aromatic compounds as well as the expected simple phenols. As Ignasiak and his coworkers have pointed out previously, many bituminous coals contain ethers that are not readily cleaved during chemical conversion reactions (37). Analysis of the available data suggests that the uncleaved ethers are heterocyclic compounds such as dibenzofuran, hydroxyaromatic compounds that readily form stable anions under the reaction conditions, and substances such as xanthene that are converted to stable aromatic anions and dianions. This contribution adds to the store of knowledge about this coal, but quite clearly much more remains to be done to characterize the phenolic and etheral compounds thoroughly.

5. ACKNOWLEDGEMENT

We are indebted to the United States Department of Energy, to the Center for Research on Sulfur in Coal of the Illinois Coal Board, and to the Gas Research Institute for the support of our research program. We are also indebted to the American Chemical Society for their permission to reproduce the figures and related information.

6. REFERENCES

(1) Stock, L.M.; Willis, R.S. J. Org. Chem. 1985, 50, 3566.

(2) Larsen, J.W.; Urban, L.O. J. Org. Chem. 1979, 44, 3219.

(3) Zielinski, M. "Supplement E, The Chemistry of Ethers, Crown Ethers, Hydroxyl Groups, and Their Sulfur Analogues"; Patai, S. Ed.; J. Wiley and Sons: New York, 1980, Chapter 10.

(4) McMurray, J.E. Org. React. 1976, 24, 187.

(5) Several recent applications are presented in the following articles: (a) Snyder, C.D.; Rapaport, H. J. Am. Chem. Soc. 1974, 96, 8046. (b) Welch, S.C.; Prakasa Rao, A.S.C. J. Org. Chem. 1978, 43, 1957. (c) McCarthy, J.; Moore, J.; Cregge, R. Tetrahedron Lett. 1978, 52, 5183. (d) Kitamura, T.; Imagaura, T.; Kiauinisi, M. Tetrahedron 1978, 34. (e) Colombo, L.; Gennari, C.; Potenza, D.; Scolastico, C.; Aragoczini, F. J. Chem. Soc., Chem. Commun. 1979, 1021. (f) Kende, A.S.; Rizzi, J.P. J. Am. Chem. Soc. 1981, 103, 4247.

(6) Node, M.; Hori, H.; Fujita, E. J. Chem. Soc., Perkin Trans. 1, 1976, 2237.

(7) Fiji, K.; Ichikawa, K.; Node, M.; Fujita, E. J. Org. Chem. 1979, 44, 1661.

(8) Van Krevelen, D.W. "Coal", Elsevier, Inc.: New York, 1961 Chapter IX.

(9) Wender, I., Heredy, L.A., Neuworth, M.B., Dryden, I.G.C. "Chemistry of Coal Utilization, Second Supplementary Volume, Elliott, M.A., Editor, J. Wiley and Sons, Inc., New York, 1980, pp. 473.

(10) Given, P.H. Coal Sci. 1984, 3, 65.

(11) Attar, A. and Hendrickson, G.G. "Coal Structure", Meyers, R.A., Editor, Academic Press, Inc.: New York, 1982, Chapter 5.

(12) Ehmann, W.D.; Koppenael, D.W.; Hamrin, C.E., Jones, W.D.; Prasad, M.N.; Tian, W.-Z. Fuel 1986, 65, 1563.

(13) Given, P.H. Fuel, 1976 55, 256.

(14) Given, P.H.; Spackman, W. Fuel, 1978, 57, 319.

(15) Given, P.H.; Yarzab, R.F. "Analytical Methods for Coal and Coal Products"; Karr, C., Ed.: Academic Press, Inc.: New York, 1978, p. 3.

(16) Zhou, P.; Dermer, O.C.; Crynes, B.L. Coal Sci. 1984, 3, 254.

(17) Dutta, P.K.; Holland, R.J. Fuel 1983, 62, 732.

(18) (a) Liotta, R.; Rose, K.; Hippo E. J. Org. Chem. 1981, 46, 277. (b) Gethner, J.S. Fuel 1982, 61, 1273.

(19) Ruberto, R.G.; Cronauer, D.C. "Organic Chemistry of Coal" Am. Chem. Soc. Symposium Series 71, Am. Chem. Soc.: Washington, D.C., 1978, p. 50.

(20) Alemany, L.B.; Grant, D.M.; Pugmire, R.J.; Stock, L.M. Fuel 1984, 63, 513.

(21) Miknis, F.P.; Sullivan, M.; Bartuska, V.J.; Maciel, G.V. Org. Geo. Chem. 1981, 3, 19.

(22) The resonances of aromatic carbon atoms substituted with hydroxy or methoxy groups in hydroaromatic alkaloids and polycyclic heterocycles appear at such low fields. Crabb, T.A. Annu. Rep. NMR Spectrosc. 1982, 13, 60.

(23) Farnum, S.A.; Messick, D.D.; Farnum, B.W. Amer. Chem. Soc., Fuel Chem. Div. Preprints, 1986, 31(1), 60.

(24) Erbatur, G., Ebatur, O.; Davis, M.F.; Maciel, G.E. Fuel 1986, 65, 1265.

(25) Erbatur, G.; Erbatur, O.; Coban, A.; Davis, M.F.; Maciel, G.E. Fuel 1986, 65, 1273.

(26) Solomon, P.R.; Carangelo, R.M. Fuel 1982, 61, 663.

(27) Snyder, R.W.; Painter, P.C.; Havens, J.R.; Koenig, Appl. Spect. 1983, 37, 497.

(28) Chu, C.J.; Cannon, S.A.; Hauge, R.H.; Margrave, J.L. Fuel 1986, 65, 1740.

(29) Chen, P.; Yang, P.W.J.; Griffiths, P.R. Fuel 1985, 64, 307.

(30) Stuart, A.D. Fuel 1986, 65, 1003.

(31) Alemany, L.B.; King, S.R.; Stock, L.M. Fuel 1978, 57, 738.

(32) Youtcheff, J.S.; Given, P.H.; Baset, Z.; Sundaram, M.S. Org. Geochem. 1983, 5, 157.

(33) Larsen, J.W.; Kovac, J. "Organic Chemistry of Coal:, Larsen, J.W., Ed.; American Chemical Society: Washington, D.C., 1978; Chapter 2 and subsequent articles in this series.

(34) Bimer, J.; Given, P.H.; Raj, S. Am. Chem. Soc. Symposium Series, 1978, 71, 86.

(35) Hayatsu, R.; Winans, R.E.; McBeth, R.L.; Scott, R.G.; Moore, L.P.; Studier, M.H. Am. Chem. Soc., Div. Fuel Chem., 1979, 24(1), 110.

(36) Wachowska, H.M.; Pawlak, W. Fuel 1977, 56, 422.

(37) Carson, D.W.; Ignasiak, B.S. Fuel 1980, 59, 757.

(38) Gilman, H.; Bradley, J. J. Am. Chem. Soc. 1938, 60, 2554.

(39) Smith, H. "Organic Reactions in Liquid Ammonia, Chemistry in Nonaqueous Ionizing Solvents", J. Wiley and Sons: New York, 1963; Volume I, Part 2.

(40) Jolly, W.L. J. Chem. Ed. 1956, 33, 512.

(41) Dutta, P.K.; Holland, R.J. Fuel 1983, 62, 732.

(42) Hayatsu, R.; Scott, R.G.; Winans, R.E. "Oxidation in Organic Chemistry, Part D"; Trahanovsky, W.D., Ed.; Academic Press, Inc.: New York, 1982; Chapter. 4.

SLOW AND RAPID PYROLYSIS OF COAL

Peter J.J.Tromp and Jacob Moulijn
University of Amsterdam
Institute of Chemical Technology
Nieuwe Achtergracht 166
1018 WV Amsterdam
The Netherlands

ABSTRACT. Coal is a heterogeneous, mainly organic, material which decomposes upon exposure to high temperatures. This process is called pyrolysis. Coal pyrolysis is the most important aspect of coal behaviour because it occurs in all major coal conversion processes. Upon pyrolysis coal is divided into a hydrogen-rich volatile fraction, consisting of gases, vapors, and tar-components, and a carbon-rich solid residue. Moreover, for certain coals decomposition can result in a temporary softening of the solid material. The pyrolysis process consists of a very complex set of reactions involving the formation of radicals. The overall decomposition of coal is open to manipulation by a variety of experimental conditions. Due to the nature of the coal conversion processes distinction should be made between slow (heating rate <1 K/s) and rapid (heating rate >1000 K/s) pyrolysis of coal. In this chapter the experimental techniques used to study the chemical an the physical changes which occur upon slow and rapid pyrolysis of coal are reviewed. Relevant experimental results are described. The influence of experimental conditions on the yield and composition of the volatile fraction, liberated upon coal pyrolysis, is considered. Some models, which describe the overall process of coal pyrolysis, are given. Finally, this chapter is concluded with a study on the thermal stability of aromatic model compounds, relevant in coal pyrolysis, to elucidate the nature and to describe the extent of primary and secondary reactions.

1. INTRODUCTION

Coal is sedimentary rock molded during a period of millions of years. It is composed of an inorganic part (minerals) and an organic part (macerals). The macerals are divided into three main groups: vitrinite, exinite and inertinite. The macerals differ in chemical composition and have distinct structural shapes, visible under an optical

Y. Yürüm (ed.), New Trends in Coal Science, 305–338.

microscope, which give an indication of the origin of the coal (1, 2).

Coal is considered to be derived from plant material which, after decease and decay by bacteria, was deposited in swamps. The heterogeneous plant material was subjected to heating and was pressurized by the weight of subsequent sediments over millions of years. As a result of these geological conditions coalification of plant material occured. More severe conditions and increasing time led to increasing coal rank. With increasing rank the various coals can be arranged in the following rank order:

peat
lignite
subbituminous coal
bituminous coal
anthracite.

Some of the rank classes are divided into subgroups. This subdivision is based upon the volatile matter content and the calorific value of the coal samples. Table I gives a classification of coals according to this rank order (3).

Table I. Classification of coals [after Neavel (3)]

ASTM Classification				International Classification	Scientific		Properties of Vitrinite				
Class	Group	Volatile Matter DMMF*	Btu/lb Moist MMF**		Stages	Rank	Elemental Carbon DAF	Volatile Matter DAF	Moisture (Saturated) wt%	MJ/kg MAF	Reflectance of Vitrinite % R_{ave}
					Peat						
~~ Undefined ~~							60		75		
Lignitic	B				Soft Brown Coal						
			6300								
	A							ca 53	35	16.8	ca 0.3
			8300								
Sub-Bituminous	C				Hard Brown Coal	Dull					
			9500								
	B			9			ca 71	ca 49	25	23	
			10500								
	A										
			11500*								
Bituminous	High Volatile C			8		Bright					
							ca 77	ca 42	8–10	29.4	ca 0.5
			13000								
	High Volatile B			7	Hard Coal (Agglomerating, Caking Coals)	Low Rank					
			14000								
	High Volatile A	>31	>14000	6							
		31		5							
	Medium Volatile			4			87	29		36.3	1.1
		22									
	Low Volatile			3		High Rank					
		14		2							
Anthractic	Semi-Anthracite										
		8					91	8		36.3	2.5
	Anthracite			1	Anthracite						
				0							

• Dry, mineral-free basis, weight percentage
•• Mineral-free, but with saturated pore moisture content
⁘ Dry, ash-free basis, weight percentage
⁘ Moist, ash-free
* Bituminous if agglomerating, sub-bituminous if non-agglomerating

A lucid introduction in coal science has been written by van Krevelen (4); his book is a must for coal scientists.

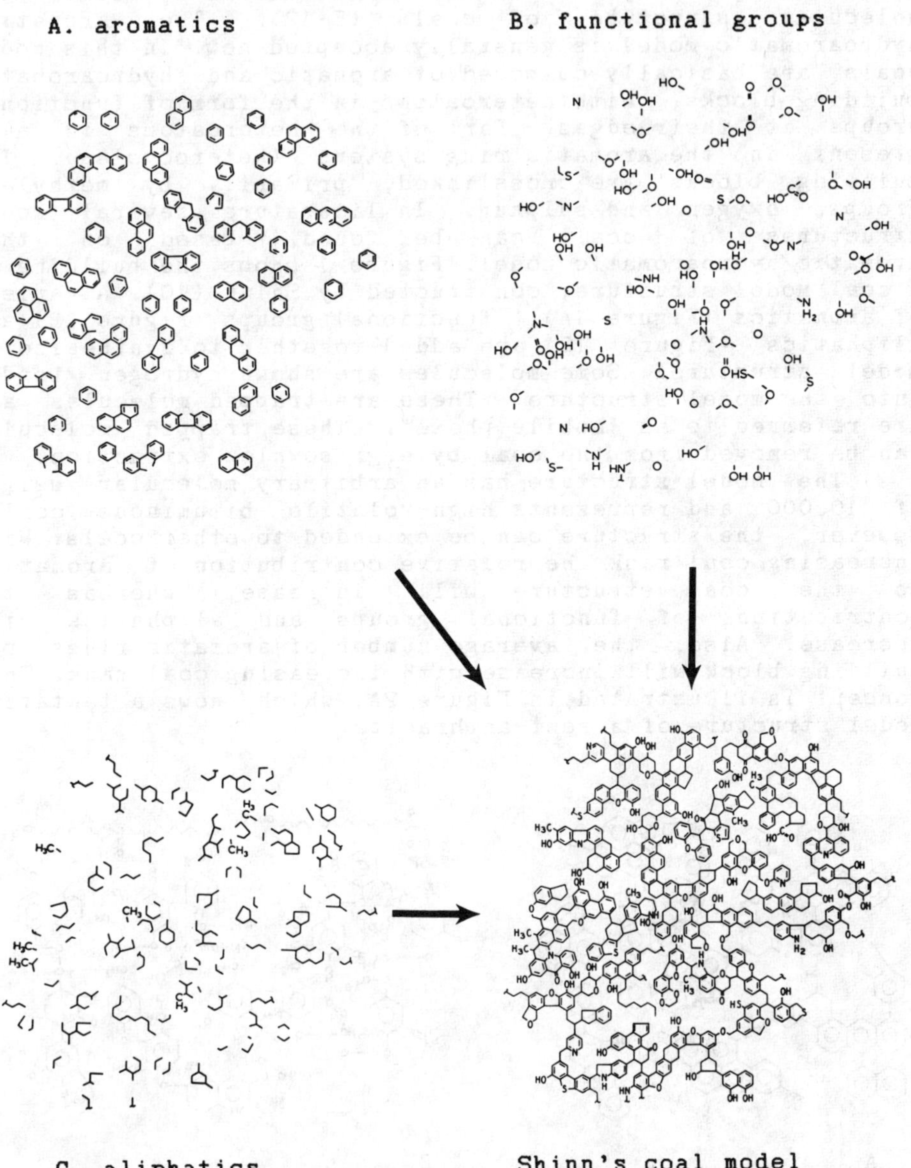

Figure 1. Construction of a coal model structure for a bituminous coal from a variety of aromatics (A), functional groups (B), and aliphatics (C) [after Shinn (10)].

1.1 Molecular Structure of Coal

In spite of the heterogeneity of coal and the large variety of coal types many efforts have been made to elucidate the molecular structure of coal (5-12). The aromatic/hydroaromatic model is generally accepted now. In this model coals are basically composed of aromatic and hydroaromatic building blocks, with heteroatoms in the form of functional groups at their edges. Part of the heteroatoms is also present in the aromatic ring systems (heterocycles). The building blocks are crosslinked, primarily by methylene groups, oxygen and sulphur. In literature several model structures of coal can be found, based on this aromatic/hydroaromatic model. Figure 1 shows the building of a coal model structure, constructed by Shinn (10). A variety of aromatics (Figure 1A), functional groups (Figure 1B) and aliphatics (Figure 1C) are added together to synthesize a model structure. Some molecules are shown hydrogen bonded into the model structure. These are trapped molecules and are referred to as "mobile phase". These trapped molecules can be removed from the coal by e.g. soxhlet extraction.

The model structure has an arbitrary molecular weight of 10,000 and represents high-volatile bituminous coals. However, the structure can be extended to other coals. With increasing coal rank the relative contribution of aromatics to the coal structure will increase, whereas the contribution of functional groups and aliphatics will decrease. Also, the average number of aromatic rings per building block will increase with increasing coal rank. This concept is illustrated in Figure 2A, which shows a tentative model structure of a semi-anthracite.

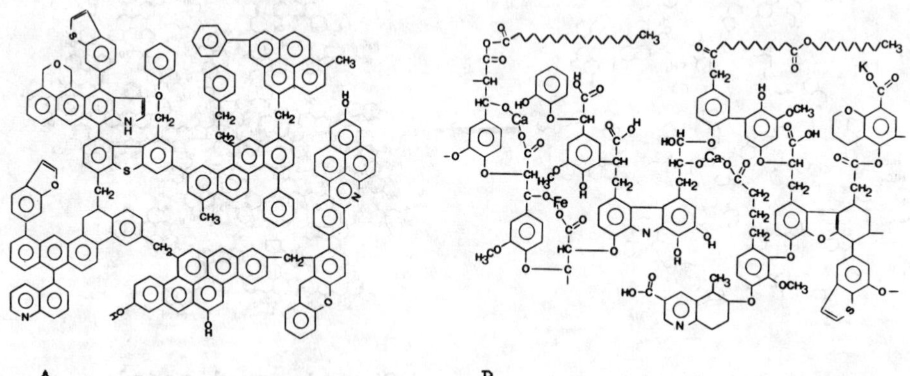

Figure 2. Possible structures of a semi-anthracite (A) and of a lignite (B). For convenience, the benzoic character of a ring system is indicated by a circle.

In Figure 2B a possible structure of a lignite is given. This structure is derived from a model structure of a guaiacyl lignin. Lignins originate from the cell walls of higher plants and are important precursors of coal. From the elementary composition of a lignite and results from analytical techniques, like Fourier Transform InfraRed spectroscopy (determination of functional groups) and Curie-point pyrolysis techniques (product distribution upon pyrolysis), the model structure of a guaiacyl lignin is adapted to form the structure as shown in Figure 2B.

2. PYROLYSIS OF COAL

The name "pyrolysis" originates from the old Greek language: pyros means fire and lysis means loose, untie. So, pyrolysis means decomposition due to exposure to high temperature.

Coal pyrolysis is the most important aspect of coal behaviour because it occurs in all major coal conversion processes, like combustion, gasification, carbonization and liquefaction.

When heated to elevated temperatures in an inert atmosphere coal decomposes and is divided into two fractions:

1. A hydrogen-rich volatile fraction. This fraction consists of gases, vapors and tar-components.
2. A carbon-rich solid residue, called char.

The overall process can be depicted as follows:

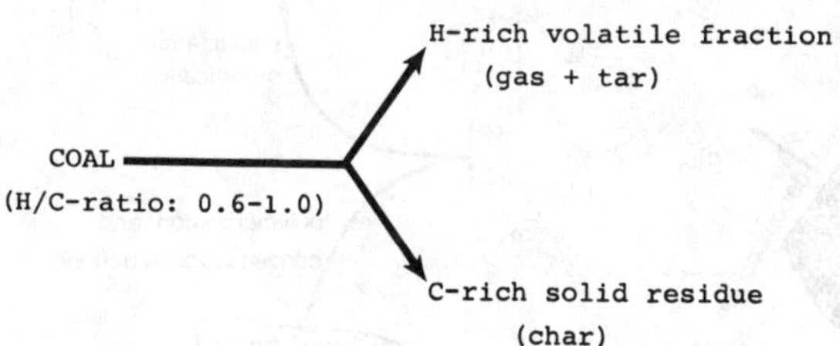

The reactions, which occur upon pyrolysis of coal, are illustrated in more detail in Figure 3. The coal structure will decompose by the breaking of the least stable bonds within the structure. The least stable bonds are the methylene-, oxygen- and sulphur-bridges between the aromatic building blocks. The result of this bond breaking is the formation of a large number of radical components, of which part is volatile under the experimental conditions. These radicals are very reactive and can undergo secondary reactions like cracking and carbon deposition, both inside and outside the coal particle. Stabilization of a radical, primarily via hydrogen addition, leads to a volatile component. Polymerization and condensation reactions, occuring via recombination of both volatile and non-volatile radical components, result in the formation of the solid char particle. The char is enriched in carbon and more aromatic in comparison with the original coal particle. The volatile products, formed after stabilization of the radical components, may also undergo secondary gas phase reactions (both cracking and condensation reactions) due to extended exposure to high temperatures.

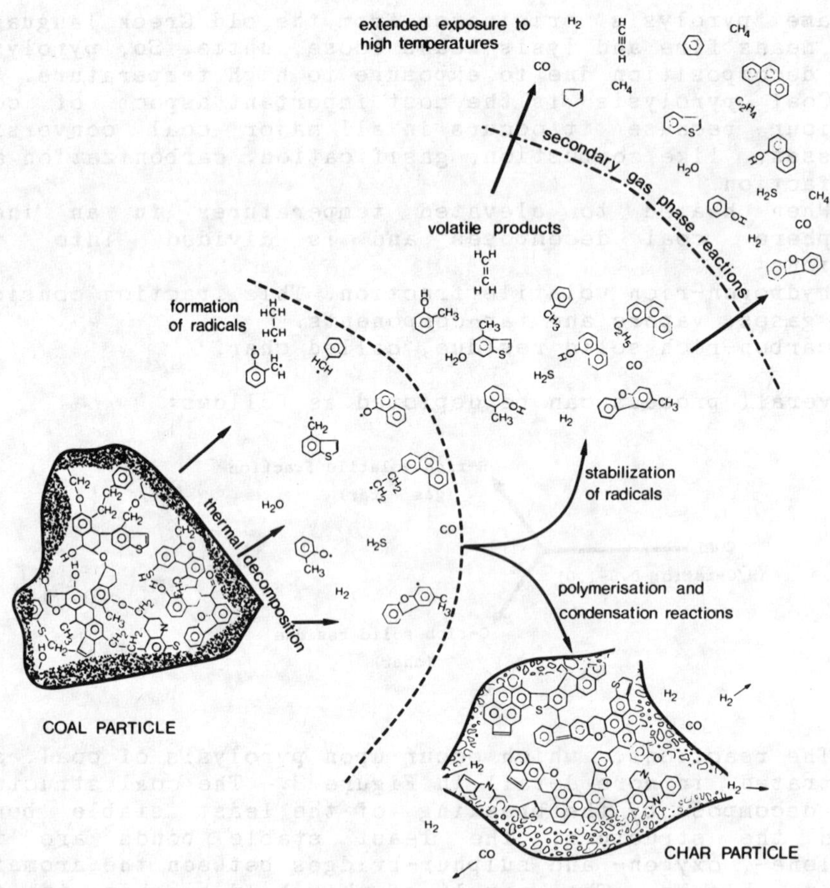

Figure 3. Reactions and processes, which occur upon pyrolysis of a coal particle.

2.1. The Softening of Coal

A coal is characterized as plastic or non-plastic, depending on whether or not it exhibits thermoplastic behaviour, i.e., it becomes viscoelastic upon heating. Generally, high temperature plasticity is observed only in bituminous coals with an ASTM volatile matter content larger than 40%. When a plastic coal is heated to progressively higher temperatures, three consecutive stages appear:

1. softenning
2. swelling
3. resolidification

The temperature at which the phenomena of softening, swelling and resolidification occur, strongly depend on the type of coal and the experimental conditions, especially the heating rate. At low heating rates (<1 K/s) thermal softening is typically observed in the temperature range 600-725 K. With increasing heating rate softening is shifted towards higher temperatures and the temperature range, over which fluidity is observed, becomes wider.

The development of fluidity in coal is still poorly understood. It has been suggested that both the number and the nature of cross-links in the coal is a key factor in the process. A macroscopic view of coal plasticity has been given by Spiro (13) and is illustrated in Figure 4.

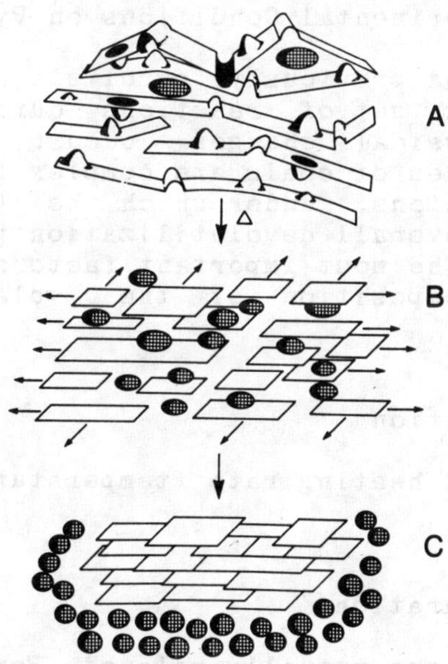

Figure 4. Macroscopic view of coal plasticity [after Spiro (13)]. See text for description of the different stages.

According to this macroscopic view a bituminous coal can be described as a network of macromolecules with trapped molecules in it (Figure 4A). Many authors distinguish a "mobile" and an "immobile" phase in coal. The mobile phase consists of (part of) the trapped molecules and fragments of the network. Upon heating the macromolecules decompose and the aromatic planar units can slide over each other, more or less freely (Figure 4B). This is a liquid state, in which the coal exhibits viscoelastic properties. The molecules, which were initially trapped, are free now and can act as lubricants. Concurrently, thermal decomposition at progressively higher temperatures leads to evolution of gases, vapors and tars. These internally generated volatile components are trapped within the viscous mass and cause swelling. This can be compared with the rise of dough. At some temperature the expanded mass solidifies (Figure 4C), due to the escape of the gaseous molecules, a lack of lubricants, and the occurrence of condensation reactions.

The thermoplastic properties of coal can cause serious operational problems in coal conversion processes like fixed or fluidized bed gasifiers by impeding the gas flow and by decreasing the hold-up. It has been reported that pre-oxidation of the coal at low temperatures (375-475 K) in air and the addition of certain inorganic compounds to three coal can reduce or even destroy the thermoplastic properties of coals (14, 15).

2.2. Effects of Experimental Conditions on Pyrolysis of Coal

As has been discussed previously, pyrolysis of coal consists of a very complex set of reactions during which both chemical and physical changes occur. The observed devolatilization rates of coals are complex functions of the experimental conditions, under which the pyrolysis process proceeds, and the overall devolatilization process is open to manipulation. The most important factors, which affect the yield and composition of the volatile fraction liberated, are:

- coal rank
- maceral composition
- particle size
- temperature and heating rate (temperature-time history)
- atmosphere
- pressure
- sample size
- reactor configuration.

Some of the factors are mutually related. For instance, when the particle size is changed, it must be realized that this also may influence the heating rate of the coal particles.

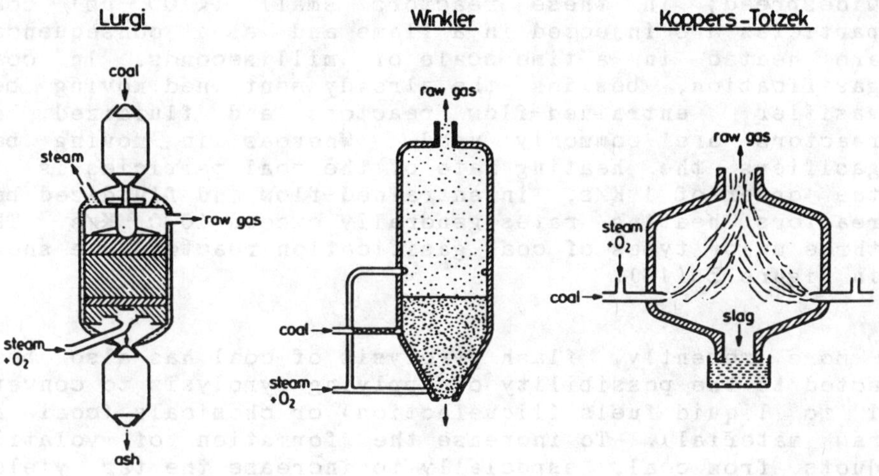

moving bed fluidized bed entrained bed

Figure 5. Three major types of industrial coal gasification reactors [after van Heek (16)]. In a moving bed gasifier various zones are present, in which a packed bed of coal (moving downwards by its own weight) is dried, devolatilized, gasified and finally burned with increasing temperature. The heating rate of the moving bed of coal is in the order of 1 K/s; the residence time of the coal in the reactor is in the order of hours. In a fluidized bed and an entrained flow reactor small coal particles are rapidly (heating rate >1000 K/s) heated in a dilute gas phase. The residence time of the coal particles in these types of reactors is in the order of seconds.

In the literature coal pyrolysis studies generally can be divided into two main areas of research:

1. slow heating (<1 K/s) of dense coal samples. Originally, this research was mainly concentrated on coke making for the steel industry. More recently, it has also been directed to the characterization of coal properties for moving bed gasifiers.
2. rapid heating (>1000 K/s) of dilute coal samples, also termed "flash pyrolysis". The aim of these studies is to create a realistic simulation of coal utilization processes like combustion and certain gasification processes. In combustion, powdered coal combustors are

widespread; in these reactors small (<100 μm) coal particles are injected in a flame and as a consequence, are heated in a time scale of milliseconds. In coal gasification, besides the already mentioned moving bed gasifier, entrained-flow reactors and fluidized bed reactors are commonly used. Whereas in moving bed gasifiers the heating rate of the coal particles is in the order of 1 K/s, in entrained-flow and fluidized bed reactors heating rates generally exceed 1000 K/s. The three major types of coal gasification reactors are shown in Figure 5 (16).

More recently, flash pyrolysis of coal has also been directed to the possibility of applying pyrolysis to convert coal to liquid fuels (liquefaction) or chemicals (coal as a raw material). To increase the formation of volatile products from coal, especially to increase the tar yield, often hydrogen is supplied; this process is referred to as hydropyrolysis.

In this chapter both slow and rapid pyrolysis of coal will be described; chemical, as well as physical changes will be considered. The influence of the factors, mentioned above, on the pyrolysis of coal will also be discussed. However, the paper is limited to the pyrolysis in an inert atmosphere. Other reviews on the pyrolysis of coal have been written by Howard (17), Gavalas (18), Berkowitz (19), Juntgen and van Heek (20), Suuberg (21), and Solomon and Hamblen (22). The paper will be concluded with a study on the thermal stability of aromatic model compounds to clarify the nature and extent of primary and secondary reactions which might occur upon coal pyrolysis.

3. SLOW PYROLYSIS OF COAL

The most commonly used method to study the progress of decomposition of coals at low heating rates is the thermogravimetric technique. In this technique the weight loss that accompanies coal pyrolysis is recorded continuously as a function of the temperature. The gas phase can be analyzed simultaneously by gas chromatography and/or mass spectrometry to identify the pyrolysis products. In Figure 6 the weight loss as a function of the temperature at a heating rate of 0.08 K/s is shown for four different types of coals, ranging from an anthracite to a lignite. From Figure 6 it is clear that with increasing coal rank thermal decomposition starts at increasing temperature.

The overall decomposition process of a certain coal can be divided into three successive stages (19):

1. Little change in weight is observed below a certain temperature T_d , the decomposition temperature of the coal.

2. Above T_d "active decomposition" takes place under evolution of methane, light volatiles and tars. When the coal is rich in chemically bound oxygen (e.g. lignites) significant amounts of carbon dioxide and water are also liberated. In general, active decomposition occurs up to 825 K.

3. At higher temperatures decomposition and condensation reactions occur, involving aromatic structures in the char matrix. These reactions result in the formation of hydrogen and carbon monoxide.

For the plastic coals the interval between initial softening and final resolidification is entirely contained in stage 2.

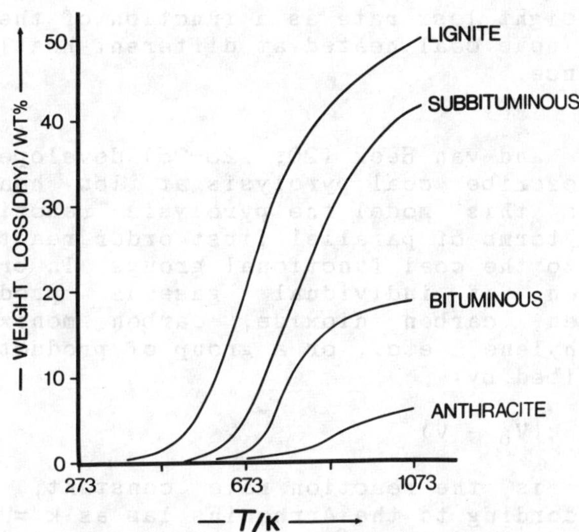

Figure 6. Weight loss as a function of the temperature for four different types of coals, heated at a heating rate of 0.08 K/s in a thermobalance.

With increasing heating rate the decomposition process is shifted towards higher temperatures. At the same time an increase in the maximum devolitilization rate is observed, as is shown in Figure 7 for a bituminous coal.

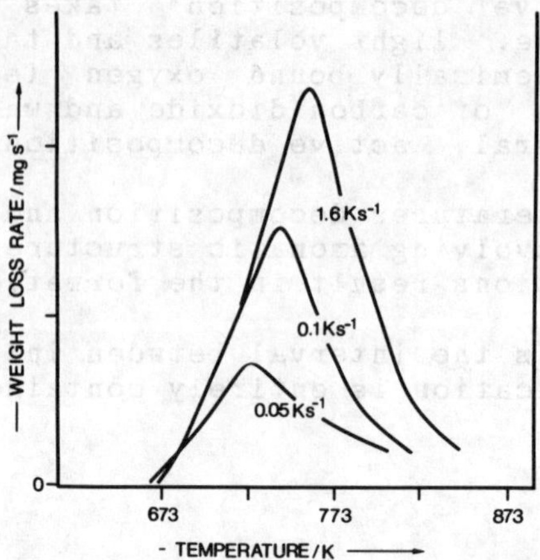

Figure 7. Weight loss rate as a function of the temperature for a bituminous coal heated at different heating rates in a thermobalance.

Juntgen and van Heek (20, 23-25) developed a kinetic model to describe coal pyrolysis at low heating rates. According to this model the pyrolysis reactions can be described in terms of parallel first order reactions, which are related to the coal functional groups. In brief the rate of formation of individual gaseous products, like e.g. hydrogen, carbon dioxide, carbon monoxide, water, methane, ethylene, etc., or a group of products, like tar, can be described by:

$$dV/dt = k(V_o - V)$$

in which k is the reaction rate constant, and can be described according to the Arrhenius law as $k = k_o \exp(-E/RT)$ with k_o= frequency factor (1/s) and E = activation energy (kJ/mole). V and V are the gas volumes (m^3 /g) released per unit weight of sample at time t and infinite time t , respectively. By introducing a constant heating rate, m = dT/dt, the rate of product formation dV/dt is converted to dV/dT. From the relationship of (dV/dT) versus (T) the activation energy and frequency factor can be estimated (24). The activation energies thus calculated from the curves of gaseous C1-C3 hydrocarbons were of the same order of magnitude as the bonding energies of bridge C-C bonds between aromatic ring systems;

C–H bonds

H_3C–H 430

C_2H_5–H 410

C–C bonds

C_6H_5–	545	350	C_6H_5–CH_2–C_6H_5
C_6H_5–C_6H_5	480	350	H_3C–CH_3
C_6H_5–CH_3	380	235	C_6H_5–CH_2–CH_2–C_6H_5
C_6H_5–CH_2–CH_3	365	210	$C_{10}H_7$–CH_2–CH_2–$C_{10}H_7$

Bond energy values of hydrocarbons (kJ/mole) [after Juntgen (24)]

From the above and the observation that the C1-C3 hydrocarbons are formed in the same temperature range as the tar components it was concluded that the coal decomposition is initiated by cracking of the bridge C-C bonds between the aromatic building blocks in the coal structure. The possible mechanism of coal pyrolysis, as suggested by Juntgen, is shown in Figure 8.

According to this mechanism at elevated temperatures there exists an equilibrium between the coal molecule and some activated form of it (stage 1). At a temperature of about 673 K this activated complex is dissociated into aromatic ring units by cracking of the bridge carbons. Hereby, radicals are formed. Recombination of the smaller radicals results in the formation of small aliphatic gas molecules, water and tar components of medium molecular weight. These volatile compounds diffuse from the interior of the coal/char particle into the bulk gas phase (stage 2). Polynuclear systems of high molecular weight are not volatile enough to escape from the solid particle into the gas phase. These compounds form the char fraction via condensation of the ring systems, with elimination of hydrogen. Hydrogen diffuses from the char particle into the bulk gas phase. At high temperatures also carbon monoxide will be produced by cracking of heterocyclic oxygen groups (stage 3). It should be noted that stage 1 to 3 as indicated in Figure 8 are equivalent with three successive stages for the overall decomposition process of a coal, as given by Berkowitz (19).

Also, the effect of the overall external gas pressure on the product formation was studied (23-25). Higher overall pressures resulted in a decrease in the tar formation and the maximum formation rate was shifted towards higher temperatures. The shifting towards higher temperatures was explained by a shift in the boiling range of the tar components with pressure. Pressure also inhibits diffusion of the volatile products out of the coal/char particle. As a consequence, these compounds are increasingly subjected to secondary cracking reactions which reduce the tar yield in favour of enhanced gas and char formation.

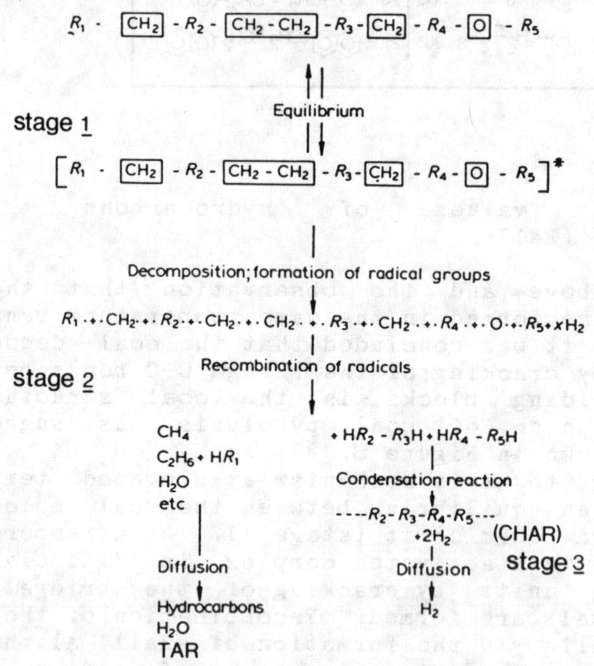

Figure 8. Possible mechanism of pyrolysis of coal [after Juntgen (24)]. ▭, C-bridges, O-bridge atoms; R, aromatic/hydroaromatic compounds with 2-5 nuclei.

On the other hand, tar formation increases with increasing heating rate. At higher heating rate more pyrolysis products are formed per unit time. As a result a larger concentration gradient and a higher internal pressure arise in the coal particle which produce a higher diffusive flow rate of the pyrolysis products out of the coal particle. Hence, a shorter residence time of the volatiles in the solid particle is obtained.

In the same way increasing particle and sample size enhance the possibility of secondary cracking reactions of the pyrolysis products on the surface of the solid particle.

3.1. Swelling and Agglomerating Properties of Coal

The thermoplastic properties, especially the agglomerating and swelling properties of bituminous coals are very important aspects in coke making and coal gasification in moving bed gasifiers. Several standard methods have been developed to determine the swelling (swelling index, dilatometry), the extent of agglomeration (variable torque plastometry), and fluidity (Gieseler plastometry) of coal upon heating at atmospheric pressure (14). However, commercial moving bed coal gasifiers work at elevated pressure (30-40 bar). Little study has been done on the thermoplastic properties of coal at elevated pressures in spite of the fact that pressure has a significant effect on the release of volatiles during pyrolysis of coal and thus on the caking and swelling of bituminous coals. Hence, recently dilatometers have been built which can operate under high pressure (26,27).

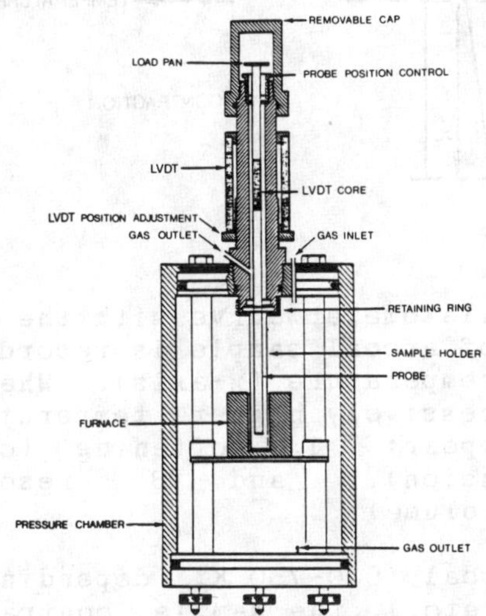

Figure 9. Details of a high pressure microdilatometer (HPMD). See text for a description of the apparatus.

In dilatometry the volume of a coal sample, pressed into a pellet, is measured as a function of the temperature. In Figure 9 details of a high pressure dilatometer are shown. In brief, the essential part of the instrument is the movable linear variable differential transformer (LVDT) core. This LVDT core is part of the dilatometer pencil, which rests on the surface of the coal sample. The coal sample (< 100 mg) in the sample holder is heated by the furnace. Any change in volume of the coal sample during heating will cause a movement of the pencil with the LVDT core and generates an electrical signal, which is proportional to the displacement of the LVDT core. This signal is displayed on the Y-axis of an X-Y recorder. The temperature of the coal sample is recorded on the X-axis. A typical dilatometer curve is shown in Figure 10.

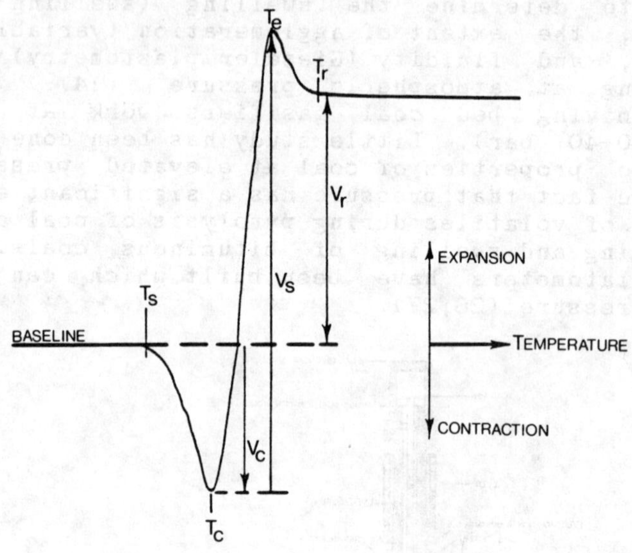

Figure 10. Typical dilatometer curve with the parameters, measured. The volume of a coal sample is recorded (y-axis) as a function of the temperature (x-axis). When a plastic coal is heated to progressively higher temperatures, three consecutive stages appear: 1. softening (contraction), 2. swelling (expansion), and 3. resolidification (contraction to final volume).

Upon softening of the coal (600-750 K, depending on type of coal, heating rate, etc.) the sample contracts because internal pores and void volume between the coal particles are filled up by the liquid coal mass. As described before, the softening stage is followed by a swelling stage which causes an expansion of the coal sample. Finally, after resolidification no volume change occurs. The following parameters are defined by the dilatometer curve (26):

1. Softening temperature (T_s)
2. Contraction temperature (T_c)
3. Maximum swelling temperature (T_e)
4. Resolidification temperature (T_r)
5. Initial contraction volume (V_c)
6. Maximum swelling volume (V_s)
7. Volume change in resolidification (V_r)

The most important parameter is the maximum volume swelling (V_S), expressed as a relative increase in the volume of the coal sample, compared with the original volume. In Figure 11, V_S is shown as a function of the applied helium pressure for different size fractions of a high volatile A bituminous coal; the heating rate is 0.16 K/s.

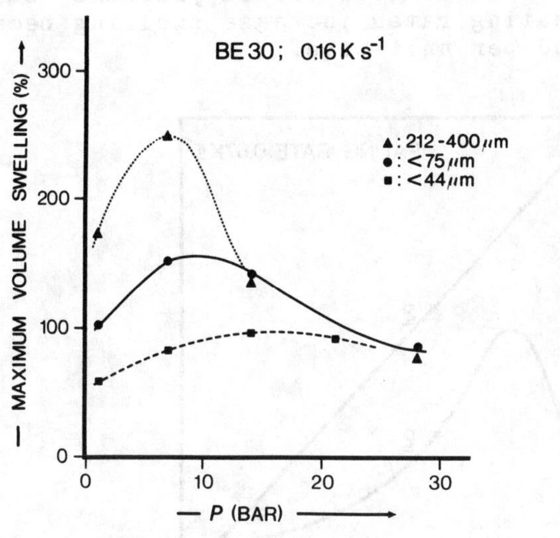

Figure 11. Maximum volume swelling (V_S), measured in a high pressure microdilatometer, as a function of the applied helium pressure for different size fractions of a high volatile A bituminous coal.

At low external pressure the maximum volume swelling increases with increasing particle size of the coal. This is explained by the decreasing surface to volume ratio of larger particles, which increases the residence time of the volatile pyrolysis products in the fluid mass of the individual coal particles. At elevated pressure, pressure itself is the dominating parameter and determines the residence time of the volatiles in the fluid mass, independent of the particle size.

Furthermore, the volume swelling shows a maximum as a function the pressure. According to Thomas (28) three types of dilatation versus pressure curves have been observed for a wide range of coals:

1. Dilatation increases with increase in pressure.
2. Dilatation increases to a maximum at 10-20 bar, followed by a decrease to a limiting value.
3. Dilatation decreases with increase in pressure.

The shapes of the three types of curves are illustrated in Figure 12. With increasing pressure two competing processes occur:

- The volume of volatile components is reduced; this reduces swelling.
- The residence time of the volatiles is increased; this enhances swelling.

The balance between the two competing processes give rise to the three types of dilatation versus pressure curves. In general, higher heating rates increase swelling because more products are formed per unit time.

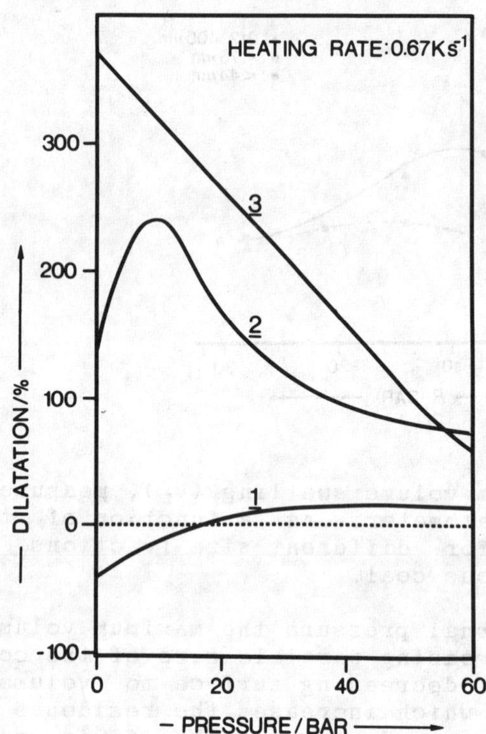

Figure 12. Three types of dilatation versus pressure curves, observed for a wide range of coals [after Thomas (28)].

4. RAPID PYROLYSIS OF COAL

As mentioned before, two classes of decomposition conditions can be distinguished (<1 K/s versus >1000 K/s), although in reality of course there is a continuous range. In this chapter the major part of the results is concerned with a heating rate of the coal particle >1000 K/s, the so called "flash pyrolysis". In flash pyrolysis secondary reactions are minimized by selecting short reaction times (1 s or less), and rapid transfer of volatiles from the point where they are generated to the bulk gas phase. Therefore, rapid pyrolysis can only be achieved with small coal particles. The larger the particles, the more time is required for heat to diffuse to the interior of the coal particle and for volatiles to reach the outer surface of the char particle. In practice, rapid pyrolysis can be experienced only by particles with a diameter in the order of 100 μm or less. In most cases the temperature of the coal particles and thus, the heating rate can not be directly measured but must be calculated from energy balances.

A wide range of experimental conditions has been employed in fundamental research on rapid pyrolysis of coal. However, two general classes of experimental techniques can be distinguished (17):

1. Captive sample techniques in which the coal sample is stationary or fixed during the experiment.
2. Continuous coal flow techniques in which the coal is continuously fed and withdrawn from the apparatus.

In the following both types of experimental techniques will be described in more detail.

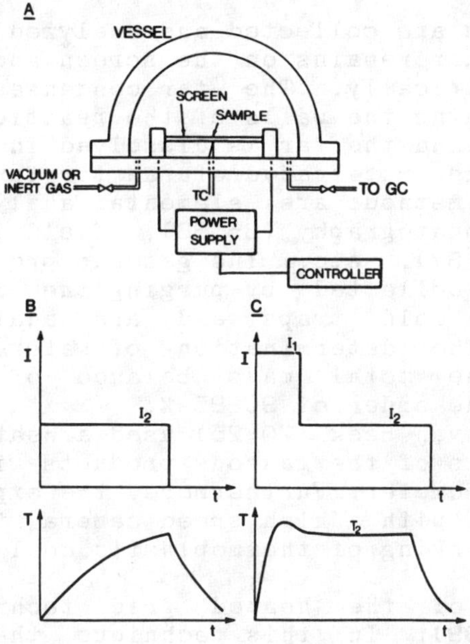

Figure 13. Heated grid system for rapid pyrolysis of coal: A. pyrolysis apparatus, B. single-pulse temperature-time response, C. double-pulse temperature-time response [after Gavalas (18)].

4.1. Captive Sample Techniques

The most widely used apparatus is the so-called heated grid system (29-33) and was introduced by Anthony et al. (34). A scheme of the experimental set-up is shown in Figure 13. A small coal sample (5-200 mg) is spread on a preweighed stainless steel screen which is then folded and connected across the terminals of a heating element. The screen is heated by an electrical current I(t) (see Figure 13 B). The screen with the coal sample is contained within a metal vessel which is closed during the experiment. The volume of the vessel is large in order to create a good dispersion of the evolved volatiles. Only the screen with sample is heated and the volatiles are rapidly quenched as they diffuse into the cold gas or condense on the cold surface of the vessel. The gas pressure can be varied from vacuum to 100 bar. A fast response thermocouple (d= 50 μm) is positioned between the folded screen to measure the temperature-time history. The temperature-time history of the coal sample does not only depend on the current input I(t), the geometry and resistance of the screen, the nature and pressure of the surrounding gas, but also on the particle and sample size. To obtain large heating rates two-step heating is applied (see Figure 13C), consisting of two consecutive pulses of constant current.

All products are collected and analyzed at the end of an experiment. Char remains on the screen and the amount is measured gravimetrically. The tar condenses on removable aluminum foils lining the walls of the reaction vessel. The foils are weighed and the tar is dissolved in solvents like methylene chloride or tetrahydrofuran for further analysis. Possible analysis methods are: elemental analysis (31), NMR, FTIR, liquid chromatography (35,36), field ionization mass spectrometry (22,37), etc. The gaseous products at room temperature are collected by purging the reactor vapors through suitable cold traps and are analyzed by gas chromatography. The determination of water causes large difficulties. The total mass balance of the products collected is in the order of 90-95 %.

Juntgen and van Heek (20,25) used a heated grid set-up with rapid analysis of the gaseous products with a time-of-flight mass spectrometer. Furthermore, the experimental set-up was equipped with a high speed camera to record the swelling and shrinking of thermoplastic coal particles upon heating.

A variant of the heated grid technique is the "pyroprobe" (38,39). In this technique the coal sample (< 3 mg) is placed in aprobe, which is directly connected to the injection port of a gas chromatograph. The pyroprobe consists of a platinum ribbon which can be heated up to 1700 K with heating rates in the order of 10,000 K/s. An inert carrier gas flushes the volatiles, generated from the decomposing coal sample, into the gas chromatograph, where they are separated on a capillary column.

Recently, analytical Curie-point pyrolysis techniques have drawn a lot of attention (40-44). In this technique the coal sample (μg scale) is fixed on a thin ferromagnetic wire which is inductively heated to its Curie-point temperature in a time scale of 0.1 s. The evolved volatile products are analyzed by gas chromatography and/or mass spectrometry. Due to the small sample size the technique gives mainly qualitative information; it is primarily used for characterization purposes. The Curie-point technique will be described in detail in another chapter of this book (44).

4.2. Continuous Coal Flow Techniques

The most widely used apparatus is the entrained flow reactor, also known as laminar flow reactor and drop tube furnace (22, 45-52), The general experimental set-up is shown schematically in Figure 14, In brief, coal particles (< 100 um) are carried by aprimary inert gas flow through a water cooled injector and are injected into a hot vertical furnace. A much larger secondary inert gas flow, which is preheated to increase the heating rate of the coal particles, flow downwards along the furnace under laminar flow conditions. The laminar flow is neccessary to avoid radial dispersion of the particles to the furnace wall and to enable the calculation of a more reliable residence time of the coal particles in the hot reactor. The pyrolysis process is stopped abruptly by collecting the particles and volatiles in a cooled probe. This collector probe is movable along the axis of the furnace. The reaction time is adjusted by the flow rate of the secondary gas streams and the position of the collector.

The temperature-time history of the coal particles must be calculated from heat transfer models and depends on parameters like the temperature of the furnace, particle size, coal feed rate, nature of the primary gas (nitrogen versus helium), etc.(22, 48-51). Reported heating times of the particles are in the range 5-200 ms, leading to heating rates in excess of 10,000 K/s. The residence time of the particles in the hot zone is calculated using both gas and particle velocity models. Particle velocities can be measured directly with a laser doppler anemometer (46).

In comparison with the heated grid technique collection and the analysis of the reaction products is much more difficult. As the collection efficiency of the char particles in the movable collector is less than 100 %, the weight loss of the coal particles can not simply be determined by weighing the char collected in the cyclone. Therefore, coal weight loss is calculated from an ash analysis of the char material using ash as a tracer. The results from this technique are independent of collection efficiency but may be subject to some error due to evaporation or decomposition of mineral matter components at the reactor temperature.

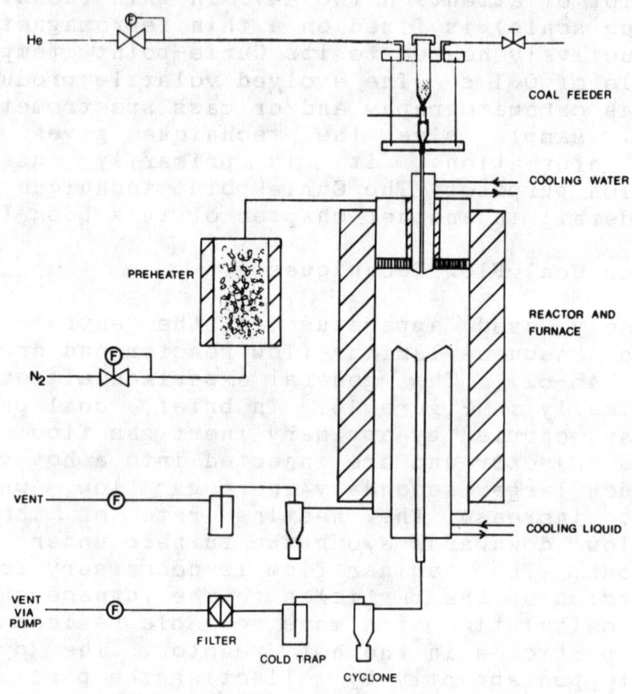

Figure 14. Schematic diagram of an entrained flow reactor system for the rapid pyrolysis of coal.

The gaseous products can be analyzed by mass spectrometry or by gas chromatography. However, to ensure high heating rates and to minimize interactions between the individual particles, the coal feed rate may be chosen so low (< 1 mg/s), that the concentration of the volatile products are below the detection limit. The tar components can not be collected quantitatively, especially when a water cooled probe is used, due to condensation on the inner wall of the collector. Also, the use of thermoplastic coals can cause problems because in the liquid state the particles stick to the collector.

The advantage of the entrained flow reactor over the heated grid technique is that it provides a more homogeneous and better defined heating of the coal particles and rapid quenching. Furthermore, the geometry of injector and reactor is a more realistic simulation of practicle coal conversion processes, like powdered coal combustion and coal gasification in entrained flow or fluidized bed reactors.

On the other hand, the captive sample technique involves much simpler instrumentation and a better overall mass balance is obtained.

Flash pyrolysis can also be obtained by injecting small coal particles into a hot fluidized bed reactor (53-55). This technique will not be discussed here. A fluidized bed reactor has disadvantages because particle residence times within the fluid bed are highly uncertain and difficult to vary.

4.3. Experimental Results from Rapid Pyrolysis of Coal

It has generally been observed that the weight losses upon rapid pyrolysis of coal particles exceed the values measured according to the standard ASTM proximate analysis for volatile content. In the ASTM method about 1 g of ground coal, loosely packed in a crucible with a close fitting capsule cover, is placed in a furnace at 1223 K and heated for 420 s (56). The measured weight loss in excess of moisture is attributed to release of volatile matter. The estimated heating rate of the sample in the standard ASTM method is 15-20 K/s. The increased volatile yields in rapid pyrolysis of small coal particles are explained as follows. Slow heating of a packed bed of coal particles favors secondary char forming reactions. In rapid heating studies small, well dispersed particles are used; these conditions reduce secondary char forming reactions of the volatile components. The extra yield of volatiles in rapid heating of coal particles is highly profitable in coal combustion and gasification because gas phase reactions are orders of magnitude faster than gas-solid reactions.

Weight loss curves as a function of the residence time of the coal particles in an entrained flow reactor are shown in Figure 15 for a Texas lignite (47). From the figure a delay in the onset of pyrolysis of the coal particles is clearly visible. This delay time is associated with the heat up of the particles. Both devolatilization rate and weight loss at a given residence time increase with increasing temperature. After a residence time of 0.4 s at 1273 K the char particles still contained more than 10 % (daf) of residual volatile matter. The maximum potential weight loss of the coal particles was a factor 1.3 higher (66.7 %, daf) than the standard ASTM volatile matter content (51.1 %, daf) of the original coal. No product distribution are known from this study (47).

The product distribution in coal pyrolysis gives important information, for two reasons. First of all, the economics of commercial processes strongly depend on the type and amount of the products. Secondly, the product distribution gives fundamental information on the processes occurring during pyrolysis. Coal rank is a very important factor in the distribution and temperature dependence of

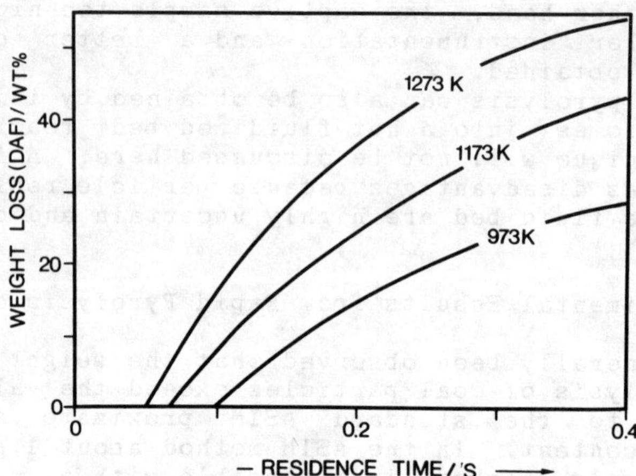

Figure 15. Weight loss curves as a function of the residence time of Texas lignite coal particles in an entrained flow reactor [after Scaroni et al. (47)].

various products. In Figure 16 the product distribution as a function of the temperature is given for a lignite (Figure 16 A) and a bituminous coal (Figure 16 B) (31). The cumulative yields of tar, water, carbon dioxide, carbon monoxide, hydrogen and gaseous hydrocarbons are shown for the two coal samples from heated grid experiments at a heating rate >1000 K/s to different peak temperatures (pressure = 1 bar helium). The results clearly illustrate that for the lignite major part of the products consist of the oxygen containing compounds; water, carbon dioxide and carbon monoxide. For the bituminous coal, however, tar is the most important product. In bituminous coals tar makes up 50-80 % of the weight loss, whereas in lignites the percentage tar is lower than 25 % of the total weight loss (18). Also the composition of the tar fraction depends on the rank of the coal. In general, the average molecular weight of the tar components from bituminous coals is higher than that from subbituminous coals and lignites. Tar from bituminous coals is a mixture of many compounds with molecular weights mainly in the range 200-1200 (22, 35, 36). At the pyrolysis temperature tar is produced as a vapor; at room temperature it becomes a viscous liquid or solid.

4.4. Mass Transfer Limitations in Rapid Pyrolysis of Coal

Gradually, researchers in coal pyrolysis more and more adopt the opinion that mass transfer processes play a prominent role in determining the amount and composition of volatiles obtained during rapid pyrolysis of coals, particularly of softening coals (21). Because of the liquid state of these coals after softening it is more logical to express the transport of volatile pyrolysis products through the coal mass in terms of liquid phase processes rather than pore diffusional processes, e.g. the evolution of tar components is viewed as an evaporation process.

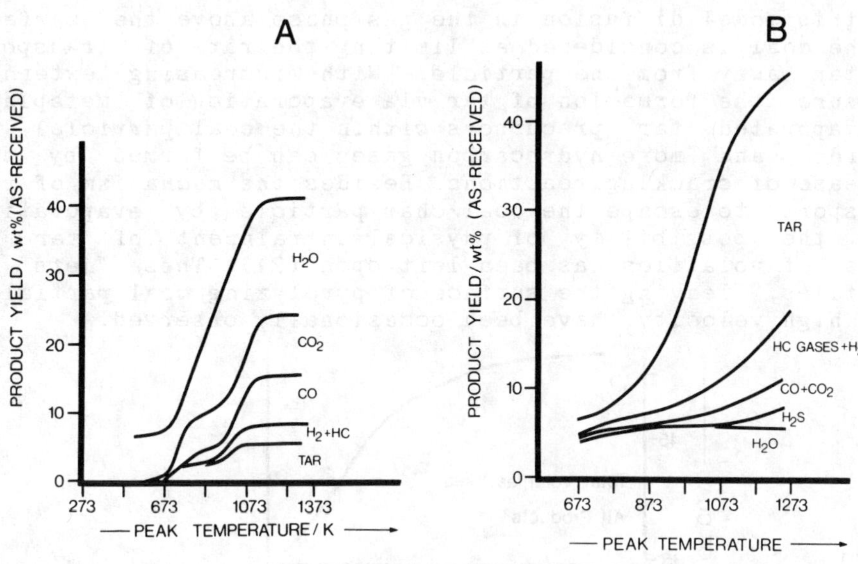

Figure 16. Pyrolysis product distribution as a function of the temperature for a lignite (A) and a bituminous coal (B), heated to different peak temperatures (heating rate 1000 K/s) in a heated grid system [after Suuberg et al.(31)].

Evidence of these mass transfer processes comes from studies in which the effect of external gas pressure on the yield of pyrolysis products is examined. The results from one of these studies are given in Figure 17 (31). In this figure the production of various compounds and classes of compounds is plotted versus external gas pressure for a Pittsburgh No. 8 bituminous coal, pyrolyzed at 1300 K for 2-10 s in a heated grid reactor (heating rate: 1000 K/s). It shows that as the external gas pressure is increased, the yield of tar is decreased. At the same time the yield of light hydrocarbon gases increases with increasing pressure. To explain these phenomena an evaporation controlled model for pyrolysis of softening coal was constructed (57).

$$\text{Coal} \longrightarrow \text{Metaplast} \longrightarrow \text{Tar (evaporation)}$$

$$\text{Metaplast} \longrightarrow H_2 \qquad \text{Coal} \longrightarrow H_2O, CO_2 \qquad \text{Metaplast} \longrightarrow \text{Hydrocarbon Gases} + CO + CO_2$$

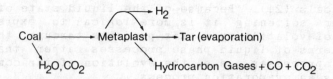

In this model diffusion in the gas phase above the surface of the coal is considered as limiting the rate of transport of tar away from the particle. With increasing external pressure the formation of tar via evaporation of metaplast (unevaporated tar precursors within the coal particle) is retarded and more hydrocarbon gases can be formed by the increase of cracking reactions. Besides the mechanism of tar transport to escape the coal/char particle by evaporation also the possibility of physical entrainment of tar by "jets" of volatiles has been left open (21). These "jets" of volatiles, leaving the surface of pyrolyzing coal particles with high velocity, have been occasionally observed.

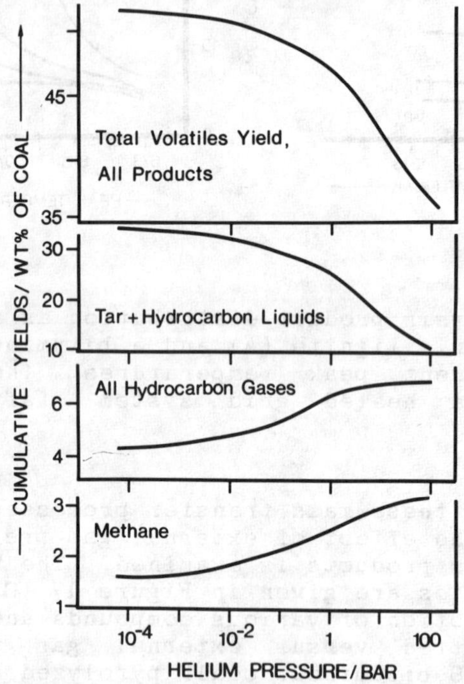

Figure 17. Effect of the external gas pressure on cumulative yields of total volatiles, tar, total hydrocarbon gases and methane from bituminous coal pyrolysis. Pyrolysis performed in a heated grid reactor at a heating rate of 1000 K/s. Maximum temperature 1300 K, with isothermal holding periods of 2-10 s [after Suuberg et al. (31)].

Another model, in which the chemistry of pyrolysis rather than the mass transfer aspects is emphasized, has been developed by Solomon et al. (22, 32). In this so called "functional group model" coal pyrolysis is viewed as a depolymerization process in parallel with thermal decomposition of the functional groups, generating gaseous components like water, carbon dioxide, carbon monoxide, hydrogen, methane, ethane, etc. All the pyrolysis products compete for the donatable hydrogen for stabilization. The kinetics of pyrolysis depend on the functional groups, but are relatively insensitive to coal rank. The important variation with the coal rank is the nature and the number of functional groups. The general structure of the functional group model is as follows:

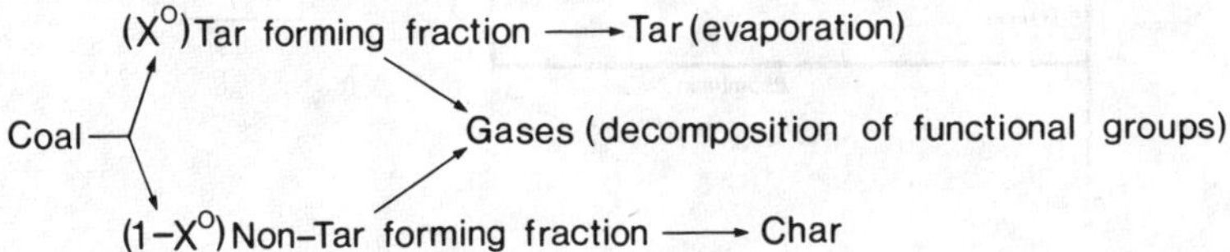

X^o = fraction of the coal, which potentially can form tar.
$(1-X^o)$ = fraction of the coal, which does not form tar.

From analysis of the starting structures in coals, based on elemental analysis and Fourier Transform InfraRed spectroscopy (FTIR), the formation of, especially, light gaseous products is predicted as a function of coal rank, temperature, residence time and particle heating rate (22).

Other pyrolysis models which include mass transfer limitations have been reviewed by Suuberg (21).

5. SECONDARY REACTIONS

The primary pyrolysis products may diffuse through the coal particle and escape, or they may react with the surface of the particle. Two kinds of secondary reactions can occur:

1. Cracking reactions forming stable, low molecular weight products, which escape and carbon which deposits on the char surface.
2. Condensation and/or polymerization reactions forming high molecular weight products, which ultimately become part of the char matrix.

The primary and secondary volatile products which escape from the char may also undergo subsequent cracking and condensation reactions, due to extended exposure to high temperatures. Three major factors determine the extent of these reactions: temperature, residence time and thermal stability of the volatile products.

A systematic study of the thermal stability and the product distribution arising from heat treatment of model compounds, relevant in coal pyrolysis, was performed (58, 59). Figure 18 gives a schematic diagram of the apparatus used.

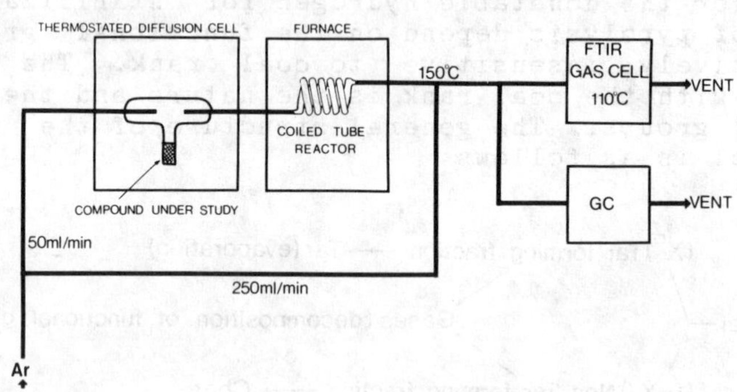

Figure 18. Schematic diagram of apparatus used to study the thermal stability of model compounds, relevant in coal pyrolysis.

The reactor was a coiled quartz tube (volume: 10 cm). The residence time of the compound in the reactor was 5 s. The temperature of the diffusion cell was chosen as such that the concentration of the model compound in the argon flow was 500 ppm. Figures 19-21 give typical results; conversion of a model compound is measured as a function of the temperature, while heating the reactor up to 1223 K at a heating rate of 0.033 K/s.

In Figure 19 the thermal stability of benzene and derivatives of benzene with different functional groups are shown. Clearly, benzene is the most stable compound, as should be expected. The presence of a functional group resulted in a lower stability of the aromatic compound. The stability depends on the type of functional group. The interpretation is rather straightforward: the lower the bonding energy, the lower the temperature at which the bond will break. E.g., the relative stability of the (alkyl)benzenes can be understood from the bonding energies:

$C_6H_5 - H$	428 kJ/mole
$C_6H_5 - CH_3$	365 kJ/mole
$C_6H_5\ CH_2 - H$	326 kJ/mole
$C_6H_5 - C_2H_5$	265 kJ/mole

From the above it can be explained that the stability decreases in the order:

benzene > toluene > ethylbenzene

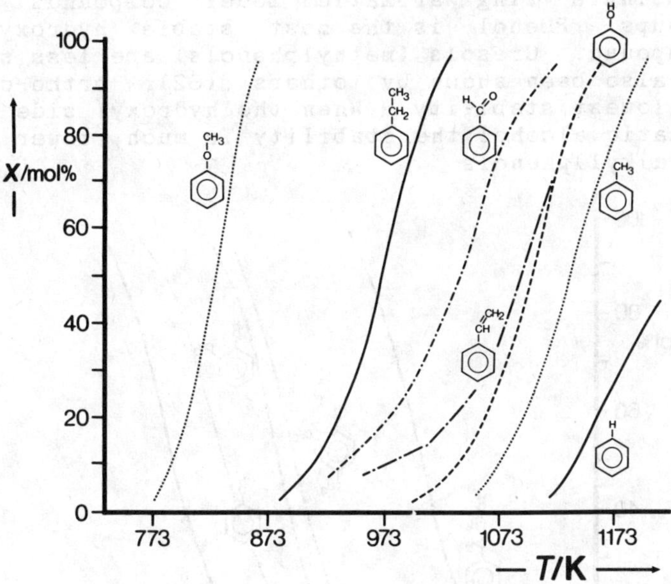

Figure 19. Thermal stability of benzene and benzene derivatives, as studied in a coiled tube reactor (residence time 5 s). Reactor was heated to 1223 K at 0.033 K/s (partial pressure of model compound 50 Pa, balance argon).

Figure 19 also shows that methoxybenzene has the lowest thermal stability. Methoxybenzene is primarily cracked into phenol and methane. Phenol itself is relatively stable and does not decompose below about 1000 K. At high temperatures phenol is converted both by cracking and polymerization reactions, as proposed by Cypres and Bettens (60):

$\rightarrow$ + CO

$\rightarrow$ 0.5 + $0.5\,H_2O$ + $0.5\,H_2$

$\xrightarrow{H_2}$ + H_2O

Coals contain many "heteroatoms" and as a consequence, also model compounds containing the most abundant heteroatoms, being O, S and N are relevant (61). Figure 20 focuses on single ring aromatic model compounds with hydroxyl groups. Phenol is the most stable hydroxylated aromatic compound. Cresols (methylphenols) are less stable and as has also been shown by others (62), ortho-cresol exhibits the lowest stability. When the hydroxyl side group is an aliphatic alcohol the stability is much lower than that of the (alkyl)phenols.

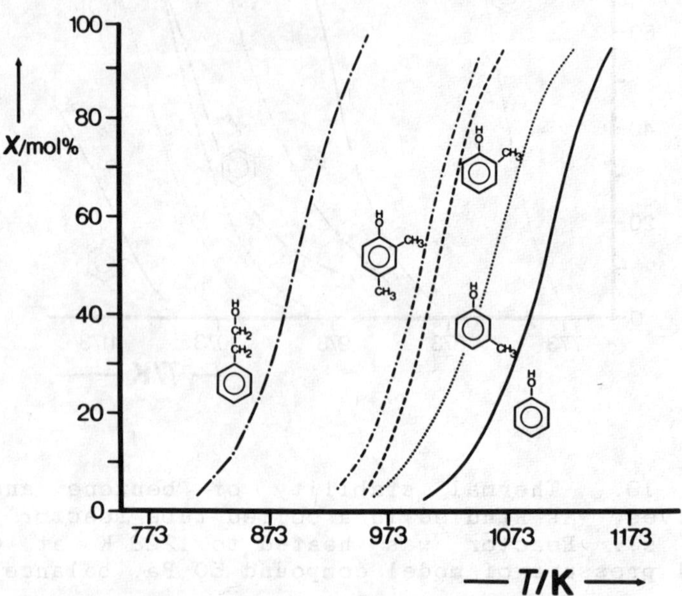

Figure 20. Thermal stability of (methylated) phenols and ethanolbenzene (partial pressure of compound 50 Pa, balance argon), as studied in a coiled tube reactor (residence time: 5 s). Reactor was heated to 1223 K at a heating rate of 0.033 K/s.

Figure 21 summarizes data on the thermal stability of single and double ring aromatic compounds, focusing on the influence of heteroatoms in the aromatic ring structure. The most stable compounds are (benzo)thiophene, benzene and naphthalene, whereas furan exhibits the lowest stability. So, considering compounds with either a S-, O- or a N- atom in the aromatic ring or just carbon atoms, the following trend in stability can be given:

S > C > N > O

An explanation can be given on the basis of aromatic stabilization of the aromatic ring structures.

The study of the thermal stabilty of aromatic model compounds, relevant in coal pyrolysis, confirm that pyrolysis of coal is initiated by bond breaking of the $-CH_2-CH_2-$ (24), and $-CH_2-O-$ (63) bridges between aromatic ring structures within the coal matrix and by decomposition of functional groups. The difference in thermal stability between the functional groups and the aromatic rings has consequences for the yield and composition of the tar fraction formed during pyrolysis of coal (64). With increasing temperatures and residence times the tar yield falls and the tar fraction becomes more and more aromatic in character.

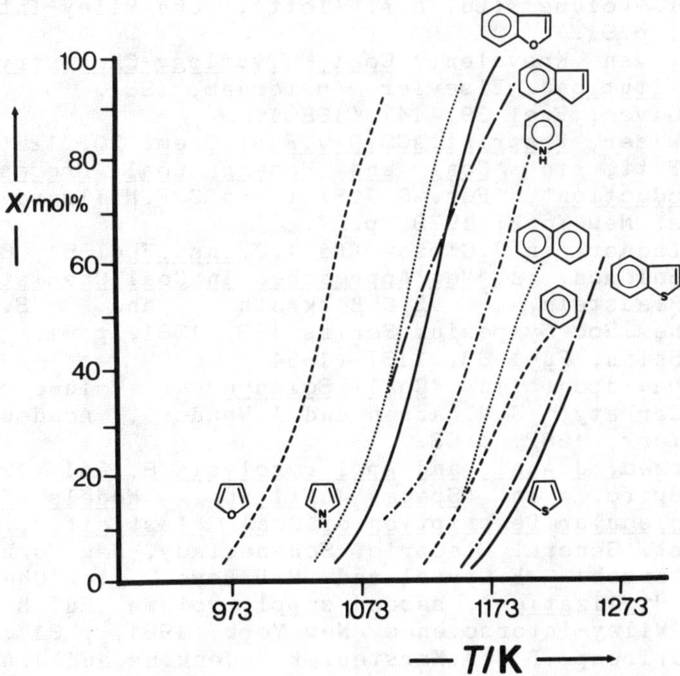

Figure 21. Thermal stability of (heterocyclic) aromatic compounds (partial pressure of compound 50 Pa, balance argon), as studied in a coiled tube reactor (residence time: 5 s). Reactor was heated to 1223 K at a heating rate of 0.033 K/s.

6. ACKNOWLEDGEMENT

This study was supported by the Netherlands Foundation of Chemical Research (S.O.N.) with financial aid from the Netherlands Organization for the Advancement of Pure Research (Z.W.O.) and by the Ministry of Economic Affairs, within the framework of the Dutch National Coal Research Programme, which is managed by the Project Office for Energy Research (PEO). Thanks are due to Prof.R.G.Jenkins of the Pennsylvania State University for using the high pressure microdilatometer and for the technical information on the atmospheric entrained flow reactor.

7. REFERENCES

1. "Stach's Textbook of Coal Petrology", Gebruder Borntraeger, Berlin, 1982.
2. F.T.C.Ting, in "Coal Structure", (Ed. R.A.Meyers), Academic Press, New York, 1982, p.7.
3. R.C.Neavel, in "Chemistry of Coal Utilization" second suppl. volume, (Ed. M.A.Eliott), John Wiley-Interscience, 1981, p.91.
4. D.W. van Krevelen, "Coal, Typology-Chemistry-Physics-Constitution", Elsevier, Amsterdam, 1981.
5. P.H.Given, Fuel 39, 147 (1960).
6. W.H.Wiser, Preprint.ACS,Div.Fuel Chem. 20, 122(1975).
7. G.J.Pitt, in "Coal and Modern Coal Processing: An Introduction", (Eds. G.J.Pitt and G.R.Millward), Academic Press, New York, 1979, p.27.
8. W.R.Ladner, in J.Gibson (Ed.),J.Inst.Fuel 51, 67 (1978).
9. P.R.Solomon, in "New Approaches in Coal Chemistry",(Eds. B.D.Blaustein, B.C.Bockrath and S.Freidman), Am.Chem.Soc.Symposium Series 169, 1981, p.61.
10. J.H.Shinn, Fuel 63, 1187 (1984).
11. R.M.Davidson, in "Coal Science", Volume 1, (Eds. M.L.Gorbaty, J.W.Larsen and I.Wender), Academic Press, New York, 1982, p.83.
12. A.Marzec, J.Anal. and Appl.Pyrolysis 8, 241 (1985).
13. C.L.Spiro, in "Space Filling Models for Coal: A Molecular Description of Coal Plasticity", Internal Report, General Electric, Schenectady, New York, 1981.
14. D.Habermehl, F.Orywal and H.D.Beyer, in "Chemistry of Coal Utilization", second suppl. volume (Ed. M.A.Elliott), John Wiley-Interscience, New York, 1981, p.317.
15. P.J.J.Tromp, P.J.A.Karsten, R.G.Jenkins and J.A.Moulijn, Fuel 65, 1450 (1986).
16. K.H. van Heek, in "Carbon and Coal Gasification", (Eds. J.L.Figueiredo and J.A.Moulijn), NATO ASI Series 105, Nijhoff Publishers, Dordrecht, The Netherlands, 1986, p.403.
17. J.B.Howard, in "Chemistry of Coal Utilization", second suppl.volume, (Ed.M.A.Elliott), John Wiley-Interscience, 1981, p.665.
18. G.R.Gavalas, in"Coal Pyrolysis", Elsevier Publ. Co., Amsterdam, 1982.
19. N.Berkowitz, in "The Chemistry of Coal", Elsevier Publ.Co., Amsterdam, 1985.
20. H.Juntgen and K.H.van Heek,Fuel Proc.Technol.2,261(1979)
21. E.M.Suuberg, in"Chemistry of Coal Conversion", (Ed. R.H.Schlosberg), Plenum Press, New York, 1985, p.67.
22. P.R.Solomon and D.G.Hamblen, in "Chemistry of Coal Conversion", (Ed. R.H.Schlosberg), Plenum Press, New York, 1985, p.121.

23. K.H. van Heek, "Druckpyrolyse von Steinkohle", Habilitationsschrift, Universitat Munster, VDI Forschungsheft 612, 1982.
24. H.Juntgen, Fuel 63, 731, (1984).
25. K.H.van Heek, Ger.Chem.Eng. 7, 319 (1984).
26. M.R.Khan and R.G.Jenkins, Fuel 63, 109 (1984).
27. P.D.Green and K.M.Thomas, Fuel 64, 1423 (1985).
28. K.M.Thomas, in "Carbon and Coal Gasification", (Eds. J.L.Figueiredo and J.A.Moulijn), NATO ASI Series No 105, Nijhoff Publishers, Dordrecht, The Netherlands, 1986, p.421.
29. D.B.Anthony, J.B.Howard, H.C.Hottel and H.P.Meisner, 15 th Symp.(Int.) on Combustion, the Combustion Institute, Pittsburgh, 1975, p.1303.
30. E.M.Suuberg, W.A.Peters and J.B.Howard, Ind. Eng. Chem. Process Des.Dev. 17(1), 37 (1978).
31. E.M.Suuberg, W.A.Peters and J.B.Howard, in "Thermal Hydrocarbon Chemistry" (Ed.Oblad), Adv. in Chem. Series No 183, Am.Chem.Soc., New York, 1979, p.239.
32. P.R.Solomon and M.B.Colket, 17 th Symp.(Int.) on Combustion, the Combustion Institute, Pittsburgh, 1979, p.131.
33. S.J.Niksa, W.B.Russel and D.A.Saville, 19 th Symp.(Int.) on Combustion, the Combustion Institute, Pittsburgh, 1982, p.1151.
34. D.B.Anthony, J.B.Howard, H.P.Meissner and H.C.Hottel, Rev.Sci.Instrum. 45(8), 992 (1974).
35. P.E.Unger and E.M.Suuberg, Fuel 63, 606 (1984).
36. E.M.Suuberg, P.E.Unger and W.D.Lilly, Fuel 64, 956 (1985).
37. P.R.Solomon and H-H King, Fuel 63, 1302 (1984).
38. D.W.Blair, J.O.L.Wendt and W.Bartok, 16 th Symp.(Int.) on Combustion, the Combustion Institute, Pittsburgh, 1977, p.475.
39. A.L.Chaffee, G.J.Perry and R.B.Johns, Fuel 62, 303 (1983).
40. G.van Graas, J.E. de Leeuw and P.A.Schenck, in "Advances in Organic Geochemistry", (Eds. A.G.Douglas and J.R.Maxwell), Pergamon Press, London, 1979, p.485.
41. H.L.C.Meuzelaar, A.M.Harper, G.R.Hill and P.H.Given, Fuel 63, 640 (1984).
42. A.M.Harper, H.L.C.Meuzelaar and P.H.Given, Fuel 63, 793 (1984).
43. P.R.Philp, T.D.Gilbert and J.Friedrich, in "Chemical and Geochemical Aspects of Fossil Energy Extraction", (Eds. T.F.Yen, F.K.Kawahara and R.Hertzberg), Ann Arbor Science, Ann Arbor, 1984, p.63.
44. P.J.J.Tromp, J.A.Moulijn and J.J.Boon, this book.
45. S.Badzioch and P.G.W.Hawksley, Ind.Eng.Chem.Process Des.Dev. 9(4), 521 (1970).

46. H.Kobayashi, J.B.Howard and A.F.Sarofim, 16 th Symp.(Int.) on Combustion, the Combustion Institute, Pittsburgh, 1977, p.411.
47. A.W.Scaroni, P.L.Walker Jr. and R.H.Essenhigh, Fuel 60, 71 (1981).
48. M.E.Morgan and R.G.Jenkins, Fuel 65, 757 (1986).
49. M.E.Morgan, in "Role of Exchangeable Cations in the Pyrolysis of Lignites", Sc.D. Thesis, Pennsylvania State University, University Park, Pa., 1983.
50. D.J.Maloney, in "Effects of Preoxidation on Rapid Pyrolysis Behaviour and Resultant Char Structure of Caking Coals", Sc.D. Thesis, Pennsylvania State University, University Park, Pa., 1983.
51. P.R.Solomon, M.A.Serio, R.M.Carangelo and J.R.Markham, Fuel 65, 182 (1986).
52. P.J.J.Tromp and J.A.Moulijn , to be published.
53. R.J.Tyler, Fuel 58, 686 (1979).
54. R.J.Tyler, Fuel 59, 218 (1980).
55. D.I.Cliff, K.R.Doolan, J.C.Mackie and R.J.Tyler, Fuel 63, 394 (1984).
56. "Methods of Analyzing and Testing Coal and Coke", Bureau of Mines, Bulletin 638, Pittsburgh, Pa., 1967.
57. P.E.Unger and E.M.Suuberg, 18 th Symp. (Int.) on Combustion, the Combustion Institute, Pittsburgh, 1981, p.1203.
58. O.S.L.Bruinsma, R.S.Geertsma, P.Bank and J.A.Moulijn, to be published.
59. O.S.L.Bruinsma, P.J.J.Tromp, H.J.J. de Sauvage Nolting and J.A.Moulijn, to be published.
60. R.Cypres and B.Bettens, Tetrahedron 30, 1253 (1974).
61. A.Attar and G.G.Hendrickson, in "Coal Structure", (Ed. R.A.Meyers), Academic Press, New York, 1982, p.131.
62. R.Cypres and B.Bettens, Tetrahedron 31, 353 (1975).
63. M.Siskin and T.Aczel, Fuel 62, 1321 (1987).
64. R.Cypres, Fuel Processing Technology 15, 1 (1987).

COAL LIQUEFACTION KINETICS

Larry L. Anderson
Department of Fuels Engineering
University of Utah
Salt Lake City, UT 84117
U.S.A.

ABSTRACT. There are similarities in both the yield of volatile products and the activation energy for hydrogenation and for dissolution of bituminous coal. Pyrolysis of the same coal, while the yield of volatile products was significantly less, gave a similar activation energy.

In all three of the processes, pyrolysis, dissolution and hydrogenation both first and second order mechanism have been assumed by different investigators. The activation energies found are usually in the range of 125-210 MJ/mole regardless of the order assumed, indicating some chemical reactions are controlling the rate. It is very difficult to generalize the kinetics and mechanism of the reactions involved in direct coal liquefaction despite the similarities pointed out. Part of the reason for this is that sufficient precise work has not been done on the same coals by different investigators and to the same extent of reaction. For example, dissolution of coal may result in nearly complete "liquefaction" up to 95% of the maf coal to products soluble in tetrahydrofuran, benzene or pyridine. However, the average molecular weight of this material is typically more than 1000. Catalytic hydrogenation, batch or continuous, results in nearly as high a yield of soluble products (80-85 weight percent of maf coal). However, the average molecular weight of such products is often 300 or less and even in heavy liquid fractions species having more than 4 rings per cluster were not found.

Until the structure for a coal used in a liquefaction process can be completely defined and a complete analysis of all products obtained by the liquefaction can be carried out, we can only speculate on the reaction mechanism. Kinetic studies give valuable, although incomplete, information on the possible rate controlling reactions. Presently these reactions appear to be diffusion or physical

Y. Yürüm (ed.), New Trends in Coal Science, 339–359.

processes at lower temperatures (<350°C) and chemical at higher temperature (>400 °C). Although kinetics and thermodynamic calculation for any reaction are considered as separate areas of study. It has been pointed out that there is usually a correlation because reactions which are thermodynamically favored proceed at a faster rate. Others have also correlated thermodynamics and kinetics. Further work will show whether this approach can be realized for a system as complex as coal liquefaction.

1. COAL LIQUEFACTION KINETICS

Kinetic studies for coal liquefaction have been made by a large number of researchers and process developers. Research by coal scientists in the area of kinetics have been used to prvide equations or expressions to describe the courses of reaction which could give reaction times, yields and plausible mechanisms. The kinetic expressions and data obtained help one to understand how reactions proceed. Because several reaction types are possible, some division of these types is necessary. For purposes of current discussion, the following processes will be examined separately before considering the general kinetics of coal liquefaction reactions:

- pyrolysis
- solvent extraction (thermal dissolution)
- hydrogenation (or hydroliquefaction).

Before discussing the theory and experimental results of these processes, some definitions must be made for the terms used in describing the extent of reaction. Kinetics results can be described in terms of the appearance of products or the disappearance of reactants. Both have been used in coal conversion kinetic studies.

Conversion to products may be based on the solubility of such products in pyridine, tetrahydrofuran, benzene, toluene, cyclohexane, hexane or other solvents. One can also base the conversion rate on the amount of insoluble material remaining at any particular reaction time. For example, Wen and Han (1) defined conversion of coal dissolution reactions as the solid organics dissolved at time, t, divided by the solid organics in the untreated coal; the rate expression was:

$$dx/dt = k\, C_{SO}(1-x)\, C/S,$$

where x : conversion (as defined above)
k : rate constant (g/hr.cm of reactor volume)
C_{SO} : weight fraction of organics in the untreated coal
C/S : coal/solvent ratio.

The determination of the conversion value of any kinetic study depends on some analytical procedure. This may be the solubility of the product in a particular solvent, the yield of a particular boiling range material or another method.

Some research studies have avoided the analysis of specific products as a method of determining the conversion. This is usually accomplished by using the total solubility of the products as an indirect measure of the "unreacted coal". The original amount of such unreacted coal before reaction is determined by the solubility of the coal feed material. Since mineral matter in the coal is essentially unaffected by conversion reactions, it can also be used to determine conversion. This is especially useful when small samples of the reacting mixture are taken during the reaction. Kang (2) used a mineral matter balance and calculated conversion by:

$$\% \text{ Conversion} = [(M_t - M_o) \times 10^4]/[M_t(100 - M_o)$$

where M_t : weight % mineral matter in the residue at a reaction time, t

M_o : weight % mineral matter in dry feed coal.

Whatever the definition of conversion used, it is important to recognize the importance of utilizing, as far as possible, a method of determination that is not appreciably affected by laboratory procedures and losses of volatile products. The method of coal conversion also impacts the choice of measuring the extent of reaction. Pyrolysis, for example, is probably the simplest coal conversion reaction to study kinetically because simple weight loss data may be used. However, in the case of simple weight loss data, the conversion values obtained are not liquefaction values but are the sum of liquefaction and gasification. The methods of determining extent of reaction are more complex in solvent extraction and hydrogenation where additional materials, such as vehicle oils or solvents, are present with the reacting coal.

1.1. Pyrolysis Kinetics

The kinetics of coal pyrolysis are important, not only for coal liquefaction, but also for related reactions of coal, including gasification and combustion. Pyrolysis of coal has been studied under isothermal conditions (where the coal is heated rapidly and held at some predetermined temperature until the weight loss rate becomes close to zero). Non isothermal pyrolysis has also been studied where the weight of the pyrolyzing coal is measured while the coal is progressively heated at a constant rate to some final temperature.

Isothermal pyrolysis studies usually monitor the weight of a charge of coal as a function of time. The final weight of a sample at any particular temperature is critical in determining the rate equation for the pyrolysis reaction at that temperature. Since the weight assymtotically approaches its final value it is impractical to carry out the experiment to completion. The final value of the weight extrapolated to infinite time has been determined by several researchers. Berkowitz and den Hertog (3) developed a procedure for extrapolation by fitting a polynomial of varying degrees to each weight loss versus time curve and computing the coefficients of each polynomial. Wiser has studied pyrolysis kinetics extensively and found that the apparent reaction order can vary considerably, depending on the value used for "a", the ultimate yield of volatiles (weight loss as a fraction of the initial weight of the sample) at a particular temperature (4). The rate of weight loss was expressed as:

$$dx/dt = k_n(a-x)^n$$

where $n \neq 1$

x : observed weight loss fraction at any time, t.

Wiser found the pyrolysis kinetics best explained as second-order reactions in isothermal studies in the temperature range 408-497°C (4).

Obviously, there is not complete agreement on even the reaction order of pyrolysis reactions from isothermal studies. The data obtained and interpreted by Wiser gives insight to possible reasons for the disagreement.

Pyrolysis kinetics are difficult to explain because so many reactions occur simultaneously. Studies of isothermal devolatilization show that both diffusion and reaction have part in controlling the rate of weight loss. Secondary reactions of primary products may also impact the results, especially when the appearance of specific volatiles is used to measure the rate of pyrolysis.

In isothermal studies some have endeavored to explain the process and kinetics by assuming that the decomposition resulting in weight loss is a series of relatively slow and overlapping reactions (5). The first-order equations used to explain the kinetics and the yields are of the form:

$$k = (1/t) \ln [a/(a - x)]$$

where, k; reaction velocity constant (of decomposition at T)

a: the ultimate yield, or maximum value of x for T

x: the cumulative amount decomposed to time, t, and is measured by the weight loss to that time.

This treatment, however, is not consistent with the numerous studies by Wiser (4) showing the reactions to be more easily described by second-order kinetics and has been criticized by Berkowitz (6) on the ground that there is an implicit assumption in the formulation that the volatile matter evolves as fast as it is formed by the decomposition of the coal. Berkowitz has advanced the view that isothermal pyrolysis is a diffusion-controlled process. Subsequent attemps to prove diffusion-control or reaction rate control have not been conclusive, since similar equations result.

Some attempts have been made to understand pyrolysis reactions and their kinetics by following the formation or evolution of individual volatile compounds such as hydrogen, water vapor, carbon monoxide or methane. These studies have not been conclusive, probably because the evolution rate for at least some of these species is a combination of primary and secondary reactions. The most likely conclusion of studies on isothermal pyrolysis is that the kinetics is a combination of reactions controlled partially by diffusion and partially by decomposition reactions.

1.2. Nonisothermal Pyrolysis Kinetics

In essentially all of the nonisothermal pyrolysis studies coal is heated at some constant rate of temperature increase while the sample weight is monitored. In principle all applicable kinetic parameters can be evaluated by a single curve (plot of weight loss versus temperature). As with the description of isothermal pyrolysis, the use of weight loss data to explain decomposition reaction rates, may be incorrect. The evaluation of the kinetic constants for nonisothermal pyrolysis follows from obtaining the rate of gas emission in terms of concentration (7).

$$dV/dt = k(V_o - V^n)$$

where, k : reaction rate constant

V : gas volume liberated at time, t

V_o : gas volume liberated at time, infinity

at a constant heating rate of dT/dt, assuming Arrhenius dependence of k, when n = 1 then:

$$dV/dt = [k_o/(dT/dt)] \exp(-E/RT) \times \{V_o^{(1+n)} + (n-1)[P_o RT^2/(dT/dt)E] \exp(-E/RT)\}^{-n/(n-1)}$$

By putting this equation into logarithmic form and determining the kinetic constants by an iterative regression analysis (7) the kinetic constants can be determined. The results for gas emission kinetics C1, C2 hydrocarbons and hydrogen and carbon monoxide) and values for coal pyrolysis give, generally, E =40-50 kcal/mole and k = 10^{10}- 10^{12} min^{-1} for n = 1. Since the order of reaction based on weight loss data, is usually closer to 2 than to 1 the kinetic values are also different. There is the same problem for nonisothermal pyrolysis kinetics as previously for isothermal studies, namely, that the data obtained are not directly pyrolysis rate measurements but rather the rate of gas emissions. For example, hydrogen evolution during linear heating compared to other reactions resulting in hydrogen formation, is too low due probably to either holdup in the coal of because the relevant hydrogen producing reactions are not in equilibrium when the coal is heated.

Whether isothermal or non isothermal the kinetics of coal pyrolysis gives, at best, specific data on the evolution of gaseous products or specific product evolution. The relation between such data and the chemical reaction sequences leading to and evolution are still to be determined.

1.3. Models of Coal Pyrolysis Kinetics

Several models have been developed for coal pyrolysis processes. These have been discussed by Gavalas (8) who also looked at earlier models. These models usually postulate reactions which result in certain products of the pyrolysis process, although these products may not be specifically identified. The postulated reactions contain parameters which are determined by comparison with experimental data. The extent to which these parameters correspond to process taking place and explain the pyrolysis chemistry and products relate to the actual reactions taking place. The way kinetic equations are applied in these models depends on the assumed structures reacting. The following functional groups and chemical reactions have been considered important for modeling the pyrolysis process:

1. functional groups
   a. aromatic nuclei with methyl and phenolic hydroxyl groups
   b. hydroaromatic structures
   c. bridging groups, including methylene and diaryl ethers
2. chemical reactions
   a. condensation of phenolic groups
   b. hydrogen abstraction and addition
   c. hydrogen exchange
   d. propagation reaction of tetralin-like structures

e. decomposition of hydroaromatic structures
f. initiation reactions to produce radicals
g. termination (recombination) reactions.

Although information on coal structure and the composition of products is still insufficient for definitive chemical modelling, the situation is improving, primarily due to more advanced analytical methods.

1.4. Extraction Kinetics

The extraction of coal by solvents can be a simple solution process when only occluded matter is taken into solution or it may be a complex process involving thermal and/or catalytic bond cleavage coupled or followed by solution and reaggreation of extracted material into colloidal or precipitated matter. Because the greatest interest in extractability has usually involved high yields, solvent extraction of coal cannot be analyzed or categorized as a process with certain characteristics. Results given here will necessarily be representative, rather than comprehensive. Such results will, hopefully, still represent the most significant work relating to the kinetics of extraction. The reactions which occur when coal is placed in a solvent depend on the coal, the solvent, the temperature and other physical conditions as well as the presence of other reactants, such as gases. Many solvents have found to disperse or dissolve coal to some extent. If the solvent is strong enough and the temperature above 400°C, essentially all of most bituminous and subbituminous coals can be dispersed, although there is disagreement whether true solutions exist. It is recognized that the action of solvents on coal at temperatures above about 350°C is accompanied by thermally initiated decomposition reactions. In this regard there appears to be many similarities in the reactions and the kinetics of the processes of pyrolysis, solvent extraction and hydrogenation of coal (9). Many investigators have studied the kinetics of coal dissolution or extraction as well as the action of solvents on model compounds. Because so many different experimental conditions, coals and methods of measuring the results have been used there is great difficulty in comparing results from all of these studies. Although the operational definitions of asphaltenes, asphaltols, oils and residue are different, it is usually because different methods or solvents were used to measure the yields of these materials. Reaction paths proposed by Squires (10) is a working model for the action of a hydrogen donor solvent on bituminous coal where the desired product is oil as shown in Figure 1.

This model is complex enough to include most of the

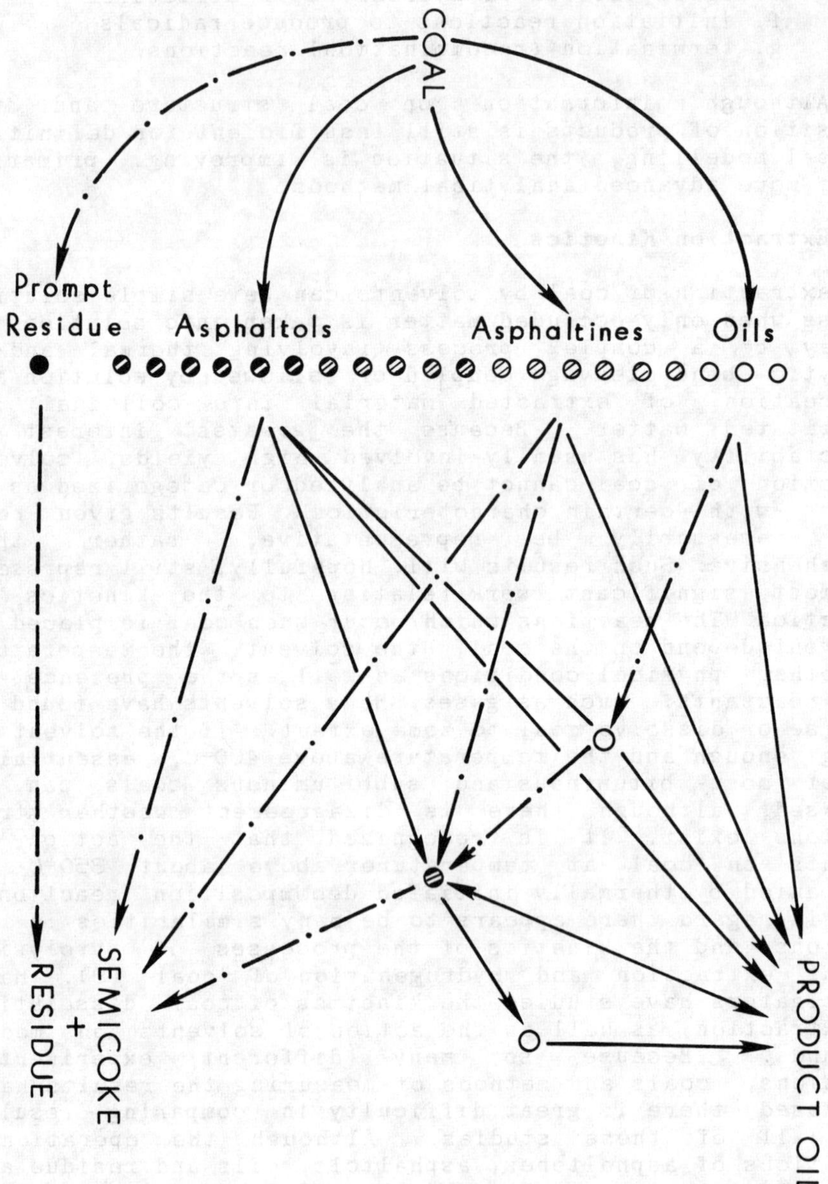

Figure 1. A working model for the action of a hydrogen donor solvent on bituminous coal [after Squires (10)].

possible reaction paths for a system which includes bituminous coal, a hydrogen donor solvent and hydrogen gas at temperatures above the softening temperature for the coal involved. The model would be even complex if all or some of the reactions were reversible. In early work by Oele (11) and others such reversibility was assumed. However, as shown by Hill et al. (12), the reactions of coal to asphaltene, asphaltols and oils are not reversible. Attempts to represent coal dissolution by more simple schemes have also been made but usually ignore obvious necessary reactions or interactions taking place. A review and analysis by Gangwer (13) of kinetic data from at least 15 publications has produced a model using reaction sequences and assumes that coal can be described as a substrate (R) with clusters of chemically bonded preasphaltenes (asphaltols) (P), asphaltenes (A), oil (L) and gas (G):

R-P-P	R-A-A	R-L-L	R-G-G
R-P-A	R-A-L	R-L-G	.
R-P-L	R-A-G		
R-P-G			

In his analysis of the results of dissolution of these structures, Gangwer postulated that preasphaltenes are formed from cleavage of the most labile bonds in the structures which have preasphaltenes as part of their makeup. The resulting cleaved preasphaltenes can appear in the products or be further converted to asphalthenes (A), oils (L) and/or gases (G). Similar possibilities as assumed for asphaltenes and oils with hydrogen capping of bonds in the cleavage process:

Coal ---> $P_{product}$ (R-P, R-A, R-L, R-G, P) + A + L + G

$P_{(intermediate)}$ ---> $A_{product}$ (R-A, R-L, R-G, A) + P + L + G

$A_{(intermediate)}$ ---> L + A + G

The values of rate constants for the formation of soluble products were given by Gangwer (13) as well as those for the formation of the specific fractions (preasphaltene, asphaltene, oil and gas). Using transition state theory, the interpretation of the data was that formation of the activated complex could involve physical as well as chemically controlled steps. While this treatment is useful in correlating the experimental data, like the other general models there are simplifications and assumptions made which are known to be incorrect. For example, in coal dissolution reactions, the hydrogen donating ability of bituminous coals is often greater than the solvents used (2,14,15). There is also no present descriptions of coal and its reactions in dissolution processes to verify that the preasphaltenes, asphaltenes, oils or gases are the result of simple cleavage reactions.

It appears that while kinetic data and correlations are useful in providing insight for the rate controlling steps in the dissolution or extraction process, the determination of exact mechanisms is not possible. The order of the reaction is similarly useful only in generally suggesting the overall process reactions and limitations. In many studies the order of reaction has not even been obtained. The data from several studies indicate that the dissolution process is first order with respect to unreacted coal and first order with respect to solvent. The latter determination requires careful evaluation at coal/solvent ratio over a range of values (12,16). While there is considerable controversy over the rate controlling step in the coal dissolution process as well as the applicable mechanism there is some arrangement that both physical and chemical processes are important in the dissolution process. The rate determining step probably changes during the course of the reaction and is also probably different for different conditions (turbulance, reaction temperature, etc.). The most probable rate controlling step at temperatures below the softening temperature of the coal (350°C) is diffusion. At higher temperatures the activation energy value indicates that chemical reactions are rate controlling. These controlling reactions could be the thermal cleavage of certain bonds in the coal (17), the transfer of hydrogen to coal radical (9) or even rehydrogenation of hydrogen donor solvent (18). It must also be recognized that coal dissolution or extraction is a complex process which includes not only dispersion, solution and depolymerization of coal components but also many secondary reactions of the solvent, such as, adduction repolymerization and isomerization which can distort the kinetic measurements obtained (19).

Since the kinetics of coal dissolution are so complex one must ask whether there is actual value in pursuing such studies or whether one can learn anything significant from the kinetic results of such studies. In general terms there has certainly been advancement in our understanding as the result of kinetic studies. At the time Oele (11) studied the process it was very difficult to measure the extent of reaction, especially during the first few minutes, when most of the dissolution took place (Figure 2). The reaction was assumed to be a zero order forward reaction and a first order backward reaction:

$$dx/dt = k_f - k_b x$$

When there was no change in the amount extracted, the

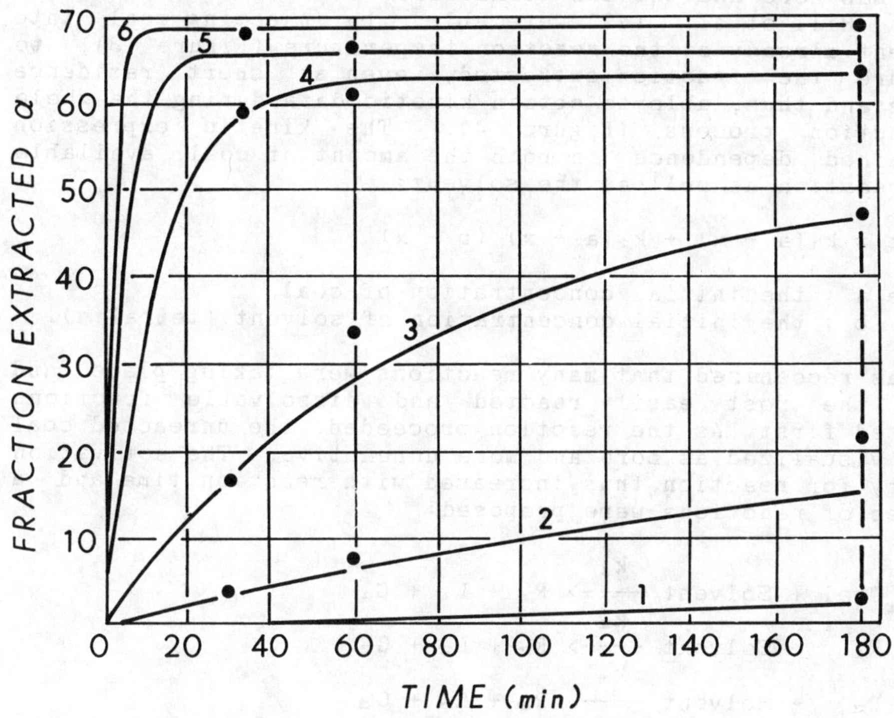

1. 165ª C
2. 185° C
3. 205° C
4. 240° C
5. 265° C
6. 290° C

<u>Figure 2. Extractive disintegration of a bituminous coal (27.2 % VM, daf) with ß-naphthol [after Oele (11)].</u>

rate was zero and the two terms were considered to be equal. Later Hill et al. (12) were able, by injecting coal into solvent already at the reaction temperature (Figure 3), to measure the fraction extracted, even at short residence times and thus, able to obtain kinetic data during the whole extraction process (Figure 4). The kinetic expression contained dependence on both the amount of coal available for reaction as well as the solvent:

$$dx/dt = k_1(a - x) + k_2(a - x)(b - x)$$

where a : the initial concentration of coal
b : the initial concentration of solvent (tetralin).

It was recognized that many reactions were taking place and that the most easily reacted and dissolvable fractions reacted first. As the reaction proceeded, the unreacted coal was visualized as more and more unreactive. The activation energy for reaction thus increased with reaction time and a series of reactions were proposed:

$$\text{Coal} + \text{Solvent} \xrightarrow{k_1} R_1 + L_1 + G_1$$

$$R_1 + \text{Solvent} \xrightarrow{k_2} R_2 + L_2 + G_2$$

$$R_2 + \text{Solvent} \xrightarrow{k_3} R_3 + L_3 + G_3$$

.

.

$$R_n + \text{Solvent} \xrightarrow{k_n} R_n + L_n + G_n$$

where R_i is the solid residue or unreacted coal, L_i is the liquid product (extract) and G_i is the gaseous product. It was assumed that

$$k_1 > k_2 > k_3 \ldots > k_n.$$

The main reaction is 1 at the beginning but shifts to 2, 3 ....n as the reaction proceeds. Thus, a first order rate expression can be defined with a variable rate constant:

$$dx/dt = k_v (1 - x).$$

It was found that k_v varied linearly with the extent of reaction (x), thus:

$$k_v = C_1 - C_2 x = k_o(1 - ax)$$

and "a" was found to be $1/x_m$ where x_m is the maximum

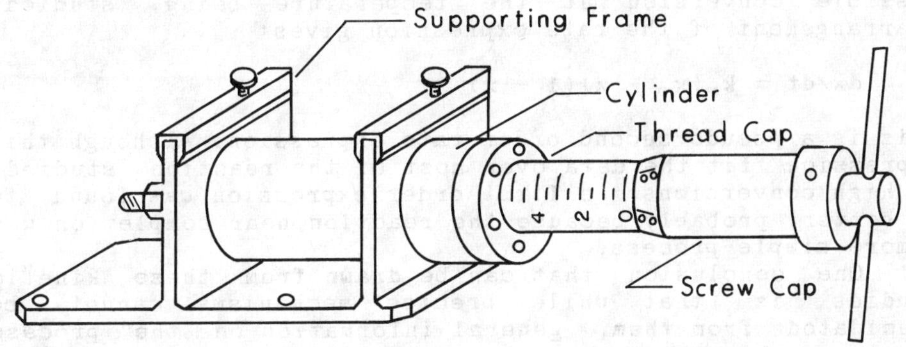

Figure 3. Coal injector used for solvent extraction [after Hill (12)].

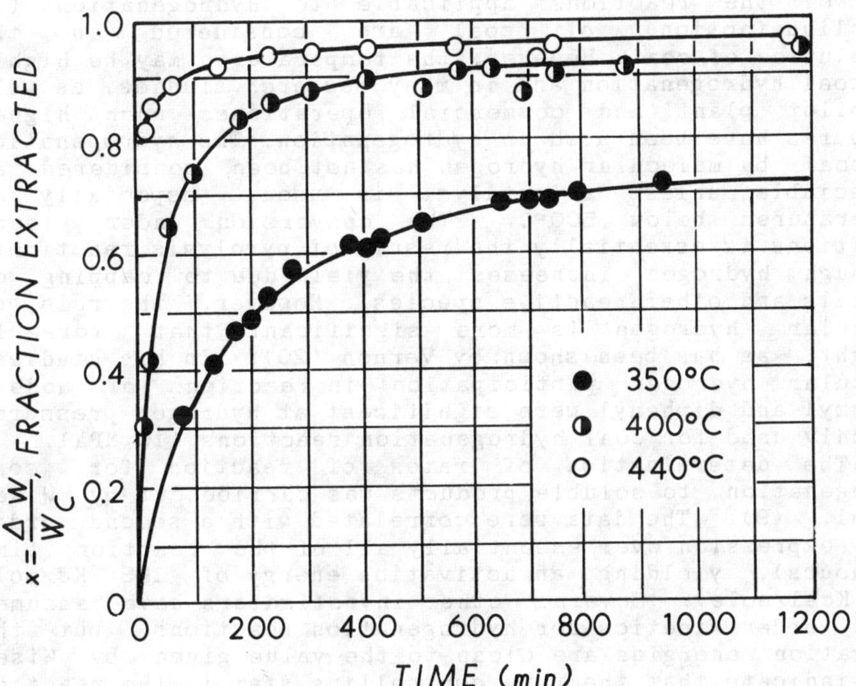

Figure 4. Time-yield curves for thermal dissolution of coal in tetralin [after Hill (12)].

possible conversion at the temperature being studied. Rearrangement of the rate expression gives:

$$dx/dt = k_m(x_m - x)(1 - x)$$

this is a pseudo second order rate expression. Although this expression fit the data over most of the reaction studied, at high conversions, a first order expression was found to be better, probably because the reaction near completion was a more simple process.

One conclusion that can be drawn from these kinetic studies is that while precise mechanisms cannot be elucidated from them, general information on the process limiting the reaction (the slow step in the dissolution process) can be. Additionally, it may be valuable to know rate constant ranges for the formation or disappearance of certain species such as preasphaltenes and asphaltenes as was determined by Gangwer (13).

1.5. Hydrogenation (Hydroliquefaction) Kinetics

Most of the reactions applicable to hydrogenation (or hydroliquefaction) of coal are considered in the dissolution of coal. However, the temperatures may be higher for coal hydrogenation and in many research studies, as well as pilot plant and commercial operations, much higher pressures have been used in hydrogenation. The hydrogenation of coal by molecular hydrogen has not been considered as appreciable unless a catalyst is used, especially at temperatures below 500°C. The conversion under these conditions is essentially the result of pyrolysis reactions, although hydrogen increases the yield due to capping of radicals and other reactive species. However, the role of molecular hydrogen is more significant that formerly thought, as has been shown by Vernon (20). In his studies, molecular hydrogen participation in reactions of models dibenzyl and diphenyl were significant at hydrogen pressures commonly used for coal hydrogenation reactions (10 MPa).

The determination of rates of reaction for coal hydrogenation to soluble products was carried out by Wiser et al. (9). The data were correlated with a second order rate expression over essentially all of the reaction time (8 hours), yielding an activation energy of 163 KJ/mole (39 Kcal/mole). Several other investigators have assumed first order kinetics for hydrogenation reactions, but the activation energies are close to the value given by Wiser and indicate that the rate controlling step in the reaction is chemical in nature. For a hydrogen reaction, one could presume that the mass transfer rate could be the controlling step. At least for some hydrogenation processes this appears not be the case (21).

Some limited data have been obtained on the kinetics of hydrocracking of coal in an ebullated bed (22). The authors assumed that the catalytic action involved only the solvent and primary liquefaction products. Hydrocracking kinetics were represented by a "pseudo" first order equation of the form:

$$F/V = kE\ (1 - \alpha)/\alpha$$

where F/V : space velocity, grams extract/hr cm^3 (sump)
k : "pseudo" first order constant (hr^{-1})
E : extract concentration, grams/cm^3 (sump)
α : conversion of extract to distillate plus gas.

Using the experimental data, the following equations resulted:

$\ln(kE) = -16{,}700/T + 23.0 \quad P_{total}=239$ atm, $P_{H_2}=197$ atm
$\ln(kE) = -20{,}000/T + 28.4 \quad P_{total}=409$ atm, $P_{H_2}=340$ atm

This study, like many others, involved reaction in the presence of a catalyst. However, it was concluded that the reactions were relatively insensitive to catalyst concentration but that thermal pyrolysis of hydrogenated extract played a major role in the hydrocracking mechanism.

When a high volatile bituminous or similar rank coal is heated in the presence of hydrogen, but in the absence of a catalyst, product yields are similar in quantity and composition to those obtained by pyrolysis. An appropriate catalyst can significantly increase the yield of soluble products and significantly changes the composition. Hydrogenation of coals with catalysts present have supplied much of the data for kinetic studies of coal hydroliquefaction. This fact alone helps us to draw some conclusions about the rate determining reactions. As pointed out by Wiser et al. (9), hydrogen reacts with thermally produced reactive fragments, which tends to lead to the production of relatively smaller molecules than are found in pyrolysis liquids. The kinetics of such catalytic hydrogenation of a bituminous coal was correlated with a second order rate expression and yielded an activation energy of 163 MJ/mole, indicating a chemical controlling slow step for the reaction. This work is typical of kinetic studies in bituminous coals.

2. COAL DEPOLYMERIZATION KINETICS AND REACTION PATHWAYS

As pointed out by Petrakis and Grandy (23) most of the coal liquefaction literature contains information about liquefaction yields or molecular structure of liquid products (and by inference, coal) with relatively little literature on liquefaction kinetics. Petrakis and grandy has summarized some results on dissolution kinetics (Table I) showing kinetic parameters by nine different research groups. These results are similar to those already discussed and show most activation energies for the dissolution process are in the range of 125-167 KJ/mole (30-40 Kcal/mole). Since these values are, in general, greater than the bond dissociation energies for C-C bonds, it has been postulated that the bonds broken in the coal liquefaction process produce radical species thata are stabilized by extensive delocalization by π electron systems of large aromatic species (24). The activation energies for such reactions would be an average of the bond dissociation energies in the chemical species involved and would vary with the rank and chemical nature (maceral composition) of the coal, if the activation energy in coal liquefaction depends on thermal bond rupture.

2.1. Model Compound Reactions

Many model compounds have been used in attempts to understand coal conversion reactions in gasification, liquefaction and combustion. The value of such studies is always suspect since the choice of a model involves an assumption about the chemical structures in coal and these are presently not well defined. Furthermore, even if precisely the same structural components are present in coal as the model compounds chosen, there are bound to be steric hindrance, mineral matter influences and other factors which may render model compound reactions invalid measure of coal reactions under the same conditions.

Having stated that the above limitation and possible invalidation of model compound studies, one is cited here to suggest a possible relation to liquefaction kinetics and to give an example of how model compound studies are used to draw conclusions about coal behaviour.

Ramanathan (25) chose 9-benzyl-1,2,3,4-tetrahydro carbazole (9-BTHC) as a model of a structural group believed to be present in bituminous coal. Pyrolysis of 9-BTHC at temperatures from 376-436ºC was studied in closed tube reactors. The pyrolysis products were analyzed and found to be mostly carbazole, 1,2,3,4-tetrahydrocarbazole, toluene and 9-benzylcarbazole. Their results indicated that the reaction was second order with an activation energy of

TABLE I

Summary of gross coal dissolution kinetics
[after Petrakis and Grandy (23)].

Coal	Solvent/Coal	T,°C	Definition of Conversion	Kinetic Parameters Kcal/mole
Utah HV	Tetralin 10:1	409-497 (pyrol.) 350-450 (liquid)	Weight loss (pyrol.) Benzene sol. (liquid)	35.6,2nd ord.intl 4.1,1st ord.latr 28.8,2nd ord.intl 13.1,1st ord.latr
Pitt Seam Ireland	Tetralin 4:1	324-387	Xylenol; cyclohexane; benzene;cresol	$E_a[K_1(fast)]=30$ $[K_2(slow)]=38$
Belle Ayr (sub)	HAO* HPH*	400-470	Pentane, benzene, pyridine	Coal-->Product E_a=16.7(HAO) 20.5(HPH)
Pitt Seam (Bruceton)	None	400-440	Benzene	Coal-->Asphaltene E_a=37
Miike	Process (330-380) MoO_3 cat	350-450	Hexane, Benzene	Coal-->Asphaltene E_a(calc)=16.4 Asphaltene Oil E_a(calc)=16
Utah Spr.Canyon	Tetralin 10:1	350-450	Benzene	ΔH= 37.2-85.5 from 0-90% reacted
Big Horn	Process 3:2	413-440	Boiling ranges heavy oil>343°C	Coal-->Oil E_a=55 Coal-->Furnace oil E_a=40.5
Bruceton	Synthoil	400	Benzene	
Makum	Tetralin 1:1	380-410	Benzene	Initial R_x E=79 Step 2 E_a=47 Step 3 E_a=35

* HAO - hydrogenated anthracene oil
HPH - Hydrogenated phenanthrene

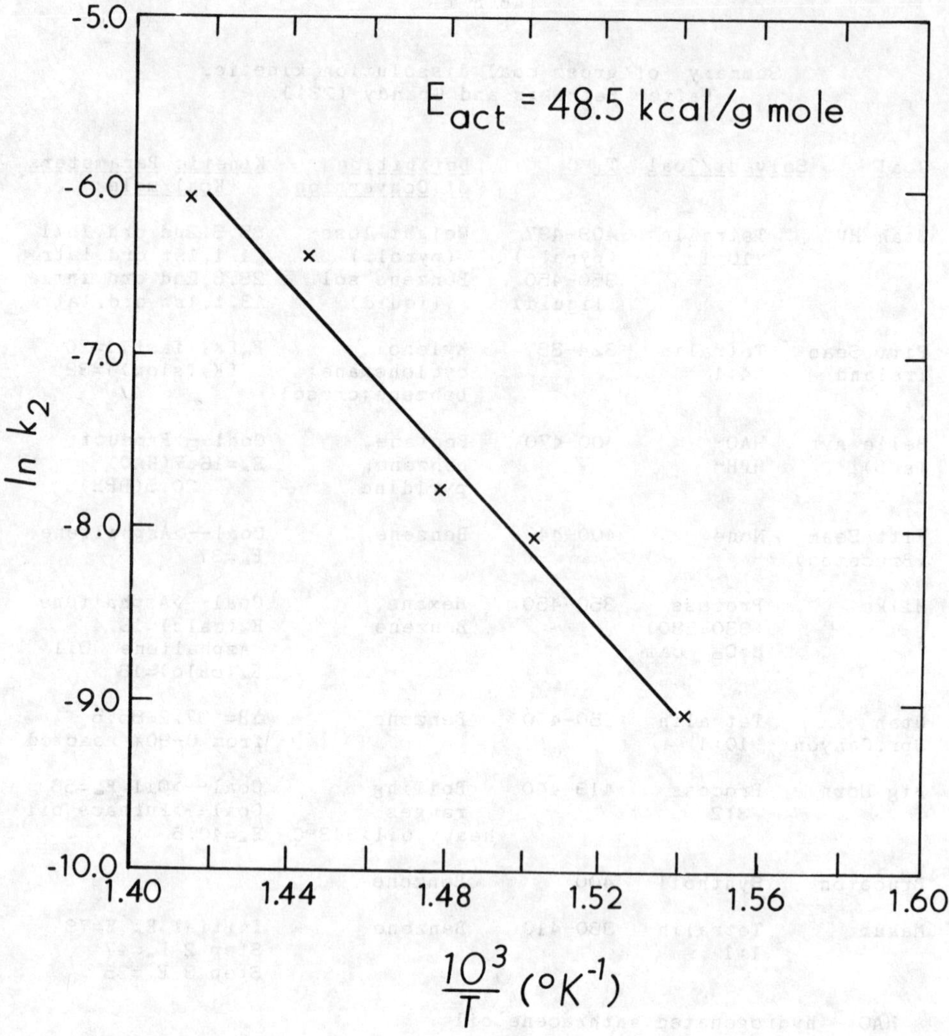

Figure 5. Arrhenius plot for second order kinetics of the 9-Benzyl 1,2,3,4-tetrahydrocarbazole (9-BTHC). The data plotted are from measurements of the amount of 9-BTHC remaining [after Ramanathan (25)].

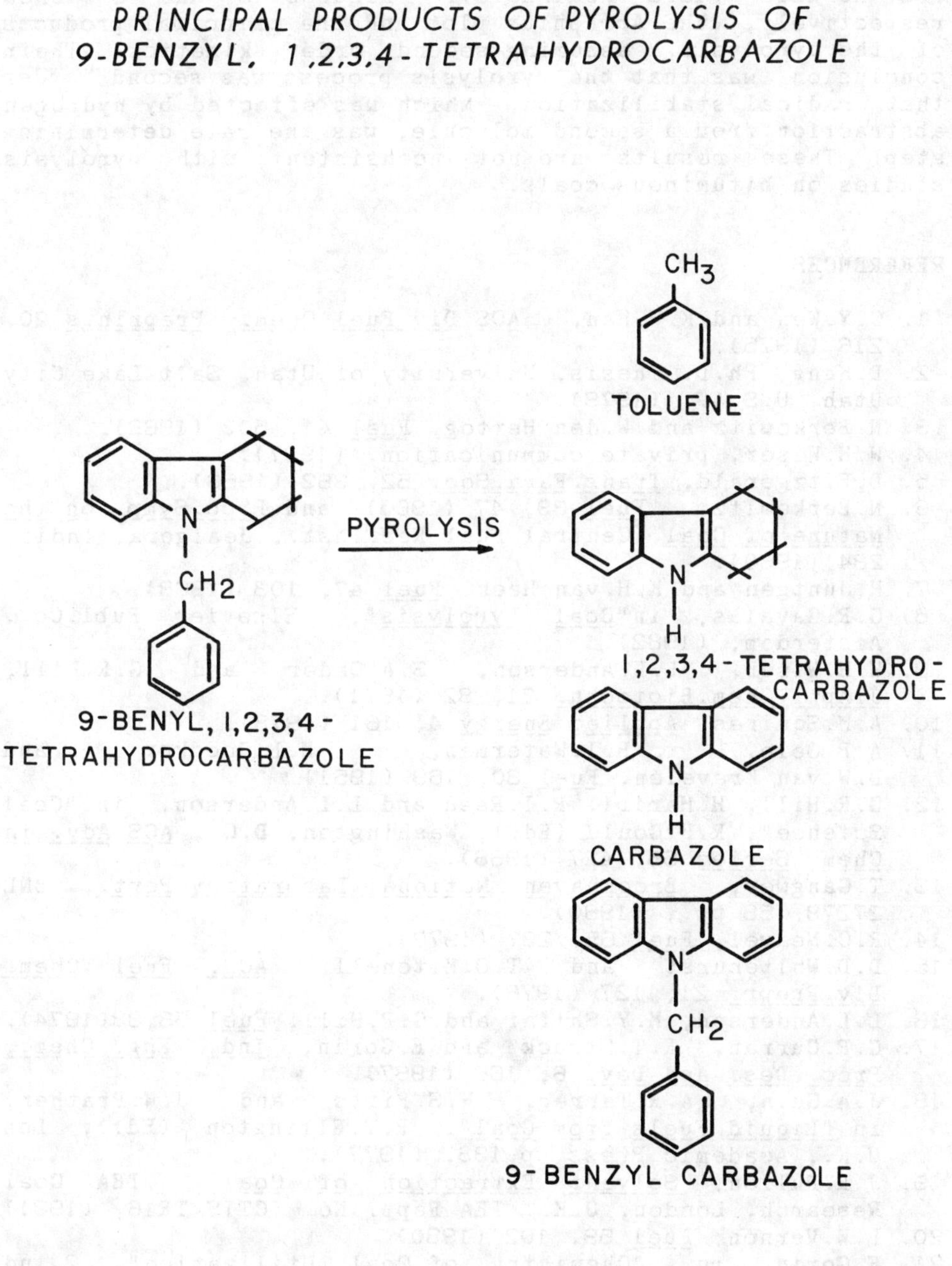

PRINCIPAL PRODUCTS OF PYROLYSIS OF 9-BENZYL, 1,2,3,4-TETRAHYDROCARBAZOLE

CH_3

TOLUENE

N

CH_2

PYROLYSIS

N

H

1,2,3,4-TETRAHYDRO-CARBAZOLE

9-BENYL, 1,2,3,4-TETRAHYDROCARBAZOLE

N

H

CARBAZOLE

N

CH_2

9-BENZYL CARBAZOLE

Figure 6. Principal products of pyrolysis of 9-Benzyl 1,2,3,4 tetrahydrocarbazole [after Ramanathan (25)].

203 KJ/mole (48.5 Kcal/mole). Figures 5 and 6 show respectively, the Arrhenius plot and the principal products of the pyrolysis, assuming second order kinetics. Their conclusion was that the pyrolysis process was second order that radical stabilization, which was effected by hydrogen abstraction from a second molecule, was the rate determining step. These results are not inconsistent with pyrolysis studies on bituminous coals.

REFERENCES

1. C.Y.Wen and K.W.Han, ACS Div.Fuel Chem. Preprints 20, 216 (1975).
2. D.Kang, Ph.D. Thesis, University of Utah, Salt Lake City, Utah, U.S.A.,(1979).
3. N.Berkowitz and W.den Hertog, Fuel **41**, 502 (1962).
4. W.H.Wiser, private communication, (1987).
5. D.Fitzgerald, Trans.Fara.Soc. 52, 362 (1956).
6. N.Berkowitz, Fuel 39, 47 (1960), and Proc.Symp. on the Nature of Coal, Central Fuel Res.Inst., Jealgora, India, 284 (1960).
7. H.Juntgen and K.H.van Heek, Fuel **47**, 103 (1968).
8. G.R.Gavalas, in"Coal Pyrolysis", Elsevier Publ.Co., Amsterdam, (1982).
9. W.H.Wiser, L.L.Anderson, S.A.Qader and G.R.Hill, J.Appl.Chem.Biotechn. 21, 82 (1971).
10. A.M.Squires, Applied Energy **4**, 161 (1978).
11. A.P.Oele, H.I.Waterman, M.L.Goedkop and D.W.van Krevelen, Fuel 30, 169 (1951).
12. G.R.Hill, H.Hariri, R.I.Reed and L.L.Anderson, in "Coal Science", R.F.Gould (Ed.), Washington, D.C., ACS Adv. in Chem. Series 55, 427 (1966).
13. T.Gangwer, Brookhaven National Laboratory Rept., BNL 27279, 58 pp., (1980).
14. R.C.Neavel, Fuel 55, 237 (1976).
15. D.D.Whitehurst and T.O.Mitchell, ACS, Fuel Chem. Div.Prepr. 21, 127 (1976).
16. L.L.Anderson, M.Y.Shifai and G.R.Hill, Fuel 53,33(1974).
17. G.P.Curran, R.T.Struck and E.Gorin, Ind. Eng. Chem., Proc. Des. and Dev. 6, 166 (1967).
18. J.A.Guin, A.R.Tarrer, W.S.Pitts and J.W.Prather, in "Liquid Fuels from Coal", R.T.Ellington (Ed.), London U.K., Academic Press, p.133, (1977).
19. J.R.Pullen, Solvent Extraction of Coal, IEA Coal Research, London, U.K., IEA Rept. No. ICTIS/TR16, (1981).
20. L.W.Vernon, Fuel 59, 102 (1980).
21. E.Gorin, in "Chemistry of Coal Utilization", 2 nd Suppl.Vol., M.A.Elliot (Ed.), Wiley Interscience, New York, N.Y., Ch. 27, 1890 p., (1977).

22. K.Anon, Pilot Sale Development of the CFS Process, U.S.D.I., OSRR&D Rept. No. 39, 4, Book 3, Consolidation Coal Co., (1971).
23. L.Petrakis and D.W.Grandy, Free Radicals in Coals and Synthetic Fuels, Coal Science and Technology 5, Elsevier Sci. Publ., Amsterdam, 211 p. (1983).
24. W.H.Wiser, Fuel 47, 475 (1968).
25. M.Ramanathan, Master of Science Thesis, University of Utah, Salt Lake City, Utah, (1979).

CATALYSIS OF COAL LIQUEFACTION

L.L.Anderson
Department of Fuels Engineering
University of Utah
306 W.C.Browning Building
Salt Lake City
Utah 84112-1183
U.S.A.

ABSTRACT. The catalysis of coal liquefaction encompasses many reactions, conditions and catalytic and support materials. Catalytic processes for direct liquefaction of coal recently piloted or demonstrated such as H-coal, have used conventional hydrotreating or hydrofining catalysts such as alumina supported Co-molybdate. There has been considerable interest in various iron compounds. Solid catalysts are recognized to act as catalysts, not with coal itself, but with liquids produced by the thermal pyrolysis of coal. Conventional liquefaction at temperatures above 400 ^{o}C have been extensively studied and almost all catalysts stable at these conditions have been applied to coal liquefaction.

Recent studies have aroused interest in possible new materials which can serve as catalysts since the conditions for initial reactions of the coal to be liquefied are lower temperatures and pressures. These processes, such as acid and base catalyzed depolymerization reactions, may make the catalytic conversion of coal to liquid fuels feasible.

1. INTRODUCTION

Catalysis is of prime importance in coal liquefaction processes whether such processes are direct, such as hydroliquefaction or solvent refining or indirect, such as Fischer-Tropsch or other synthesis processes. The catalysts used in a particular process depends upon, among other factors, the processing temperature and the physical state of the reacting species. Several reviews have been prepared to summarize the experience of research and process development in applying catalysts to coal liquefaction. The problems of summarizing the experiences and developing an

Y. Yürüm (ed.), New Trends in Coal Science, 361–381.

understanding of catalysis for coal liquefaction are compounded by the number of reaction types involved as well as such considerations as poisoning by components in coal and coal derived products, the presence of mineral matter and ash, the heterogeneity of coal and difficulties associated with the use of conventional solid catalysts and supports with solid reactants.

The hydrogenation or hydroliquefaction of coal is often termed, when a catalyst is used, as "catalytic liquefaction of coal". However, it is recognized that the pyrolytic or thermolytic breakup of the coal skeleton or macrostructure cannot be accelerated (or, in other words, catalyzed). One should not even expect that a solid reactant (coal) and a solid catalyst could achieve the contact required for conventional catalytic reactions [see Tables I and II, (1)]. The function of solid catalysts, which are considered as heterogeneous catalysts, is in the upgrading of solvent, or other hydrogen donors in the system, and in such reactions as hydrodesulfurization, hydrodeoxygenation and reduction of reactive, unsaturated compounds (2).

Catalysts used for coal liquefaction have included transition metals, oxides and sulfides, (mostly Ni, Mo, Fe) but also Sn and Zn compounds, hydrogen sulfide, hydrogen chloride and many varieties or combinations of these. Most recent work has been concentrated on catalysts designed to optimize specific processes for the selectivity desired. Catalysis may be applied to the conversion reactions or to one or more of the reactants. Studies of the volatility, acidity and chemical stability have been related to catalysts but much is still to be learned about the reactions and their susceptibility to catalysis.

Objectives of a catalyst application center around either reduction of molecular weight, heteroatom reduction or removal or both.

Since the types of reactions, the conditions of reaction and other factors make the direct liquefaction of coal so different from indirect methods, the emphasis here will be on direct liquefaction. This reasoning is also justified in that the catalysis of indirect liquefaction of coal does not, at any stage, actually involve catalytic coal liquefaction. In indirect liquefaction the initial process is gasification of coal to a synthesis gas while the subsequent process is synthesis of products from the synthesis gas produced in the gasification process.

2. HISTORICAL PERSPECTIVE

In order to understand why certain materials have been used as catalysts for hydrogenation or other direct liquefaction processing, it is helpful to know something about the

TABLE I

Comparison of Catalysts for LEFCO Process Using 1/8 in. (ID)x30 in. long and Last Chance (Kaipoarowitz) Coal*, (1)

Catalyst	Conversion, %	Catalyst	Conversion, %
$ZnBr_2$	58.5	Sn (powder)	7.9
ZnI_2	46.3	$SnCl_2.2H_2O$	7.6
$ZnCl_2$	41.1	$FeCl_3.6H_2O$	7.2
$SnCl_2.2H_2O$	40.5	Zn (powder)	7.0
$SnCl_4.5H_2O$	25.6	$ZnSO_4.7H_2O$	5.4
LiI	16.6	$(NH_4)_6Mo_7O_{24}.4H_2O$	5.4
$CrCl_2$	12.3	$FeCl_2$	3.3
$Pb(C_2H_3O_2)_2.3H_2O$	11.7	$CaCl_2.H_2O$	no reaction
NH_4Cl	11.0	$Na_2CO_3.H_2O$	no reaction
$CdCl_2.2.5H_2O$	7.9		

* P = 2000 psi, T = 650°C. Where no catalyst was present, essentially no reaction took place. Catalyst concentration used was 10% of the coal.

TABLE II

Comparison of Catalysts for LEFCO Process Using 5/16 in. (ID)x54 ft long, Kaiparowitz Coal* (1)

Catalyst	Conversion, %
$ZnCl_2$	65.9
Molecular sieves	21.5
Cobalt molybdate on silica-alumina	17.5
No catalyst added	19.1
WS_2	21.5
Fe_2O_3	17.7

* P = 1800 psi, T = 482°C. Catalyst concentration used was 5.8% of the coal weight.

history of catalytic liquefaction, especially in Germany during the period 1913-1945. A personal summary of this period has been given from first hand knowledge by Donath and Hoering (3) and some facts from their account will be used here.

The conditions which led investigators to look for hydrogenation catalysts for coal processing grew out of the work of Matthias Pier who discovered selective, oxidic catalysts that were not poisened by sulfur in the methanol synthesis (1923). Before this, even though Bergius had demonstrated that coal could be liquefied with hydrogen at high pressures, the situation seemed hopeless since essentially all coals contain sulfur, oxygen and nitrogen. At the time sulfur alone was considered to be a prohibitive poison for all hydrogenation catalysts.

The first sulfur resistant coal hydrogenation catalysts prepared by Pier in 1924 were sulfides and oxides of molybdenum, tungsten and the iron group metals. The use of these catalysts made it possible to add hydrogen, split carbon-carbon bonds and eliminate S, N and O from coals and oils.

The work and time spent to develop processes and suitable catalysts were major investments by Badische Anilin und Soda Fabrik (BASF) in Germany. The initial catalysts used for coal conversion were those found to be resistant to sulfur poisoning in the methanol synthesis. Later as a result of thousands of experiments, both small and large scale, and in depth discussions and collaboration with others, M.Pier and associates developed other catalysts and processes. From the discovery of molybdenum-zinc oxide catalysts in 1925 to the discovery of dual function catalysts (1930 and later) the work was greatly assisted by the operation of large scale plants. At the same time more than 1500 catalysts were tested in high pressure flow reactors and about five times as many catalysts were actually prepared though many were not tested at high pressure due either to "unpromising composition" or unsatisfactory physical properties. This extensive testing was only a part of the catalyst development since the catalysts mentioned here were vapor phase catalysts.

After the discovery of Mo-ZnO catalyst, which gave 100% yield of gasoline from coal tar, it was found that the throughputs were too small and the hydrogen recycle rates too high for commercial feasibility. At higher throughputs the catalyst deactivated severely and M.Pier and coworkers concluded that the hydrogenation process needed to be conducted in two stages:

1. A "liquid phase" where coal and liquid residues were converted to a middle oil (boiling below 325°C) using a suspended catalyst,

and

2. A "vapor phase" where the middle oil was converted into motor fuels using a fixed bed catalyst.

Liquid phase catalysts at first were molybdenum or ammonium molybdate. Because of cost and supply problems with molybdenum other catalysts were developed, including low cost iron ores and especially Bayermasse. With such low cost catalysts or catalysts used in small concentrations, these could be discarded after passing through the reactor, greatly simplifying this part of the process.

Vapor phase catalysts at first were hydrorefining types and the requirements for a fixed bed catalyst were obviously quite different from the dispersed liquid phase catalysts, even in physical properties such as physical strength, shape, size and melting point. In 1927 a catalyst was adopted which consisted of equimolar amounts of MoO_3, ZnO and MgO (catalyst 3510). Later (1930) a superior catalyst (catalyst 5058) composed of tungsten disulfide was adopted. This catalyst, a dual function catalyst, performed well at full scale and gave conversion rates of middle oil to gasoline three times as great as did catalyst 3510. Results are shown for catalyst 5058 in Table III.

TABLE III

Hydrogenation with Tungsten Disulfide Catalyst (3).

Feed	Product	Reaction Conditions Temperature °C	Pressure atm	Rate, kg/l.h	Conv. %
Diisobutane	Isooctane	216	250	2.0	99
Naphthalene	Decaline	336	200	0.9	90
Brown coal	Prehydrogenated M.O.	380	200	1.0	
Middle oil					
Phenols 18%					99.5
N-bases 4%					99.5
Sulfur 2.5%					99.5
n-Butane	Isobutane	400	200	0.5	35
Paraffins 260-320 °C	Gasoline 180 °C EP	408	220	1.0	90

One problem with 5058 was that in the production of gasolines the aromatic content and thus, the octane number, were low. In 1934 a montmorillonite clay (Terrana) was found which, when treated with hydrochloric acid and used with 10% catalyst 5058, gave gasolines with higher octane number but the same yield at the same temperature as catalyst 5058.

Further developments on vapor phase catalysts were necessary because of the presence of nitrogen compounds and their effect on some catalysts. The vapor phase process was eventually divided into two steps, "prehydrogenation" and "splitting hydrogenation or benzination" (hydrocracking). Catalysts such as metal sulfides were developed and much work was done on the effects of acidic and basic support materials.

Much of the work on the catalysis of coal hydrogenation have found more application in the petroleum refining industry. The work by coal scientists laid a foundation for many of the catalysts and processes presently used in hydrorefining and hydrocracking. It still remains to be seen if these catalysts and processes will be used for direct liquefaction of coal on a commercial scale.

3. CATALYTIC REACTIONS OF IMPORTANCE IN COAL HYDROGENATION

From the early Germany work and subsequent research and development on direct liquefaction of coal it because obvious that heterogenous catalysis, where the reaction takes place on a solid catalyst surface, involved liquid products from coal decomposition or thermal reactions. The reactions taking place after the initial coal decompostion are numerous.This variety has been summarized by Cusumano, et al. (4) and is shown in Table IV.

Some of the general reactions and products listed in Table IV are only "expected" since coal liquefaction, except utilizing the Fischer-Tropsch (Sasol) process is not presently commercial.

In addition to the general reactions and products from coal conversion to liquids there are a multitude of effects and interactions which can affect the catalytic process involved. Some of these are catalyst particle size, effects of catalyst-support interactions, surface effects of catalysts, poisoning, regeneration and recovery of catalyst materials. Cusumano et al (4) have also summarized (Table V) some of the significant advances of importance to coal conversion catalysis.

4. CATALYSTS USED IN DIRECT LIQUEFACTION

The least expensive catalyst available for coal conversion is the mineral matter in the coal. The catalytic effects of mineral matter on the SRC processes and the Exxon Donor Solvent process have been investigated by several researchers, including Gangwer et al. (5), Mukherjee and Chowdhury (6) and Gray (7).

TABLE IV

Catalytic Reactions in Coal Conversion (4).

Process	General Reactions	Specific Reactions/ Products
Direct liquefaction	Hydrogenation Cracking Hydrofining	Aromatic liquids Hydrodesulfurization (HDS) Hydrodenitrogenation (HDN)
CO/H_2 synthesis	Fischer-Tropsch	Methane Hydrocarbon liquids Alcohols Chemicals
Water-gas shift		Hydrogen
Direct gasification	Hydrogasification Oxidative gasification	Methane Synthesis gas
Liquids refining and upgrading	Cracking Reforming Hydroforming	Hydrogenation Dehydrogenation Dehydrocyclization Isomerization Hydrogenolysis Hydrodesulfurization (HDS) Hydrodenitrogenation

Both homogeneous and heterogenous catalysts have been used for coal liquefaction. Homogenous catalysts for coal and coal derived liquids have included hydrogenation transition metal complexes of cobalt, ruthenium, rhodium, palladium and platinum (2,8). Examples of such complexes are dicobalt octacarbonyl (9) and the rhodium complex of N-phenylanthranilic acid (10). A major problem with these homogeneous catalysts is recovery from the product mixture. The catalysts mentioned above are expensive and consequently, anything less than close to complete recovery results in a process too expensive to be feasible. Additionally, some of these catalysts are quickly poisoned by heteroatoms, especially sulfur.

Heterogenous catalysts which have been tested for coal hydrogenation, hydrocracking or hydrogenolysis include almost all solid catalytic materials. Over the years

TABLE V

Advances of Importance to the Catalytic Conversion of Coal (4).

Subject	Representative areas of impact
Multimetallic catalysis	-Upgrading of coal liquids. Methanation. Fischer-Tropsch synthesis.
Catalyst-support interactions	-Thermal and chemical stabilization of catalysts and supports.
Catalyst characterization	-Improvement of coal liquefaction catalysts. Upgrading of coal liquids. Determination of catalyst intrinsic activity.
Catalyst preparation	-New catalyst formulations. Controlled variations in catalyst properties. Higher surface area catalysts.
Poisoning and regeneration	-Effect of sulfur in coal conversion catalysis. Preventation of carbon deposition, removal of carbon.
Mechanism and surface chemistry	-Better understanding of the important steps in coal conversion. Identification of rate limiting processes, directions for process improvement.
Reaction engineering and catalyst testing	-Development of effective catalyst testing procedures. Data interpretation.
Inorganic chemistry	-New catalytic materials and compositions with improved poison tolerance and activity.
Materials science	-Novel support materials and structures. Novel refractory compounds. Understanding of sintering phenomena (catalyst deactivation).
Surface science	-New characterization techniques.

essentially all solid metals and metal compounds have been at least considered. The most often tested and utilized in process developments have been sulfided nickel/molybdenum catalysts supported on alumina, molybdenum catalysts or iron sulfides. Many other materials have been tested as catalysts which are directly related to the transition metal catalysts listed above. In a recent survey of the literature 1977-1986 on catalysts and catalytic liquefaction of coal the catalysts listed in Table VI were cited.

As indicated earlier the reactions desirable in coaliquefaction are primarily molecular weight reduction (depolymerization), heteroatom removal or both. Catalysts applied to the liquefaction process are usually directed mainly to depolymerization since upgrading (heteroatom removal) can be carried out in a subsequent process without the complications of the presence of solid residues and mineral matter. Molecular weight reduction or depolymerization by hydrocracking can be effectively catalyzed by such catalysts as $ZnCl_2$ for either coal or coal derived liquids (11,12). The hydrocracking action of metal halides, where smaller organic molecules are produced by the thermal cleavage (pyrolysis) of larger entities in the coal structure assisted by a catalyst and using hydrogen, takes place at lower temperatures (400-430 ^{o}C) with $ZnCl_2$. Some other metal halides are effective catalyst for hydrocracking as shown by Wood et al. (13) and probably act as homogenous catalysts at the temperature applied, as shown by their vapor pressures. The use of $ZnCl_2$ and other halides results in high conversions but some, such as $AlCl_3$, are too active, producing smaller molecules which are gases instead of liquids. The metal halides, particularly Zn and Sn, have been extensively tested. Sn is far too expensive for application unless 100 percent can be removed. $ZnCl_2$, which is approximately one order of magnitude less expensive than tin chloride, has received more attention but can also be too expensive for application without high recovery rates (14). Although considerable data are available using $ZnCl_2$ by Wood et al. (15) utilizing small concentrations of catalyst (5%) much of the work reported used massive amounts of catalyst (1:1 catalyst:coal ratios). This leads to major problems of recovery especially since during the hydrocracking of coal some zinc chloride is converted to zinc sulfide and other species. These compounds must then be converted back to zinc chloride in order to recycle the catalyst. There are also significant engineering problems associated with handling the zinc chloride which is corrosive (2).

The most commonly utilized heterogenous catalyst for direct coal liquefaction is sulfided cobalt molybdate on alumina and the current literature still includes more

TABLE VI

Coal Liquefaction Catalysts (1977-1986)

Catalyst	Support	Form of Preparation
MoO_3	-	Impregnated in coal
	α-alumina, γ-alumina, zirconia, silicon dioxide, refractory oxides, clays, silicate sludge	Sulfided
$(NH_4)_6Mo_7O_{24}$	-	Micro encapsulated
Mo naphthenate		
MoS_3		Impregnated in coal
Co-Mo, Co-W	alumina, silicon dioxide	
Ni-Mo, Ni-W		
Sn-Mo	alumina	
Mo-Ti		
Ni-Mo-Ti		
Sn-Fe-Mo		
Sn-Co-Mo	alumina	Stabilized with La-Si, La-Zr
Sn-Co-Ti-Mo		
Sn		Molten
$SnCl_2$		Molten
$SnCl_2$	alumina	
SnO_2	alumina	
ZnO	alumina	
$ZnCl_2$		
$ZnO-TiO_2$		Mixed with coal, molten
Bi_2O_3	alumina	
Ga_2O_3	alumina	
Se		Mixed with coal
Pd hydrous titanate		
TiO_2		
MnO		
Mo, Co, Ni, Mn, Sn (oil sol.catalyst)		With ligands n-octanoic acid or naphthenic acid
Co boride		From $CoBr_2 + NaB_5H_8$
Ni boride		From $NiBr_2 + NaB_5H_8$
Ni		Electroplated on coal

Table VI continued

Material	Treatment
SRC mineral residue	
SRC mineral residue ash	
Fe	Impregnated on coal, Mol. dispersed
Fe dust	From coal gasification and electric furnace
Fe ores	
magnetite, geothite, limonite, hematite bauxite, laterite red mud	Oxided, sulfided, reduced with CO
Fe sulfides	
FeS	Adsorbed (colloidal from $FeSO_4$+ Na_2S(aq)
Pyrite (FeS_2)	
Pyrite + H_2S	
Troilite (FeS)	
Pyrrhotite ($Fe_{1-x}S$)	
Iron III sulfide (Fe_2S_3)	
Fe_7S_8	
Fe oxides	
FeO	Treated with CO, H
Fe_2O_3	
Fe_2O_3 + S powder	
Fe_2O_3-TiO_2	
Fe_2O_3-MoO_3	
$FeSO_4$	Impregnated in coal
H_2S	

papers on this catalyst. Co/Mo on alumina is an active hydrogenation catalyst that is not easily deactivated by the heteroatoms in coal derived liquids. Deactivation is usually due to clogging of the catalyst pores by mineral matter in the coal. The best results (yields) are obtained when the catalyst is used in combination with hydrogen and a donor solvent (16).

According to Gun et al.(16), the donor solvent gives up hydrogen to free radicals formed from the coal thermolysis reactions to stabilize molecules. Then the dehydrogenated solvent goes to the catalyst surface to react with dissociated hydrogen and regain its hydrogen donating ability.

5. MILD TEMPERATURE CATALYTIC REACTIONS

Especially since about 1962 major efforts of many investigators who have been interested in coal liquefaction, have been directed at "depolymerization of coal". Coal does not have a simple repeating structure so the term depolymerization does not imply the production of monomers from a polymeric substance. The main objective of many coal liquefaction reactions is to cleave the minimum numbers of chemical and physical bonds which will result in the maximum number of liquid range products. One of the obstacles to systematic studies to accomplish this objective has been the conditions, especially temperature, at which liquefaction reactions have been studied. This dilemma has prompted many studies during the last two decades where catalysts were designed to attack certain linking groups or bonds thought to be present in coal. In order to avoid the indiscriminate thermal bond ruptures which occur above the softenining temperature of bituminous and subbituminous coals (350 ^{o}C) many experiments have now been conducted on catalytic liquefaction of coal at temperatures below 300 ^{o}C. These conditions have opened whole families of new catalyst possibilities since many materials stable below 300 °C decomposed, became unstable or vaporized at ordinary coal liquefaction conditions (>400°C). Representative studies and their catalysts will now be discussed which show the types of catalysts, reactions and products from these lower temperature (often called mild conditions) studies.

As far back as 1962 Heredy and Neuworth (17), showed that coal could be "depolymerized" and rendered soluble by reaction with phenol and boron trifloride at 100 ^{o}C. The reactions of this application have been studied using model compounds. The following reactions are visualized for coal depolymerization (18-22):

$$(AR)_1-CH_2-(AR)_2-CH_2-CH_2-(AR)_3 + 2\ C_6H_5-OH$$

$$\longrightarrow C_6H_5-CH_2-(AR)_1 + C_6H_5-CH_2-CH_2-(AR)_2H + H(AR)_3$$

$$(AR)_1-CH_2-C_6H_5 + C_6H_5OH \longrightarrow HO-C_6H_4-CH_2-C_6H_4-OH + H(AR)$$

These depolymerization reactions dramatically affect the solubility of the components of coal as shown in Tables VII and VIII the original coals (being less than 5 percent soluble in single solvents such as benzene and even tetralin at its normal boiling point). The depolymerization reactions listed are clearly not the only cleavage reactions taking place when coals are treated with phenol/boron trifloride since it has been shown that at least some C-O, C-N and C-S bonds are also affected (23).

TABLE VII

Coals Used in 'Depolymerization' Studies, [After Heredy and Neuworth (17)].

Coal	% C dmmf	Composition/100 C atoms	Petrogr. comp. V	E*	I**
1. N.Dakota	70.6	$C_{100}H_{84.38}O_{23.82}N_{1.45}S_{0.30}$	79	2	19
2. Colorado	76.7	$C_{100}H_{80.78}O_{18.88}N_{1.78}S_{0.18}$	87	3	10
3. W.Virgn. Ireland S.	82.4	$C_{100}H_{83.88}O_{8.02}N_{1.38}S_{0.77}$	83	3	14
4. Pennsyl. Montour S.	85.1	$C_{100}H_{81.21}O_{8.91}N_{1.91}S_{0.22}$	76	5	19
5. W.Virgn. Powelton S.	85.8	$C_{100}H_{78.16}O_{8.95}N_{1.80}S_{0.13}$	73	9	19
6. W.Virgn.	90.7	$C_{100}H_{80.43}O_{2.40}N_{1.04}S_{0.28}$	82	-	18

*: Including resinite
**: Comprised of micrinite and fusinite.

TABLE VIII

'Depolymerized' Products from Reaction with Phenol/BF_3 at 100°C, [after Heredy and Neuworth (17)].

Coal No	% Soluble in C_6H_6	C_6H_6/MeOH	Phenol	Total % Sol.	Blank*	Net % Sol.	% Phenol in prod**
1.	6.6	48.1	20.5	75.2	2.4	72.8	41.2
2.	6.3	15.1	2.1	23.4	6.3	17.1	32.8
3.	9.1	24.8	13.5	47.4	18.0	29.4	16.3
4.	7.0	14.7	7.1	28.8	11.7	17.1	12.4
5.	5.6	-	19.4	25.0	6.5	18.5	13.0
6.	4.4	-	5.4	9.8	1.1	8.7	15.5

*: Phenol solubles in original coal, net gain solubility = total - blank.
**: Total phenol content of reaction product mixture (including insoluble residue).

Other investigators (24-28) have applied similar reactions to catalyze the depolymerization of coals, usually with the objective of rendering a large fraction of the coal soluble so that more information may be obtained on the chemical structures present. Ouchi (24), for example, was able to solubilize all but a few coals having greater than 85 percent carbon to more than 90 percent using p-toluene sulfonic acid/phenol at 185°C (Table IX).

Although the depolymerization reactions cited here are referred to as "acid catalyzed depolymerization" (23), they must be considered as chemical reactions involving the acid components as well as phenol. This could also be said of some other solubilization reactions recently applied to coal, including reductive alkylation. However, for catalytic considerations, acid materials must be recognized as facilitating many reactions at mild conditions which may be applied for solubilization but might also serve as liquefaction reactions. These reactions would include non reductive alkylation (29), Friedel-Crafts alkylation and acylation. Similarly, based-catalyzed depolymerization reactions might serve as liquefaction steps; hydrolytic depolymerization (30), hydrogenolysis (31-34).

In looking at the total range of acid- and base-catalysed reactions of coal at mild conditions, Anderson and Miin looked at over 100 materials (35). Tables IX through XV give results for Bronsted acids and a whole range of compounds which may act as acid or base catalysts. Their conclusions included recognition of the different classes of acids and their properties which could be related to the extent of coal conversion. Only one property was shown to be directly related to the catalyst's ability to catalyze the production of liquids from a high volatile bituminous coal. That property was the HSAB hardness of the acid material. Only borderline catalysts gave more than 34 percent conversion of coal to liquids and gases (mostly liquids).

Although these mild temperature catalytic reactions of coal are being investigated vigorously by some, the practical application of the catalysts and conditions have not been demonstrated. This area of coal processing has not only opened up new possibilities for catalytic coal liquefaction but also made possible more exact identification of the bonds and structures involved in the liquefaction reactions. The application of organic chemistry principles to coal depolymerization reactions in a major liquefaction scheme has been demonstrated by Shabtai et al. (36) and consists of the following steps, (Figure 1):

1) Intercalation of coal with a Lewis acid catalyst material, zinc chloride, ferric chloride or other metal halide, followed by mild hydrotreatment (HT);

TABLE IX

Phenolation of Japanese Coals with Phenol/p-Toluene Sulphonic Acid at 185°C [After Ouchi (24)].

% C dmmf	Composition/100 C atoms	% Soluble in pyridine
69.7	$C_{100}H_{88.9}O_{26.2}N_{0.6}S_{0.2}$	90+
70.9	$C_{100}H_{78.0}O_{24.6}N_{1.0}S_{0.3}$	90+
75.8	$C_{100}H_{73.9}O_{18.0}N_{0.9}S_{0.2}$	90+
78.0	$C_{100}H_{88.3}O_{13.9}N_{0.9}S_{trace}$	90+
81.9	$C_{100}H_{85.9}O_{9.6}N_{2.2}S_{0.2}$	90+
83.1	$C_{100}H_{87.6}O_{6.2}N_{1.9}S_{1.0}$	90+
84.6	$C_{100}H_{80.2}O_{6.7}N_{1.7}S_{0.04}$	90+
86.2	$C_{100}H_{81.6}O_{x}N_{y}S_{z}$	90
89.6	$C_{100}H_{65.2}O_{2.6}N_{1.9}S_{0.3}$	30
92.9	$C_{100}H_{43.6}O_{x}N_{y}S_{z}$	10

TABLE X

Yields of Reaction Products (THF soluble + gases), from hydrotreatment of Hiawatha (Utah) hvbb Coal at P=13.5 MPa-H_2 and T=300°C for 3 hours. Catalyst:Transition metal compounds

Catalyst	Yields (wt %)		
	Gas	Liquid	Total
$RuCl_3.nH_2O$	13	19	32
$CoCl_2.6H_2O$	1	23	24
CoS	1	23	24
NiS	1	16	17
$NiCl_2.6H_2O$	3	23	26
Cu_2Cl_2	2	20	22
$Cu_2Cl_2.2H_2O$	3	17	20

TABLE XI

Yields of Reaction Products (THF soluble + gases), wt %, from Hydrotreatment of Hiawatha (Utah) hvbb Coal at P=13.5 MPa-H_2 and T=300°C for 3 hours. Catalysts: Group VA compounds.

Catalyst	Yields (wt %)		
	Gas	Liquid	Total
$SbCl_3$	2	32	34
BiF_3	3	19	22
$BiCl_3$	4	55	59
$BiBr_3$	8	64	72
BiI_3	9	34	43

TABLE XII

Yields of Reaction Products (THF soluble + gases), wt %, from Hydrotreatment of Hiawatha (Utah) hvbb Coal at P=13.5 MPa-H_2 and T=300°C for 3 hours. Catalysts: Group IVA compounds.

Catalyst	Yields (wt %)		
	Gas	Liquid	Total
$GeCl_4$	4	82	86
SnF_2	3	17	20
$SnCl_2.2H_2O$	9	78	87
$SnBr_2$	13	81	94
SnI_2	13	80	93
SnO	3	13	16
SnF_4	12	31	43
$SnCl_4.5H_2O$	2	73	75
$SnBr_4$	4	77	81
SnI_4	4	93	97
PbF_2	2	19	21
$PbCl_2.5H_2O$	1	21	22
$PbBr_2$	2	46	48
PbI_2	3	38	41

TABLE XIII

Yields of Reaction Products (THF soluble + gases), wt %, from Hydrotreatment of Hiawatha (Utah) hvbb Coal at P=13.5 MPa-H_2 and T=300°C for 3 hours. Catalysts: Iron compounds and hard acids.

Catalyst	Yields (wt %)		
	Gas	Liquid	Total
FeS	1	17	18
FeF_2	2	19	21
$FeCl_2.4H_2O$	2	27	29
$FeBr_2$	2	45	47
FeI_2	3	41	44
$FeCl_3$	14	19	33
$AlCl_3$	13	17	30
$CrCl_2.6H_2O$	14	26	40
$CeCl_3.7H_2O$	1	4	5

TABLE XIV

Yields of Reaction Products (THF soluble + gases), wt %, from Hydrotreatment of Hiawatha (Utah) hvbb Coal at P=13.5 MPa-H_2 and T=300°C for 3 hours. Catalysts: Group IIB compounds.

Catalyst	Yields (wt %)		
	Gas	Liquid	Total
ZnF_2	1	15	16
$ZnCl_2$	8	71	79
$ZnBr_2$	12	82	94
ZnI_2	12	73	85
$ZnSO_3$	2	32	34
$ZnSO_4.H_2O$	2	18	20
Zn_3P_2	1	23	24
$Zn_3(PO_4)_2$	1	19	20
$Zn(OAc)_2$	1	25	26
ZnO	1	15	16
CdF_2	2	26	28
$CdCl_2.2.5H_2O$	1	15	16
$CdBr_2$	1	29	30
CdI_2	3	18	21
$HgCl_2$	1	23	24

TABLE XV

Yields of Reaction Products (THF soluble + gases), wt %, from Hydrotreatment of Hiawatha (Utah) hvbb Coal at P=13.5 MPa-H_2 and T=300°C for 3 hours. Catalysts: Bronsted acids

Catalyst	Yields (wt %)		
	Gas	Liquid	Total
Benzoic acid (anhydrous)	3	11	14
Benzoic acid (2.7 N)	7	19	26
Phenol	2	19	21
HOAc (glacial)	7	13	20
HOAc (15 N)	6	20	26
H_3PO_4 (85%)	2	18	20
HCl (4.9 N)	12	38	50
HBr (4.9 N)	18	42	60
HI (4.9 N)	45	43	88

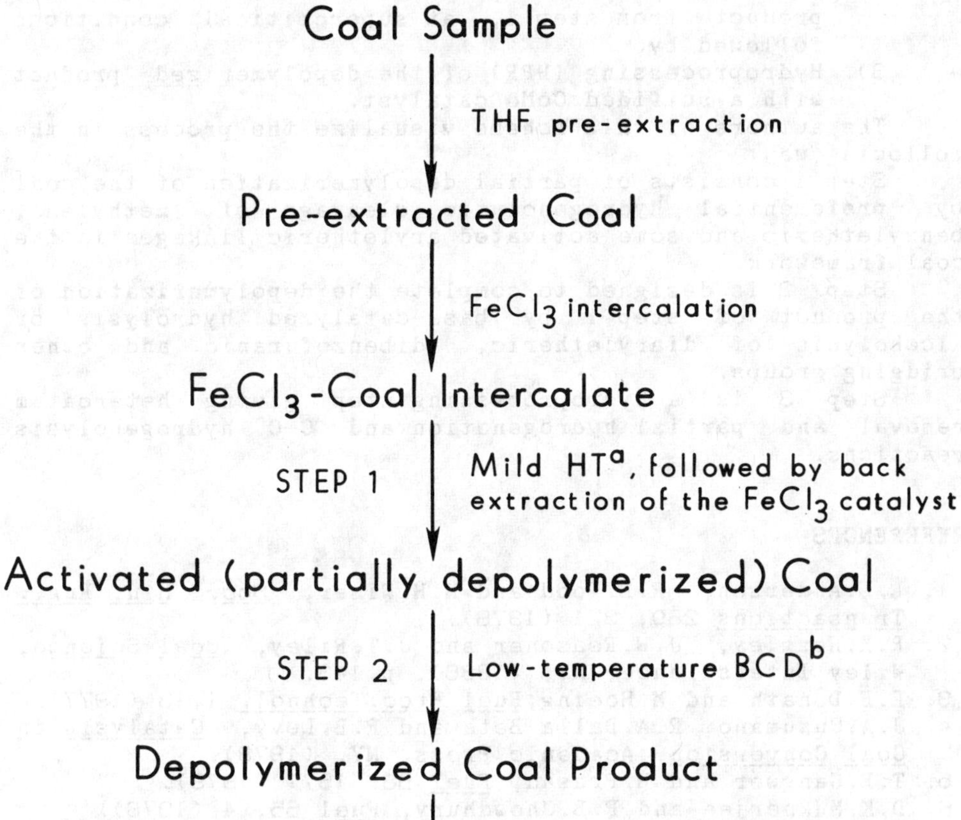

Coal Sample

THF pre-extraction

Pre-extracted Coal

$FeCl_3$ intercalation

$FeCl_3$-Coal Intercalate

STEP 1 — Mild HT[a], followed by back extraction of the $FeCl_3$ catalyst

Activated (partially depolymerized) Coal

STEP 2 — Low-temperature BCD[b]

Depolymerized Coal Product

STEP 3 — Hydroprocessing (HPR)[c] with sulfided catalysts

Low Molecular Weight Hydrocarbons (oils)

Figure 1. Procedure for production of low molecular weight hydrocarbons by base-catalyzed-depolymerization [after Shabtai (36)].

a. HT: hydrotreatment (250-275°C; H pressure, 1000-1500psig)

b. BCD: base-catalyzed depolymerization (250-275°C, initial N pressure, 1000 psig)

c. HPR hydroprocessing (350-370°C, H pressure, 2700 psig)

2) Base-catalyzed depolymerization (BCD) of the product from step 1, at supercritical conditions followed by,

3) Hydroprocessing (HPR) of the depolymerized product with a sulfided CoMo catalyst.

The authors of this scheme visualize the process in the following way:

Step 1 consists of partial depolymerization of the coal by preferential hydrogenolytic cleavage of methylene, benzyletheric and some activated aryletheric linkages in the coal framework.

Step 2 is designed to complete the depolymerization of the product of step 1 by base-catalyzed hydrolysis or alcoholysis of diaryletheric, dibenzofuranic and other bridging groups.

Step 3 is a hydroprocessing step giving heteroatom removal and partial hydrogenation and C-C hydrogenolysis reactions.

REFERENCES

1. L.L.Anderson, R.E.Wood and W.H.Wiser, Soc. Min. Eng., Transactions 260, 321 (1976).
2. R.K.Hessley, J.W.Reasoner and J.T.Riley, Coal Science, Wiley Interscience, NY, (1986), p.143.
3. E.E.Donath and M.Hoeing,Fuel Proc.Technol. 1, 3 (1977).
4. J.A.Cusumano, R.A.Dalla Beta and R.B.Levy, Catalysis in Coal Conversion, Academic Press, NY, (1978).
5. T.E.Gangwer and H.Prasad, Fuel 58, 1577 (1979).
6. D.K.Mukherjee and P.B.Chowdhury, Fuel 55, 4 (1976).
7. D.Gray, Fuel 57, 213 (1978).
8. J.L.Cox, S.A.Wilcox and G.L.Roberts, in 'Organic Chemistry of Coal',John Larsen (Ed.), ACS Symp. Series 71, ACS, Washington, D.C., (1978), p.186.
9. S.Friedman, S.Metlin, A.Swedi and I.Wender, J. Org.Chem. 24, 1287 (1959).
10. N.T.Holy, N.Nalesnik and S.McClanahan, Fuel 56, 47 (1977).
11. C.W.Zielke, R.T.Struck, J.M.Evans, C.P.Costanza and E.Gorin, Ind. Eng. Chem., Process Des. Devel. 5, 158 (1966).
12. C.W.Zielke, R.T.Struck, J.M.Evans, C.P.Costanza and E.Gorin, Ind. Eng. Chem., Process Des. Devel. 5, 151 (1966).
13. R.E.Wood, W.H.Wiser, L.L.Anderson and A.G.Oblad, U.S.Pat. 4,134,822, (1977).
14. A.Hill, Zinc Chloride Catalyst Recovery from Char, Masters Thesis, University of Utah, Salt Lake City, Utah, (1976).

15. R.E.Wood, L.L.Anderson and G.R.Hill, Quarterly of the Colorado School of Mines, Golden, CO. 65, 201, (1970).
16. S.R.Gun, S.K.Sama, P.B.Chowdhury, S.K.Mukherjee and D.K.Mukherjee, Fuel 58, 171 (1979).
17. L.A.Heredy and M.B.Neuworth, Fuel 41, 221 (1962).
18. L.A.Heredy, A.E.Kostyo and M.B.Neuworth, Fuel 42, 182 (1963).
19. L.A.Heredy, A.E.Kostyo and M.B.Neuworth, Fuel 43, 414 (1964).
20. L.A.Heredy, A.E.Kostyo and M.B.Neuworth, Fuel 44, 125 (1965).
21. L.A.Heredy, A.E.Kostyo and M.B.Neuworth, Adv. in Chem. Series No55, 493 (1966).
22. J.W.Larsen and E.Kuemmerle, Fuel 55, 162 (1976).
23. N.Berkowitz, 'The Chemistry of Coal', Elsevier Sci. Publ., Amsterdam, The Netherlands, (1985), p.182.
24. K.Ouchi, K.Imuta and Y.Yamashita, Fuel 44, 205 (1965).
25. Y.Yurum, I.Yiginsu and B.Zumreoglu, Fuel 58, 75 (1979).
26. Y.Yurum and I.Yiginsu, Fuel 60, 1027 (1981).
27. Y.Yurum, Fuel 60, 1031 (1981).
28. Y.Yurum and I.Yiginsu, Fuel 61, 1138 (1982).
29. B.S.Ignasiak, D.Carson, A.J.Szladow and N.Berkowitz, ACS Symp. Series No 139, 97 (1980).
30. I.G.C.Dryden, in 'Chemistry of Coal Utilization' Suppl. Vol. (H.H.Lowry, Ed.), J.Wiley, NY, (1963), p.232.
31. L.Reggel, C.Zahn, I.Wender and R.Raymond, U.S. Bu. of Mines Bull. No 615 (1965).
32. M.Makabe and K.Ouchi, Fuel 58, 43 (1979).
33. K.Ouchi, K.Fuwata, M.Makabe and H.Itoh, ACS, Fuel Chem. Preprints 24, 185 (1979).
34. L.L.Anderson, K.E.Chung, R.G.Pugmire and J.S.Shabtai, in 'ACS Symp. Series', No. 169, (1981), p.223.
35. L.L.Anderson and T.C.Miin, Fuel Proc. Technol. 12, 165 (1986).
36. J.S.Shabtai, T.Skulthai and I.Saito, ACS Fuel Chem. Div. Preprints 31, 15 (1986).

SOLUBILIZATION OF COAL

Aral Olcay
Department of Chemical Engineering,
Ankara University, Ankara, Turkey

ABSTRACT. Recent literature on solvent extraction of coals has been reviewed and discussed.

1. INTRODUCTION

The solubilization of coal has been studied for the determination of coal structure. Several methods have been applied so far for the solubilization of coal either partly or almost completely. Solvent extraction being the most convenient technique to be applied to coal was widely studied in the early days of coal chemistry. Wheeler and his colleagues (1) used pyridine, while Fischer (2) used benzene under pressure and obtained substantial yields of extracts. However, the aim of early attempts of coal solubilization was not the identification of coal structure but to separate the mixture of compounds which were responsible for caking properties of good coking coals and incorporate them into poor coking coals to upgrade them. Both groups of researchers (1,2) observed that the coking properties of the coals diminished as extraction proceeded. They also found out that to improve the coking properties of coal, incorporation of the extracts into the coal was not sufficient, the characteristics of the coal were also important. However, the early solvent extraction studies led to the better understanding of the action of a wide range of solvents on coals.

From the 1920's on, increasing attention was directed to solvent extraction under conditions in which coal undergoes active thermal decomposition.

In the 1930's Pott and Broche (3) demonstrated that anthracene oil and tetralin at 400 ^{o}C could dissolve almost all the coal substance.

Investigation of the chemistry of coal-solvent interactions is difficult because the molecular structure of coal is not clearly understood. Attempts have been made to identify the properties of pure solvents which make them effective for coal extraction. Solvents with different chemical and physical properties have different effects on coal.

In 1951 Oele and his colleagues (4) distinguished four types of solvent in respect of their effect on coal: non specific, specific, degrading and reactive solvents.

Non specific solvents are believed to extract, at temperatures below 100 ^{o}C, resins and waxes of the original plant material that are

Y. Yürüm (ed.), New Trends in Coal Science, 383–399.

present as macroscopic flakes or occluded in the coal matrix (5). The extract yield never exceeds a few percent of the coal. Non specific solvents are not important in the study of coal chemistry. Common organic solvents with low boiling point such as ethyl alcohol, benzene, acetone are in this class.

Specific solvents dissolve 20-40 % of the coal substance at temperatures below 200 °C. The process is physical and the nature of the extract is similar to that of the coal therefore specific solvents are nonselective solvents. Specific solvent extraction can supply information on coal structures. Nitrogen and oxygen compounds are generally good solvents due to the presence of an unshared electron pair on the nitrogen or oxygen atom in the molecule which makes the solvent polar (6). Nitrogen compounds are generally better solvents than oxygen compounds. Extract yields, as high as 35-40 % can be obtained by extraction with pyridine, certain heterocyclic bases and primary aliphatic amines. Secondary and tertiary aliphatic amines are less effective because more than an alkyl group on the amine presents steric hindrance to interaction between the solvent and coal.

Degrading solvents are effective solvents when used for coal extraction at temperatures in the range of 200-400 °C. After extraction the solvents can be recovered from the extract almost totally and substantially unchanged. Their action depends on the thermal degradation of the coal to form smaller and more soluble fragments. Phenanthrene and diphenyl are examples of degrading solvents. Since the extent and the characteristics of the degradation process of coal depend on temperature, the yield and, to some extent, composition of the extract show a variation with the extraction temperature. At temperatures above 400 °C pyrolytic condensation reactions in the coal make it progressively less soluble leading to sharp decrease in the extract yield.

Heredy and Fugassi (7) using tritium and ^{14}C labelled phenanthrene in coal extraction, have concluded that degrading solvents also act as hydrogen transfer agents or hydrogen shuttlers beside dispersing smaller and more soluble fragments of thermally degraded coal: a thermally produced coal radical abstracts hydrogen from phenanthrene yielding a phenanthrene radical which in turn abstracts a hydrogen from another part of coal. In this way hydrogen is shuttled between different parts of the coal. In this respect, these solvents should be considered as reactive solvents although they are recovered more or less unchanged.

Tar oil fractions containing a variety of chemical compounds are also used as degrading solvents. They are not always recovered unchanged from the coal extract. When the solvent recovered from the extract is used again in coal extraction, a steady decrease in the solvent power has been observed. This may indicate the presence of traces of reactive compounds which are consumed during the first cycles by reacting with coal (5).

Reactive solvents dissolve coal by reacting with it at temperatures around 400 °C. The composition of the extracts obtained with reactive solvents differ markedly from those obtained by degrading solvents. Chemical reaction between coal and solvent can be estimated from chemical changes in the recovered solvent. The extract and the residue usually weigh more than the original coal indicating that some chemical reaction had taken place between coal and solvent (5). While Oele et al. (4) termed this process "extractive chemical disintegration", the terms "solvolysis" (8) or "thermal dissolution" are more often used. Almost all work on reactive solvents relates to systems involving thermal degradation of coal to soluble fragments and hydrogen transfer reactions. The hydrogen donor power of a reactive solvent depends on its molecular structure. Solvents containing hydroaromatic rings such as tetralin give higher coal conversion than their aromatic analogue and the hydroaromatic member of a heterocyclic pair (such as indoline) is more effective than the aromatic member (indole). These solvents can act as hydrogen donor or as hydrogen transfer agents (5).

Clark et al. (9) have studied the reactivities of cycloalkanes either alone or as a part of a solvent mixture during the solvent extraction of coal at 430 ^{o}C. When used with polyaromatic compounds having 3 or more rings as solvents, cyclohexanes donate hydrogen in the presence of coal-derived radicals, giving rise to high extraction yields.

Volker and coworkers (10) have studied the influence of o-cresol on conversions of Bruceton coal to THF extractable material over the full range of composition for mixtures of o-cresol with either tetralin, phenanthrene, anthracene or pyrene. The % conversion is plotted against % cresol content of the mixture. As can be seen from Fig.1, in the

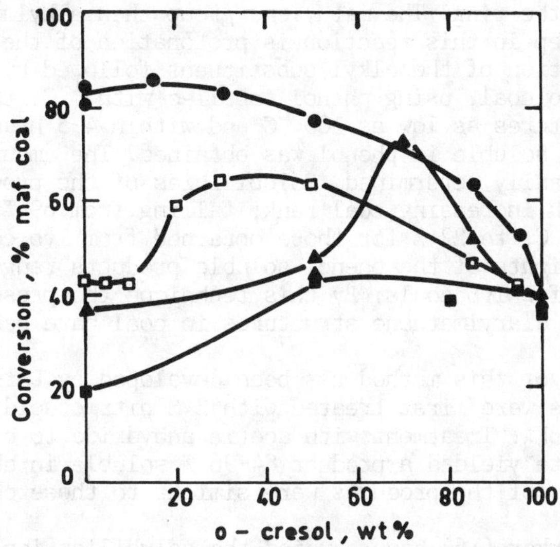

Figure 1 Conversion of Bruceton coal to THF extract in various solvents as a function of added o-cresol. Tetralin ● ;pyrene ▲ ; anthracene ■ ; phenanthrene □ .

absence of cresol, conversion of coal to THF extract is much smaller, in anthracene, pyrene or phenanthrene than in tetralin.

Addition of cresol to polynuclear compounds causes an increase in coal conversion while addition of cresol to tetralin did not result in an increase. This indicates that o-cresol can act as a hydrogen donor in the presence of nondonor polynuclear aromatics; which is in agreement with an earlier report (11). While tetralin has been extensively used as a hydrogen donor agent in coal conversion and is still used in coal research. Moschopedis (12) has studied the use of oil sands bitumen as a hydrogen donor vehicle.

Isopropyl alcohol and methyl alcohol can act as hydrogen donors to coal. In contrast to tetralin, however, the transfer of hydrogen by the alcohol can be promoted by the presence of either potassium isopropoxide or KOH (13). The amount of acetone formed during the reaction was equal to the amounts of hydrogen transferred to coal. With such mixtures it was possible to have a 96 % pyridine soluble product at 335 ^{o}C in 90 min. Model compound studies with anthracene and diphenylether indicated that while the aromatic ring was hydrogenated, the ether linkage was unchanged. Isopropyl alcohol was found to be unstable in the presence of bases at temperatures above 335 ^{o}C, but the methanol/KOH system was sufficiently stable at 400 ^{o}C, which made the latter system more suitable because of the higher activity of coal at this temperature.

Heredy and Neuworth (14) tried to solubilize coal by using it as an alkylating agent. Experiments carried out with model compounds containing methylene chains joining two aromatic groups and using phenol as solvent and borontrifluoride as catalyst, indicated the cleavage of methylene groups at the ring. The methylene group then alkylated the phenol. The first step in this reaction is protonation of the aromatic compound at the position of the alkyl substituent followed by alkylation of phenol. Applied to coal, using phenol labelled with ^{14}C, the reaction proceeded at temperatures as low as 100 ^{o}C and within 4-5 hr a product which was up to 75 % soluble in phenol was obtained. The amount of phenol in the product was easily determined. Solubilities of the products in phenol decreased with increasing coal rank; falling from 69 % in the case of lignite with 71 % C, to 22 % for those obtained from lvb coals with 85 % C. Molecular weights of the phenol soluble products ranged from 300 for lignites to 930 for lvb coals. By this technique the presence of isopropyl groups and diarylmethane structures in coal have been demonstrated (14).

A modification of this method has been developed by Darlage and coworkers (15). Coals were first treated with 2 M nitric acid, then reacted with phenol-BF_3. Treatment with acetic anhydride to convert the phenolic OH to acetate yielded a product 84-96 % soluble in chloroform. The molecular weights of the products were similar to those observed by Neuworth (14).

Ouchi and coworkers(16) have studied the solubilization of some Japanese coals by using phenol and p-toluene sulphonic acid as the catalyst. The yield of pyridine soluble products was 80-90 % for coals having a carbon content in the range of 70-80 %, the yield decreasing with increasing coal rank. Solubilization of coal was explained by cleavage of short methylene groups between aromatic units in coal yielding low molecular weight compounds. More recent studies by Ouchi et al. (17) also indicated that the yield of lower molecular weight products increased with repeated reactions in contrast to Larsen and coworkers (18), claims about the presence of polymeric units in the reaction mixture.

2. SOLVENT CHARACTERISTICS - SOLVENT EFFECTIVENESS

Solvent extraction of coal at temperatures below the onset of active thermal decomposition is a useful method for the separation of materials with different chemical structure and different solubilities. The process is a physical process and much of the original coal structure is retained in the extracts. Therefore researchers interested in the study of the constitution of coal have investigated coal dissolution by specific solvents. Several comprehensive reviews of coal extraction can be found in the literature (5, 19-23).

Much research has been carried out in an attempt to correlate the physical properties of specific solvents with their ability to dissolve coals. No discernible relations between solvent power and the dielectric constant, dipole moment or surface tension of the solvent have been observed (19). To explain coal solubility, a thermodynamic approach was made by Van Krevelen (24).

According to Van Krevelen coal should be viewed as a cross-linked polymer. Coalification is considered to be the progress of a polymerization reaction with progressive loss of functional groups, the insoluble product being the three dimensional macromolecular network and extractable substances being the monomers or oligomers filling the pores of the network. The amount of the extractable substances dissolved by a particular solvent is determined by the solubility parameter of the solvent. The solubility parameter is defined as the positive square root of the cohesive energy density.

$$\delta = C^{1/2}$$

The cohesive energy density or specific cohesive energy was defined as

$$C = \frac{\Delta U_v}{V_1}$$

Where ΔU_v is the energy change for complete isothermal vaporization of the saturated liquid to the ideal gas state and V_1 is the molar volume of the liquid. The unit of the solubility parameter in the cgs system is $(cal/cm^3)^{1/2}$ or "hildebrands". In SI the most appropriate and convenient unit is $MPa^{1/2}$ which is numerically equal to $J^{1/2}\ cm^{-3/2}$.

Solubilization occurs when the free energy of mixing $\Delta G = \Delta H - T\Delta S$ is negative. Since the entropy of mixing, ΔS, is always positive, the sign of ΔG was determined by the sign and magnitute of the heat of mixing, ΔH. For reasonably non-polar molecules and in the absence of hydrogen bonding, ΔH is positive. According to the Hildebrand-Scatchard (25) theory, the heat of mixing is given by

$$\Delta H_m = v_1 v_2\ (\delta_1 - \delta_2)^2$$

where v_1 and v_2 are the volume fractions and δ_1 and δ_2 are the solubility parameter of the solvent and solute respectively. Therefore when $\delta_1 = \delta_2$ the total heat of mixing will be small or zero and the free energy change negative.

Angelowich and his coworkers (26) showed that the solvent effectiveness was a function of the non-polar solubility parameter for a series of pure and coal tar solvents. Van Krevelen (24) calculated theoretical values of δ for coals as a function of rank by summing the energy contributions of atoms and constitutional characteristics. When a cross-linked polymer is soluble in a solvent, the solvent causes it to swell. The polymer is assigned the δ value of the liquid providing the maximum swelling. Kirov et al. (27) determined d values from swelling experiments in various solvents. It was also shown that compounds which are as chemically different as pyridine ($\delta = 21.9\ J^{1/2}cm^{-3/2}$), phenol ($\delta = 22.1\ J^{1/2}cm^{-3/2}$) and thionyl chloride ($\delta = 21.1\ J^{1/2}cm^{-3/2}$) are all reasonably good solvents for coal.

Often a mixture of two solvents, one having a δ value higher and the other having a δ value lower than that of the solute is a better solvent than the two solvents themselves.

When two miscible liquids are mixed, the solubility parameter of the mixture is given by

$$\delta_m = v_1\delta_1 + v_2\delta_2$$

where v_1 and v_2 are the volume fractions of the two solvents.

Hombach (28) studied a series of ternary systems coal-solvent$_1$-solvent$_2$ and the systems were treated as quasi-binary systems. The solubility parameters of six coals were determined. In Fig.2 are presented the solubility parameters of coals versus rank. A minimum was found for

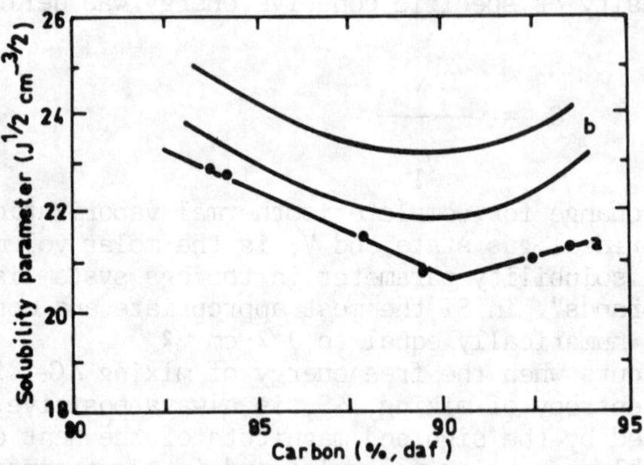

Figure 2 Solubility parameter of coals versus rank (28). Comparison of: (a) measured values; (b)calculated area (24).

coals having a carbon content of 89 % (daf). The minimum can be explained as a result of adding two independent functions: 1- the decrease of the solubility parameter with increase in rank due to the loss of polar groups e.g. hydroxyl groups, 2- the increase of the solubility parameter with increasing rank due to increasing aromaticity and increasing cross-linking. For a completely cross-linked product, the solubility parameter would be meaningless.

Weinberg and Yen (29) have studied the swelling properties of a hvA bituminous coal before and after removal of some extractable matter.

The extractability spectrum of the untreated coal using mixed solvent systems with solubility parameters ranging from 14.3 to 47.9 $MPa^{1/2}$ is shown in Fig.3. The swelling is given by $(W_s - W_o)/W_o$ where W_s and W_o are the swollen and initial volume of the coal respectively. As can be seen in Fig.3, there are two maxima in the spectrum at 22.5 and 28.6 $MPa^{1/2}$. For removal of soluble material coal has been extracted with pyridine in a soxhlet extractor. The swelling spectrum of the pyridine extracted coal is shown in Fig.4. As is seen, while the peak at 28.6 $MPa^{1/2}$ disappeared, the peak at the lower parameter remained. From this they

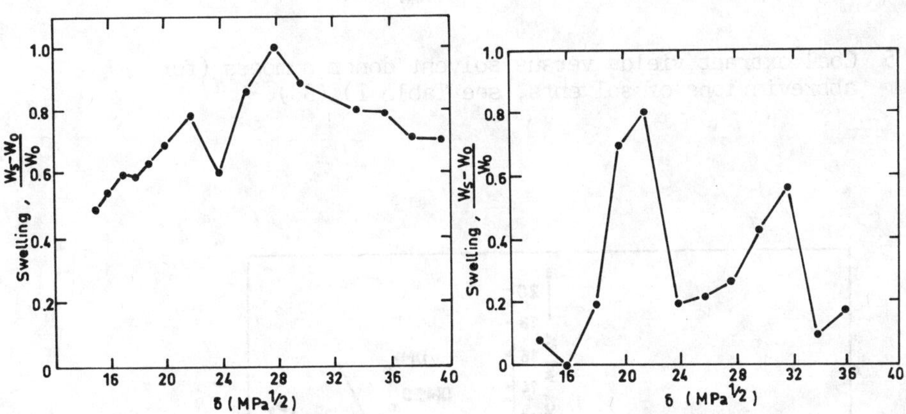

Figure 3 Swelling spectrum of the dried untreated PSOC coal using the mixed-solvent system.

Figure 4 Swelling spectrum of pyridine-extracted PSOC 102 coal using the mixed-solvent system.

concluded that the solvent must interact with the soluble matter in coal for maximum swelling to occur. When the soluble matter is removed swelling is suppressed.

The extraction of a hvA bituminous coal at ambient temperature is considered in terms of donor-acceptor interactions between coal and solvent (30). Donor number (DN) describes the nucleophilic behaviour and acceptor number (AN) describes the electrophilic behaviour of the solvent.

In Table 1 solvent effectiveness is presented in terms of extract yields and electron donor acceptor numbers.

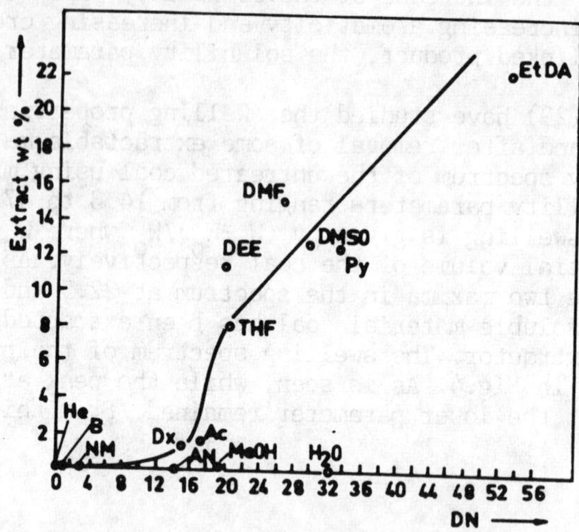

Figure 5 Coal extract yields versus solvent donor numbers (for abbreviations of solvents, see Table 1) (30).

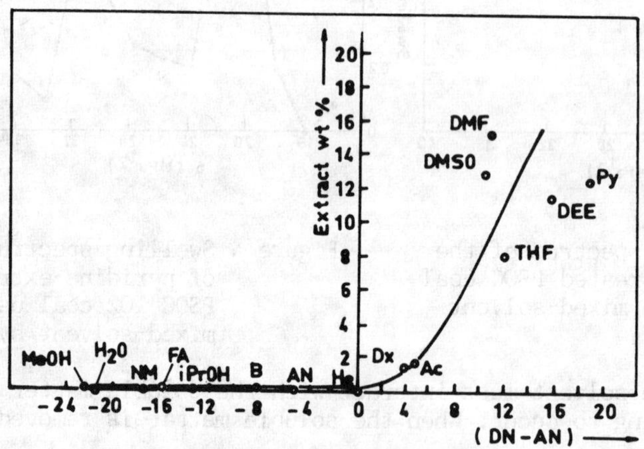

Figure 6 Coal extract yields versus DN-AN of solvents (for abbreviations of solvents, see Table 1) (30).

Table 1 Yields of coal extracts and electron donor, DN, and acceptor, AN, numbers of solvents (30).

SOLVENT	Abbreviation used	YIELD OF COAL EXTRACT (wt.% daf coal)	DN	AN	DN-AN
n-Hexane	He	0.0	0	0	0
Water	H_2O	0.0	33.0	54.0	-21.8
Formamide	FA	0.0	24.0	39.8	-15.8
Acetonitrile	AN	0.0	14.1	19.3	-5.2
Nitromethane	NM	0.0	2.7	20.5	-17.8
Isopropanol	iPrOH	0.0	20.0	33.5	-13.5
Acetic acid	AA	0.9	-	52.9	-
Methanol	MeOH	0.1	19.0	41.3	-22.3
Benzene	B	0.1	0.1	8.2	-8.1
Ethanol	EtOH	0.2	20.5	37.1	-16.6
Chloroform	$CHCl_3$	0.35	-	23.1	-
Dioxane	DX	1.3	14.8	10.8	+4.0
Acetone	Ac	1.7	17.0	12.5	+4.5
Tetrahydrofuran	THF	8.0	20.0	8.0	+12.0
Diethyl ether	DEE	11.4	19.2	3.9	+15.3
Pyridine	Py	12.5	33.1	14.2	+18.9
Dimethylsulphoxide	DMSO	12.8	29.8	19.3	+10.5
Dimetyhlformamide	DMF	15.2	26.6	16.0	+10.6
Ethylenediamine	EtDA	22.4	55.0	20.9	+34.1
1-Methyl-2-pyrrolidinone	MP	35.0	27.0	13.3	+14.0

As can be seen from Fig.5, extract yield increases with increasing electron donor number of the solvent; which is in agreement with Dryden's (6) observations on solvents containing unshared electron pairs. Although there is no correlation between extract yield and the acceptor number of the solvent, a correlation has been observed between extract yield and DN minus AN values of the solvent, Fig.6.

The model of coal extraction has been based on the concept that coal consist of a macromolecular three dimensional network with extractable substances filling pores of the network. High extract yields obtained from coal extraction at temperatures low enough not to provoke thermal cleavage of the bonds, lead to the conclusion that the extractable content of coal is high and the substances in the pores are less strongly bonded than the matrix material. Therefore the coal macromolecular network and the pore substances are considered separately on the coal extraction model (Fig.7).

DN_N and AN_N denote electron donor and electron acceptor centres occuring in the coal network.

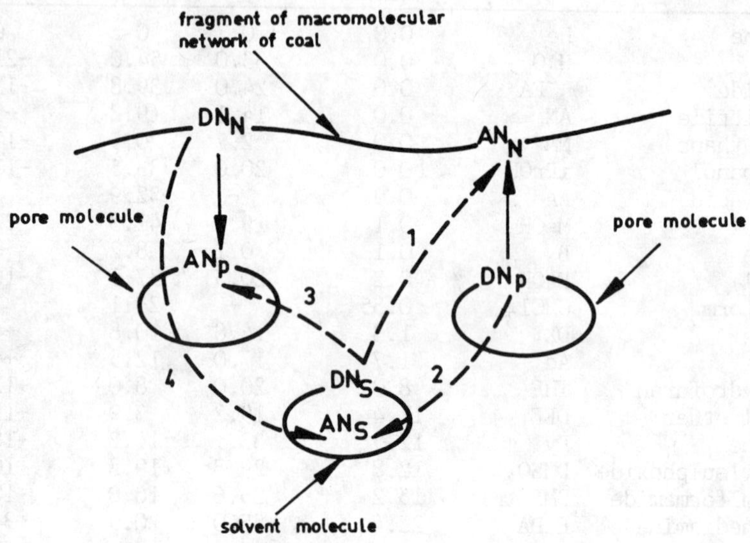

Figure 7 Coal extraction model showing donor-acceptor bonds occuring in coal. 1,2,3,4 represent possible routes of donor-acceptor bond formation between solvent molecules and electron-donor or -acceptor centres of coal (30).

DN_p and AN_p denote electron donor and electron acceptor centres occuring in pore material. Electron donor centres DN_p and DN_N are pyridine rings, oxygen functional groups, π rich heteroatomic rings. Electron acceptor centres (AN_p and AN_N) are phenol protons, pyrrole-NH protons and π deficient heteroatomic groups.

In the coal model, the presence of two types of donor-acceptor bonds $DN_N \dashrightarrow AN_p$, $DN_p \dashrightarrow AN_N$ binding together pore substances and the macromolecular network is assumed. These bonds may be hydrogen bonds, coordinate bonds or charge transfer links. Extraction will occur if the following new bonds are formed during extraction.

$$DN_S \dashrightarrow AN_p$$

$$AN_S \leftarrow DN_P$$

$$DN_S \rightarrow AN_N$$

$$AN_S \leftarrow DN_N$$

Donor-acceptor bond energy can be calculated from the donor and acceptor number

$$\Delta H = DN \times AN / 100$$

The formation of new bonds will be possible if their bond energies are higher than those of the intermolecular bonds of the solvent molecules and the bonds present in coal. Therefore extraction is assumed, in principle to be a substitution reaction; pore substances are replaced by a solvent molecule in their $D_N \rightarrow A_P$ or $D_P \rightarrow A_N$ bonds that bind together structural elements of the original coal (30).

Maximum and minimum values for acceptor and donor numbers for coal are assigned on the basis of the functional groups present.

The group of solvents characterized either by low DN and AN values or by considerably higher values of DN and AN cannot fulfil the above mentioned requirement therefore they do not yield any extract.

The group of solvents with DN higher than 19 and AN lower than 21 are effective solvents in coal extraction. Diethyl ether, tetrahydrofuran, pyridine and ethylendiamine are examples of effective solvents in coal extraction.

Szeliga and Marzec (31) have used twenty different solvents characterized by electron-donor and electron-acceptor number and solubility parameter to study the swelling of a hvb coal (Table 2).

The swelling ratio is plotted versus solvent donor number and solubility parameter, in Figs.8 and 9 respectively. As can be seen from the figures, swelling and solvent electron donor number shows a reasonably good correlation while swelling and solubility parameter does not. The solvents with DN ranging from 0 - 16 swell the coal only slightly. The solvents having DN between 16 - 30 swell coal to a great extent. From this, it was concluded that electron donor sites in coal macromolecules have donor numbers of 16 - 30 approximately.

Marzec and Pajak (32) have also studied the effect of preswelling on the extract yield of the hvb coal. Swelling of the coal by the vapour of the solvents pyridine, tetrahydrofuran and dimethylformamide was carried out at 25°C. Vapour saturated coals were extracted with THF, DMF, benzene and diethylether. Fresh coals were extracted both with the same solvent that was used for extraction of the swollen sample and with the same solvent together with the same amount as present in the swollen coal of the solvent used for preswelling.

The results indicated that preswelling brings about a decrease or an increase in the extract yields compared with the yields derived from fresh coal. For example preswelling of coal in pyridine vapour resulted in a decrease in THF and DMF extracts yields but in an increase of benzene or diethyl ether extract yields. From these results it was

Table 2 Swelling of a high bituminous coal and characteristics of solvents (31).

Solvent	Solvent number	Swelling ratio, Q	Solvent Characteristics	
			Donor number	Solubility parameter ($cal^{1/2}cm^{-3/2}$)
Benzene	1	1.0	0.1	9.2
Nitrobenzene	2	1.1	4.4	11.1
Isopropyl alcohol	3	1.14	20.0	11.44
Acetonitrile	4	1.15	14.1	11.8
Diethyl ether	5	1.15	19.2	7.4
Dioxane	6	1.16	14.8	9.8
Nitromethane	7	1.18	2.7	11.0
Methanol	8	1.19	19.0	12.9
Propan-1-ol	9	1.23	-	10.2
Ethanol	10	1.25	20.5	11.2
Ethyl acetate	11	1.26	17.1	8.6
Acetone	12	1.30	17.0	9.4
Methyl acetate	13	1.32	16.5	9.2
Methyl ethyl ketone	14	1.49	-	9.45
Tetrahydrofuran	15	1.59	20.0	9.1
1,2-Dimethoxyethane	16	1.60	24.0	-
Dimethylformamide	17	1.69	26.6	11.5
Dimethylsulphoxide	18	2.04	29.8	12.8
Pyridine	19	2.08	33.1	10.4
Ethylenediamine	20	2.08	55.0	-
1-Methyl-2-pyrrolidinone	21	2.38	27.3	-

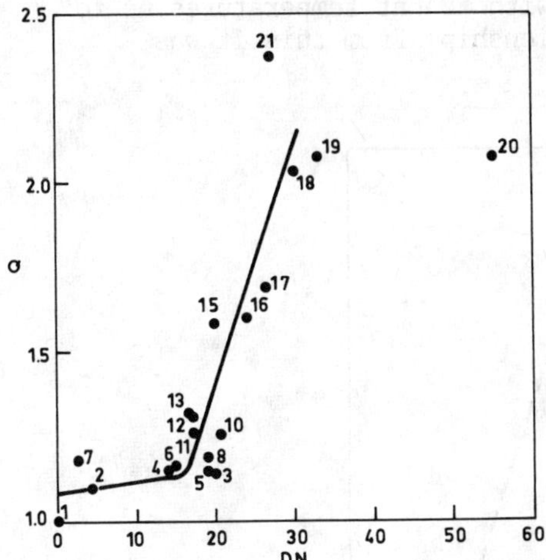

Figure 8 Swelling ratio, Q, of hvb coal versus electron-donor number, DN, of solvents. Numbers indicate solvents listed in Table 2.

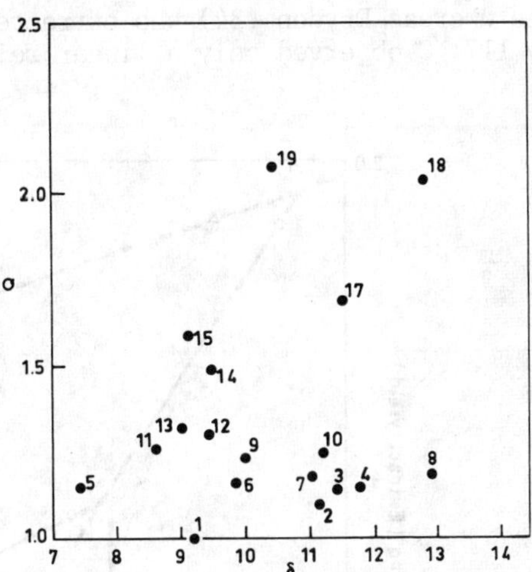

Figure 9 Swelling ratio, Q, of hvb coal versus solubility parameter, , of solvents. Numbers indicate solvents listed in Table 2.

concluded that some solvents block pathways to some, although not all, sites that were accessible to other solvents when acting singly on fresh coal (32). Solvents less powerful such as benzene and diethyl ether could extract a little more from the swollen coal due to their access to new sites, rather than to their successful competition with the solvents used for preswelling.

If coalification is to be considered the progress of polymerization with progressive loss of functional groups, low rank coal should consist almost entirely of non-cross-linked oligomers and should therefore be quite soluble in suitable solvents. Ethanol having a δ-value between those of ethylenediamine and monoethanolamine does not extract coal to any appreciable extent. Thus, chemical factors also play an important role in coal extraction. Van Bodegom et al. (33), starting from the above reasoning, have examined the dissolution behaviour of brown coals having a carbon content in the range of 67-75 % daf coal in pyridine, primary amines and aqueous KOH.

In amine solvents the dissolution process reached an appreciable rate at 140 ^{o}C and in aqueous KOH at 80 ^{o}C. As seen in Fig.10, a plot of log extract yield versus 1/T, there are changes in the slope of the lines whereas Dryden (34) who extracted with EDA at temperatures up to only 117 °C observed only a linear relationship. From this it was

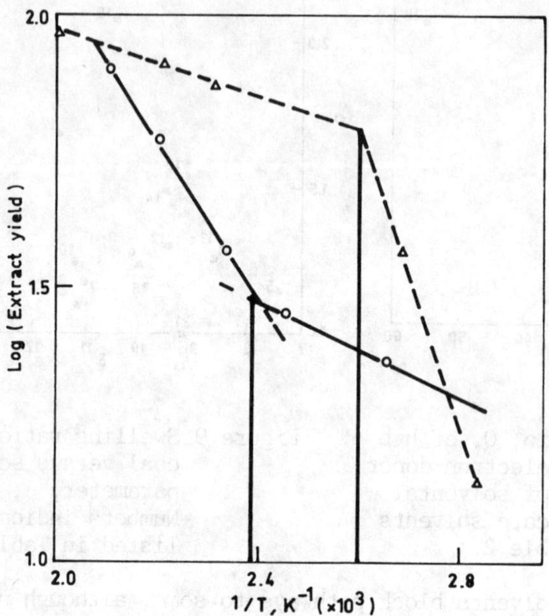

Figure 10 Plot of log (extract yield) versus reciprocal temperature. o , Morwell/pentylamine; Δ , Wyoming/aq.KOH

concluded that a chemical reaction does take between coal and solvent. As to the nature of chemical reaction, it was proposed that it is the breaking of ester bonds present in lower rank coals:

$$RCOOR' + R''NH_2 \longrightarrow RCONHR' + R'OH$$

since pyridine is unable to break ester bonds the solubility in pyridine should not depend on temperature and that was indeed the case. Pretreatment of the brown coals with sodium ethanolate in ethanol or with aqueous-HCl caused an increase in solubility due to the ester bond breaking. The coal fragments, having acid groups, are too polar to be soluble in apolar amines, but are easily soluble in more polar solvents such as MEA.

In Fig.11 the percent residue is plotted against the alkyl chain length of the alkyl amines used as solvents. As is seen, the solubility of Morwell brown coal at 180 °C increases with increasing alkyl chain length in the amine in contrast to one's expectation.

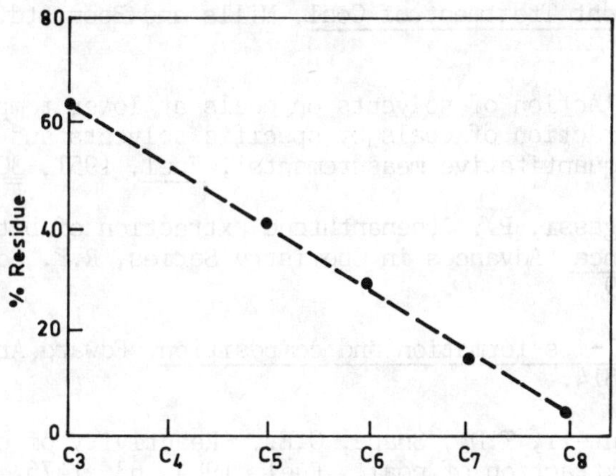

Figure 11 Plot of percentage residue as a function of alkyl-chain length; Morwell/n-alkyl amines, 180 °C.

This is explained by the formation of an amide or a salt from the reaction of amine with the acidic groups of coal. Therefore the longer the alkyl chain the more close the solubility parameter of the solvent will be to that of the reaction products resulting in an increase in solvent power.

To break the ester bonds and solubilize the acidic coal fragments, good solvents for brown coals are strong bases and they should be considered as reactive solvents. Complete recovery of amine solvents are not possible.

REFERENCES

1. Wheeler, R.V., Burgess, M.J., 'The volatile constituents of coal', J. Chem.Soc., 1911, 649;

   Jones, D.T., Wheeler, R.V., 'The composition of coal', J.Chem.Soc., 1915, 313; 1916, 707;

   Cochram, C., Wheeler, R.V., 'Studies in the composition of coal. The resolution of coal by means of solvents', J.Chem.Soc., 1927, 700.

2. Fischer, F., Gluud, W., 'Die ergiebigkeit der kohlenextraktion mit benzol', Ber.dt.Chem.Ges., 1916, 49, 1460.

3. Pott, A., Broche, H., 'The solution of coal by extraction under pressure - the hydrogenation of the extract', Fuel, 1934, 13(3), 91-6.

4. Oele, A.P., Waterman, H.I., Goedkoop, M.L., Van Krevelen, D.W., 'Extractive disintegration of bituminous coals', Fuel, 1951, 30, 169.

5. Wise, W.S., Solvent Treatment of Coal, Mills and Boon Ltd., London, 1971, p.58.

6. Dryden, I.G.C., 'Action of solvents on coals at lower temperatures mechanism of extraction of coals by specific solvents and the significance of quantitative measurements', Fuel, 1951, 30, 145-158.

7. Heredy, L.A., Fugassi, P., 'Phenanthrene extraction of bituminous coal', Coal Science, Advances in Chemistry Series, R.F. Gould, Ed., 1966, 55, 448-459.

8. Francis, W., Coal-its formation and composition, Edward Arnold, London, 1961, p.814.

9. Clarke, J.W., Rantell, T.D., Snape, C.E., 'Reactivity of cycloalkanes during solvent extraction of coal', Fuel, 1984, 63, 1476.

10. Volker, E.J., Bockrath, C.B., 'Effect of cresol as a co-solvent in coal liquefaction and product analysis', Fuel, 1984, 63, 285-287.

11. Whitehurst, D.D., Mitchell, T.D., Farcasiu, M., The Chemistry and Technology of Thermal Processes, Academic Press, N.Y., 1980.

12. Moschopedis, S.E., 'Use of oil sands bitumen and its derivatives as hydrogen donors to coal', Fuel, 1980, 59(1), 67.

13. Ross, D.S., Blessing, J.E., 'Alcohols as H-donor media in coal conversion. 1. Base promoted H-donation to coal by isopropyl alcohol', Fuel, 1979, 58(6), 433-437.

14. Heredy, L.A., Neuworth, M.B., 'Low temperature depolymerization of bituminous coal', Fuel, 1962, 41, 221-231.

15. Darlage, L.J., Weidner, J.P., Block, S.S., 'Depolymerization of oxidized bituminous coals', Fuel, 1974, 53(1), 54-59.

16. Ouchi, K., Imuta, K., Yamashita, Y., 'Catalysts for the depolymerization of mature coals', Fuel, 1973, 52, 156.

17. Ouchi, K., Yokoyamo, Y., Katoh, T., Htironori, I., 'Depolymerization of Yallourn brown coal. 1. Reduction of molecular weight with repeated reaction', Fuel, 1987, 66, 1115.

18. Larsen, J.W., Sams, T.L., Rodgers, B.R., 'Internal rearrangement of hydrogen during heating of coals with phenol', Fuel, 1981, 60, 335-341.

19. Kiebler, M.W., 'The action of solvents on coal', Chap.19 of Chemistry of Coal Utilization, Vol.1, H.H. Lowry, Ed., J. Wiley and sons, New York, 1945, p.677-760.

20. Dryden, I.G.C., 'Chemical constitution and reactions of coal', Chap.6 of Chemistry of Coal Utilization, Suppl.Vol., H.H.Lowry, Ed., John Wiley and sons, New York, 1963.

21. Lahaye, Ph., Decroocq, D., 'Coal solubilization in an organic medium', Revue de l'Institute Francis du Petrole, 1976, 31, 99.

22. Roy, J., Banarjee, P., Singh, P.N., 'Action of dipolar aprotic solvents on coal', Indian Journal of Technology, 1976, 14, 289.

23. Pullen, J.R., Solvent Extraction of Coal Report No. IC-TIS/TR16, IEA Coal Research, London, 1981.

24. Van Krevelen, D.W., 'Chemical structure and properties of coal. XXVIII. Coal constitution and solvent extraction', Fuel, 1965, 44, 229.

25. Hildebrand, J.H., Scott, R.L., The Solubility of Nonelectrolytes, 3 rd. Ed., Reinhold Pub.Corp., New York, 1950.

26. Angelovich, J.M., Pastor, G.R., Silver, H.F., 'Solvents used in the conversion of coal', Ind.Eng.Chem., Process Des.Dev., 1970, 9(1), 106.

27. Kirov, N.Y., O'Shea, J.M., Sargent, G.D., 'The determination of solubility parameters for coal', Fuel, 1967, 46, 415.

28. Hombach, H.P., 'General aspects of coal solubility', Fuel, 1980, 58(7), 465-470.

29. Weinberg, V.L., Yen, T.F., 'Solubility parameters in coal and coal liquefaction products', Fuel, 1980, 59(5), 287-289.

30. Marzec, A., Juzwa, M., Betlej, K., Sobkowiak, M., 'Bituminous coal extraction in terms of electron-donor and -acceptor interactions in the solvent/coal system', Fuel Process.Technol., 1979, 2(1), 35-44.

31. Szeliga, J., Marzec, A., 'Swelling of coal in relation to solvent electron-donor numbers', Fuel, 1983, 62, 1229.

32. Pajak, J., Marzec, A., 'Influence of pre-swelling on extraction of coal', Fuel, 1983, 62, 979.

33. Van Bodegom, B., Van Veen, J.A.R., Van Kessel, G.M.M., Sinnige-Nijssen, M.W.A., Stuiver, H.C.M., 'Action of solvents on coal at low temperatures. 1. Low rank coals', Fuel, 1984, 63, 346.

34. Dryden, I.G.C., 'Action of solvents on coals at lower temperatures. IV. Characteristics of extracts and residues from the treatment of coal with amine solvents', Fuel, 1952, 31, 176.

SUPERCRITICAL GAS EXTRACTION OF COAL

Aral Olcay
Department of Chemical Engineering,
Ankara University, Ankara, Turkey

ABSTRACT. The fundamentals of supercritical gas extraction is briefly described, and a review of the recent investigations of the factors controlling the supercritical gas extraction of coal is presented.

1. INTRODUCTION

Supercritical gas (SCG) extraction of coal is one of the promising techniques developed a few years ago by the Coal Research Establishment of the National Coal Board at Stoke Orchard.

As Paul and Wise (1) pointed out, supercritical gas extraction is analogous to both solvent extraction and distillation. Whereas leaching of a solid by a stream of liquid is a form of liquid extraction, vaporization of a substance into a stream of carrier gas is a form of distillation. However, below its critical temperature, the solvent will be liquid, while above its critical temperature it will be gaseous whatever the pressure. Thus both processes can be realized with the same solvent. In supercritical extraction the carrier gas is a supercritical gas and under the conditions which are realized it is difficult to distinguish physically between the two processes.

Investigation of Fig.1 in which the density of propane is plotted versus pressure at different temperatures, gives information about the optimum conditions for supercritical gas extraction.

At temperatures well above its critical temperature, the density of propane increases almost linearly with pressure according to the ideal gas law. Only at higher pressures are deviations noticeable.

The closer the temperature to its critical temperature the more marked the deviations from linearity are. However at temperatures below its critical temperature, the density-pressure curve shows inflexions due to the vapor-liquid phase transitions. The curve AB represents density changes of the propane vapor with pressure. At B, propane liquid appears and the line BC represents the conversion of vapor to liquid at the constant saturated vapor pressure of propane. At C there is only one liquid phase. Line CD represents the density changes of liquid propane with pressure.

Y. Yürüm (ed.), New Trends in Coal Science, 401–415.

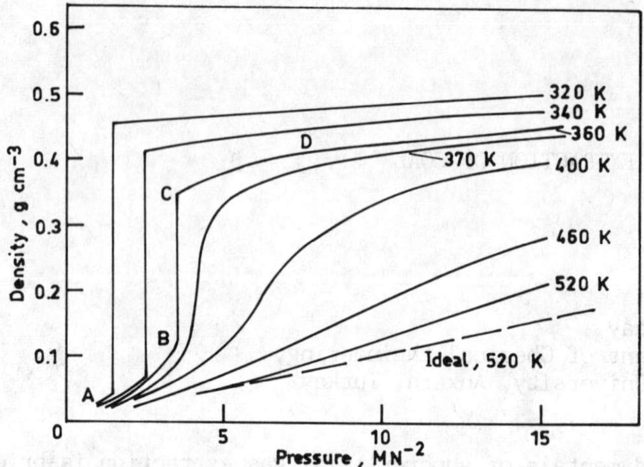

Figure 1 The density of propane as a function of pressure at different temperatures (1).

At 360 K and at its saturated vapor pressure liquid with a density of C and vapor with a density of B are in equilibrium. The liquid of a density C has a certain solvent power. If a propane gas at 400 K had the same density then it should have a similar solvent power. This can be accomplished by compressing the propane at 400 K to 5.5 MPa. The higher the temperature of propane gas, the higher the pressure that is required to match the liquid density. From this it may be concluded that for optimum extraction, the supercritical gas should have a high density at the extraction temperature. The greatest density for a given temperature will be obtained at higher pressure; for a given pressure the greatest density will be obtained at temperatures close to the critical temperature of the solvent.

In supercritical gas extraction, an organic solvent under supercritical conditions is used for the extraction of coal. The process involves the selective extraction of hydrogen rich compounds of the coal and is based on the ability of a solid substance to vaporize in the presence of a compressed supercritical gas. The products obtained upon heating coal to 400 ^{o}C cannot be separated by distillation from the parent coal because of the low temperature, whilst at higher temperatures, polymerization of the products takes place, which is undesirable. However, when compressed solvent vapors are used near their critical temperature the heavy organic coal components acquire a high volatility. Under favourable conditions an increase of 10000 times in volatility can be achieved which permits the extraction of low volatile compounds at temperatures considerably lower than their boiling points.

Supercritical gas extrcation is either carried out in a stirred batch or in a semicontinuous autoclave. In semicontinuous processes the autoclave is charged with coal and solvent and preheated. Then a stream of supercritical gas is pumped through the heated bed of coal. The gas phase containing the coal constituents is withdrawn through a valve into a condenser and collected in a flask at atmospheric pressure causing the coal extract to be precipitated. By carrying out the precipitation in two stages it is possible to fractionate the extract into a low melting solid and a viscous liquid, thus avoiding the troublesome filtration problems encountered in solvent extraction of coal. Semicontinuous operations are reported to give higher conversion than batch operations since they prevent secondary condensation reactions. Since the reactivity of coal is appreciable at 400 ^{o}C, most supercritical gas extraction of coal has been carried out at 350-450 ^{o}C and at a pressure of 10-20 MPa.

Among the several solvents studied for the SCG extraction of coal, such as pyridine, toluene, paracresol and dodecane, toluene seems to be most suitable due to its stability under extraction conditions, although higher yields of extract can be obtained with some other solvents. A recovery of toluene between 98.0 % to 99.6 % has been reported.

Whilst in batch autoclave extraction solvent/coal ratios ranging from 7/1 to 10/1 are usually used, the value of this ratio is higher in semicontinuous operations.

Due to the high diffusivity of the supercritical gas, coal particle size below 1.6 mm has only a small effect on the conversion (2).

For supercritical toluene extraction of coal at 10 MPa and at 400-440 ^{o}C the residence time in the reactor is usually 30 min. The gas residence time is kept as short as possible to prevent side reactions of the compounds present in the gas phase. However, the structures of the identified compounds in the supercritical gas-extracts and the small amount of water and gas formed during extraction show that fluid-phase thermal decomposition does not occur. The absence of olefins which are thought to arise by degradation of straight chain alkanes supports this conclusion (3).

Using supercritical toluene at about 400 ^{o}C and 10 MPa 25-30 % of a bituminous coal can be extracted. Table 1 indicates the typical composition of a hvb coal and the SCG extraction products obtained.

Table 1 Analyses of a supercritical extract, residue and feed coal (4)

	Coal	Extract	Residue
C wt %	82.7	84.0	84.6
H wt %	5.0	6.9	4.4
O wt %	9.0	6.8	7.8
N wt %	1.85	1.25	1.9
S wt %	1.55	0.95	1.45
H/C (atomic)	0.72	0.98	0.63
VM	37.40	-	25.00

The residue is a porous char and usually contains two-thirds of the volatile matter of the feed coal in contrast to the char obtained from pyrolysis which contains practically no volatile matter.

The extract is free of ash and solvent and its hydrogen content is significantly higher than that of the coal. As is seen in Table 1, when no hydrogen is used, the decrease in oxygen, nitrogen and sulphur, due in practise to pyrolitic reactions in the coal whilst it is heated to the extraction temperature, is moderate.

For the analysis of the extract, it is first divided into three fractions; petroleum ether solubles (oils), benzene solubles (asphaltenes) and benzene insolubles (preasphaltenes). After the extract has been refluxed with benzene and filtered to separate benzene insolubles, the filtrate is concentrated and 10 times its volume of petroleum ether is added to precipitate the benzene solubles, leaving oils in solution. Oils are usually further fractionated into paraffins, aromatics and polar compounds by adsorption chromatography on silica gel (5). The ratio of the three fractions of the extract does change with coal rank; oil yield decreases with increasing carbon content; the yield of asphaltenes shows a maximum, while that of benzene insolubles shows a minimum at about a carbon content of 71 % daf (6).

Gas chromatography-mass spectrometry of the simpler molecules indicates the presence of n-alkanes, alkyl benzenes, alkyl phenols and alkyl naphthalenes (7,3).

Structural analysis based on elemental analysis, molecular mass and nmr of the more complex compounds present in SCG extracts indicated the presence of an open chain polynuclear hydroaromatic structure linked by ether linkages. An average structure for benzene solubles constructed to fit the analysis is given in Fig.2

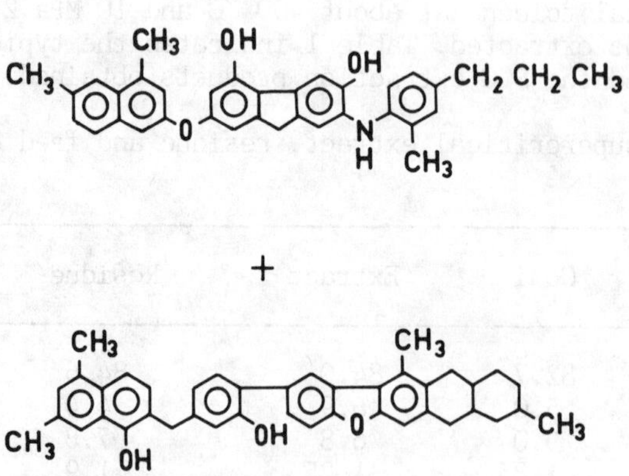

Figure 2 Average structure of the benzene solubles of a coal SCG extract (8).

2. INVESTIGATIONS OF THE FACTORS CONTROLLING SUPERCRITICAL GAS EXTRACTION OF COAL

2.1. Effect of Coal Properties :

Williams et al. (4) have investigated the influence of the volatile content of coal on the extract yield and found a linear relationship for the supercritical toluene extraction of different coals at 350 °C and 10 MPa (Fig.3).

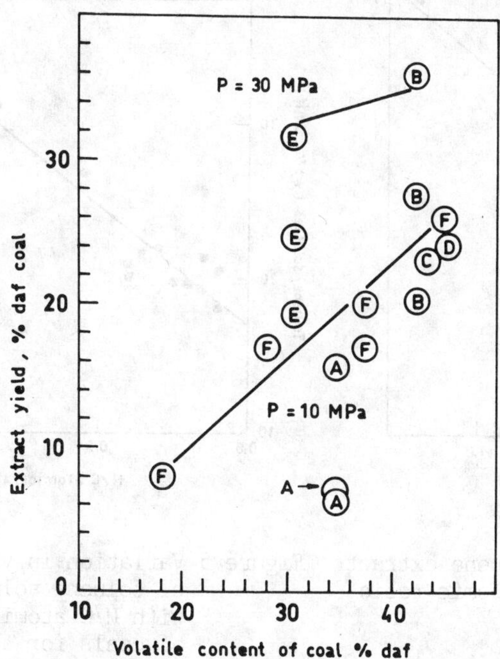

Figure 3 Variations in extract yield with the volatile content of the coal. Solvent (9) : Supercritical toluene. A) Illinois no.6, T = 400 °C, P = 5.89; 9.89; 16.0 MPa (10); B) Illinois no.6, T = 400 °C, P = 8.82; 10.1; 35.7 MPa (11); C) lignite Tunçbilek, T = 400 °C, P = 16 MPa (7); D) lignite Elbistan-Afşin, T = 400 °C, P = 17 MPa (7); E) Zonguldak, T = 360 °C, P = 11.6; 14.2; 29.4 MPa (3); F) different coals, T = 350 °C, P = 10 MPa (4).

In the same figure are also plotted the SC toluene extract yields of different coals obtained from the literature. Scattering of the points may be due to the different operational parameters used in the experiments.

Kershaw et al. (12) have investigated the supercritical toluene extraction of fifteen Victorian brown coals at 400 °C and 10 MPa. As is seen in Figs 4 and 5 good correlations were obtained between both the extract yield and toluene solubles (oils + asphaltenes) and the H/C atomic ratio of coal. A decrease in aromaticity and increase in the

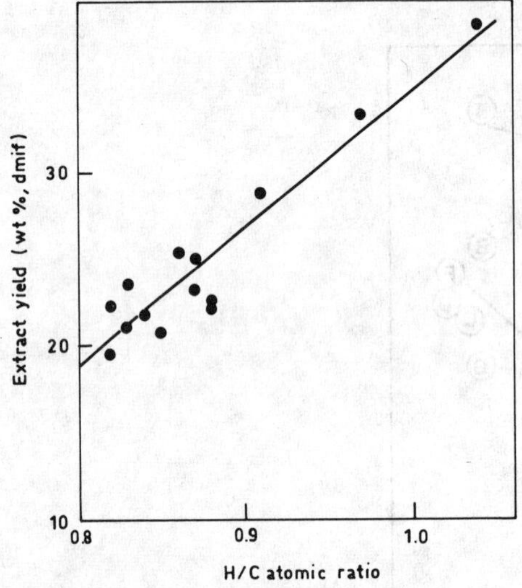

Figure 4 Variation in toluene extract yield with H/C atomic ratio of coals (12).

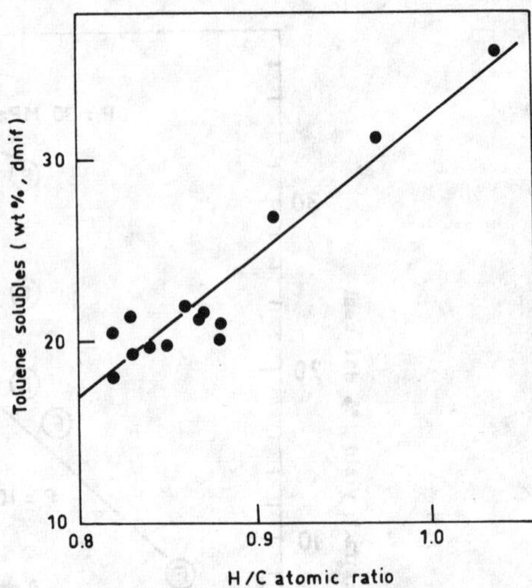

Figure 5 Variation in yield of toluene solubles with H/C atomic ratio of coals for toluene extraction (12).

molecular weight with increasing H/C atomic ratio of the coal have also been reported for toluene solubles (12).

No remarkable effect of inorganic constituents on the conversion or the extract yield of a coal has been observed.

2.2. Effect of Process Variables :

2.2.1. Solvent to Coal Ratio :

Amestica et al. (10), investigated the effect of solvent to coal ratio on the extract yield by using the same volume of toluene and varying the weight of coal. The results indicate that the extract yield increases with the increasing solvent/coal ratio. However, to minimize energy requirements this ratio should be kept as low as possible.

2.2.2. Temperature and Pressure :

Tufano and coworkers (9) used the data present in the literature and plotted the SC toluene extract yields obtained from different coals using batch or semicontinuous reactors against pressure at different temperatures (Fig.6).

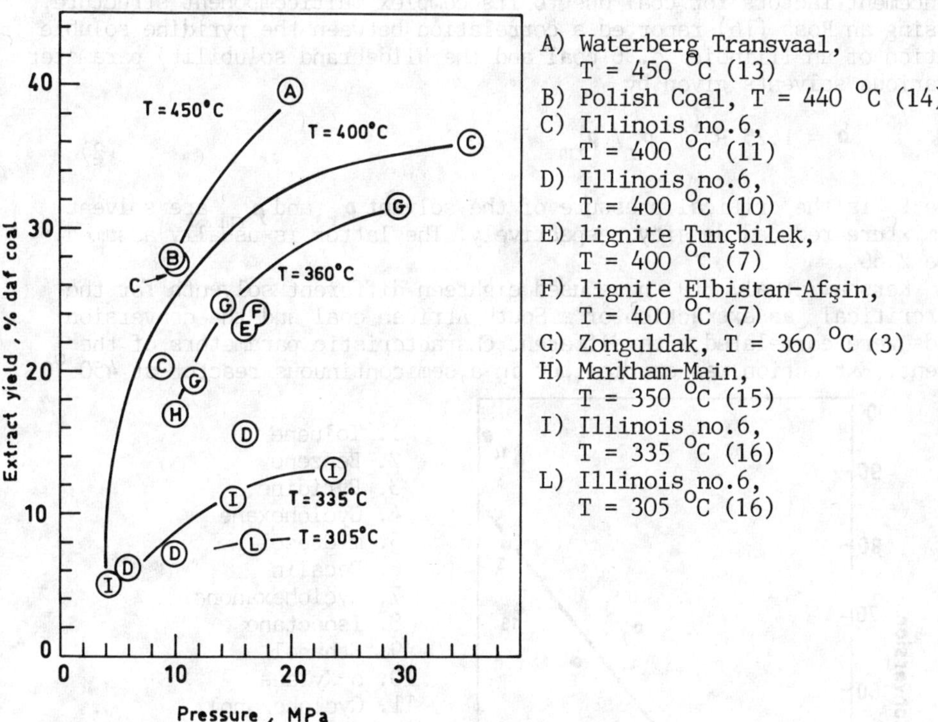

Figure 6 Variations in SC toluene extract yield of different coals with pressure (9).

Fig.6 shows a noticeable increase in the extract yield at the supercritical pressure of toluene (4.16 MPa). At higher pressure the effect of change in pressure on the extract yield seems to be small. The extract yields increase with increasing temperature, due to the combined effect of phase behaviour and thermolysis reactions of the various dissolved fractions.

2.3. Effect of Solvent Parameters :

For a single component system the solubility of a solid substance in a supercritical gas is directly proportional to an enhancement factor and the solid partial pressure (17). The enhancement factor is given by

$$\ln \alpha = (v_s - 2B_{12})/v \qquad (1)$$

where B_{12} is the cross virial coefficient and represents the interaction between the molecules 1 and 2. v_s is the volume of the solute, v is the total molar volume of the system. B_{12} can be estimated using the method proposed by Prausnitz (18). However, it is not possible to compute the enhancement factors for coal due to its complex multicomponent structure. Blessing an Ross (16) reported a correlation between the pyridine soluble fraction of an Illinois No. 6 coal and the Hildebrand solubility parameter of various solvents given by

$$\delta = 1.25 P_c^{1/2} \rho_r / \rho_{rm} \qquad (2)$$

where P_c is the critical pressure of the solvent ρ_r and ρ_{rm} are solvent and mixture reduced density respectively. The latter is usually assumed to be 2.66 .

Kershaw et al. (19) have used eighteen different solvents for the supercritical gas extraction of a South African coal and the conversion yields were correlated with different characteristic parameters of the solvent. Extraction was carried out in a semicontinuous reactor at 450 °C

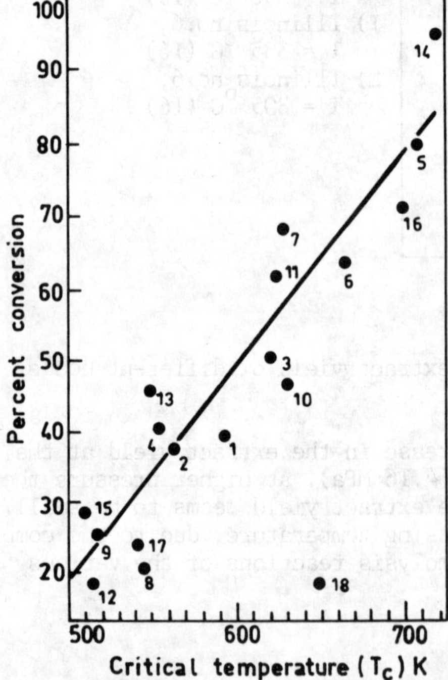

1. Toluene
2. Benzene
3. Pyridine
4. Cyclohexane
5. m-Cresol
6. Decalin
7. Cyclohexanone
8. Isooctane
9. Ethanol
10. o-Xylene
11. Cyclohexanol
12. Methanol
13. Isobutanol
14. Tetralin
15. Isopropanol
16. Aniline
17. n-Heptane
18. Water

Figure 7 Conversion versus critical temperature for extraction of Waterberg coal at 450 °C and 20 MPa (19).

and 20 Mpa so that the solvents used were in the supercritical state.

In Fig.7 the % conversion is plotted against the critical temperature of the solvents. As can be seen from Fig.7 tetralin gave higher conversions than estimated from its critical temperature due to its hydrogen donor capacity. A similar relationship is observed with % conversion-density of the solvent or solubility parameter of the solvent under experimental conditions (Fig.8).

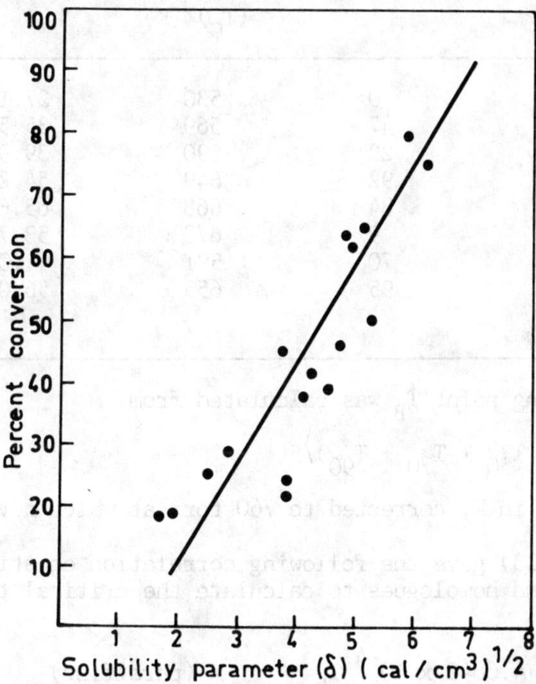

Figure 8 Conversion versus Hildebrand solubility parameter for extraction of Waterberg coal at 450 °C and 20 MPa (19).

Alcohols gave higher conversion than hydrocarbons having similar densities or Hildebrand solubility parameters. Wolf et al. (10) have also reported higher yields with supercritical ethanol due to its partial reaction with coal. Indeed the recovery of ethanol ranged from 73 % to 85 %. Ross and Blessing (20) have suggested that alcohols act as hydrogen donors by hydride transfer of their hydrogen. Makabe et al. (21) have also found that during supercritical ethanol extraction of model compounds splitting of ether bonds, ethylation of the benzene nuclei and hydrogenation of aromatic nuclei did occur.

Kershaw (22) has also investigated the supercritical extraction of the same coal under the same conditions using various commercially available paraffinic and aromatic hydrocarbon mixtures containing less than 5 % of cyclohexanes (Table 2).

Table 2 Characteristics of hydrocarbon solvent mixtures used and conversions obtained (22)

Solvent	Average boiling point T_B (K)	Aromatic content (mass %)	Calculated critical temperature (T_C) (K)	Conversion (mass % daf)
1	363	0	530	27.1
2	332	47	569	35.5
3	412	22	590	39.9
4	447	92	649	54.2
5	472	84	668	65.6
6	509	9	678	53.7
7	394	70	591	41.0
8	452	95	655	56.0

The average boiling point T_B was calculated from

$$T_B = (T_{10} + T_{30} + T_{50} + T_{70} + T_{90})/5 \qquad (3)$$

where T_n is temperature in K, corrected to 760 torr at which n vol % of the solvent distilled.

Ambrosse et al. (23) gave the following correlation equations for paraffins and benzene and homologues to calculate the critical temperature.

$$T_b/T_c = 0.51453 + 4.6842 \times 10^{-4}\, T_b \quad \text{(paraffins)} \qquad (4)$$

$$T_b/T_c = 0.415 + 6.05 \times 10^{-4}\, T_b \quad \text{(benzene and homologues)} \qquad (5)$$

For the computation of critical temperature of the mixture of paraffins and aromatic hydrocarbons in which the weight fraction of aromatic compounds is x, the following equation was obtained

$$T_b/T_c = (0.515+4.68\text{x}10^{-4}T_b)(1-x)+(0.415+6.05\text{x}10^{-4}T_b)x \qquad (6)$$

where T_b is the average boiling point of the solvent mixture.

In Fig.9 the % conversion is plotted versus the calculated values of critical temperature of the solvent mixture.

As can be seen from Fig.9 the points lie close to the line obtained for pure solvents. From this it is concluded that conversion data can also be estimated from the critical temperature of the solvent mixtures.

As is seen in equation 2, the Hildebrand solubility parameter depends on the density at the experimental conditions and therefore on

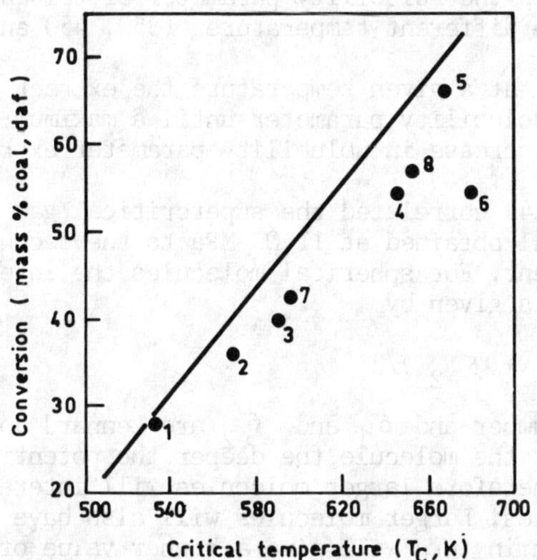

Figure 9 Conversion versus critical temperature for extraction of Waterberg coal at 450 °C and 20 MPa mixed solvents; the line is for pure solvents (21).

the pressure and temperature. At a given temperature by changing the pressure, the solubility parameter of the solvent can be changed. Fig. 10 shows the change in the supercritical toluene extract yields of a coal

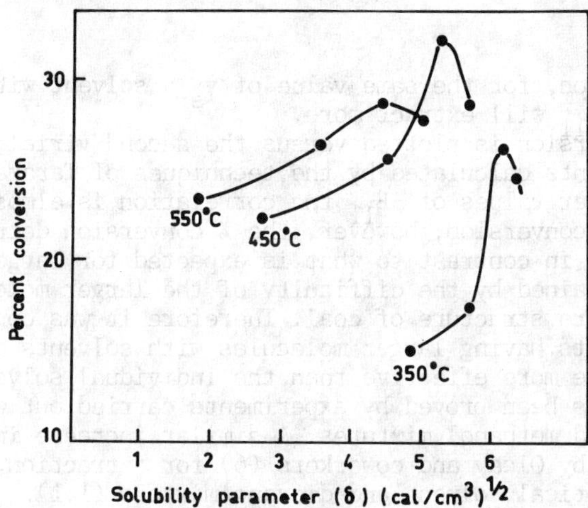

Figure 10 Conversion versus Hildebrand solubility parameter for toluene extraction of New Wakefield coal (19).

containing 79 % C (daf) with the solubility parameter of toluene (at 10, 20, 30 and 40 MPa) at three different temperatures (350, 450 and 550 ^{o}C) (19).

As is seen in Fig.10, at a given temperature the extract yield increases with increasing solubility parameter until a maximum is reached. After that with a further increase in solubility parameter extract yields decrease.

Fong and coworkers (24) correlated the supercritical gas extract yields of Illinois No.6 coal obtained at 11.03 MPa to the second virial coefficient of the SC solvent. For spherical molecules the interaction second virial coefficient is given by

$$B_{12} = - 2/3\, \pi N_o\, (\sigma_1 + \sigma_2)/2 \qquad (7)$$

where N_o is the Avagadro number and σ_1 and σ_2 are Lennard Johnes parameters (17). The larger the molecule the deeper the potential well of the molecule will be. Therefore larger molecules will interact more strongly with other molecules. Larger molecules will also have a more negative value of B_{12} which in turn will give a higher value of the enhancement factor. If B_{12} is taken approximately to be

$$B_{12} = (B_{11} + B_{22})/2 \qquad (8)$$

then the equation 1 will have the following form

$$\ln \alpha = \frac{(v_s - B_{22}) - B_{11}}{v} \qquad (9)$$

According to this equation, for the same value of v_s a solvent with a more negative value of B_{11} will extract more.

In Fig.11, % conversion is plotted versus the second virial coefficient of the solvents calculated by the techniques of Tarakad and Danner (25). For the lower values of $-B_{11}$ the correlation is almost linear. After a maximum conversion, however, the % conversion decreases at higher values of $-B_{11}$ in contrast to what is expected for larger molecules. This was explained by the difficulty of the larger molecules in moving through the pore structure of coal. Therefore it was concluded that a mixture of solvents having larger molecules with solvents of smaller molecules will be more effective than the individual solvents.

This assumption has been proved by experiments carried out with supercritical toluene and methanol mixtures. A similar increase in yield has been observed by Olcay and coworkers (6) for extraction of lignites with a supercritical toluene and dioxane mixture (1:1).

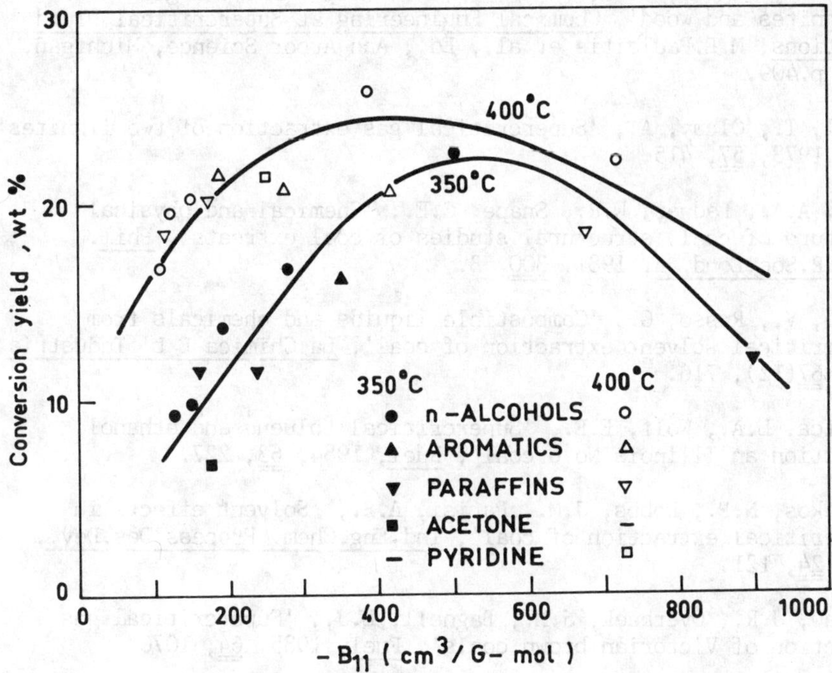

Figure 11 Variations in % conversion with the second virial coefficients of the solvents (24).

REFERENCES

1. Paul, P.F.M., Wise, W.S., The Principles of Gas Extraction, Mills and Boon Ltd., London, 1971.

2. Maddocks, R.R., Gibson, J., Williams, D.F., 'Coal processing technology: Supercritical extraction of coal', CEP, 1979, 49.

3. Ceylan, R., Olcay, A., 'Supercritical gas extraction of Turkish coking coal', Fuel, 1981, 60, 197.

4. Whitehead, J.C., Williams, D.F., 'Solvent extraction of coal by supercritical gases', J.Inst.Fuel, 1975, 182.

5. Bartle, K.D., Jones, D.W., Martin, T.G., Wise, W.S., 'Characterization of the neutral oil from a low temperature coal tar', J.Appl.Chem., 1970, 20, 197.

6. Olcay, A., Tuğrul, T., Çalımlı, A., 'The supercritical gas extraction of lignites and wood', Chemical Engineering at Supercritical Fluid Conditions, M.E.Paulaitis et al., Ed., Ann Arbor Science, Michigan, 1983, p.409.

7. Tuğrul, T., Olcay, A., 'Supercritical gas extraction of two lignites', Fuel, 1978, 57, 415.

8. Herod, A.A., Ladner, W.R., Snape, C.E., 'Chemical and physical structure of coal, structural studies of coal extrcats', Phil. Trans.R.Soc.Lond. A, 1981, 300, 3.

9. Tufano, V., Russo, G., 'Combustible liquids and chemicals from supercritical solvent extraction of coal', La Chimica E L' Industria, 1985, 67(12), 716.

10. Amestica, L.A., Wolf, E.E., 'Supercritical toluene and ethanol extraction an Illinois No.6 coal', Fuel, 1984, 63, 227.

11. Vasilakos, N.P., Dobbs, J.M., Parisi, A.S., 'Solvent effects in supercritical extraction of coal', Ind.Eng.Chem. Process Des.Dev., 1985, 24, 121.

12. Kershaw, J.R., Overbeek, J.M., Bagnell, L.J., 'Supercritical gas extraction of Victorian brown coals', Fuel, 1985, 64, 1070.

13. Kershaw, J.R., Barrass, G., Gray, D., Jezko, J., 'Process effects on the nature of coal liquefaction products', Fuel, 1980, 59, 413.

14. Slomka, B., Rutkowski, A., 'A kinetic study of the supercritical toluene extraction of coal at 9.8 MPa', Fuel Process.Technol., 1982, 5, 247.

15. Bartle, K.D., Martin, T.G., Williams, P.F., 'Chemical nature of a supercritical gas extract of coal at 350 °C', Fuel, 1975, 54, 226.

16. Blessing, J.E., Ross, D.S., 'Supercritical solvents and the dissolution of coal and lignite', ACS Symposium Series, 1978, 71, 171.

17. Rowlinson, J.S., Richardson, M.J., 'Solubility of solids in compressed gases', Adv.Chem.Phys., 1959, 2, 85.

18. Prausnitz, J.M., 'Fugacities in high-pressure equilibria and in rate processes', J.Am.Inst.Chem.Eng., 1959, 5(1), 3.

19. Jezko, J., Gray, D., Kershaw, J.R., 'The effect of solvent properties on the solvent extraction of coal', Fuel Process.Technol., 1982, 5, 229.

20. Ross, D.S., Blessing, J.E., 'Alcohols as H-donor media in coal conversion. 1. Base promoted H-donation to coal by isopropyl alcohol, 2. Base promoted H-donation by methyl alcohol', Fuel, 1979, 58, 433 and 438.

21. Makabe, M., Ouchi, K., 'Solubility increase of coals by the reaction with alcohol-elucidation of reaction mechanism using model compounds', Fuel Process.Technol., 1982, 6, 307.

22. Kershaw, J.R., 'Solvent effects on the supercritical gas extraction of coal, the role of mixed solvents', Fuel Process.Technol., 1982, 5, 241.

23. Ambrose, D., Cox, J.D., Townsend, R., 'The critical temperatures of forty organic compounds', Trans.Faraday Soc., 1960, 66, 1452.

24. Fong, W.S., Chan, P.C.F., Pichaichanarong, P., Corcoran, W.H., 'Experimental observations on a systematic approach to supercritical extraction of coal', Chemical Engineering at Supercritical Fluid Conditions, M.E.Paulaitis et al., Ed., Ann Arbor Science, Michigan, 1983, p.377.

25. Tarakad, R.R., Danner, R.P., 'An improved corresponding states method for polar fluids: Correlation of second virial coefficients', AIChE Journal, 1977, 23, 685.

COAL DESULFURIZATION. THE PROTODESULFURIZATION OF NAPHTHALENE THIOLS AND SULFIDES

Leon M. Stock and Ryszard Wolny
Department of Chemistry
University of Chicago
5735 S. Ellis Avenue
Chicago, Illinois
USA

ABSTRACT. Basic information concerning the chemistry of organosulfur compounds is needed for the design of new coal desulfurization reactions. A study of the nucleophilic replacement of sulfur by oxygen in aryl thiols and sulfides led to the discovery of an acid-catalyzed reductive desulfurization reaction that proceeds readily under surprisingly mild conditions. Specifically, the treatment of 1- and 2-naphthalenethiol, 1- and 2-naphthyl alkyl sulfides, and di-1- and di-2-naphthyl sulfide with triflic acid in methylene chloride at 40°C leads to the formation of naphthalene as the principal organic product. The scope and limitations of the reaction have been explored and the reaction mechanism has been investigated. Several lines of evidence indicate that the process is best described as an electrophilic protodesulfurization reaction that yields naphthalene and oxidized sulfur compounds.

1. INTRODUCTION

The inorganic and organic sulfur compounds in coal restrict the utilization of this important world resource. Consequently, many laboratories are now concerned with the definition of the structures and reactions of the sulfur compounds in coal in a search for new chemistry that would permit its facile desulfurization. The chemical issues can be illustrated by a brief discussion of a representative bituminous coal from the Illinois Basin. One of the coals selected for study by the Illinois Coal Board comes from the No. 6 seam, it contains 3.6% sulfur (1). The sulfur is distributed among organic compounds, 2.6%, pyrite, 0.8%, sulfate, 0.1%, and elemental sulfur, 0.1% (1,2). Elemental and sulfatic sulfur are not generally detected in pristine coal samples and it seems safe to conclude that these substances are formed after the coal has been exposed to the earth's environment by oxidative or microbiological processes (2). The sulfur-containing minerals have been quite intensively studied and a great deal is known about

Y. Yürüm (ed.), New Trends in Coal Science, 417–431.

their occurrence in coals (1). Considerably less information is available concerning the nature and the composition of the organosulfur compounds. Attar and his coworkers suggested that the organosulfur compounds in an Illinois No. 6 coal were distributed among many different kinds of compounds with 7% in aliphatic thiols, 15% in aromatic thiols, 18% in aliphatic thioethers, 2% in aryl thioethers, and 58% in heterocycles (3). Further work has cast doubt on the reliability of the indirect methods used by Attar (4), but a very recent investigation of the pyrolysis of another Illinois No. 6 coal suggests that the organic sulfur compounds can be divided into two distinct classes one of which may be aliphatic compounds and the other of which may be heterocyclic compounds (5).

The environmentally effective utilization of coal requires the removal of the sulfur, the sulfur minerals and the organosulfur compounds. Washing and other physical methods can be used successfully for the elimination of elemental sulfur and the minerals during the preparation of coals for commercial usage. Unfortunately, these methods have no significant impact on the organosulfur compounds. Their removal requires chemical disruption of the coal macromolecules.

Nucleophilic substitution reactions are foremost among the chemical methods that could be used for the replacement of the sulfur atoms in the divalent organosulfur compounds that are apparently so abundant in coal. The simple hydrolysis reaction is dazzlingly attractive and constitutes a goal for basic research. Unfortunate-

$$\text{CoalSCoal} + 2H_2O \rightarrow 2\text{CoalOH} + 2H_2S \qquad (1)$$

ly, the energy requirements for this conversion are very large and the direct transformation is difficult to accomplish. We, therefore, undertook a study of two alternative approaches that exploit indirect catalytic methods to obtain the same goal. These methods depend on quite different concepts. One reaction is appropriate for basic environments and the other in acidic environments. Both involve novel procedures that have not been considered in discussions of desulfurization reactions. Both methods are, in principle, suitable for aromatic organic compounds and sulfur-containing heterocycles. Both methods center on chemical conversions that should produce hydrogen sulfide or sulfur as the utlimate reaction products. Neither method involves extreme reaction conditions such as high temperature or high pressure.

The approach for desulfurization in basic media that we have elected to study exploits anion radical chemistry (6). The idea is illustrated by the replacement reaction of a thiol with a nucleophile, Nuc^-, as shown in equation 2.

$$\text{ArSH} + e \rightarrow [\text{ArSH}]^{\dot{-}} \qquad (2a)$$

$$[\text{ArSH}]^{\dot{-}} \rightarrow \text{Ar}\bullet + \text{SH}^- \qquad (2b)$$

$$\text{Ar}\bullet + \text{Nuc}^- \rightarrow \text{ArNuc}^{\dot{-}} \qquad (2c)$$

$$ArNuc^{\overset{\cdot\cdot}{-}} + ArSH \rightarrow ArSH^{\overset{\cdot\cdot}{-}} + ArNuc \qquad (2d)$$

The overall reaction is shown in equation 3. Reactions of this kind

$$ArSH + Nuc^- \rightarrow ArNuc + SH^- \qquad (3)$$

are plausible for many of the organosulfur compounds that are thought to be present in coal and we are examining the scope and limitations of such tactics.

Desulfurization in acidic media was undertaken on the basis of the idea that sulfur-oxygen interchange could be accomplished via a hemithioketal intermediate as illustrated in equation 4 for a simple

$$C_6H_5SR + H_3O^+ \rightarrow \text{[cyclohexadiene intermediate bearing H, H, OH, SR]} \rightarrow C_6H_5OH + RSH_2^+ \qquad (4)$$

for a simple benzene derivative. Our attempts to accomplish reactions of this kind and the subsequent even more fruitful research are described in this article.

2. MATERIALS AND METHODS

2.1 Materials

Quality commercial grades of trifluoroacetic acid, trifluoroacetic acid-d, heptafluorobutyric acid, trifluoromethanesulfonic acid, chloroform-d, and benzene-d_6 were sufficiently pure for use without further purification. The thiols, thioethers, disulfides, and heterocycles needed in this study were obtained in a variety of ways. Many were available from commercial sources and were purified as necessary. For example, 1-naphthalenethiol was purified by vacuum distillation and 2-naphthalenethiol was purified by recrystallization from ethanol.

Many compounds had to be synthesized. In the interests of brevity, we shall only note that all these compounds exhibited physical and spectroscopic properties in full accord with their structures. Methyl 2-naphthyl sulfide, ethyl 2-naphthyl sulfide, methyl 1-naphthyl sulfide and ethyl 1-naphthyl sulfide were prepared by the method of Nakayama and coworkers (7). Typically, 50 mmol of the arenethiol was stirred with 55 mmol of potassium hydroxide in 20 ml of water. After a few minutes, 0.5g of tetrabutylammonium chloride, which served as a phase transfer catalyst, was added and then an alkyl iodide was added in one portion. The mixture was stirred for one hour. Each reaction yielded about 95% of the desired product which was crystallized two times from petroleum ether or distilled under vacuum.

Another method based on the work of Migota and his associates (8) was used for the synthesis of di-1-naphthyl sulfide, di-2-naphthyl sulfide, 1-naphthyl 2-naphthyl sulfide, phenyl 1-naphthyl sulfide, phenyl 1-naphthyl sulfide, and butyl 2-naphthyl sulfide. Metallic sodium (1.5g) was dissolved under nitrogen in absolute alcohol (100 ml). The naphthalenethiol (25 mmol) was added to this solution together with 25 mmol of the corresponding iodonaphthalene, iodobenzene or iodobutane in 50 ml of absolute ethanol and then 0.3 mmol of tetrakis-(triphenylphosphine)palladium(0). The mixture was refluxed for 12 hours under nitrogen. Then the ethanol was evaporated using a rotary evaporator. The residue was crystallized three times from ethanol. Liquid products were purified by vacuum distillation. These reactions provided about 90% of the desired materials.

Di-1-naphthyl disulfide and di-2-naphthyl disulfide were prepared by titration of the corresponding thiols in ethanol solution by iodine in methanol solution. Both compounds precipitated during the reaction. The solid products were collected on a filter and the disulfides were washed with a small portion of ethanol and then water. Pure products were obtained by recrystallization of the crude products from absolute ethanol. The yields in both reactions were over 95%.

1-Naphthyl and 2-naphthyl trifluoroacetate and their sulfur analogues were synthesized by refluxing the corresponding naphthol or naphthalenethiol with trifluoroacetic anhydride. Excess anhydride was removed using a rotary evaporator and the product was crystallized from ethanol.

The benzothianthrene perchlorates and thianthrene perchlorate were obtained using the procedure of Shine and Silber (9).

In addition, several deuterium labeled compounds had to be prepared. For example, diphenylmethane-d_2 was prepared by the procedure of Chen and coworkers (10). Two exchanges were sufficient to raise the deuterium content at the benzylic position to over 95%. Triflic acid-d was prepared by refluxing equimolar amounts of triflic acid anhydride with heavy water for 1 hour. The reaction mixture was protected from the atmosphere to prevent the introduction of moisture. The proton content was assessed by NMR spectroscopy.

2.2 Reaction Procedures

The desulfurization reactions and other experiments were performed over a range of conditions as reported in the next section. Two procedures are described here to illustrate the approach.

The reactions of the naphthalenethiols and sulfides with nucleophilic reagents in trifluoroacetic acid were carried out with 1-naphthalenethiol, 2-naphthalenethiol, di-2-naphthyl sulfide, 2-naphthyl methyl sulfide and nucleophiles such as water, acetate ion, 1-naphthalenethiol, methyl sulfide, and 2-naphthalenethiol. The reactions were carried out for 10 hours at 70°C and then poured into water and extracted with petroleum ether. The products were analyzed by magnetic resonance, mass spectroscopy and gas chromatography.

The reactions with trifluoroacetic acid and trifluoromethanesulfonic acid were carried out at high acidity using 10 parts trifluoroacetic acid and 1 part trifluoromethanesulfuric acid. For 2-naphthalenethiol, 2:1 and 20:1 solvent compositions were also used. The reactions were carried out for 10 hours at 70°C. The products were analyzed by magnetic resonance and mass spectroscopy.

The reactions of the thiols and sulfides with ethanethiol and aluminum-chloride were performed in dichloromethane with the sulfide or thiol at a concentration of 0.146 M with 2.5 and 5.0 molar equivalents of aluminum chloride and ethanethiol, respectively. The reaction was allowed to proceed for 6 hours in most cases. Other reaction conditions are specified in the next section. The reaction products obtained after the completion of the reaction were washed five times with water and then dissolved in an organic solvent and dried with molecular sieves A4 at least 24 hours prior to spectroscopic gas chromatographic analyses.

2.3 Equipment

Gas chromatography-mass spectra were recorded using a Hewlett-Packard Model 5790A chromatograph equipped with a capillary column containing a crosslinked methyl silicone (12 m). The chromatograph was used in conjunction with a VG 70-250 mass spectrometer. Mass spectra were also obtained using the direct insertion probe of the same spectrometer.

Gas chromatographic analyses were also performed with a Perkin Elmer Sigma 3B instrument in the temperature range 60 to 295°C using a 3 m column packed with 3% OV-101 on Chromasorb W (80-100 mesh) and in the temperature range from 60 to 350°C using a 3 m column packed with 3% DEXSIL 300 GC on chromosorb W-HP (100-120 mesh).

The nuclear magnetic resonance were recorded using the Bruker HX-270, Chicago 500 MHz, and Varian XL 400 MHz spectrometers.

3. RESULTS AND DISCUSSION

Several considerations suggested that it would be more appropriate to investigate naphthalene derivatives rather than to examine elementary benzene derivatives or heterocycles. Accordingly, we first undertook a study of the sulfur-oxygen interchange in naphthalenethiol and its derivatives using trifluoroacetic acid as the key reagent. It was first established that the naphthalenethiols and their derivatives readily underwent deuteration and exchange in trifluoroacetic acid

$$2\text{-}C_{10}H_7SR + CF_3CO_2D \rightarrow [\text{1-D,1-H-naphthalenium-2-SR}]^+ + CF_3CO_2^- \quad (5a)$$

$$CF_3CO_2^- + [\text{1-D,1-H-naphthalenium-2-SR}]^+ \rightarrow \text{1-D-2-SR-naphthalene} + CF_3CO_2H \quad (5b)$$

(eq 5) thereby insuring the formation of the arenonium ion necessary for the formation of the necessary hemithioketal shown in the equation. When the reaction was carried out with very reactive nucleophiles such as 1- and 2-naphthalenethiol, it was found that these compounds were converted to the corresponding naphthyl trifluoroacetates (eq 6). The same reaction takes place in heptafluorobutyric

$$\text{NaphthaleneSH} + CF_3CO_2H \rightarrow \text{NaphthaleneSCOCF}_3 + H_2O \quad (6)$$

acid at 120°C and in trifluoroacetic acid containing triethylsilane at 72°C. The thioesters are significantly less reactive than the thiols in electrophilic reactions. Consequently, the rates of formation of the arenonium ions are greatly decreased and there is no opportunity for the generation of the needed hemithioketal. This difficulty was circumvented by the addition of water to the reaction mixtures. Water (eq 7) hydrolyzes the thioester to the thiol. However,

$$\text{NaphthaleneSCCF}_3 + H_2O \rightarrow \text{NaphthaleneSH} + CF_3CO_2H \quad (7)$$

neither of the naphthalenethiols are converted to the corresponding naphthols under these conditions. Subsequent work on the reactions of di-2-naphthyl sulfide and 2-naphthyl methyl sulfide with mildly nucleophilic reagents such as water and acetic acid and with very nucleophilic reagents such as naphthalenethiol did not take place in trifluoroacetic acid at 72°C or in heptafluorobutyric acid at 120°C. These sulfides were recovered unchanged from the reaction mixtures. Attempts to intercept the arenonium ion intermediates with reducing reagents such as triethylsilane (eq 8) were also unsuccessful.

$$[\text{arenonium ion: H, H, SR, +}] + (C_2H_5)_3SiH \rightarrow [\text{dihydronaphthalene: H, H, H, SR}] + (C_2H_5)_3Si^+ \quad (8)$$

These observations suggested that the concentration of the arenonium ion in the fluorinated acid solvents was too small to permit efficient reactions of this essential intermediate with the nucleophilic reagents. Hence, the necessary hemithioketal could not be formed under these conditions.* Accordingly, we turned our attention to the selection of reaction conditions that would be more favorable for the

---

*It is important to note that reaction 4 proceeds very readily from left to right. Indeed, many suitable methods for the preparation of aryl thiols and thioethers exploit this chemistry (11). Even though we have turned our attention to other promising tactics for chemical desulfurization, the scheme outlined in equations 1 and 4 clearly deserves more research effort.

formation of the necessary arenonium ion intermediates. Elementary considerations prompted us to add a very strong acid, trifluoromethane sulfonic acid, triflic acid, to trifluoroacetic acid in order to increase the concentration of arenonium ions with thiol, alkylthiol, and arylthiol substituents. Reactions were carried out with 1- and 2-naphthalenethiol, di-1- and di-2-naphthyl sulfide, 1-naphthyl 2-naphthyl sulfide, and 2-naphthyl trifluorothioacetate in a solution of triflic acid in trifluoroacetic acid in a 1:10 mole ratio. The reactions of most of the sulfur compounds occurred rapidly as the solutions darkened. Preliminary studies of the reaction products revealed that the starting materials were converted to naphthalene as shown in Table I. Typical rate data are shown in Table II.

Table I. Conversions of Sulfides and Thiols to Naphthalene During Reactions with Trifluoroacetate Acid-Triflic Acid.

Compound	Naphthalene, %
1-Naphthalenethiol	54
Di-1-naphthyl sulfide	47
1-Naphthyl butyl sulfide	43
2-Naphthyl butyl sulfide	48
2-Naphthyl trifluorothioacetate	5
2-Naphthalenethiol	50

Table II. The Kinetics of the Reaction of 2-Naphthalenethiol with Trifluoroacetic Acid and Triflic Acid at 70°C.

Time, hrs.	2-Naphthyl thiotrifluoroacetate, %	Naphthalene, %
2	53	7
4	37	20
6	25	29
12	5	50
24	5	50

The observations indicate that naphthyl trifluoroacetate and naphthalene are both formed under the experimental conditions. It is evident that naphthyl trifluoroacetate is also converted to naphthalene. However, pure 2-naphthyl trifluorothioacetate gave only 5% of this compound under comparable conditions. It appeared that the differences in the experimental results arose from the presence of competitive nucleophiles in the reaction mixture. We found that the addition of even a mild competitive nucleophile, such as water, to the reaction system would enhance the rate of formation of naphthalene from the thioacetate ester. For example, when the reaction of the ester (0.15M) was carried out with water (0.15M), there was a sevenfold increase in the yield of naphthalene.

During the early phases of the study, it was also observed that the reactions of some of the organosulfur compounds were rather complex and that insoluble compounds were also formed. Mass spectroscopy and NMR spectroscopy established that 2-naphthalenethiol derivatives were converted into two benzothianthrenes.

S S S S

The facility with which the naphthalenethiols and their derivatives are converted into naphthalene in an acidic environment is surprising. Study of the literature revealed, however, that reactions of this kind were not unprecedented. Alper and Blais reported the molybdenum(VI) hexacarbonyl reacted with naphthalenethiol in acetic acid to produce naphthalene (12). Although these workers proposed that the molybdenum(II) was the crucial reagent, it seems equally plausible an acidic molybdenum(V) compound may be responsible for the reaction. Luh and his coworkers have used this reaction (13), a Japanese group (14) found that ethanethiol and aluminum chloride reduced many substituted benzenes including benzenethiol to hydrocarbons (eq 9).

$$ArSR \xrightarrow{C_2H_5SH, AlCl_3} ArH \qquad (9)$$

These observations appeared to be related to our findings and prompted an examination of the reactions in more detail because, among other reasons, the simplicity of the reaction renders it especially attractive for coal desulfurization. Consequently, we have undertaken a broad investigation of the scope of the reaction and the conditions under which it proceeds most readily. The results presented in the following paragraphs constitute a progress report on this continuing investigation.

3.1 1-Naphthyl Derivatives

The reaction of 1-naphthalenethiol with triflic acid produces about 85-90% naphthalene (eq 10). The other products include di-1- and di-

$$1\text{-}C_{10}H_7SH + CF_3SO_3H \xrightarrow[25°C,\ 6\ hrs]{CH_2Cl_2} \begin{array}{l}\text{Naphthalene}\\ \text{Dinaphthyl sulfides}\\ \text{Binaphthyls}\end{array} \quad (10)$$

2-naphthyl sulfide, 1-naphthyl 2-naphthyl sulfide, 1,1'-binaphthyl, 1,2'-binaphthyl, 2,2'-binaphthyl, and two isomers of benzothianthrene. Elemental sulfur is also present among the reaction products.

Several kinetic experiments were performed to determine the reaction velocity and to establish the relative yields of the different products at early and late stages of the reaction. Naphthalene appears to be formed from 1-naphthalenethiol in dichloromethane at 24°C in the presence of 0.15M triflic acid with a rate constant of 4.8×10^{-4} per second. The product distribution is shown in Table III.

Table III. The Product Distribution for the Reaction of 1-Naphthalenethiol with Triflic Acid in Dichloromethane.

	Concentration, %				
Time min.	$C_{10}H_8$	$(1\text{-}C_{10}H_7)_2S$	$(1\text{-}C_{10}H_7)(2\text{-}C_{10}H_7)S$	$(1\text{-}C_{10}H_7)_2S_2$ [a]	Dithianes [b]
3	67.0	19.8	4.8	8.3	-
10	75.4	18.6	5.4	0.6	-
30	77.8	16.1	5.6	8.3	0.05
60	86.1	9.7	1.0	-	0.08
180	89.0	5.4	4.5	0.3	0.12
360	89.4	3.0	4.2	1.3	2.05

[a] These values are somewhat less certain than the other entries in the Table. At early reaction times, however, the concentration of the disulfide can be accurately measured.

[b] Several different isomers are formed as discussed in the text.

These observations suggest that several rather complex reactions occur under the experimental conditions. Although naphthalene is clearly the principal product of the reaction, the thiol is also converted to an array of sulfides, disulfides and dithianes. The relative yields of these products change as the reaction proceeds toward completion. This finding implies that some of the products, for example the dithianes are formed in secondary reactions. Certain of the sulfides are formed in simple substitution reactions, but others can only be formed through more deepseated rearrangement reactions. It is pertinent that the oxidation products include disulfides and dithianes.

To gain further insight concerning this chemistry, we examined the reaction of 1-naphthalenethiol with aluminum chloride and ethanethiol. The major product of the reaction was naphthalene but many other compounds including 1-naphthyl ethyl sulfide, 1-naphthyl ethyl disulfide, diethyl sulfide, diethyl disulfide, and diethyl trisulfide were produced (eq 11).

$$\text{1-Naphthalenethiol} \xrightarrow{AlCl_3, C_2H_5SH} \begin{array}{l}\text{Naphthalene}\\ \text{Naphthyl ethyl sulfide}\\ \text{Naphthyl ethyl disulfide}\\ \text{Aliphatic disulfides}\\ \text{Aliphatic trisulfides}\end{array} \quad (11)$$

Rate data for the reaction of 1-naphthyl butyl sulfide were obtained to define the course of the reaction more completely. The results are shown in Figure 1.

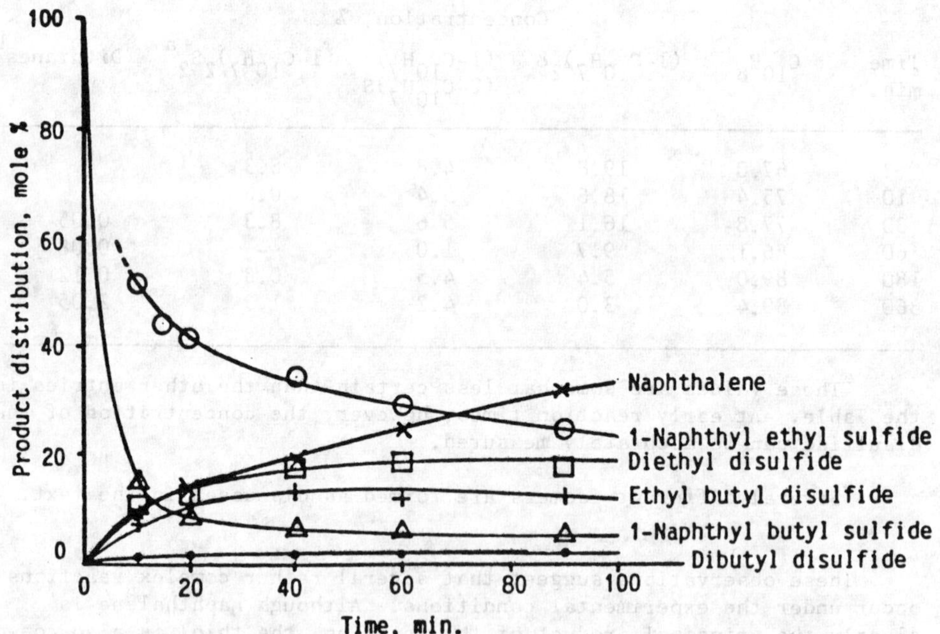

Figure 1. The product distribution for the desulfurization of 1-naphthyl ethyl sulfide using aluminum chloride and ethanethiol.

The kinetic information reveals that 1-naphthyl butyl sulfide is converted rapidly into 1-naphthyl ethyl sulfide which is slowly converted to naphthalene, the only other aromatic compound in the reaction mixture. As already mentioned, aliphatic sulfides and disulfides such as diethyl sulfide, dibutyl sulfide, ethyl butyl sulfide, diethyl disulfide and dibutyl disulfide are also formed. Smaller quantities of aliphatic trisulfides are present.

The richness of this chemistry prompted us to examine the reaction between 1-naphthyl butyl sulfide and triflic acid in the presence of ethanethiol. Quite similar results were obtained with the butyl thioether converted rapidly to the ethyl thioether. Overall, the rate of formation of naphthalene increased significantly.

A representative diaryl thioether, di-1-naphthyl sulfide, was examined next, it is rapidly transformed to naphthalene by triflic acid in dichloromethane (eq 12). Other major products of this reaction include sulfur and the binaphthyls which are formed in different rela-

$$(1\text{-}C_{10}H_7)_2S + CF_3SO_3H \rightarrow \text{Naphthalene} \qquad (12)$$

tive concentrations at different times during the reaction (Table IV).

Table IV. The Desulfurization of Di-1-Naphthyl Sulfide Using Triflic Acid in Dichloromethane.

	Product distribution			
	Binaphthyl			Naphthalene / Binaphthyl (total)
Time, hrs.	1,1'	1,2'	2,2'	
0.5	80.5	19.5	0.0	15.1
1.5	49.2	27.2	23.6	4.5
3.0	17.1	16.9	66.0	3.1
6.0	1.5	5.5	93.0	3.1
10.0	2.5	6.9	90.6	3.3

Such results imply that kinetic factors govern the course of these reeactions at short reaction time, but that the initial products isomerize to other more thermodynamically stable products at longer reaction times.

3.2 2-Naphthyl Derivatives

The main products in the reaction of 2-naphthalenethiol with triflic acid are naphthalene (65%), di-2-naphthyl sulfide, di-2-naphthyl disulfide and two isomers of benzothianthrene (eq 13).

$$2\text{-}C_{10}H_7SH + CF_3SO_3H \rightarrow \begin{array}{l}\text{Naphthalene}\\ \text{Binaphthyls}\\ \text{Dinaphthyl sulfides}\\ \text{Dinaphthyl disulfides}\\ \text{Benzothianthrenes}\end{array} \quad (13)$$

The yield of the benzothianthrenes obtained from 2-naphthalenethiol is much greater than the yield realized with 1-naphthalenethiol. The addition of ethanethiol (5.0 equivalents) to a reaction mixture consisting of 2-naphthalenethiol and triflic acid (2.5 equivalents) completely changed the reaction pathway. As before, 2-naphthyl ethyl sulfide is formed rapidly. This compound is then very slowly converted into naphthalene. For example, after 6 hours at room temperature 80% of naphthyl ethyl sulfide remains and only 2% naphthalene is produced. The reaction with aluminum chloride and ethanethiol gave similar results.

Inasmuch as 2-naphthyl ethyl sulfide appears to be an important intermediate in this reaction. The chemistry of its analogue, 2-naphthyl 1-butylsulfide was investigated. It turned out to be the most unreactive compound among the substances examined in this study. In the presence of triflic acid or in the presence of aluminum chloride and ethanethiol, less than 1% of the material is converted to naphthalene. Only a small amount of dibutyl disulfide, and two new substances are formed. Mass spectroscopy suggests that one of these compounds is 1,2-di-(butylthio)naphthalene.

The reaction of di-2-naphthyl sulfide was also studied. The reaction of this compound with triflic acid gave more than 30% naphthalene, eq (14). Benzothianthrene, di-2-naphthyl disulfide, binaphthyl and two unidentified compounds of mass 206 and 208 were also formed in lesser yields. More than 50% of the starting material was recovered after 6 hours. During the reaction of this sulfide with aluminum chloride and ethanethiol, 2-naphthyl ethyl sulfide formed quite slowly. The concentration of di-2-naphthyl sulfide and 2-naphthyl ethyl sulfide decreased as the concentration of naphthalene increased during the course of the reaction, but the yield of naphthalene was only 9% after six hours.

$$\text{Di-2-Naphthyl sulfide} + CF_3SO_3H \xrightarrow[25^\circ C,\ 6\ \text{hours}]{} \begin{array}{l}\text{Naphthalene}\\ \text{Di-2-naphthyl disulfide}\\ \text{Benzothianthrenes}\\ \text{Binaphthyls}\end{array} \quad (14)$$

In summary, the results indicate that the reactions begin via the interconversion of sulfides and thiols to produce an array of compounds, eventually the sulfur compounds react with acids to give reduction products and oxidation products rather than simple substitution products. This is a novel situation.

3.3 Reaction Pathway

The unusual course of the reaction prompted an investigation of its mechanism. The involvement of cation radicals was tested first. ESR measurements clearly established that radicals were present in the reaction mixtures, but it was unclear whether or not these radicals were intermediates in the reaction or merely produced by the oxidation of products, such as the dithianes.

A determination of the source of the hydrogen atom in the naphthalene appeared to offer a reasonably direct approach for the definition of the reaction pathway because a cation radical would produce naphthyl radical which would abstract a hydrogen atom from the solvent or another donor (eq 15).

$$1\text{-}C_{10}H_7SR \rightarrow 1\text{-}C_{10}H_7SR^{+\cdot} \rightarrow 1\text{-}C_{10}H_7\cdot + RS^+ \qquad (15a)$$

$$1\text{-}C_{10}H_7\cdot + \text{H-Donor} \rightarrow C_{10}H_8 \qquad (15b)$$

On the other hand, an electrophilic protodesulfurization reaction would involve only the proton from the acidic reagent (eq 16).

$$1\text{-}C_{10}H_7SR + H^+ \rightarrow [\text{1-H,1-SR-naphthalenium cation}] \rightarrow 1\text{-}C_{10}H_8 + RS^+ \qquad (16)$$

Accordingly, we determined the source of hydrogen atom that is incorporated into the aromatic hydrocarbons formed from the sulfides and thiols. Experiments were performed in labeled dichloromethane with triflic acid, and in dichloromethane with triflic acid-d (eqs 17 and 18). The nmr and mass spectra of the naphthalene and other products formed in the reaction provided definitive evidence. Specifically,

$$\text{Dinaphthyl sulfide} + CF_3SO_3H \xrightarrow{CD_2Cl_2} \text{Naphthalene, Other Products} \qquad (17)$$

$$\text{Dinaphthyl sulfide} + CF_3SO_3D \xrightarrow{CH_2Cl_2} \text{Naphthalene-d, Other Products} \qquad (18)$$

the spectroscopic studies established that the naphthalene which is produced in the reactions of di-1-naphthyl, 1-naphthyl 2-naphthyl, and di-2-naphthyl sulfide with triflic acid in labeled dichloromethane does not contain deuterium. In contrast, experiments performed in undeuterated dichloromethane with triflic acid-d show a large incorporation of deuterium into the naphthalene obtained from these sulfides. Identical results were obtained when 1- and 2-naphthalenethiol were treated with triflic acid-d in unlabeled dichloromethane.

Kinetic experiments performed with 1-naphthalenethiol and triflic acid-d in dichloromethane also revealed that proton-deuterium exchange in the naphthalene nucleus is very fast at the 2 and 4 positions.

To test the involvement of naphthyl radical in another way, we examined the reaction of a highly reactive hydrogen atom donor, diphenylmethane, with the sulfides and thiols and triflic acid. Diphenylmethane-d_2 (5 molar equivalents relative to the sulfide) did not transfer any significant amount of deuterium into the reaction products. These exchange experiments clearly establish that the reactions of the aromatic organosulfur compounds with triflic acid are not taking place through free naphthyl radicals generated from a cation ion radical intermediate. The observation that the naphthalene formed in the reaction in the presence of diphenylmethane-d_2 yields only unlabeled naphthalene definitively excludes the involvement of radicals of this kind.

4. CONCLUSIONS

A novel desulfurization reaction involving a strong acid as the key reagent has been uncovered. Our work indicates that this reaction successfully removes sulfur from thiols, aryl alkyl sulfides and diaryl sulfides. The sulfur is distributed principally between disulfides, trisulfides, sulfur and small amounts of organosulfur compounds such as the dithianes. The reaction pathway has not been completely established, but cation radicals do not appear to be important intermediates in the reactions. Rather the reaction appears to be reasonably well described as an electrophilic process in which sulfur interchange reactions and thioether isomerization reactions occur rapidly and precede the protodesulfurization reaction which leads to naphthalene, and formally the positively charged sulfur species (eq 19).

$$C_{10}H_7SR + C_2H_5SH \xrightarrow{H^+} C_{10}H_7SC_2H_5 + RSH \qquad (19a)$$

$$C_{10}H_7SC_2H_5 + H^+ \rightarrow [\text{1-H,1-}SC_2H_5\text{-naphthalenium cation}] \rightarrow C_{10}H_8 + C_2H_5S^+ \qquad (19b)$$

These oxidized sulfur intermediates are presumably the precursors of the disulfides and trisulfides formed in the reactions. It should be pointed out that, while cation-radicals do not appear to be important intermediates for the formation of naphthalene, they may be significant in the formation of binaphthyls and related kinds of products.

5. ACKNOWLEDGEMENT We are grateful to Dr. S. Ray Mahasay for many fruitful discussions of the reaction mechanism and to the Center for Research on Sulfur in Coal of the Illinois Coal Board for their support of this investigation.

6. REFERENCES

(1) Kruse, C.W. 1985 Annual Report, Center for Research on Sulfur in Coal, Champaign, Illinois, p. 26.

(2) Duran, J.; Ray Mahasay, S.; Stock, L.M. Fuel, 1986, 65, 1167.

(3) Attar, A.; Dupuis, F. Coal Structure, ACS Adv. Chem. Ser. 1981, 192, 239 and related publications in this series.

(4) Nichols, D. Private communication, 1985.

(5) Calkins, W.H. Energy and Fuels 1987, 1, 59.

(6) Rossi, R.A.; de Rossi, R.H. Aromatic Substitution by the $S_{RN}1$ Mechanism, Am. Chem. Soc. Monograph 178, Washington, 1983.

(7) Nakayama, J.; Fujita, T.; Hoshino, M. Chem. Lett. 1983, 249.

(8) Migota, T.; Shimizu, T.; Asario, Y.; Shiobane, J.; Kato, Y.; Kosugi, M. Bull. Chem. Soc. Jap., 1980, 1385.

(9) Silber, J.J; Shine, H.J. J. Org. Chem. 1971, 36, 2923.

(10) Chen, T.S.; Wolinska-Mocydiarz, J.: Leitch, L.C. Journal of Labelled Compounds 1971, 6, 285.

(11) Furman, F.M.; Thelin, J.H.; Heiro, D.W.; Haely, W.B. J. Am. Chem. Soc 1960, 82, 1450.

(12) Alper, H.; Blais, C. J. Chem. Soc., Chem. Commun. 1980, 169.

(13) Luh, T.Y.; Wang, C.S. J. Org. Chem. 1985, 50, 5413.

(14) Node, M.; Nishide, K.; Ohta, K.; Fujita, E. Tetrahedron Letters 1982, 23, 689.

KINETICS OF COAL GASIFICATION

K. J. Hüttinger
Institut für Chemische Technik
Universität Karlsruhe
Kaiserstr. 12
D-7500 Karlsruhe

ABSTRACT. Coal gasification consists of coal pyrolysis and gasification of the pyrolysis products. During gasification gas/gas-, liquid/gas- and solid gas reactions can occur. All three types of reactions are treated, discussion of the heterogeneous solid/gas reaction between coal char and the gasification media (CO_2, H_2O, H_2) is restricted to the surface reactions, transport phenomena are widely neglected. Some remarks on experimental procedures for studying coal gasification are additionally given.

1. INTRODUCTION

Coal gasification is defined as a process, in which coal is converted to gases. This definition is understandable, but it does not describe, what really occurs in coal gasification processes. When coal is fed into a gasification reactor and thereby heated up, it primarily undergoes pyrolysis reactions, whereby gases, volatile or tar products and char are formed.

$$\text{Coal} \longrightarrow \text{gases } (CH_4, C_2H_4, C_2H_6, CO, CO_2, H_2) + \text{volatiles (tar)} + \text{char} + H_2O \qquad (1)$$

The intensity of the pyrolysis reactions increases with decreasing rank, high volatile bituminous coals can form more than 30 % and brown coals more than 40 % gaseous and volatile products. Pyrolysis reactions are extremely fast. Studies using the CURIE-point-technique for shock heating showed that they are finished within two to five seconds (1,2).

At high heating rates, which also exist in fluidized or entrained bed reactors, pyrolysis reactions can be so strong,

Y. Yürüm (ed.), New Trends in Coal Science, 433–452.

that the gasification agent has nearly no chance to interact with the coal particle before devolatilization fades away. It means, that pyrolysis as the primary step of coal gasification is only insignificantly influenced by the gasification medium. But the sudden formation of large gas volumina can disturb the hydrodynamics of the bed.

A high pressure can alter the situation. Pressure does not only shift the evaporation equilibria, but it also lowers the evaporation kinetics (3). As a consequence, a decrease of tar and an increase of char yield is observed in the pyrolysis step (2,4). If pyrolysis is performed in a hydrogen atmosphere at increased pressure, this gasification agent can react with coal already during pyrolysis, a process which is termed as hydropyrolysis. One reason is the high diffusion coefficient of this molecule.

In other words, not coal - anthracite may be an exception - but volatile products and a solid residue or char are gasified. Depending on pressure, three main types of gasification reactions can occur in a gasification reactor, namely

- homogeneous gas/gas reactions
- heterogeneous liquid/gas reactions
- heterogeneous solid/gas reactions.

Homogeneous gas/gas- and liquid/gas reactions summarize the reactions between volatile products and the gasification medium. Temperature and pressure control which type of reaction is predominant. These reactions may pose severe problems, because some aromatics like naphthalene are extremely stable, do not completely react and are deposited in tubes and plant units behind the reactor. The heterogeneous solid/gas reaction between the solid pyrolysis residue and the gasification medium is, what we generally describe or study as coal gasification. In conclusion, coal gasification is not only a complex process as generally known, but it is additionally composed of various types of reactions.

Even if we concentrate on the so-called coal gasification reaction, namely between the solid pyrolysis residue or char and the gasification medium, it is extremely difficult to present a definite picture about mechanisms and kinetics. It is well known from reactivity studies of cokes and carbons, that reactivity strongly depends not only on the precursor material but also on the history of its formation, for instance heating rate, pressure, atmosphere and final temperature. It means, that any kind of carbon cannot completely be characterized by physical measurements or properties. In relation to coal chars, which are formed in gasifiers by pyrolysis under a certain atmosphere, the reactivity may be quite different even if we use the same coal. Another point of great importance represent the minerals in coal. Content and composition change from coal to coal. Some cations are extremely catalytically active, depending on the gasification atmosphere (5-7). Earth alkali metals like calcium or alkali metals like sodium and potassium are active under oxidizing conditions, provided that they are not inactivated by silicates. Iron, on the other hand, exhibits its catalytical activity only under reducing conditions, provided that it is not poisoned by sulfur (8-10). This element can completely be responsible for the gasification kinetics of any coal or coal char if the gasification temperature is beyond 880 °C and the atmosphere a reducing one (9,10).

Obviously, all these problems cannot be treated in a single paper. Fortunately, we find in literature of the last decades and also the last years some excellent reviews on gasification kinetics, namely by WALKER et al.(11), ERGUN (12), LAURENDEAU (13), KAPTEIJN and MOULIJN (14) and JOHNSON (15). It does not seem to be necessary to reinterprete these reviews. Consequently, this paper will be concentrated on the main basics of gasification mechanisms and kinetics, which are more or less accepted, but it has to be added, still in discussion. New ideas about mechanisms and kinetics will be treated in more detail. Furthermore, it seems worthwhile to discuss some details of approved experimental and especially newer experimental techniques for analyzing the mechanisms and kinetics of coal gasification. Unfortunately, the expectations being combined with techniques like isotope tracer studies or transient kinetic experiments have not been fulfilled. KAPTEIJN and MOULIJN accordingly concluded that the investigation of one problem poses several new problems (14). We are far away to understand the mystery of coal or carbon gasification. This problem is still a challenge for scientists all over the world.

2. HOMOGENEOUS GAS/GAS AND HETEROGENEOUS LIQUID/GAS REACTIONS

Studies on coal gasification have mainly been focussed on char gasification. As a consequence, literature on the reactions of volatile products with various gasification agents either in condensed or gaseous state is rare. Liquid/gas reactions are only important at high pressures and low or medium temperatures in a gasification reactor. Gas/gas reactions between volatile products and the gasification medium play a role at all pressures. Only at gasification temperatures decisively above 1000 °C, these reactions are probably so fast that they have not to be considered separately.

Volatiles or tars are composed of hundreds or more single compounds, mainly aromatics. Larger amounts of aliphatics are only formed in decomposition of brown coals. Little attention has been given on tar reactivity in gasification. Because of the complex composition of tars, studies with well-defined model compounds like non-substituted, alkyl-substituted and hetero-aromatics are helpful. This concept was successfully applied for both types of reactions by HÜTTINGER and KIRMANN (liquid/gas reactions) (16), GRÄBER and HÜTTINGER (gas/gas reactions, normal pressure) (17,68) and NELSON and HÜTTINGER (gas/gas reactions, increased pressure) (18), but only for the case of hydrogasification. Reaction schemes for all types of aromatics have been proposed. The correlation between the molecular structure and reactivity is evident. The primary bond rupture decisively influences the further reaction path. The reactivity increases with decreasing aromaticity. Heteroatoms weaken ring systems. Hydrogen partial pressure can change the reaction mechanism. Dibenzofurane is a good example. At low pressure, decarbonylisation is the primary reaction, which effects a ring cleavage (17,18). At increased hydrogen pressure dehydroxylation is predominant, whereby the extremely stable biphenyl is formed (18). Although the heterogeneous solid/gas reaction in char gasification is generally regarded as the rate limiting step in coal gasification, further work should be done on the reactions of tar or tar components with various gasification media.

3. HETEROGENEOUS SOLID/GAS REACTIONS (CHAR GASIFICATION)

3.1 Fundamentals

Char gasification is a heterogeneous reaction which can be influenced or controlled by mass transport effects, namely pore and/or boundary film diffusion. This problem will not be treated in this chapter which only deals with the surface reactions in char, coke or carbon gasification. Various catalytical effects possibly arising from the minerals of coal will also be neglected.

It seems to be established nowadays that the different reactivities of chars, cokes or carbons are based on similar or equal mechanisms, the differences in gasification rate per g or mol carbon result from different active surface areas. Some authors favour rates related on the total specific surface area (TSA) (22-24), some others rates related on the active surface area (ASA) (19-21). Some confusion still exists about TSA determinations by carbon dioxide or nitrogen adsorption. The reason is that adsorption of nitrogen at 77 K in micropores is diffusion-controlled (25). Therefore, equilibrium is not often obtained and the TSA values are lower than those received by carbon dioxide adsorption at 300 K. The ratio between ASA and TSA (ASA/TSA) is the fraction of active surface area. C_t, the active site density or the concentration of active sites per total surface area (numbers or mol per total surface area) is (ASA/TSA) divided by A*, the area of an active site ($A^* = 8.3 \cdot 10^{-20}$ m^2). During gasification one part of the active sites is occupied $[C_o]$ and the other one is free $[C_f]$.

$$[C_t] = [C_o] + [C_f] \qquad (2)$$

Treatment of gasification kinetics on the basis of active site theory is elegant, the problem is the determination of the active sites, which can only be explained by themselves. Adsorption of oxygen, as proposed in literature, does not seem to be a reliable method (25) because oxygen can adsorb at centers which are not accessible to carbon dioxide or water vapour (26). The same holds for adsorption of carbon monoxide or hydrogen. Transient kinetic experiments or desorption of frozen-in surface groups also seem to be questionable. KAPTEIJN and MOULIJN suggest that the only available method to obtain the true active site density is by applying transient kinetic methods, but this technique still has to be proved (14). The ASA/TSA ratio and thus $[C_t]$ changes from carbon to carbon, but also during gasification. Normally an increase is observed. Values of the ASA/TSA ratio scatter from 0.003 (carbon black, (19)) to 0.15 (coke, (12)), but also values of 0.00019 have been reported (26).

Using the active site theory, the surface gasification rate r_S can be expressed as follows :

$$r_S = [C_t] \cdot f(k_i, p_i) \qquad (3)$$

$[C_t]$ is the surface concentration of active sites, k_i are the rate constants of the individual surface reactions and p_i the partial pressures of the reaction gases at the reaction surface. The function $f(k_i, p_i)$ is the intrinsic site conversion rate, it can be received from LANGMUIR-HINSHELWOOD kinetics, if a homogeneous surface is assumed.

If the concentration of active sites is defined as the number of active sites per mass of carbon $= [C_t] \cdot TSA = [C_t] \cdot A_g$, eq. (3) becomes :

$$r_S = m_C \cdot [C_t] \cdot A_g \cdot f(k_i, p_i) \qquad (4)$$

m_C = mass of carbon present (g).

m_C can be expressed by the initial amount of carbon $m_{C,i}$ and the degree of carbon conversion X.

$$m_C = m_{C,i}\ (1 - X) \qquad (5)$$

With coals liberation of volatiles $m_{C,V}$ has to be taken into account:

$$m_C = (m_{C,i} - m_{C,V})\ (1 - X) \qquad (6)$$

Using eq.(5) two normalized rate laws can be formulated:

$$r_{S,n} = \frac{r_S}{m_{C,i}} = \frac{dX}{dt} = (1-X) \cdot C_t \cdot A_g \cdot f(k_i, p_i) \qquad (7)$$

or

$$r_{S,N} = \frac{r_S}{m_C} = \frac{dX}{dt}\,\frac{1}{(1-X)} = C_t \cdot A_g \cdot f(k_i, p_i) \qquad (8)$$

Both types of equations are used. As can be seen, $r_{S,n}$ in related to the initial amount of carbon, whereas $r_{S,N}$ takes into account the consumption of carbon. Eq.(8) gives a better fit of experimental data (30).

For practical purposes, the following rate law is sometimes used:

$$\frac{dX}{dt} = k \cdot (1 - X)^n \qquad (9)$$

This and similar semi-empirical rate equations taking into account the change of internal surface area with carbon conversion were investigated by several authors (24,27-31). A detailed treatment of the problem is given in another paper of this conference (70).

In the following, the function $f(k_i, p_i)$ will be derived for the most common models of char-, coke- or carbon gasification with carbon dioxide, water vapour and hydrogen.

The gasification with carbon dioxide will be treated first:

$$C + CO_2 \rightleftharpoons 2\,CO \qquad (10)$$

The advantage of this reaction is the selectivity, which is one. Selectivity problems exist with water vapour gasification. The common models only account for the reaction:

$$C + H_2O \rightleftharpoons CO + H_2 \qquad (11)$$

Three consecutive reactions have to be considered, namely

the water gas shift reaction

$$CO + H_2O \rightleftharpoons CO_2 + H_2 , \qquad (12)$$

methanation of carbon monoxide

$$CO + 3\,H_2 \rightleftharpoons CH_4 + H_2O , \qquad (13)$$

and hydrogasification

$$C + 2\,H_2 \rightleftharpoons CH_4 . \qquad (14).$$

Eqs.(12) to (14) only describe stoichiometric reactions, but formation of carbon dioxide or methane can occur at the carbon surface in different ways:

$$C(O) + H_2O \rightleftharpoons CO_2 + H_2 \qquad (15)$$

$$C(H_2) + H_2O \rightleftharpoons CH_4 + C(O) \qquad (16)$$

or

$$C(H_2) + H_2 \rightleftharpoons CH_4 + C_f \qquad (17)$$

Uncatalyzed hydrogasification according to eq.(17)

$$C + 2\,H_2 \rightleftharpoons CH_4 \qquad (18)$$

is very slow as compared with water vapour or carbon dioxide gasification. Sufficient rates are only achieved with very reactive chars at high hydrogen pressures. Thermodynamic plays a decisive role especially in this reaction, which is the only exothermic one (Table 1).

Table 1 - Heats of reaction ΔH_R^O and temperatures at which $K_p > 1$ of gasification reactions

Reaction	$\Delta H^O_{R,298}$, kcal·mol^{-1}	$T(K_p > 1)$, K
$C + CO_2 \rightleftharpoons 2\,CO$	41.2	> 950
$C + H_2O \rightleftharpoons CO + H_2$	31.4	> 950
$C + 2\,H_2 \rightleftharpoons CH_4$	- 17.9	< 820

3.2 Carbon dioxide gasification

The most common mechanism of carbon dioxide gasification is the oxygen exchange, originally proposed by REIF (32), later on supported by ERGUN (32,33,12), WALKER (11,34)

and others.

$$C_f + CO_2 \underset{k_{-1}}{\overset{k_1}{\rightleftarrows}} C(O) + CO \quad (19)$$

$$C(O) \overset{k_2}{\rightarrow} CO + C_f \quad (20)$$

According to LANGMUIR-HINSHELWOOD the rates of the oxygen exchange and desorption reaction are given by :

$$r_{(19)} = k_1 \cdot (1-\Theta) \cdot p_{CO_2} - k_{-1} \cdot \Theta \cdot p_{CO} \quad (21)$$

$$r_{(20)} = k_2 \cdot \Theta \quad (22)$$

Θ represents the fraction of occupied sites = $[C_o] / [C_t]$ or $C(O) / [C_t]$. Under steady state conditions

$$r_{(21)} = r_{(22)} \quad (23)$$

which yields :

$$\Theta = \frac{k_1 \cdot p_{CO_2}}{k_1 \cdot p_{CO_2} + k_{-1} \cdot p_{CO} + k_2} \quad (24)$$

Substituion of Θ in eq. (21) or eq. (22) gives the intrinsic site conversion rate :

$$f(k_i, p_i) = \frac{k_1 \cdot p_{CO_2}}{\frac{k_1}{k_2} \cdot p_{CO_2} + \frac{k_{-1}}{k_2} \cdot p_{CO} + 1} \quad (25)$$

The global rate is

$$r_S = [C_t] \cdot k_2 \cdot \Theta \quad (26).$$

The oxygen exchange mechanism suggests that carbon monoxide inhibition occurs not by adsorption of carbon monoxide from the gas phase but rather by the equilibrium of the oxygen exchange reaction. The equilibrium constant $K_1 = \frac{k_1}{k_{-1}}$ was determined by ERGUN (12) as 23 kcal·mol^{-1} and later confirmed by STRANGE and WALKER (34).

Some other authors found (35-37), that the oxygen exchange mechanism is not applicable in the case of high CO/CO_2

ratios and postulated a third step, namely adsorption of carbon monoxide from the gas phase :

$$C_f + CO \underset{k_{-3}}{\overset{k_3}{\rightleftharpoons}} C(CO) \qquad (27)$$

The oxygen exchange reaction (eq. (19)) is simultaneously assumed to be irreversible. WALKER et al. (11) have shown that this model leads to a similar rate equation as given above.
It underlines that care has to be taken if it is concluded from kinetics to a mechanism. Further mechanisms, derived from increased pressure experiments, were reported by KEY (38), BLACKWOOD and INGEME (39) and SHAW (40). Rate laws derived by these authors have an additional second order term of carbon dioxide partial pressure in the indicator. A detailed discussion of these kinetics is given elsewhere (13).

In recent papers it was suggested by several authors (26, 41, 42) that desorption of the C(O) surface complex is based on a two-step mechanism. Simulation of such a desorption mechanism revealed no confirmation (43), especially of the experimental results of FREUND (26).

All mechanisms discussed above are based on a single site adsorption. As shown by LAURENDEAU et al. (44, 45), a dual site adsorption can be predominant, if gasification is performed with pure carbon dioxide, i.e. in the absence or at low levels of carbon monoxide. In this case, the rate is proportional to a half order dependence on the carbon dioxide partial pressure instead of a first order dependence. These results confirm earlier observations of TURKDOGAN and VINTERS (23, 46). According to these authors, the order changes from 0.5 to 1 depending on the gas composition. A dual site chemisorption was also postulated from desorption measurements in the laboratory of the author (43). Inhibition of carbon dioxide gasification by carbon monoxide is generally accepted. But BIEDERMANN and WALKER (47) found, that traces of hydrogen in the ppm-range also strongly inhibit the carbon dioxide gasification. Due to these results the authors suggest, that a true gasification rate with carbon dioxide has never been measured and is possibly not measurable.

The rate-limiting step is a general discussion in gasification kinetics. In carbon dioxide gasification most authors are convinced, that this is desorption of the C(O) surface complex. Values of activation energies scatter over a rather wide range, but most trustworthy results for the desorption step are in the range of 60 kcal·mol^{-1} (13, 14). A similar value is found for the global kinetics of carbon dioxide gasification (Fig.3), but the activation energy of the forward reaction of the oxygen exchange is only slightly smaller, namely 55 kcal·mol^{-1} (13).

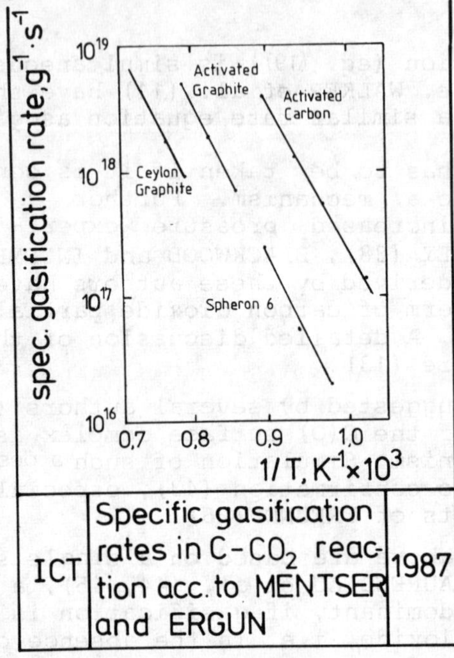

Figure 3

3.3 Water vapour gasification

As already pointed out, a mechanism of water vapour gasification being valid over wide ranges of pressure, temperature and gas composition must a priori be more complicated than of carbon dioxide gasifications. If the consideration is restricted to the water-gas reaction, a similar oxygen exchange mechanism like in carbon dioxide gasification has to be favoured (12).

$$C_f + H_2O \underset{k_{-1}}{\overset{k_1}{\rightleftharpoons}} C(O) + H_2 \tag{28}$$

$$C(O) \overset{k_2}{\rightarrow} CO + C_f \tag{29}$$

Analogous to carbon dioxide gasification a second similar mechanism was proposed for this reaction in which inhibition is caused by hydrogen adsorption from the gas phase and the oxygen exchange reaction is assumed to be irreversible (48-50). Reasons why the first mechanism has to be preferred were given by LAURENDEAU (13).

According to mechanism one the intrinsic site conversion rate $f(k_i, p_i)$ is :

$$f\ (k_i, p_i) = k_2 \cdot \Theta = \frac{k_1\ p_{H_2O}}{\frac{k_1}{k_2}\ p_{H_2O} + \frac{k_{-1}}{k_2}\ p_{H_2} + 1} \qquad (30)$$

For the global surface rate results :

$$r_s = [C_t] \cdot k_2 \cdot \Theta \qquad (31)$$

Several further mechanisms have been proposed (see Ref.(13)). YANG and YANG (51) found by TEM- and SEM studies the following expression for the site conversion rate at monolayer edges of Ticonderoga graphite:

$$f(k_i, p_i) = \frac{k_1\ p_{H_2O}}{\frac{k_1}{k_2}\ p_{H_2O} + \frac{k_{-1}}{k_2}\ p_{H_2}^n + 1} \qquad (32)$$

whereby n is 0.5 at 700, 0.85 at 800 and 1.0 at 900 °C. For multilayer edges n = 0.5 was found at all temperatures. The latter rate function was also reported in an earlier paper by GIBERSON and WALKER (52).

In some other papers, dissociation of steam to OH and H is supposed (49, 50, 53-55). More important are mechanisms, which fit data over larger ranges of pressure and thus take into account the selectivity of water vapour gasification. Methane and carbon dioxide for instance can be formed by the following surface reactions :

$$C(H_2) + H_2O \rightleftharpoons CH_4 + C(O) \qquad (33)$$

$$C(O) + H_2O \rightleftharpoons CO_2 + H_2 + C_f \qquad (34)$$

Including these reactions, the following global rate results (53) :

$$r_S = \frac{k \cdot p_{H_2O} + c \cdot p_{H_2O}^2 + d \cdot p_{H_2} \cdot p_{H_2O}}{1 + a \cdot p_{H_2} + b \cdot p_{H_2O}} \qquad (35)$$

MÜHLEN (56) showed that a better fit of experimental data is achieved if eq. (35) is extended by a second order hydrogen term accounting for methane formation by hydrogasification:

$$C(H_2) + H_2 \rightleftharpoons CH_4 + C_f \tag{36}$$

With no water vapour this extended surface rate law is identical to a possible one of hydrogasification as will be shown later.

According to all presented global surface rates of water vapour gasification, carbon monoxide does not inhibit the reaction. Inhibition by hydrogen is very strong and it decisively surpasses carbon monoxide inhibition in carbon dioxide gasification. Results with various inert carrier gases (He, Ar, Kr) reveal that also helium as compared to argon or krypton strongly reduces the rate of water vapour gasification, but only at elevated pressure (57). No similar effect is found in carbon dioxide gasification. Diffusion effects may be excluded. It is therefore assumed that strong hydrogen inhibition is partially caused by the effect of the small molecule.

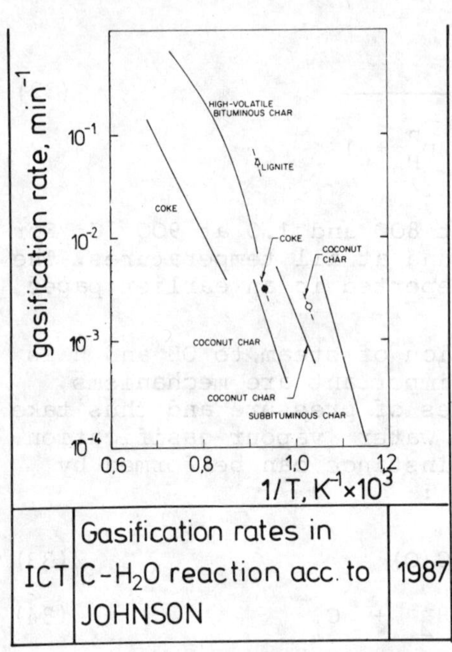

Figure 4

A different mechanism of water vapour gasification of carbon was proposed by HERMANN and HÜTTINGER (59). Based on TPD measurements, which showed extremely stable bound oxygen surface groups, these authors concluded that the oxygen is heterocyclicly bound as ether. Calculations of the concentration of the active sites density supported this conclusion, because a constant value after gasification in various H_2/H_2O mixtures was only received by assuming that oxygen is bound to two carbon atoms. This result needs confirmation by a direct prove of the proposed ether structure.

The rate-limiting step in water vapour gasification is

probably the same like in carbon dioxide gasification, namely desorption of the carbon oxygen surface complex. The uncertainty in kinetic data and the activation energy is higher than in carbon dioxide gasification (13). But the activation energies found for the oxygen exchange step are clearly lower than those found for the desorption step, whereas the latter values are in the same range like the activation energies found for the global gasification kinetics (Fig.4). Most values scatter around 60 - 70 kcal·mol^{-1} but higher values were also reported (11).

3.4 Hydrogasification

Hydrogasification of coals or organic matter in general is quite different from carbon dioxide or water vapour gasification, especially at increased pressure which is a prerequisite for thermodynamic and kinetic reasons. Hydrogen is able to hydrogenate radicals formed in the pyrolysis step by thermal bond cleavage. At sufficiently high hydrogen pressure, polyaromatics can be stabilized up to 800 °C (57), as will be shown later. In respect to coke or carbon gasification, the selectivity of this reaction is one, because methane is the only product. However, with coal the selectivity of hydrogasification is extremely complex and mainly depending on the heating procedure.

Only few mechanistic studies are available on hydrogasification of char which is generally termed as slow hydrogasification (3). SHAW (60) using the experimental data of BLACKWOOD et al. (61, 62) showed that they misinterpreted their data. He concludes that methane-forming steps can be simplified to :

$$C_f + H_2 \underset{k_{-1}}{\overset{k_1}{\rightleftharpoons}} C(H_2) \tag{37}$$

$$C(H_2) + H_2 \underset{k_{-2}}{\overset{k_2}{\rightleftharpoons}} CH_4 + C_f \tag{38}$$

Applying conventional LANGMUIR-HINSHELWOOD kinetics to this mechanism, the steady-state condition yields for the intrinsic site conversion rate :

$$f(k_i, p_i) = \frac{k_{-1} \cdot k_{-2} \cdot (K \cdot p_{H_2}{}^2 - p_{CH_4})}{(k_1 + k_2) \cdot p_{H_2} + k_{-2} \cdot p_{CH_4} + k_{-1}} \tag{39}$$

K is the equilibrium constant of the overall reaction, i.e. $k_1 \cdot k_2 \,/\, k_{-1} \cdot k_{-2}$.

The intrinsic global surface rate is

$$r_S = [C_t] \cdot f(k_i, p_i) \tag{40}$$

With eq.(41) SHAW received an excellent fit of BLACKWOOD's data with coconut char after the initially rapid reaction has died away.
Most authors use a simple first order dependence on hydrogen partial pressure.

$$\bar{r}_S = k \cdot p_{H_2} \tag{41}$$

The same relationship results from SHAW's rate law if

$$p_{CH_4} \ll K \cdot p_{H_2}^2 \text{ and } (k_{-2} \cdot p_{CH_4} + k_{-1}) \ll (k_1 + k_2) \cdot p_{H_2} \tag{42}$$

Another simplified rate law as compared to SHAW's model was used by ZIELKE and GORIN (63) and FEISTEL et al. (64).

$$\bar{r}_S = \frac{k \cdot p_{H_2}^2}{1 + a \cdot p_{H_2}} \tag{43}$$

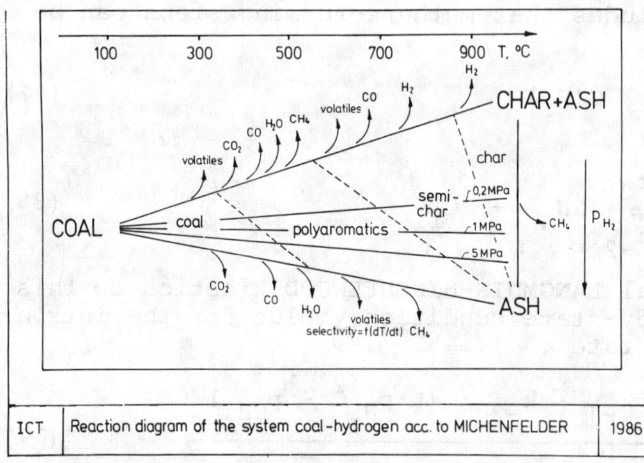

ICT | Reaction diagram of the system coal-hydrogen acc. to MICHENFELDER | 1986

Figure 5

All results indicate that hydrogasification of chars at high pressure is proportional to the hydrogen partial pressure. In this property, hydrogasification differs from carbon dioxide and water vapour gasification, where the gasification rate levels out at 2 (CO_2) or between 1 and 1.5 MPa (H_2O). The most often found value of the activation energy for char gasification is about 40 kcal·mol^{-1} (13), and thus lower than of carbon dioxide and water vapour gasification. Activation energies for the elementary reactions were reported by SHAW (60).

In gasification of coal and especially of brown coals and lignites, which due to their high reactivities are the preferred raw materials for hydrogasification, the first step, the so-called rapid hydrogasification or hydropyrolysis, can control the overall reaction. This is the case at high hydrogen partial pressures of about 5 MPa where the coal is nearly completely 'gasified' under conditions of hydropyrolysis (57, 65, 66). This reaction path is presented in the triangular diagram shown in Fig.5. The upper axis describes the pyrolysis path, which is equivalent to low hydrogen pressure. With increasing temperature, coal is converted to polyaromatics, semi-char and char. The lower axis is for high hydrogen pressure. Under this condition, the polyaromatic system is maintained up to high temperatures. The heating rate determines the selectivity of the reaction. At slow heating, volatiles are the main product, but at fast heating methane selectivity amounts to 0.7 or more.

A severe problem in hydrogasification represents inhibition by moisture (67). Only 1 % water vapour drastically lowers the reactivity. As low rank coals (brown coal, lignite), even well dried, liberate substantial amounts of water by decomposition of the carboxyl groups, this problem cannot be easily overcome in a technical reactor. Although the mechanism of inhibition is not fully clear, it is probable that water forms extremely stable surface groups (ethers like in water vapour gasification) which block active sites. MICHENFELDER (57) has shown that these groups are stable in humid hydrogen up to 750 °C. The same author showed that 950 °C and 5 MPa are not sufficient to overcome the problem of water vapour inhibition completely (57).

4. REMARKS ON EXPERIMENTAL TECHNIQUES

Kinetic studies on coal, char or carbon gasification are performed in fixed or fluidized bed reactors, normally operated batch-wise and under differential reactor conditions. The latter condition is guaranteed if product gas concentration remains below 1 % of the gasification medium. Influence of boundary film diffusion is negligible if the gasification rate does not further increase by increasing gas flow rate. Influence of pore diffusion can be tested by experiments with decreasing particle size. Elimination of pore diffusion is achieved if the rate does not further increase with decreasing particle size provided that the rate is not controlled by micropore diffusion. At high pressure and temperature smaller particles have to be used.

The best analytical technique for studying gasification kinetics represents measurement of mass loss by using a thermobalance, but at increased pressure, because modern coal gasification processes use pressures at least up to 2 MPa. Care must be taken on diffusion effects, special sample holders are necessary (27). The advantage of this technique is that the rate of carbon gasification is directly measured by mass loss. Evolution of tar products from coal is identified additionally. For following selectivity in complex gasification reactions an additional gas analysis is necessary (see below).

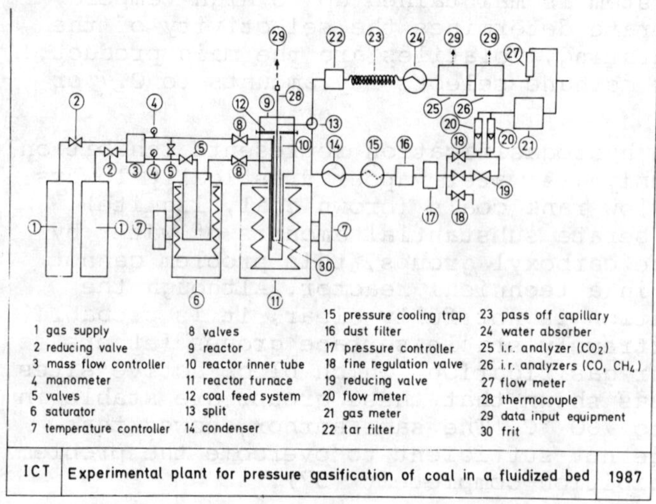

ICT | Experimental plant for pressure gasification of coal in a fluidized bed | 1987

Figure 6

Analysis of product gases for following the gasification rate poses several problems. Studies on inhibition by product gases fed with the gasification medium to the reactor are nearly impossible because the concentration changes are too low to be analyzed. In the case of coal gasification formation of volatile products is not identified, therefore the initial amount of char at the beginning of gasification is unknown. The amount of volatile products can only be determined after gasification by a carbon balance. Gas analysis itself represents a problem in non-steady state experiments. Gas formation peaks can be extremely broadened during flowing of the gas through condensator, drying units etc. This problem can be solved by using a by-pass, whereby the gas is expanded with the aid of an extremely thin expansion capillary (0.1 to 0.2 inner diameter). Without this technique, a deconvolution of the measured gas formation peaks is necessary, but it is only possible if a master curve is found for peak broadening under all experimental conditions (59).

A thermobalance for pressures up to 10 MPa is described by MÜHLEN (56), a gasification apparatus with a fixed or fluidized bed reactor for 5 MPa equipped with the above mentioned technique for gas analysis is shown in Fig.6. As an example, gas peaks measured with the expansion capillary and at the outlet of the apparatus resulting from $CaCO_3$ decomposition at 0.5 MPa and 900 °C are represented in Fig.7. Special problems arise with carbon dioxide because it is adsorbed at tube walls, in water condensors, at silica gel etc. Pressure enhances the problem.

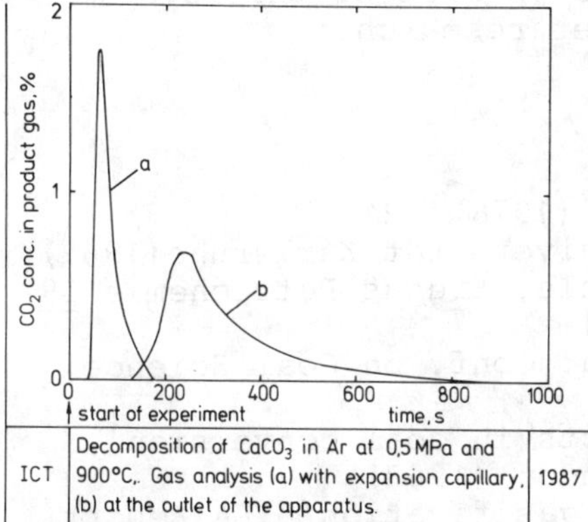

Decomposition of $CaCO_3$ in Ar at 0,5 MPa and 900°C,. Gas analysis (a) with expansion capillary, (b) at the outlet of the apparatus.

Figure 7

Kinetic studies can be used for testing a supposed mechanism. Rearrangement of the rate equation shows, which parameter must be varied in order to receive the desired information. TEM- and SEM studies originally applied by HENNIG (69) and in recent years by YANG and YANG (51) are very useful for studying mechanisms. The same holds for isotope tracer studies, as shown for instance by ERGUN (12) and MOULIJN et al. (14). Transient kinetic experiments as favoured by KAPTEIJN and MOULIJN can be helpful in some special cases (14). Desorption measurements performed either under isothermal conditions (26) or at linear heating (TPD of frozen-in surface groups) (59) are doubtful (43), because only extremely stable bound oxygen surface groups are detected and their role in gasification is not yet clear.

A fundamental problem represent active sites. It seems, that the available or tested methods are not satisfying. More research seems to be necessary in this field. The same holds for the development of internal pore structures during gasification and their relevance to gasification kinetics. Care has to be taken with inert carrier gases, especially if helium is used (58). The inhibition effect of this molecule and also of hydrogen need further research.

5. CONCLUSION

A lot of information on the gasification behaviour of coals, chars or carbons is available. The complex composition of coals, but also the complex structures (X-ray fine structure, micro- and macrostructure) of chars or carbons complicate self-consisting treatments of gasification kinetics. The concept of active sites can be a tool to solve the problem, but a reliable technique is needed for detection of active sites. Lateral interactions of chemisorbed species, not considered in the past, can be responsible for some contradictions.

In this paper, only some selected subjects could be treated. Nevertheless, it is hoped, that both beginning and more advanced researchers can take profit of this overview and are stimulated for further research.

6. REFERENCES

(1) Anthony, D.B. et al., FUEL 55 (1976) 121
(2) Sperling, R., Dissertation, Universität Karlsruhe (1987)
(3) Hüttinger, K.J., Erdöl und Kohle, Erdgas Petrochemie 39 (1986) 495; ibida 40 (1987) 21
(4) Wanzl, W. et al., Proc. Intern. Conf. on Coal Science, Sydney (1985) 899
(5) Davidson, R.M., 'Mineral effects in coal conversion', IEA Report No. ICTIS/TR 22, London (1983)
(6) Pullen, J.R., 'Catalytic coal gasification' IEA Report No. ICTIS/TR 26, London (1984)
(7) Hüttinger, K.J., Erdöl und Kohle, Erdgas, Petrochemie 39 (1986) 261
(8) Adler, J. et al., FUEL 64 (1985) 1215
(9) Hüttinger, K.J., FUEL 62 (1983) 166
(10) Hüttinger, K.J. et al., 60 (1981) 93; ibida 61 (1982) 291
(11) Walker, P.L., Jr. et al., Adv. Catalysis 11 (1959) 133
(12) Ergun, S. and Mentser, M. in 'Chemistry and Physics of Carbon', M. Dekker, N.Y., Vol. 1 (1966) 203
(13) Laurendeau, N.M., Prog. Energy Combust. Sc. 4 (1978) 211

(14) Kapteijn, F. and Moulijn, J.A., in 'Carbon and Coal Gasification' (J.L. Figueiredo, J.A. Moulijn, eds.), M. Nijhoff Publ., Boston (1986) 291
(15) Johnson, D.L., 'Kinetics of Coal Gasification', J. Wiley, N.Y. (1979)
(16) Hüttinger, K.J. and Kirmann, H., Erdöl und Kohle, Erdgas, Petrochemie 35 (1982) 17
(17) Gräber, W.D. and Hüttinger, K.J., Erdöl und Kohle, Erdgas, Petrochemie 32 (1979) 26; ibida 32 (1979) 519; ibida 33 (1980) 416
(18) Nelson, P. and Hüttinger, K.J., FUEL 65 (1986) 354
(19) Laine, N.R. et al., J. Phys. Chem. 67 (1963) 2030
(20) Radovic, L.R. et al., FUEL 62 (1983) 849
(21) Garcia, X. and Radovic, L.R., FUEL 65 (1986) 292
(22) Otto, K. and Sheleff, M., Proc. 6th Intern. Congress on Catalysis, London (1976), paper B47
(23) Turkdogan, E.T. and Vinters, J.V., Carbon 7 (1969) 101
(24) Adschiri, T. and Furasawa, R., FUEL 65 (1986) 927
(25) Rodriguez-Reinoso, F., in 'Carbon and Coal Gasification' (J.L. Figueiredo, J.A. Moulijn, eds.), M. Nijhoff Publ., Boston (1986) 601
(26) Freund, H., FUEL 65 (1986) 63
(27) Johnson, J.L., Adv. Chem. Series 131 (1974), ACS, Washington, D.C., 145
(28) Chornet, E. et al., FUEL 58 (1979) 395
(29) Mahajan, O.P. et al., FUEL 57 (1978) 643
(30) Jüntgen, H., Carbon 19 (1981) 167
(31) Tomkow, K. et al., FUEL 56 (1977) 101
(32) Reif, A.E., J. Phys. Chem. 56 (1952) 785
(33) Ergun, S., J. Phys. Chem. 60 (1956) 480
(34) Strange, J.F. and Walker, P.L., Jr., Carbon 14 (1976) 345
(35) Gadsby, J. et al., Proc. Roy. Soc. 193A (1948) 357
(36) Löwe, A., Carbon 12 (1974) 335
(37) Grabke, H.J., Carbon 10 (1972) 587
(38) Key, A., Gas Research Board, Comm. 40 (1948) 36
(39) Blackwood, J.D. and Ingeme, A.J., Aust. J. Chem. 13 (1960) 194
(40) Shaw, J.T., FUEL 56 (1977) 134
(41) Mc Carthy, D.J., Carbon 24 (1986) 652
(42) Moulijn, J.A. and Kapteijn, F., Erdöl und Kohle, Erdgas, Petrochemie 40 (1987) 15
(43) Hüttinger, K.J., Erdöl und Kohle, Erdgas, Petrochemie 40 (1987) in press
(44) Koenig, P.C. et al., Carbon 23 (1985) 531
(45) Koenig, P.C. et al., FUEL 65 (1986) 412
(46) Turkdogan, E.T. and Vinters, J.V., Carbon 8 (1970) 39
(47) Biedermann, D.L. et al., Carbon 14 (1976) 351
(48) Gadsby, J. et al., Proc. Roy. Soc. A 187 (1946) 129

(49) Long, F.J. and Sykes, K.W., Proc. Roy. Soc. A 193 (1948) 377
(50) Johnstone, J.F. et al., Ind. Engg. Chem. 44 (1952) 1564
(51) Yang, R.T. and Yang, K.L. Carbon 23 (1985) 537
(52) Gibersen, R.C. and Walker, P.L., Jr., Carbon 3 (1966) 521
(53) Blackwood, J.D. and Mc Grory, F., Aust. J. Chem. 11 (1958) 16
(54) Blackwood, J.D. and Mc Taggart, F.K., Aust. J. Chem. 12 (1959) 533
(55) Wehrer, A. et al., J. Chem. Phys. Physicochim. Biol. 70 (1966) 664
(56) Mühlen, H.-J., Dissertation, Universität Essen (1983)
(57) Michenfelder, A., Dissertation, Universität Karlsruhe (1986)
(58) Hüttinger, K.J. et al., Proc. 17th Bienn. Conf. on Carbon, Worcester, USA (1987)
(59) Hermann, G. and Hüttinger, K.J., Carbon 24 (1986) 705
(60) Shaw, J.T., Proc. Intern. Conf. on Coal Science, Düsseldorf (1981) Glückauf, 209
(61) Blackwood, J.D., Aust. J. Chem. 12 (1959) 14; ibida 15 (1962) 397
(62) Blackwood, J.D. and Mc Carthy, D.J., Aust. J. Chem. 19 (1966) 797
(63) Zielke, C.W. and Gorin, E., Ind. Engg. Chem. 47 (1955) 820
(64) Feistel, P.P. et al., Am. Chem. Soc. Div. Fuel Chem., preprints 22 (1) (1977) 53
(65) Hüttinger, K.J. and Michenfelder, A., Erdöl und Kohle, Erdgas, Petrochemie 40 (1987) in press
(66) Hüttinger, K.J. and Michenfelder, A., Proc. Intern. Conf. on Coal Science, Maastricht (1987)
(67) Hüttinger, K.J. and Michenfelder,A., FUEL 64 (1985) 1723
(68) Hüttinger, K.J. and Gräber, W.D., FUEL 61 (1982) 499; ibida 61 (1982) 505; ibida 61 (1982) 509
(69) Hennig, G.R., in 'Chemistry and Physics of Carbon' (P.L. Walker, Jr., ed.), M. Dekker, N.Y., Vol. 2 (1966) 1
(70) Hüttinger, K.J., 'Transport and other effects in coal gasification', this conference.

TRANSPORT AND OTHER EFFECTS IN COAL GASIFICATION

K. J. Hüttinger
Institut für Chemische Technik
Universität Karlsruhe
Kaiserstr. 12
D-7500 Karlsruhe

ABSTRACT. The paper summarizes the kinetics of coal char gasification excepted surface reactions (mechanisms). The following subjects controlling coal char gasification are treated: Coal as the raw material, coal char formation and coal char properties, catalytic effects by minerals, changes of particle size and/or pore structure during gasification, transport phenomena. Numerous experimental results and theoretical models are available. Nevertheless, confirmations of the various models or a preferred model for a broad spectrum of gasification conditions is still missing.

1. INTRODUCTION

In the chapter on 'Kinetics of Coal Gasification' (1) it was shown, that the overall process consists of

- coal pyrolysis, whereby gases, volatile and non-volatile tar products and char as solid residue are formed in a very rapid reaction,
- homogeneous gas/gas reactions of volatile tar products,
- heterogeneous liquid/gas reactions of non-volatile tar products,
- heterogeneous solid/gas reactions of the char

with the gasification media. In the case of the heterogeneous char gasification, only the surface reactions were treated. Effects on reactivity caused by

- coal rank, which implies composition of organic matter (carbon and hydrogen content, heteroatoms), chemical and physical structure and thus pyrolysis behaviour,
- inorganic matter or minerals (amount, composition, chemical state and distribution),
- history of char formation (initial particle size, atmosphere, pressure, heating rate, temperature),

Y. Yürüm (ed.), New Trends in Coal Science, 453–480.

- properties of chars as predetermined by their origin (coal) and history of formation,and
- the changes of char properties during gasification

were widely neglected. Some of these parameters or properties can influence the surface gasification rate (intrinsic reactivity), some others like pore structure of chars the overall rate of gasification, namely by mass transport limitation. Many variables depend on each other, partially in a very complicated manner, as will be shown by a simple example. Highly dispersed iron formed from pyrite or marcasite can enhance the gasification rate to such an extent, that the reaction becomes diffusion-controlled. On the other hand, diffusion control can effect in an oxidizing atmosphere that iron minerals are reduced to metallic, catalytically active iron.

Before discussing transport phenomena in coal gasification, some fundamental parameters influencing char gasification will be considered. In some cases, it is necessary to restrict the discussion on simple statements, either for not extending the paper too much or for a lack of evident experimental data.

2. FACTORS OTHER THAN MASS TRANSPORT INFLUENCING REACTIVITY

2.1 Coal rank

Brown coals, lignites and high volatile bituminous coals are preferred coal types for gasification. Porosity of these low rank coals is dominated by macropores ($d_p > 50$ nm). With increasing rank, pore size distribution is shifted to smaller pore diameters. Medium rank coals ($\sim$ 80 % carbon content) exhibit a significant high content of mesopores ($2 \text{ nm} < d_p < 50$ nm), high rank coals of micropores ($d_p < 2$ nm). (2). During pyrolysis of low rank coals (isothermal, fast heating), more than 30 % of volatiles escape within a few seconds. The resulting chars are porous materials with porosities in the range of 0.5, whereby macro- and mesopores are important for gasification as feeder pores.

Reactivity of coals and chars formed thereof decreases with increasing rank, a drastic change occurs at transition from non-caking (thermosetting) coals to caking coals (3). For the higher rank coals with more than 75 to 80 % carbon content, the initial gasification rate in carbon dioxide or water vapour $(dX/dt)_{t=0}$ is proportional to the carbon content of the coals, for the lower rank coals with less than 75 % to 80 % carbon content $(dX/dt)_{t=0}$ was found to be proportional to the initial rate of carbon dioxide formation during pyrolysis. Correlations of the initial gasification rates with other coal properties (ultimate and proximate analysis, reflectance, metal contents in char, surface area of char or the amount of oxygen trapped in the char) of 25 investigated coals failed. Other authors (4,5) correlated the char reactivity with the oxygen content of the coal.

In water vapour gasification, lower rank coals also yield decisively higher CO_2/CO ratios than higher rank coals. This is attributed to a catalytic effect by minerals (K^+, Na^+, Ca^{++}) (3). According to own observations with model cokes, which have no minerals (sugar coke, polyvinylchloride coke) an additional effect by the oxygen of the coke or char must be assumed. Catalytic effects with caking coals are small or even negligible, because the minerals are encapsulated by carbon (6). The low CO_2/CO ratios with higher rank coals can also result from transport limitations leading to a lack of water vapour in the pore system.

Even if transport phenomena are negligible, which is more probable with low rank coals, a clear prediction of char reactivity by coal properties is still not possible. Determination and use of initial gasification rate as performed by HASHIMOTO et al. (3) seems to be a promising way, because initial rate includes the conditions of char formation.

2.2 Inorganic matter

The content of inorganic matter of coals varies with rank and additionally from seam to seam. Some coals exhibit up to 30 % mineral matter. Such high contents mainly result from silicon containing compounds, which are catalytically inactive. Most important cations are Si, Al, Ca, Mg, Fe, in some special cases K and Na. During pyrolysis, minerals are converted to oxides (7). Some cations (Ca^{++}, K^+, Na^+) can also influence pyrolysis of organic matter of coal, namely if they are present in an active form. This is the case again with low rank coals, where these cations are partially exchanged. A decrease of tar formation and a simultaneous decrease of weight loss is normally observed (8,9).

Catalytically active metals present in coals are well known (10,11), namely Ca $\gg$ Mg, K $\approx$ Na and Fe. Calcium is known to be only active, if it is present in a finely dispersed state, preferentially exchanged. This is the case with brown coals and lignites due to the high content of carboxyl groups (12). For the activity of potassium and sodium the anion is extremely important (13,14); these metals can develop their catalytic activity only if they are not bound in silicates. The reaction between silica or silicates with alkali metals or alkali metal oxides is not only thermodynamicly favoured, but also because of the high mobility of the alkali metals and their oxides. This problem does not exist with iron. The activity of this element is determined by sulfur, which poisons iron by formation of extremely stable surface sulfides (15). Agglomeration during progressive gasification reduces activity additionally (16, 17).

Calcium, potassium and sodium are active in carbon dioxide and water vapour gasification. The oxidic form is extremely important. No activity is found with these metals or their oxides in hydrogasification. But potassium and also calcium can catalyze the water gas shift reaction (18, 19), potassium also the methanation of carbon monoxide, but only at elevated pressure (20). Iron catalyzes hydro- and water vapour gasification, but the latter one only under reducing conditions (16, 17, 21), which normally exist in gasifiers. Iron catalyzed hydrogasification is strongly favoured by pressure. A summary of catalytic effects by the various metals is shown in Fig.1.

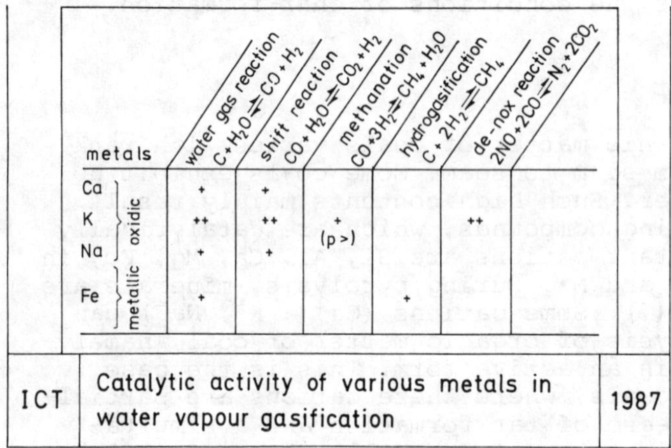

Figure 1

In a simplified manner, the catalytic activity of all metals in gasification can be described by an oxygen exchange mechanism, as shown for water vapour gasification.

$$Me + H_2O \rightleftharpoons Me(O) + H_2 \qquad (1)$$

$$Me(O) + C \rightleftharpoons Me + CO \quad (2)$$

The true mechanisms are more complex, especially with calcium and the alkali metals (10, 22).

According to the simplified mechanism it is evident, that the catalysts preferentially catalyze dissociation of carbon dioxide and water vapour, but not desorption of the intermediate carbon-oxygen surface complex. This means, that catalytically active metals increase the concentration of dissociation centers (active sites), but they do not change the rate-limiting step. Activation energies of non-catalyzed and catalyzed gasification are in the same order of magnitude, although gasification rates can differ by orders of magnitude.

In relation to the catalytic effect of mineral matter in coal gasification it is important that the necessary amounts of iron are extremely low, i.e. far below 1 %. Dispersion is the decisive factor. Small iron contents are normally found in all coals. At gasification temperatures above 880 °C, where the poisoning effect of sulfur obviously decreases, this element is supposed to be responsible for the gasification rate of low rank coals in technical gasifiers (6). On the contrary, remarkable catalytic effects with calcium, potassium or sodium require at least 1 % or higher percentages of the metal. Such alkali metal contents of coals represent an exception (23).

As shown, catalytic effects as caused by minerals depend on many parameters. Amount, distribution, chemical state and possible poisons have to be considered. These criteria are not sufficient, availability of porosity and accessibility of porosity by the gasification medium are of similar importance. These preconditions for catalytic activity are again fulfilled by low rank coals, but not by caking coals.

As far as gasification rate is concerned, the kinetics of non-catalyzed gasification should be applicable. Numerous investigations support this conclusion. However, care has to be taken in calculation or representing gasification rates related on the internal surface area.

2.3 History of char formation

The history of char formation also influences char activity. Considering particle size, it is known that larger amounts of tar are released from smaller particles, because secondary reactions in the coal particle are diminished by reduced mass and heat transport effects (24). A similar effect is achieved by fast heating, whereby it is assumed, that evaporation is kinetically favoured in comparison to pyrolysis (condensation) reactions. The opposite effect results from increased pressure, namely by limitation of tar devolatilization (24, 25, 26).

Pyrolysis reactions cause dimension changes of the coal particles, which can be observed directly with the aid of a high speed camera (27) or by measuring the swelling behaviour (26, 28). No correlation was found between coal properties and swelling behaviour at normal pressure on the one hand and swelling behaviour at increased pressure on the other hand (26, 28). Resolidification plays a major role for the final dimension changes of a particle, caused by swelling and subsequent shrinkage. Only if volatiles may escape before resolidification occurs, a pronounced shrinkage is observed with increasing temperature or time. This is the case with small particles at low pressure (27). The pore volume of macro- and mesopores of chars exhibits a minimum between 0.5 and 2.5 MPa (29). At about 2.8 MPa swelling was found to be independent of particle size and heating rate (29).

Minerals have an additional influence. Active metal cations, especially exchangeable cations like calcium, potassium and sodium reduce the yield of volatiles or weight loss during pyrolysis (9). As a consequence, dilatation of coal particles is reduced by these metals (30), but only at low pressure (28). Influence of the atmosphere (CO_2, H_2O, H_2) in the pyrolysis step is small as far as catalytic effects by minerals are neglected (31). Minerals present in lignites diminish weight loss, as already pointed out, but the superimposed influence of the various gasification atmospheres is small.

Finally, it is noteworthy that the influence of the raw coal or the history of char formation on reactivity of the resulting char is diminished with increasing pyrolysis or heat treatment temperature and time, because the size of the graphitic layers is enlarged and defects are progressively annealed (32-34 , 4). Pore size simultaneously decreases (35) and reactivity decreases, too (34, 36). Maximum porosity is available at 700 °C. BLACKWOOD et al. (37) and JOHNSON (38) concluded, that the reactivity correlates better with heat treatment temperature than with coal type already at pyrolysis temperatures above 700 °C (rate $\sim$ exp $(1/T_p)$, T_p = pyrolysis temperature $>$ 700 °C). In view of catalytic effects generalization of this conclusion is doubtful.

2.4 Char properties

As shown in the previous chapter, char properties depend in a complicated manner on the original coal and the history of char formation. As far as we are not able to predict char properties in relation to the gasification behaviour by the properties of coal and the procedures of char formation, it is necessary to characterize chars. As char gasification is a heterogeneous reaction, porosity, pore structure and pore surface area are the most important parameters. Techniques for determination of these properties by pycnometry, gas adsorption, molecule probes, mercury porosimetry and small angle X-ray diffraction are available (39). For more advanced treatments of the heterogeneous reaction, development of pore structure models is necessary, as shown for instance by SIMONS et al. (40, 41), who investigated the relevance of the pore tree model. The same authors showed that assumption of pore branching according to the pore tree model is not sufficient. For a better description of pore structure, pore combination has to be considered, as pointed out by WHEELER (42) and HASHIMOTO et al. (43) in earlier years and others (44, 45) in recent years.

2.5 Changes of char properties during gasification and their influence on gasification kinetics

Carbon and also chars are often supposed to exist of domains with different reactivity. This means that the more reactive parts are primarily gasified and the reactivity per volume unit of carbon decreases. A further lowering of reactivity can result from a thermal annealing of defects during gasification. If such effects and possible activations or deactivations of catalytically active cations of the minerals during gasification are neglected, the most important fact is the change of internal surface and porosity. According to several authors (46-50), the internal surface area at first increases and then decreases with progressive carbon conversion, provided that gasification is performed in the chemically reaction-controlled regime.

Maximum internal surface area exists at medium carbon conversion, whereas porosity continuously increases. This general behaviour is found with all gasification media (O_2, CO_2, H_2O, H_2). For example, results of TOMKOW et al. (49) are presented in Fig. 2, showing the relative surface areas for three different pore size ranges during oxygen gasification of two chars prepared at 500 and 900 °C.

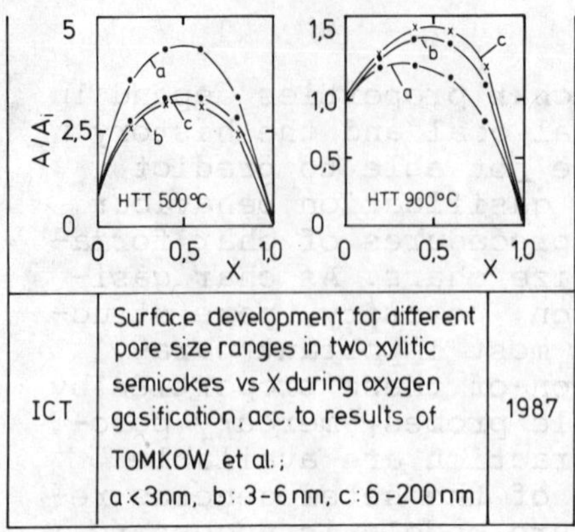

Figure 2

Surface developments as shown for example in Fig.2 are explained by pore growth, initation of new pores and coalescence or combination of adjacent pores with extent of gasification. For consideration of internal surface and corresponding by porosity changes different models were developed. They can be classified as follows:

(1) Empirical, semi-empirical or order of reaction models
(2) Particle or grain models
(3) Pore models.

2.51 Order of reaction and related models

The order of reaction models assume that the rate of mass change is proportional to the mass of carbon present:

$$\frac{dm}{dt} = - k \cdot m^n \tag{3}$$

Substitution of m by the initial mass of carbon m_o and carbon conversion X yields:

$$m = m_o \; (1 - X) \tag{4}$$

$$\frac{dX}{dt} = k \cdot m_o^{\,n-1} \; (1 - X)^n \tag{5}$$

The term m_o^{n-1} is often neglected, which is incorrect if $n \neq 1$. For the probable case $n = 1$ follows from eq.(5) that the rate of gasification is proportional to the mass of unconverted carbon or :

$$\frac{dX}{dt} \cdot \frac{1}{1-X} = k \tag{6}$$

JÜNTGEN (51,52) received a rather good fit of experimental data of coal and low temperature char gasification with steam at 1 MPa and 810 °C by eq.(6). ADSCHIRI et al. (53, 54) found that the surface area of high-porous chars (porosity > 0.6) produced in a fluidized bed decreases linearly with increasing conversion in carbon dioxide gasification at normal pressure :

$$A_X / A_{X=0} = 1 - X \tag{7}$$

These results explain the findings of JÜNTGEN (51,52). It means that a plot of dX/dt versus X gives straight lines with intercepts dX/dt = 0 at X = 1 :

$$\frac{dX}{dt} = k \cdot (1 - X) \tag{8}$$

The slope of the straight lines k for the various coals corresponds to the initial gasification rate at X = 0:

$$k = (dX/dt)_{X=0} \tag{9}$$

On the other hand, a plot of $\frac{dX}{dt} \cdot \frac{1}{1-X}$ according to eq.(6) would give horizontal lines, whereby the height is equal to k or $(dX/dt)_{X=0}$. The results envisualize that the initial rate in isothermal gasification has a fundamental meaning, because it is a fingerprint of the coal and the 'history' of char formation.

In this connection some remarks are necessary concerning the mode of char preparation. In many experimental studies chars were prepared by slow pyrolysis in inert atmosphere. Such chars exhibit a much lower porosity or internal surface area than chars prepared at fast heating in gasification atmosphere. Most research on internal surface area or porosity development in gasification as mentioned above was performed with chars produced at slow heating and additional soaking times at the final temperature. The relevance of such studies on coal char gasification is not proved.

The concept of initial reaction rate was also applied by HASHIMOTO et al. (55,56) in catalyzed steam gasification of coconut shell carbon- and carbon black pellets, prepared with a phenolic resin binder and pretreated for 15 min at 850 °C. The following results were received:

$$\frac{dX}{dt} = \left(\frac{dX}{dt}\right)_{X=0} \cdot (1 - X)^n \tag{10}$$

n = 0 : K, Na

n = 2/3 : Ni

n = 1 : pure, Ca, Fe

An exponent n = 0 means that the gasification rate is independent of carbon conversion. Similar results have been reported by several authors for catalytic gasification with potassium (57-59, 52).

An exponent $n = 2/3$ corresponds to the shrinkage particle or grain model. This result has also been received for non-catalyzed gasifications in earlier investigations, for example by JOHNSON (60). The single particle can be regarded as to be a solid sphere (shrinking particle) which is rather improbable with a porous particle being gasified in the reaction-controlled regime. It is more probable to assume a particle being composed of several smaller spheres. Both models are based on a surface-related reaction rate. The exponent n describes the decrease of the external surface with decreasing sphere volume. As k is proportional to the reciprocal initial sphere radius a test of the mechanism is easy. Exponents n of eq.(10) others than 0, 2/3 or 1 are meaningless, because the carbon gasification rate would depend on the initial amount of carbon (eq.(5)).

Further semi-empirical models based on a dimensionless time parameter

$$\tau = t / t_{X=0.5} \tag{11}$$

were reported by GARDNER et al. (61) for catalyzed hydrogasification (eq.(12)) and JOHNSON (62) for steam gasification (eq.(13)).

$$\frac{dX}{d\tau} = k \cdot \exp(-aX) \cdot (1 - X) \tag{12}$$

$$\frac{dX}{d\tau} = k \cdot \exp(-bX^2) \cdot (1 - X)^{2/3} , \tag{13}$$

and CHORNET et al. (63) using data of WALKER et al. (48) on gasification of coal chars in oxygen, carbon dioxide, water vapour and hydrogen (eq.(14)):

$$\frac{dX}{d\tau} = k \cdot X^{0.5} \cdot (1 - X) \tag{14}$$

Fig. 3 shows a plot of eq.(14). CHORNET et al. (63) received a good fit of the experimental data of WALKER et al. (48). JÜNTGEN (51, 52) showed that the shape of $dX/d\tau$ versus X is similar to the dependence of the relative specific surface area $A_X / A_{X=0}$ on carbon conversion X. This also follows from the results of TOMKOW et al. (49) shown in Fig. 2.

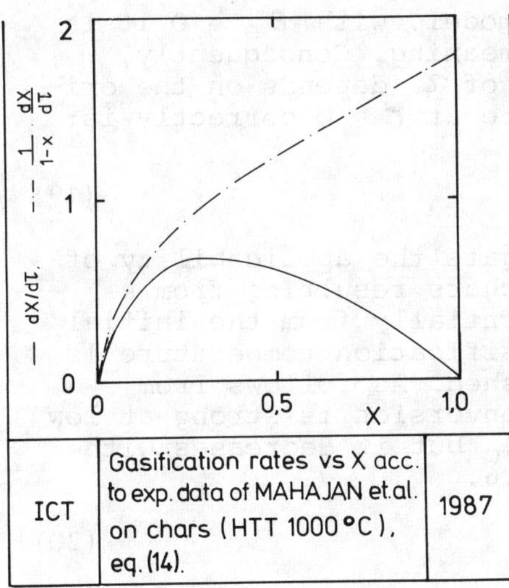

ICT	Gasification rates vs X acc. to exp. data of MAHAJAN et.al. on chars (HTT 1000 °C), eq. (14).	1987

Figure 3

However, the applicability of these correlations for coal gasification is questionable for two reasons:
(1) The studies are generally based on chars produced at slow pyrolysis, which means that the initial porosity is small.
(2) According to eq. (14) the gasification rate for $X \rightarrow 0$ is zero, which is in contradiction to experimental findings or only possible in the case of non-porous chars.
The models are therefore limited to cases, where the initial porosity and the initial surface areas are vanishing small.

SIMONS (64) pointed out that all mentioned semi-empirical models have no physical meaning. He showed that the increase of internal surface area by pore growth and the decrease with pore combination can be described for the reaction-controlled regime as follows (65) :

$$A \sim \varepsilon^{0.5} \cdot (1 - \varepsilon) \tag{15}$$

ε = porosity

Assuming that the gasification rate is directly proportional to the internal surface area gives :

$$\frac{dX}{d\tau} = \varepsilon^{0.5} \cdot (1 - \varepsilon) \tag{16}$$

The carbon conversion X is related to ε by

$$X = \frac{\varepsilon - \varepsilon_o}{1 - \varepsilon_o} . \tag{17}$$

With eq. (17) the carbon conversion rate may be expressed as follows :

$$\frac{dX}{d\tau} = k \cdot \left[X + \varepsilon_o \cdot (1 - X)\right]^{0.5} \cdot (1 - X) \tag{18}$$

Eq.(18) is similar to CHORNET's model, with $\varepsilon_o = 0$ it is identical, but it has a physical meaning. Consequently, carbon conversion X as function of τ depends on the original porosity (Fig.4) and the rate at $\tau = 0$ correctly is:

$$\frac{dX}{d\tau} = k \cdot \varepsilon_o^{0.5} \tag{19}$$

It would be worthwhile to investigate the applicability of eq.(18) for gasification of coal chars resulting from a more realistic production, preferentially from the initial step of gasification after the gasification temperature is reached and devolatilization finished. As follows from eq.(18) the influence of carbon conversion is strong at low and medium initial char porosity ε_o but it decreases with high initial char porosity $\varepsilon_o > 0.6$.

$$\varepsilon_o + (1 - \varepsilon_o) \cdot X \approx \varepsilon_o. \tag{20}$$

In this case eq.(18) is similar to eq.(8), which was successfully applied for gasification of chars with $\varepsilon_o > 0.6$ for example by ADSCHIRI et al. (50).

2.52 Single particle models

Another possibility to describe heterogeneous solid-gas reactions represent single particle models (66,67). Depending on the solid and the reaction products, two cases can be distinguished, namely

- constant size (spherical) particle models
- shrinking (spherical) particle models.

To the first group of models belong

1. the homogeneous, continuous or progressive reaction model
2. the heterogeneous, shrinking or unreactive core model
3. the crackling core model.

According to the homogeneous model the gas enters and reacts throughout the particle at all times and the solid is converted continuously and progressively throughout the particle. Precondition for application of this model is a porous particle and the formation of a solid reaction product, in coal gasification for example ash, which remains as a skeleton.

According to the heterogeneous model the reaction front moves into the solid particle leaving behind an ash layer.

In both models the ash skeleton or the ash layer can theoretically be replaced by unreactive unconverted carbon. Both models are idealized models. For example, the reaction front between the ash layer and the unreacted core may be rather diffuse than sharp, thus giving a behaviour intermediate between the shrinking core and the continuous reaction model. This problem was considered by WEN (68) and ISHIDA and WEN (69). Despite this and other complications WEN (68) and ISHIDA et al. (69,70) conclude that the shrinking core model is the best representation for the majority of reacting solid-gas reactions.

Several extensions to these simple models have been proposed, namely the so-called porous pellet model (71) and the grainy porous pellet model (72). In both cases porous particles are assumed. For initially non-porous particles PARK and LEVENSPIEL (73) introduced the so-called 'crackling' core model. According to this model a non-porous material forms a grainy intermediate, which then reacts away to the final product according to the shrinking core model. 'Crackling' which is the formation of the grainy intermediate, may occur by a pure physical change or by a thermal decomposition reaction.

The shrinking spherical particle model was already treated in chapter 2.51. It is the simplest of all models, nevertheless, it was successfully applied to coal char gasification kinetics (62).

Most of the models are represented in textbooks, monographs (66,67) and review articles (68,74). On the other hand, pore models and especially random pore models seem to be more suitable for describing coal gasification kinetics. Therefore, no further discussion of single particle models is given. The following equations give some time/conversion relationships for the constant size and shrinking sphere model for the case of chemical reaction control (66).

Constant size sphere:

$$\frac{t}{\tau} = 1 - (1 - X)^{1/3} \qquad (21)$$

Shrinking sphere:

$$\frac{t}{\tau} = 1 - (1 - X)^{1/3} \qquad (22)$$

t = time, τ = time for complete conversion ($X = 1$).

The influence of particle diameter on τ is as follows:

$$\tau \neq f\,(d_p) \qquad \text{(homogeneous)} \qquad (23)$$

$$\tau \sim d_p = d_{P,O} \qquad \text{(heterogeneous)} \qquad (24)$$

$$\tau \sim d_{P,O} \qquad \text{(shrinking sphere)} \qquad (25)$$

Differentiation of eq. (21) or (22) gives

$$\frac{dX}{dt} \sim (1 - X)^{2/3} \qquad (26)$$

This equation was already discussed in chapter 2.51.

2.53 Pore models

Pure order of reaction models and also particle or grain models predict for the chemical reaction-controlled regime a continuously decreasing reaction rate dX/dt with increasing conversion X. This dependence follows from eqs.(8) and (10) and also from the semi-empirical eqs.(12) and (13). A zero order reaction (n = O) according to eq.(10) is an exception. For the particle models eq.(26) leads to the same conclusion. The reason is evident, because the reacting surface of each particle or grain is receding with progressive gasification.

In coal or coal char gasification in the reaction-controlled regime such a behaviour may be expected with initially non-porous materials (shrinking particle), but also with porous materials, if it is assumed that the particle is composed of several smaller spheres (grainy pellet), as already pointed out in chapter 2.51. With initially less porous materials the gasification rate can go through a maximum during the early or intermediate stages of gasification. This is attributed to an increase of the reacting surface area (46-52, 75, 76) as shown in exemplary manner in Fig.2.

Changes of internal surface and pore structure were considered in the empirical model of CHORNET et al. (63) and the theoretical model of SIMONS (64). A further exception taking into account an intermediate increase of the reacting surface represents the crackling core model of PARK and LEVENSPIEL (73).

Pore models need assumptions about the pore structure and pore structure development. In an early model PETERSEN (77) assumed that (1) all pores are cylindrical and of uniform radius and (2) no new intersections occur as the reaction surfaces within the solid particle grow. The model is based on two parameters, the resulting time-conversion relationship is complex. Extensions of this model were performed by SZEKELY et al. (78, 67), CALVELO and CUNNINGHAM (79) and RAMACHANDRAN and SMITH (80).

A more realistic pore model was developed by HASHIMOTO and SILVESTON (43). These authors adopted a population balance technique and accounted for the distribution in pore sizes as well as the collapse of pore walls into each other as reaction proceeds. These assumptions lead to a two parameter model as PETERSEN's model (77).

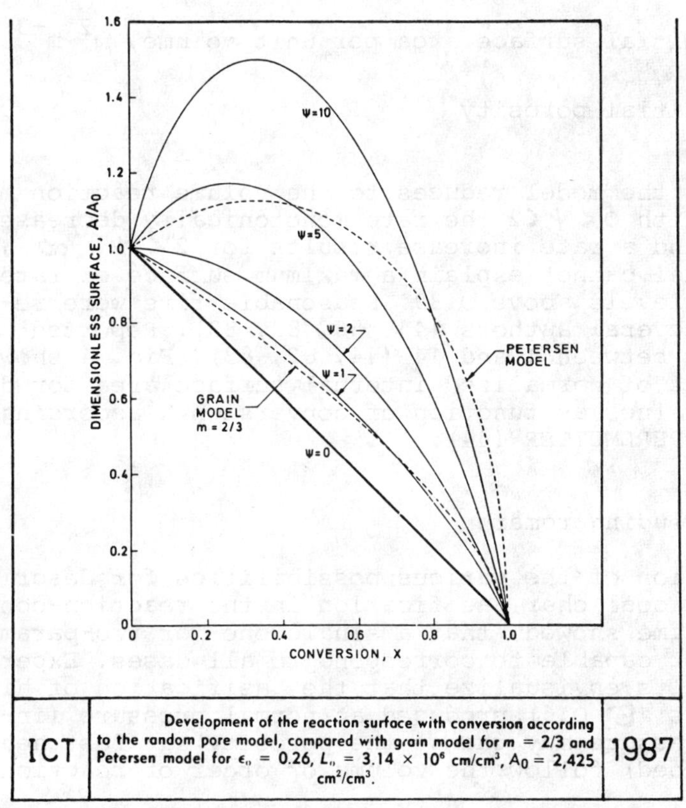

Development of the reaction surface with conversion according to the random pore model, compared with grain model for $m = 2/3$ and Petersen model for $\epsilon_0 = 0.26$, $L_0 = 3.14 \times 10^6$ cm/cm^3, $A_0 = 2{,}425$ cm^2/cm^3.

Figure 4

A completely different pore model was developed by BHATIA and PERLMUTTER (44). This model accounts for the random overlap of reacting surfaces as they grow. It is based on a theory originally developed by AVRAMI (81) for the analysis of the geometry of crystal aggregates. The advantages of this model are that (1) no assumption is needed as to the actual shape of the pores and (2) only one parameter is required to predict the rate-conversion relationship.

$$r \sim \left[1 - \psi \ln (1 - X)\right]^{1/2} \cdot (1 - X) \qquad (27)$$

$$\psi = \frac{4\pi L_O \cdot (1 - \varepsilon_O)}{A_O^2} \quad \text{(initial pore structure parameter)} \qquad (28)$$

L_O = initial pore length per unit volume, $m \cdot m^{-3}$

A_O = initial surface area per unit volume, $m^2 \cdot m^{-3}$

ε_O = initial porosity

With $\psi = 0$ the model reduces to the volume reaction model (eq. (8)). With $0 < \psi < 2$ the rate monotonically decreases. A surface and a rate increase results for $2 < \psi < \infty$, but the model cannot explain a maximum surface or rate at conversion levels above 0.39. Reasonable fits were received by several authors (44, 56, 82, 83). Reported values vary between 1 and 14 (44, 82, 83). Fig. 4 shows calculations of normalized internal surface area for different ψ values as function of conversion X according to BHATIA and PERLMUTTER (44).

2.5.4 Concluding remark

The discussion of the various possibilities for describing kinetics of coal char gasification in the reaction-controlled regime showed, that a single one- or two-parameter model is not capable to correspond to all cases. Experimental results envisualize that the gasification of highporous chars ($\varepsilon > 0.6$) produced at normal pressure (increasing pressure may change the porosity) at fast heating (fluidized bed) follows the volume or order of reaction model:

$$\frac{dX}{dt} = k \cdot (1 - X) \tag{8}$$

with $$k = (dX/dt)_{X=0} \tag{9}$$

Catalytic active species may alter the kinetics (K, Na : n = O; Ni : n = 2/3). An exponent n = 2/3 corresponding to the shrinkage particle, shrinking core or grainy pellet model gives a significant change of the rate-conversion curve (Figs. 5,6).

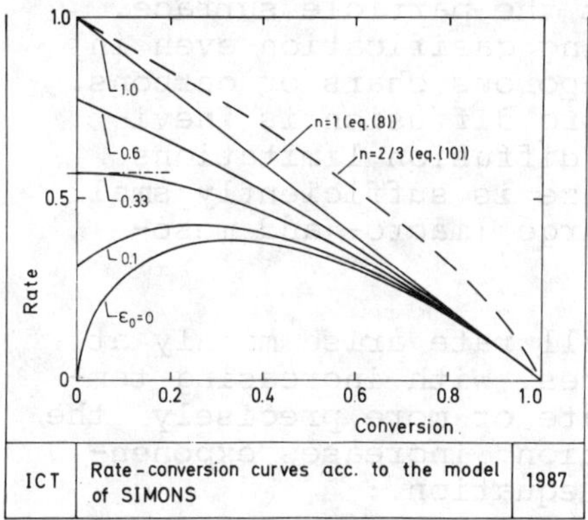

Figure 5

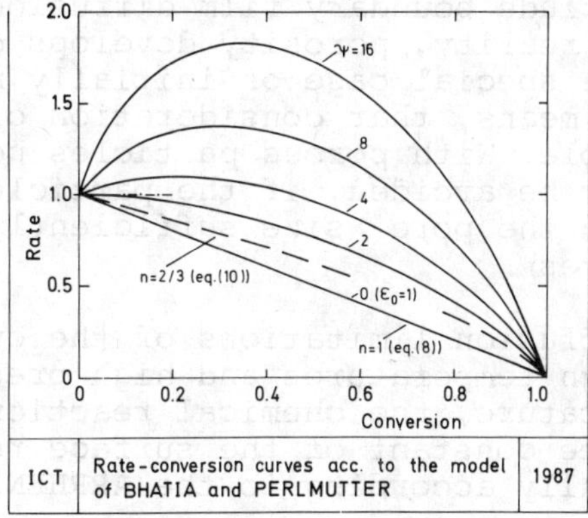

Figure 6

The results with n = 1 according to eq.(8) with high-porous chars are not surprising, because the pore models of SIMONS (64) and BHATIA and PERLMUTTER (44) yield nearly linear rate-conversion relationships for $\varepsilon >$ 0.6 as follows from Figs.5 and 6. A comparison between these models shows that the latter one is more flexible. It allows a higher increase of the rate for high ψ values (low porosities) and the limitation of the maximum rate is at X = 0.39 as compared to 0.33 in the SIMONS model. The main disadvantage of the SIMONS model is, that it is only based on porosity, which can be related to internal surface area in different ways. It means, that influences of pore size and pore size distribution are neglected.

Finally, it should be reminded, that the rates in the particle and grain as well as in the pore models are related to surface area. A fit of experimental data by such a model must not necessarily mean, that the surface area changes according to the model. It may also be, that the concentration of active sites occasionally changes as given by the model parameters.

3. THE ROLE OF MASS TRANSPORT

In chapter 2 possible influences of mass transport on the overall kinetics of char gasification were neglected. This assumption is valid in gasification of really non-porous char particles, if the flow rate of the gas at a given temperature and pressure is sufficiently high in order to exclude boundary film diffusion at the particle surface. In reality, porosity develops during gasification even in the special case of initially non-porous chars or carbons. It means, that consideration of pore diffusion is inevitable. With porous particles pore diffusion limitations may be avoided, if the particle size is sufficiently small and the pore size sufficiently large (macro- and mesopores).

Diffusion limitations of the overall rate arise mainly at high temperatures and high pressures. With increasing temperature, the chemical reaction rate or more precisely the rate constant of the surface reaction increases exponentially according to the ARRHENIUS equation :

$$k_S = k_O \cdot \exp(-E/RT_S) \qquad (29)$$

On the contrary, the diffusion coefficient for bulk or continuum diffusion increases with temperature only to the power of 1.5 (eqs.(30) - (32)), for KNUDSEN diffusion in narrow pores only to the power of 0.5 (eq.(33)) :

$$D_B = \frac{1}{3} \cdot \overline{l} \cdot \overline{w} \qquad (30)$$

$$\overline{l} = k \cdot T/4.44 \cdot d_M^{\,2} \cdot p \qquad (31)$$

$$\overline{w} = (8kT/\pi M)^{0.5} \qquad (32)$$

with $\overline{l}$ = mean free path

$\overline{w}$ = mean molecular speed

k = BOLTZMANN constant

T = temperature

d_M = molecular diameter

p = pressure

M = molecular mass

$$D_K = \frac{1}{3} d_p \bar{w} \tag{33}$$

with d_p = pore diameter.

An exception represents surface diffusion in submicropores, which is a thermally activated process :

$$D_S = const \cdot \exp(-Q/RT) \tag{34}$$

with R = gas constant.

The rate of surface diffusion is small as compared to bulk or KNUDSEN diffusion. In the case of film diffusion, not the diffusion coefficient D but the mass transport coefficient ß is the limiting value, which has a temperature dependence only to the power of $\leqslant 1$. Binary diffusion coefficients may be calculated using the CHAPMAN-ENSKOG or similar formulas (84), mass transfer coefficients ß using the SHERWOOD number.

With increasing pressure, the reaction rate increases, in water vapour gasification up to approx. 1.5 MPa, in carbon dioxide gasification up to approx. 2 MPa and in hydrogasification continuously, approx. to the power of 1. As follows from eqs.(30) to (32) the bulk diffusion coefficient decreases linearly with pressure, which means, that transport limitation problems strongly increase at elevated pressure. The KNUDSEN diffusion coefficient is independent of pressure, but the mean free path $\bar{l}$ (eq.(31)) decreases. Therefore, KNUDSEN diffusion may be enhanced, but it cannot compensate for diminished bulk diffusion.

A schematic presentation of the mass transport effects in char or carbon gasification is shown in Fig.7 according to WALKER et al. (85). Three main ranges I, II and III and two transition reasons a and b are distinguished. At low temperatures chemical reaction controls (range I). With increasing temperature pore diffusion becomes effective at first (range II) provided, that the gas flow rate is sufficiently high. At very high temperatures, film diffusion finally limits the gasification kinetics. Particle size, pore structure, intrinsic char reactivity, the gasification medium and pressure can shift the transition temperatures from one range to the other. Fig.7 also shows the pore effectiveness factor η according to DAMKOEHLER or THIELE, which is one in range I and nearly zero in range III. Note that the apparent activation energy E_a in range II is one half of the true activation energy of the surface reaction E_t.

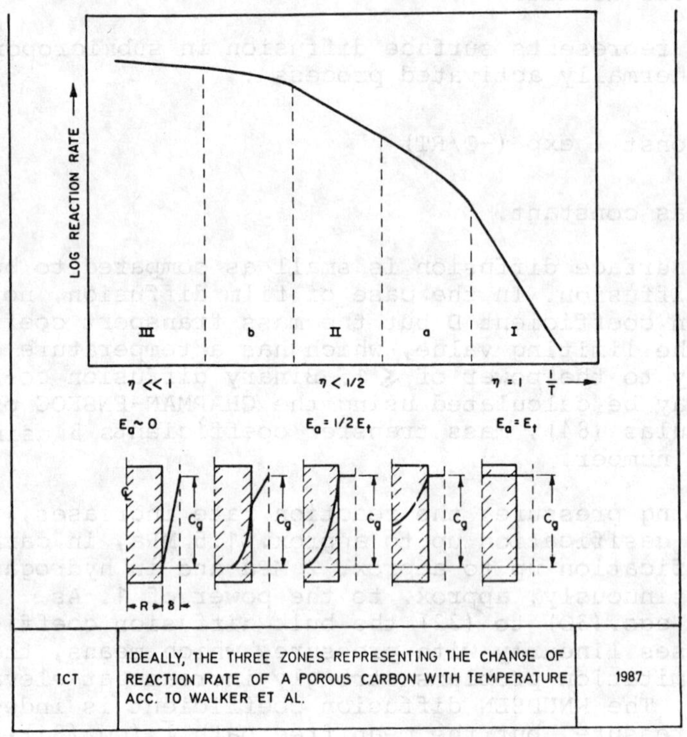

Figure 7

The following quantitative treatment of the problem is based on the assumption that the surface reaction of the gasification medium follows an irreversible first order reaction as shown in the previous sections. The external mass transfer rate of the gasification medium from the fluid to the partial surface per unit of time and external surface A_{ex} of the particle is:

$$j = \dot{n}_T/A_{ex} = -\beta\cdot(c_F - c_S) \tag{35}$$

ß = mass transfer coefficient

c = concentration of the gasification medium

The amount of gasification medium, which is converted at the external and internal surface (A_{ex} and A_{in}) per unit of time is given by :

$$\dot{n}_R = -(k_S\cdot A_{ex}\cdot c_S + k_S\cdot A_{in}\cdot \bar{c}) \tag{36}$$

The gasification rate per unit of external surface $r_{A_{ex}}$ is :

$$r_{A_{ex}} = \dot{n}_R/A_{ex} = -k_S\cdot(c_S + \frac{A_{in}}{A_{ex}}\cdot \bar{c}) \tag{37}$$

$\bar{c}$ represents the average concentration of the gasification medium in the pores. It may be expressed by the aid of the pore effectiveness factor η :

$$\eta = \frac{k_S\cdot\bar{c}}{k_S\cdot c_S} = \frac{\bar{c}}{c_S} \tag{38}$$

or

$$\bar{c} = \eta\cdot c_S \tag{39}$$

With eq.(39), eq.(37) yields:

$$-r_{A_{ex}} = -k_S\cdot c_S\cdot(1+\eta\cdot\frac{A_{in}}{A_{ex}}) \tag{40}$$

The unknown concentration c_S at the external surface results from the steady-state condition eq.(41):

$$j = -r_{A_{ex}} \tag{41}$$

$$-\beta(c_F - c_S) = -k_S\cdot c_S\cdot(1+\eta\,\frac{A_{in}}{A_{ex}}) \tag{42}$$

$$c_S = \frac{\beta}{\beta + k_S\cdot(1+\eta\cdot\frac{A_{in}}{A_{ex}})} \tag{43}$$

With eq. (43) the effective gasification rate per unit of external surface is :

$$(r_{eff})_{A_{ex}} = \frac{1}{\frac{1}{\beta} + \frac{1}{k_S(1+\eta\frac{A_{in}}{A_{ex}})}} \cdot c_F \tag{44}$$

$$(r_{eff})_{A_{ex}} = (k_{eff})_{A_{ex}} \cdot c_F \tag{45}$$

$(k_{eff})_{A_{ex}}$ may also be related to the mass unit of the solid (carbon, char) :

$$(k_{eff}) = (k_{eff})_{A_{ex}} \cdot (A_{ex})_m \tag{46}$$

For a spherical particle $(A_{ex})_m$ is:

$$(A_{ex})_m = \frac{A_{ex}}{m} = \frac{A_{ex}}{V\cdot\rho} = \frac{3}{R\cdot\rho} \tag{47}$$

ρ = apparent density of the solid.

The effective gasification rate per unit of mass is as follows :

$$(r_{eff})_m = (k_{eff})_m \cdot c_F = \frac{3}{R\cdot\rho} \cdot \frac{1}{\frac{1}{\beta} + \frac{1}{k_S\cdot\left[1+\eta\cdot\frac{R}{3}\cdot(A_{in})_V\right]}} c_F \tag{48}$$

$(A_{in})_V$ = internal surface per volume unit.

For mathematical calculations of r_{eff} the pore effectiveness parameter η is required. It can be calculated with the aid of the dimensionless THIELE modulus $\emptyset$. Solutions $\eta = f(\emptyset)$ for particles of various shape (plate, cylindrical rod, sphere) are available in textbooks (86-88). The relationship for a spherical particle is :

$$\eta = \frac{3}{\emptyset} \cdot \left(\frac{1}{\tanh\emptyset} - \frac{1}{\emptyset}\right) \tag{49}$$

$$\emptyset = (k_S \cdot (A_{in})_V \cdot L^2/D_e)^{0.5} = \tag{50}$$

THIELE modulus for an isothermal first order reaction.

For a spherical particle :

$$\emptyset = L(2k_S/D_e \cdot r)^{0.5} . \quad (51)$$

D_e = effective pore diffusion coefficient

L = pore length

r = pore radius

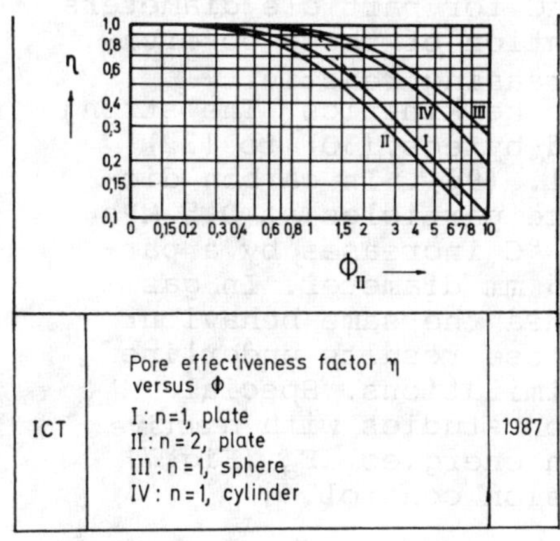

Figure 8

Fig.8 shows a plot of n versus Ø for particles of different shape and different reaction orders. Solutions for non-isothermal particles or reactions with changing volume are available in literature (86-88). For carbon or char particles L may be substituted by the particle radius R

$$L = \lambda \cdot R \quad (52)$$

λ = tortuosity factor.

Three ranges of Ø may be distinguished :

(1) $\emptyset \ll 1 : \eta = 1$ (range I in Fig.7) (53)

(2) $\emptyset > 3 : \eta = 3 \cdot (\emptyset - 1)/\emptyset^2$ (54)

(3) $\emptyset \gg 1 : \eta = 3/\emptyset$ (range II in Fig.7; $E_a = 0{\cdot}5\ E_t$) (55)

For a more detailed treatment of pore diffusion in heterogeneous solid-gas reactions monographs are recommended (89, 90). A special problem in calculating pore diffusion control represents determination of Ø , because k_S is not known. For this purpose several methods have been developed, for example the WEISZ-PRATER criterion (91) or the ROBERTS-SATTERFIELD approach (92). A summary of the methods is given in (89, 90, 93).

Most of the models discussed in chapter 2 are extended to diffusion limitation. This holds for external or boundary film diffusion and ash or product layer diffusion in particle models (66, 67), and also for pore diffusion in pore models (94, 95).

Experimental results on the influence of pore diffusion were presented by WALKER et al. (85) and TURKDOGAN et al. (96) for carbon dioxide gasification and by ERGUN and MENTSER (97) for water vapour gasification. According to all authors pore diffusion control is negligible at normal pressure below approximately 1000 °C for particle diameters up to approx. 2 or 3 mm. Generalization of these observations is not possible, because increasing reactivity and decreasing pore size may alter the kinetics. The strong influence of pressure, as predicted by eqs.(30) to (32) was demonstrated by TURKDOGAN et al. (96). In carbon dioxide gasification of electrographite particles at 0.5 MPa it was found that the rate at 1000 °C increases by a particle size reduction even below 0.5 mm diameter. In gasification of a coconut char at 0.5 MPa the same behaviour was observed already at 800 °C. These results underline the importance of pore diffusion limitations. Special care has to be taken in gasification studies with reactive coal chars. Low apparent activation energies E_a (Fig.7) may easily be caused by pore diffusion control.

For technical purposes it is necessary not only to measure char gasification but also to model gasifiers. This was done for example by YOON et al. (98) and BLIEK (83) for moving bed gasifiers and by CARAM and AMUNDSON (99) for fluidized bed gasifiers.

4. CONCLUDING REMARK

The present paper represents an attempt to briefly summarize the main problems of coal char gasification, which are related to the raw material coal, char formation, char properties, catalytic effects of minerals, change of particle size and/or pore structure during gasification and transport phenomena. The literature contains numerous more experimental results and theoretical models, but it would be hard, if not impossible to cover all the literature in this field in a single review article. Furthermore, more research is necessary to elucidate the merits of the various particle and pore models. It is hoped that contributions of future research will help to make the complex coal char gasification kinetics more understandable.

REFERENCES

1. Hüttinger, K.J.: 'Kinetics of Coal Gasification', this conference
2. Gan, H. et al.: FUEL 51 (1972) 272
3. Hashimoto, K. et al.: FUEL 65 (1986) 1516
4. Blackwood, J.D. and Mc Taggart, F.K.: Aust. J. Chem.12 (1959) 533
5. Dutta, S. and Wen, C.Y.: Ind. Eng. Chem. Proc. Des. Dev. 16 (1) (1977) 20; ibida 16 (1) (1977) 31
6. Hüttinger, K.J. and Krauss, W.: FUEL 60 (1981) 93; ibida 61 (1982) 291
7. Gluskoter, H.J.: Adv. in Chem. Ser., ACS 141 (1975) 1
8. Tyler, R.J. and Schafer, N.A.: FUEL 59 (1980) 487
9. Morgan, M.E. and Jenkins, R.G.: FUEL 65 (1986) 757; ibida 65 (1986) 765
10. see References 5, 6, 7 in Reference (1)
11. Proceedings International Symposium on 'Fundamentals of Catalytic Coal and Carbon Gasification', Amsterdam (1983) and Rolduc, Netherlands (1986)
12. Hüttinger, K.J. and Michenfelder, A. : Erdöl und Kohle, Erdgas, Petrochemie 40 (1987) 166; see also FUEL 66 (1987) in press
13. Lang, R.J.: in (11), (1986) 9
14. Hüttinger, K.J. and Minges, R.: FUEL 65 (1986) 1112; ibida 65 (1986) 1122
15. Adler, J. et al.: FUEL 64 (1985) 1215
16. Hüttinger, K.J. : in (11), (1983) 113
17. Nishiyama, Y.: in (11), (1986) 210
18. Hüttinger, K.J. et al.: FUEL 65 (1986) 932
19. Kapteijn, F. et al.: Proc. 4. Intern. Conf. on Carbon, DKG, Baden-Baden (1986) 605
20. Hüttinger, K.J. et al.: Chem.-Ing.-Tech. 58 (1986) 409
21. Hermann, G. and Hüttinger, K.J.: Carbon 24 (1986) 705
22. Moulijn, F.A. and Kapteijn, F.: in 'Carbon and Coal Gasification' (J.L. Figueiredo, J.A. Moulijn, eds.), M. Nijhoff Publ., Boston (1986) 181
23. Kyotani, T. et al.: in (11), (1986) 337
24. Kaiser, M. et al.: Proc. Intern. Conf. on Coal Sc., Sidney (1985) 899
25. Anthony, D.B. et al.: FUEL 55 (1976) 121
26. Khan, R.M. and Jenkins, R.G.: FUEL 65 (1986) 725
27. Löwenthal, G. et al.: FUEL 65 (1986) 346
28. Trompp, P.J.J.: in (11), (1986) 289
29. van Heek, K.H.: Habilitationsschrift, Universität Münster
30. Bexley, K. et al.: FUEL 65 (1986) 47
31. Jenkins, R.G. and Morgan, M.E.: FUEL 65 (1986) 769
32. Jenkins, R.G. et al.: FUEL 52 (1973) 288

33. Hippo, E. and Walker, P.L., Jr.: FUEL 54 (1975) 245
34. Blake, J.H. et al.: FUEL 46 (1967) 115
35. Nandi, S.P. et al.: Carbon 2 (1964) 199
36. Mühlen, H.J. et al.: FUEL 65 (1986) 591
37. Blackwood, J.D. et al.: Aust. J. Chem. 20 (1967) 2525
38. Johnson, J.L.: ACS, Div. Fuel Chemistry, Preprints 520 (4) (1975) 85
39. Gregg, S.J. and Sing, K.S.W.: 'Adsorption, Surface Area and Porosity', Academic Press, London (1967)
40. Simons, G.A. and Finson, M.L.: Comb. Sci. Tech. 19 (1979) 217; ibida 19 (1979) 227
41. Simons, G.A.: Comb. Sci. Tech.20 (1979); ibida 20 (1979) 117
42. Wheeler, A., Adv. Catalysis 3 (1951) 249
43. Hashimoto, K. and Silveston, P.L. : AIChE J. 19 (1973) 259; ibida 19 (1973) 268
44. Bhatia, S.K. and Perlmutter, D.D. : AIChE J. 26 (1980) 379
45. Gavalas, G.R.: AIChE J. 26 (1980) 577
46. Wicke, E. and Hedden: K., Z. Elektrochemie 57 (1953) 636
47. Kawahata, M. and Walker, P.L., Jr.: Proc. 5th Carbon Conf., Pergamon Press, London (1962) 251
48. Mahajan, O.P. et al.: FUEL 57 (1978) 643
49. Tomkow, K. et al.: FUEL 56 (1977) 101
50. Adschiri, T. and Takehiko, F.: FUEL 65 (1986) 927
51. Jüntgen, H.: Carbon 19 (1981) 167
52. Jüntgen, H. and v.Heek, K.H.: 'Kohlevergasung',Verlag Karl Thiemig, München (1981) 80
53. Adschiri, T. et al.: The Inst. of Chem. Engineers Symp. Ser. 87
54. Adschiri, T. et al.: FUEL 65 (1986) 1688
55. Hashimoto, K. et al.: FUEL 65 (1986) 407
56. Hashimoto, K. et al.: FUEL 65 (1986) 1516
57. Veraa, M.J. and Bell, A.L.: FUEL 57 (1978) 194
58. Walker, P.L., Jr. et al.: FUEL 58 (1979) 338
59. Tomita, A. et al.: FUEL 58 (1979) 614
60. Johnson, J.L.: Adv. in Chem. Series 131 (1974) 145
61. Gardner, N.C. et al.: Adv. in Chem. Ser. 131 (1970) 217
62. Johnson, J.L.: Adv. in Chem. Ser. 137 (1974) 145
63. Chornet, E. et al.: FUEL 58 (1979) 395
64. Simons, G.A.: FUEL 59 (1980) 143
65. Simons, G.A.: Comb. Sci. Tech. 19 (1979) 227
66. Levenspiel, O., 'Chemical Reaction Engineering', J. Wiley & Sons, N.Y. (1972)
67. Szekely, J. et al.: 'Gas-Solid Reactions', Academic Press, N.Y. (1972)
68. Wen, C.Y.: Ind. Eng. Chem. 60 (1968) 34

69. Ishida, M. et al.: Chem. Eng. Sci. 26 (1971) 1031
70. Ishida, M. et al.: Chem. Eng. Sci. 26 (1971) 1043
71. Ishida, M. and Wen, C.Y. : AIChE J. 14 (1968) 311
72. Sohn, H.Y. and Szekely, J.: Chem. Eng. Sci. 27 (1972) 763
73. Park, J.Y. and Levenspiel, O.: Chem. Eng. Sci. 30 (1975) 1207
74. Rodriguez, A.E.: in 'Carbon and Coal Gasification' (J.L. Figueiredo, J.A. Moulijn, eds.), M. Nijhoff Publ., Boston (1986) 361
75. Dutta, S. et al.: Ind. Eng. Chem. Proc. Des. Dev. 16 (1977) 20
76. Dutta, S. and Wen, C.Y.: Ind. Eng. Chem. Proc. Des. Dev. 16 (1977) 31
77. Petersen, E.E.: AIChE J. 3 (1957) 413
78. Szekely, J. and Evans, J.W.: Chem. Eng. Sci. 25 (1970) 1091
79. Calvedo, A. and Cunningham, R.E.: J. Catalysis 17 (1977) 1
80. Ramachandram, P.A. and Smith, J.M.: AIChE J. 23 (1977) 353
81. Avrami, M.: J. Chem. Phys. 8 (1940) 212
82. Hashimoto, K. et al.: J. Energy Dev. in Japan 7 (1985) 297
83. Bliek, A.: Ph.D. Thesis, Twente University, The Netherlands (1984)
84. Bird, R.B. et al.: 'Transport Phenomena', J. Wiley & Sons, New York (1960)
85. Walker, P.L., Jr. et al.: Adv. Catalysis 11 (1959) 133
86. Smith, J.M.: 'Chemical Engineering Kinetics', Mc Graw Hill, N.Y. (1970)
87. Fitzer, E. and Fritz, W.: 'Technische Chemie', Springer-Verlag, Berlin (1975)
88. Froment, G.F. and Bischoff, K.B.: 'Chemical Reactor Analysis and Design', J. Wiley & Sons, N.Y. (1979)
89. Aris, R.: 'The Mathematical Theory of Diffusion and Reaction in Permeable Catalysts', Vol. I, Clarendon Press, Oxford (1975)
90. Satterfield, C.N.: 'Mass Transfer in Heterogeneous Catalysis', MIT Press, Cambridge, Mass. (1970)
91. Weisz, P.B. and Prater, C.D.: Adv. Catalysis 6 (1954) 143
92. Roberts, G.W. and Satterfield, C.N.: Ind. Eng. Chem. Fundamentals 4 (1965) 288
93. Emig, G. and Hofmann, H.: Chem. Ing. Tech. 47 (1975) 637; 717; 889
94. Hashimoto, K. and Silveston, P.L. : AIChE J. 19 (1973) 268

95. Simons, G.A.: Comb. Sci. Tech. 20 (1979) 107
96. Turkdogan, E.T. et al.: Carbon 6 (1968) 462
97. Ergun, S. and Mentser, M.: in 'Chemistry and Physics of Carbon', M. Dekker, N.Y., Vol. 1 (1966) 203
98. Yoon, H. et al.: AIChE J. 24 (1978) 885
99. Caram, H.S. and Amundson, N.R.: Ind. Eng. Chem. Proc. Des. Dev. 18 (1979) 80

FLUIDIZED BED COMBUSTION OF COAL

N. Selçuk
Department of Chemical Engineering
Middle East Technical University
Ankara 06531, Turkey

ABSTRACT. Features of fluidized bed combustion of coal, design considerations associated with atmospheric fluidized bed boilers and types of fluidized bed combustors are briefly summarized.

1. INTRODUCTION

Fluidized combustion is the most promising new method of direct coal utilization for steam raising, for process heat generation or for power generation in gas turbine cycles.

Solid fuels are burned in steam generators mainly on travelling grates or by pulverized fuel firing, depending upon the capacity involved. The travelling grate stoker is for the combustion of coarse coal particles which have a relatively small surface area for the combustion reaction and burn only slowly. Pulverized fuel firing on the other hand, uses coal which has been ground to a very fine size by an extensive pulverizing process and which enables good mixing of combustion air and fuel, thus allowing quick burn-up to be achieved.

The furnace or radiant chamber size of such a steam generator is also determined by the condition which has to be met for safe operation and stipulates that prior to entering the convection section, the flue gases are to be cooled down to a temperature below the ash-softening point. Furthermore, the ash-sintering temperature has to be considered for the burner arrangement and the furnace dimensions.

Fluidized bed combustion is regarded as a solution, as it comprises a fluidized bed of inert substances in which a variety of fuels are burned.

Fluidized bed technology has been known for more than 50 years. In the process engineering sector, it is used for chemical processes, for drying and cooling as well as for coal gasification.

Due to the increasing emphasis on energy conservation and air pollution in the past two decades, the use of fluidized bed as a combustor for burning coal in industrial and utility boilers has gained interest and many experimental rigs have been built. The technology may now be regarded as commercial, as confirmed by the recent rapid increase

Y. Yürüm (ed.), New Trends in Coal Science, 481–494.

in the amount of installed coal-burning FBC capacity, see Fig.1.

Fluidized combustion involves burning of coal particles in a hot fluidized bed of inert solids (Fig.2). Coal particles are normally in millimeter size range and account for less than 2% of the total bed solids, the rest being inerts (e.g. ash) and material for sulfur retention such as limestone or dolomite. The fluidizing air is passed through a perforated distributor plate and the coal is fed through either one or a series of feed points and is rapidly dispersed through the bed as it burns. Heat exchange tubes (usually steam or water coils) are immersed in the bed to control the bed temperature which is usually in the range 750-950 ^{o}C. The lower and upper limits are determined by the combustion stability and ash-melting point respectively. Some of the finer solids are elutriated with the combustion products and the solid is removed from the bed through overflow pipes. In order to increase the carbon combustion efficiency the elutriated particles are collected in a cyclone and reinjected into the bed.

Fluidized combustors offer many advantages over the conventional coal firing systems:

i. Higher average volumetric heat release rates are obtained in fluidized combustors.
ii. Higher heat transfer coefficients between the bed material and immersed surfaces are achieved, so that although the temperature difference is lower, overall heat flux is larger. Therefore less immersed surface is required which leads to compact size.
iii. They are characterized by better mixing between the particles and the gas which in turn makes the operation at lower excess air levels possible.
iv. Fluidized combustors are flexible in coping with fuels of variable quality. They are capable of burning low grade coal regardless of its volatile content and ash swelling characteristics. The coal need not be cleaned and pulverized but merely crushed thus reducing coal preparation costs considerably.
v. They are also characterized by uniform combustion temperatures which are lower than those of conventional combustors. Due to low combustion temperatures, the ash particles are kept below their melting temperature and slagging and fouling on heating surfaces are reduced.
vi. Emission of combustion-generated pollutants is lower in fluidized combustors. Low operation temperature reduces the emission of nitrogen oxides and vapourization of toxic elements such as alkali metal sulfates or chlorides. By feeding limestone or dolomite additives to the bed, in situ desulfurization of combustion products is possible.

Inevitably there are also disadvantages.

i. Operating rates are limited to within the range over which the bed can be fluidized.
ii. The cost of the pumping power to fluidize the bed may be excessive and particularly so with deep beds.
iii. There is a limit to the size and type of particle that can be handled by this technique.

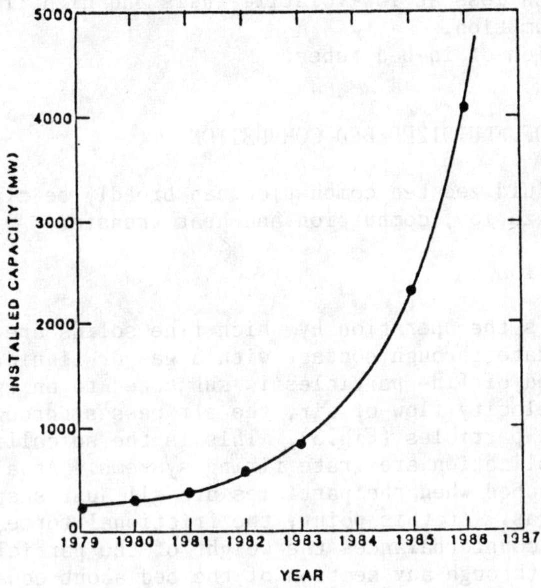

Figure 1. Installed Capacity for AFBC |1|

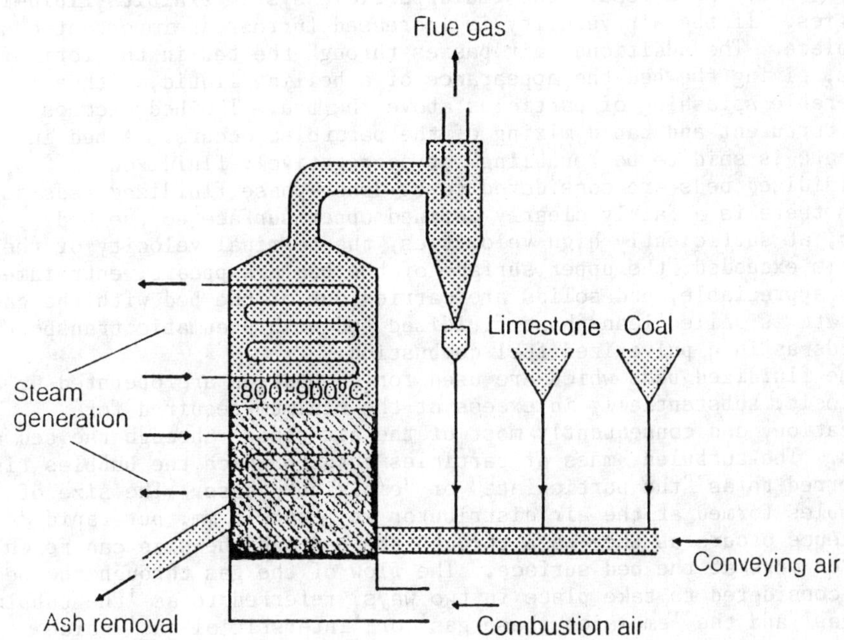

Figure 2. Schematic Diagram of a Fluidized Bed Combustor.

iv. Carbon loss at low-volatile fuels and high limestone consumption.
v. Erosion of in-bed tubes.

2. FEATURES OF FLUIDIZED BED COMBUSTION

Features of fluidized bed combustion can broadly be classified into 3 groups; fluidization, combustion and heat transfer.

2.1. Fluidization

Fluidization is the operation by which fine solids are transformed into a fluidlike state through contact with a gas or liquid.

When a bed of fine particles is subjected to an evenly distributed upward, low-velocity flow of air, the air passes through the bed without disturbing the particles (Fig.3). This is the so called fixed bed. A well known application are grate firing systems. At a higher velocity, a point is reached when the particles are all just suspended in the upward flowing gas. At this point, the frictional force between a particle and the fluid counterbalances the weight of the particle, and the pressure drop through any section of the bed about equals the weight of fluid and particles in that section. The bed is considered to be just fluidized and is referred to as an incipiently fluidized bed or a bed at minimum fluidization. This leads to a continuous change in location of the particles. As a result the fluid/particle system exhibits fluid-like properties. If the air velocity is increased further an important change takes place. The additional air passes through the bed in the form of bubbles, giving the bed the appearance of a boiling liquid, with a considerable splashing of particles above the bed. The bed becomes highly turbulent and rapid mixing of the particles occurs. A bed in this state is said to be "bubbling" or aggregatively fluidized.

Fluidized beds are considered to be dense-phase fluidized beds as long as there is a fairly clearly defined upper surface to the bed. However, at sufficiently high velocities, the terminal velocity of the solids is exceeded, the upper surface of the bed disappears, entrainment becomes appreciable, and solids are carried out of the bed with the gas. This state is called lean-phase fluidized bed with pneumatic transport of solids as in a pulverized fuel combustion.

The fluidized beds which are used for combustion are operated at an air velocity substantially in excess of the minimum required for fluidization, and consequently most of the air passes through the bed as bubbles. The turbulent mass of particles through which the bubbles rise is referred to as "the particulate" or "emulsion" phase. The size of the bubbles formed at the air distributor is about 15 mm, but rapid coalescence occurs as they pass through the bed and the size can reach 300 mm or more at the bed surface. The flow of the gas through the bed can be considered to take place in two ways, referred to as 'the bubble phase gas' and the 'emulsion phase gas' or 'interstitial gas' which passes through the particles relatively slowly. The rate of gas interchange between the two phases is rapid for the small bubbles at the

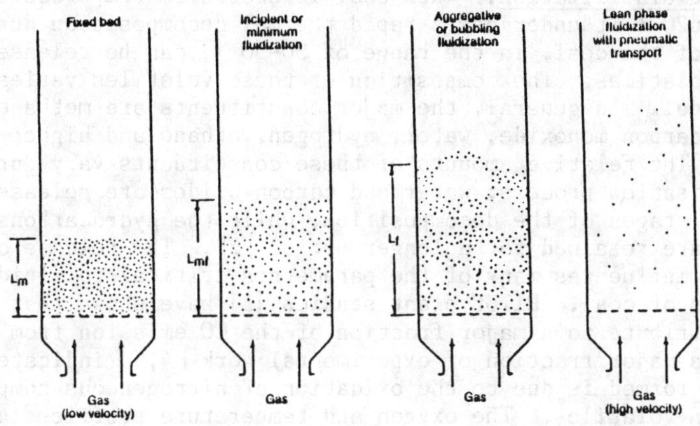

Figure 3. Fluidization Phenomena.

bottom of the bed, but reduces as the bubble size increases.

2.2. Combustion

Consider the combustion of a coal particle in a fluidized bed. The important combustion and related processes (Fig.4) are:

i) Drying
ii) Devolatilization
iii) Volatiles combustion
iv) Char combustion
v) Fuel/gas mixing
vi) Heat transfer local to the particle

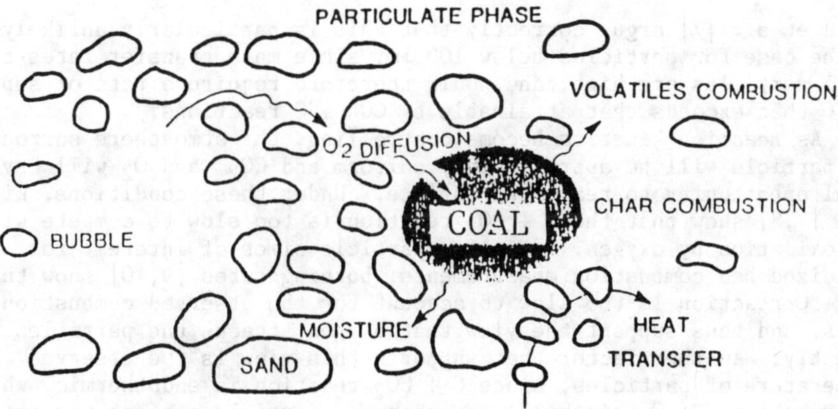

Figure 4. Processes Inovolved During the Combustion of Coal.

2.2.1. Devolatilization. When coal is heated to temperatures in excess of about 620 K it undergoes a rapid thermal decomposition during which a fraction of the coal, in the range of 30-50 %, can be released in the form of volatiles. The composition of these volatiles varies with the type of coal. In general, the major constituents are methane, carbon dioxide, carbon monoxide, water, hydrogen, ethane and higher hydrocarbons. The relative amounts of these constituents vary during the devolatilisation process; water and carbon oxides are released mainly in the early stages of the decomposition, while the hydrocarbons and hydrogen are retained for a longer period |2|. The release of the volatiles influences many of the parameters critical to fluidized bed combustion of coal. Pilot plant studies |3| have shown that the volatiles contribute to a major fraction of the CO emission from fluidized beds and a major fraction of experimental work |4,5| indicates that most of the NO formed is due to the oxidation of nitrogeneous compounds present in volatiles. The oxygen and temperature profiles in the bed are also affected by the volatile release, since a strong evolution in some particular section of the bed would deplete the oxygen there, and possibly cause hot spots. Finally, due to their low ignition temperature, the volatiles effectively extend the stable range of operation of fluidized beds to lower temperatures.

2.2.2. Char combustion. Burning of the char occurs mainly due to a carbon/oxygen reaction yielding varying amounts of CO and CO_2. Under conditions of external diffusion control of the combustion process, it has been maintained |6| that it is possible for the rate of the reaction

$$2CO + O_2 \rightarrow 2CO_2$$

to be fast enough to consume all the O_2 before it reaches the carbon surface, the CO then being supplied by the reduction reaction

$$CO_2 + C \rightarrow 2CO$$

Field et al. |7| argue correctly that this is particularly unlikely to be the case for particles below 100 μm, since mass transfer rates to small particles are high, and would therefore require a rate of supply of CO that exceeds that attainable by CO_2 + C reactions.

As reaction kinetics become controlling, the atmosphere surrounding the particle will be approximately uniform and CO_2 and O_2 will have equal opportunity to reach the surface. Under these conditions, kinetic data |7,8| show that the C + CO_2 reaction is too slow to compete with the oxidation by oxygen. Even for particle sizes of interest for fluidized bed combustion measurements, burning rates |9,10| show that CO_2 + C reaction is too slow to account for the observed combustion rates, and thus support the view that oxygen attacks the particles directly. Another factor that supports this view is the observed temperature of particles; since C + CO_2 reaction is endothermic, while the reaction C + O_2 is highly exothermic, particle surface temperatures will be expected to fall below or exceed bed temperature for these respective reactions. Temperature readings of burning particles give surface temperatures which are substantially higher than the bed temperatures, indicating oxidation by oxygen. The current view is that at temperature relevant to FBC the main product of this reaction is CO |8| , the subsequent oxidation of CO by oxygen is assumed to occur fast enough to convert most of the CO to CO_2 within the boundary layer of particles, so that for the purpose of oxygen balance the reaction $C + O_2 \rightarrow CO_2$ can be considered to occur. Detailed calculations of the CO reactions in the boundary layer have been performed as a function of bed temperature, particle size, and the partial pressure of water vapor which show that the CO oxidation may under certain conditions such as very small particle size, low temperature and low oxygen or OH concentrations occur far removed from the particle surface |11|. The order of the heterogeneous reaction

$$2C + O_2 \rightarrow 2CO$$

is uncertain. Reaction orders from experiment vary from 0.5 to 1.0 . Howard and Essenhigh |12| suggest that the order must depend on temperature and oxygen partial pressure, with low temperatures favouring zero order kinetics and low partial pressure favouring a first order process.

Char combustion reactions take place within the pores as well as at the surface of the char. It seems at present an exceedingly complicated task to model accurately the structural changes which take place in the char particle during combusiton.

The wide spread in pore size distribution, the complexity of the intersecting passages in the coal matrix, varying swelling in the coal matrix, varying swelling and softening behaviour of different coals when heated, are factors that need to be better understood.

3. HEAT TRANSFER

In fluidized bed boilers, tubes are immersed in the bed to evaporate water or superheat steam. The rate of heat transfer can be correlated by an equation of the form

$$Q = \underbrace{hA(T_b - T_t)}_{(1)} + \underbrace{\varepsilon \sigma A(T_b^4 - T_t^4)}_{(2)}$$

where the first term is the heat transfer by convection and the second term is that by radiation. The convective heat transfer represents about 80% of the total and is therefore the controlling mechanism. The effective heat transfer coefficient decrease with increasing height above the expanded bed level and finally approach the value of the coefficient for forced gas convective heat transfer at the operating velocity. Temmink and Menlink |13|, for example, report an approximately linear decrease in bed-to-tube coefficient from 330 W/m^2K for an in bed tube to 150 W/m^2K for a tube at a height of 30 cm above the expanded bed surface with coal <25 mm.

4. DESIGN CONSIDERATIONS ASSOCIATED WITH AFFBs.

Operating experience with the demonstration and commercial units has indicated some areas of concern that continue to challenge system designers. While it is true that most are with auxilary subsystems, they do influence the efficiency and reliability of the whole system.

4.1. Combustion Efficiency

Operating efficiency of a majority of FBBs has been found few percent lower than the design efficiency. The problem has many possible sources:

i) Insufficient freeboard height. Enough freeboard volume must be provided to completely burn up carbon remaining in the solids.
ii) High fines content. If the fines content in the fuel is inconsistent with its design point, excessive carryover will result.
iii) Poor bed conditions. The turbulence of the bed and the height of the bed must be sufficient to burn coarse fuel particles completely. Some designers feel that bubbling beds fluidized only from below may not give enough turbulence to compensate for unavoidable variations in fuel feed sizes.

To improve combustion efficiency without making changes in basic designs, two immediate solutions are apparent: to provide more refractory and height in the freeboard area to reduce heat and particle losses, or to install a recycle loop and reinject all or a portion of the solids.

4.2. Limestone Properties

3 major parameters are important in assuring adequate SO_2 control:

i) particle size of limestone
ii) type of limestone
iii) Ca/S mole ratio

Tests show varying removal rates with different limestones, but most boilers require a Ca/S ratio of 2-4 for 90% SO_2 removal, though the final value depends on the S content of the fuel. If the particles are too coarse, limestone utilisation is poor. But if the particles are too fine, they are elutriated from the bed again reducing utilization, but also overloading the reinjection mechanism.

4.3. Fuel Feeding

The question here is where in the bed to feed the fuel. Some systems employ underbed feed. To give proper distribution, a number of feed points are needed. Systems range from 1 feed point every 9 sq.ft of bed area to one every 20 sq.ft. But as the combustor volume increases to accomodate higher output, the problem of evenly distributing fuel to the full area of the bed becomes difficult.

Some installations (Georgetown) employ overbed feeding systems through either a conventional spreader-stoker system or gravity feed pipes. In these systems, fuel must be sized within close tolerances, or fines will elutriate, reducing combustion efficiency. Overbed feeding system also causes poor limestone utilization and hence higher SO_2 emissions as a result of lower residence times.

4.4. Air Distribution

This parameter is important in maintaining uniform fluidization across the bed. When the particles are large emough to dampen the fluidizing air velocity, partial slumping of the bed results and creates local hot spots with poor combustion characteristics. Depending on the size of the air distributor's openings, the apparatus may be prone to plugging. It is also susceptible to thermal degradation because it must handle a weighty mass of hot material, especially in slumped-bed conditions; thus some form of cooling should be designed into the distributor.

Most of the designs used today range from simple perforated plates to multiple layers of horizontal sparger tubes. One design makes use of T-shaped pipes welded onto the distributor plate. This allows a layer of ash to build up on the plate, to insulate the metal from the hot solids.

4.5. Erosion/Corrosion

In many boiler designs, especially larger sizes, heat transfer surface is provided in the bed. However, severe erosion of in-bed tubes are reported.

The degree of erosion depends on the orientation of the surfaces, the abrasiveness of bed material and the fluidization velocity. Experience indicated that vertically oriented tubes are less prone to erosion than horizontal ones. As a solution, some designs eliminated the use of in-bed tubes.

Although the temperature in AFBBs is low enough to prevent alkali metal deposits, it is argued that this problem still should not be discounted. Lignites, for example, tend to have high alkali content. Because alkalis tend to deposit on particles, the ash can grow larger, causing clinkers and upsetting fluidization.

4.6. Load Control

A heat transfer tube immersed in a fluidized bed will experience about 4 times the heat flux of a tube above the bed. Thus, one simple way to control turndown is to vary the bed height so that more tubes are exposed. Bed material can be added or removed to adjust the height, or the superficial velocity can be regulated. As in conventional combustors, fuel feed can also be varied. In practice, all three methods are generally used to provide the control.

Using a zoned air-distribution plenum is practical in the smaller sizes but as the boiler size increases, the method becomes cumbersome.

4.7. Fuel Flexibility

Be vary when you hear statements like "AFBBs can burn anything you feed them". While fuel flexibility is an inherent advantage of these systems, they are not "dispose-alls" for solid liquid and gaseous fuels at least not without major sacrifice in boiler efficiency and overall performance. Like any other boiler a FBB must be designed for a specific fuel. However, there is much more latitude within a specific fuel than with a stoker or PC-fired unit.

4.8. Ash Removal

Both bottom ash and fly ash removal are serious considerations in FBB technology. Spent bed material must be removed at a controlled rate to maintain optimum bed conditions. If this is not performed carefully and with heat removal, a significant thermal loss can occur to the bottom ash handling system, resulting in materials-related problems.

To meet the emission standards for particulates, flyash remaining after the cyclones must be picked up by a fabric filter, electrostatic precipitator or other dust collector. But this ash has different characteristics than ash from conventional boiler. It has larger quantities of carbon & higher concentration of calcium compounds. Electrostatic precipitators encounter resistivity problems and fabric filters are more prone to fire hazards and have shorter life.

As a solution to these problems, several new configurations have been developed. These are:

i) Combustor and convective sections have been separated and hence in-bed tubes have been eliminated.
ii) Circulating FBCs have been developed.
iii) Multiple-bed FBCs have been developed.

5. TYPES OF FLUIDIZED BED COMBUSTORS

5.1. Bubbling Fluidized Bed Combustors (BFBC)

Figure 5 shows a bubbling bed fluidized combustor. Steady combustion can take place in such a system by adding fuel when the bed is above the ignition temperature. The fluidity of the system ensures rapid mixing of the fuel throughout the bed.

5.2. Circulating Fluidized Bed Combustors (CFBC)

If the gas velocity in the bubbling fluidized bed is increased still further, more and more of the particles are entrained in the gas stream and leave the vessel. Eventually the transport velocity for most of the particles is reached. Above this value the vessel can quickly empty of solids unless additional particles are fed to the base of the reactor. If the solids leaving the vessel are returned through an external collection system, then a circulating or fast fluidized bed system (CFB) results. The system is characterized by clusters of particles streaming upward in the reactor at solids concentrations well above that for dilute phase transport.

Figure 6 also shows a schematic diagram of a CFBC system.

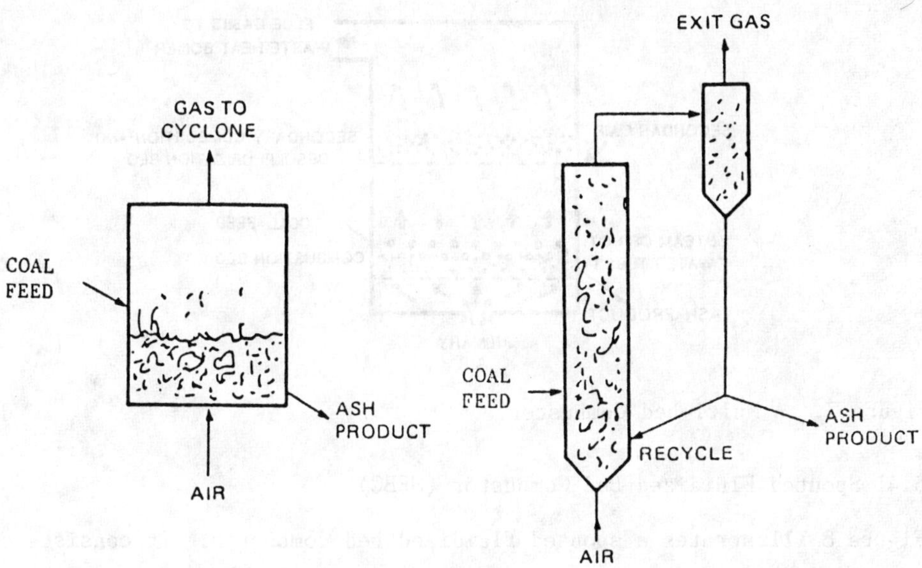

Figure 5. A Bubbling Fluidized Bed Combustor.

Figure 6. A Circulating Fluidized Bed Combustor.

Typical operating conditions for bubbling and circulating bed fluidized combustors are given in Table 1.

Table 1. Typical Operating Conditions for BFBCs and CFBCs.|1|

	Bubbling beds	Circulating beds
Top feed size (mm)	50	10
Bed particle size (mm)	0.1–4.0	0.05–1.0
Fluidizing velocity (m/s)	1–3	4–12
Combustion efficiency (%)	90–99	95–99

5.3. Multi-bed Combustion Systems (MBC)

Variants on BFBC system that involve multiple beds (usually two), are being developed (see Fig.7). In the lower or primary bed combustion takes place while further combustion may occur in the upper or secondary bed. The secondary bed also can serve as a predrying stage or an emission control bed.

Commercial units, such as CFBC superimposed on BFBC for wide particle size ranges of bed material burning coal have been installed (Battelle in U.S.A.).

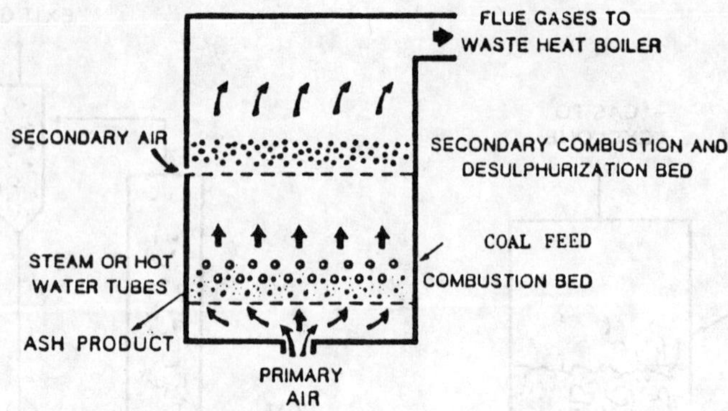

Figure 7. A Multi-bed Combustor.

5.4. Spouted Fluidized Bed Combustor (SFBC)

Figure 8 illustrates a spouted fluidized bed combustor. It consists of

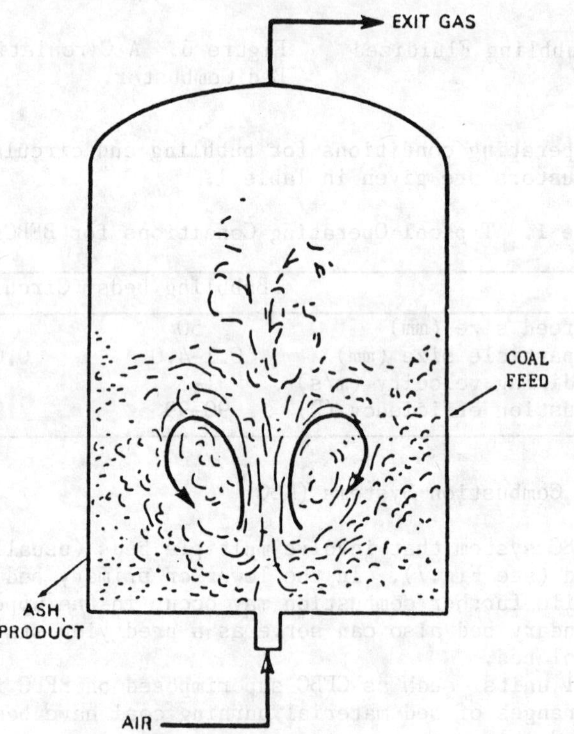

Figure 8. A Spoute/Fluidized Combustor.

a bed of coarse particles partly filling the vessel which is provided with a relatively large central opening at its base. Gas is injected through this aperture. With sufficient flow of gas, the particles in the gas can be made to rise in a fountain at the center of the bed and impart a cyclic motion on the bed as a whole.

6. CONCLUSION

In conclusion, fluidized bed combustors overcome the two fundamental limitations of the more established combustion techniques by allowing the combustion of a range of coal with widely varying ash and moisture contents and by providing a means of control for SO_x and NO_x emissions. So more and more of high capacity systems are expected to be seen in the future.

NOMENCLATURE

A	Area of heat transfer surface in the bed (m^2)
h	Convective heat transfer coefficient (W/m^2-K)
Q	Rate of heat transfer (W)
T_b	Temperature of fluidized bed (K)
T_t	Temperature of heat transfer surface (K)
ε	Emissivity
σ	Stefan-Boltzmann constant (W/m^2K^4)

REFERENCES

1. La Nauze, R.D., 'A Review of the Fluidized Bed Combustion of Biomass', J. Inst. of Energy, 60, No. 443, p.66, 1987.

2. Howard, J.B., 'Fundamentals of Coal Pyrolyses and Hydropyrolysis', Chemistry of Coal Utilization Second Supplementary Volume (M.A. Elliot, Ed.), Wiley, New York, p.665, 1981.

3. Gibbs, B.M. and Beer, J.M., 'Concentration and Temperature Distributions in a Fluidized Bed Coal Combustor', Combustion Institute European Symposium., (Ed. F.J. Weinberg), Academic Press, London, 1973.

4. Gibbs, B.M., Pereira, F.J. and Beer, J.M., 'Coal Combustion and NO Formation in an Experimental Fluidized Bed', Inst. of Fuel Symposium Series No. 1, D6-1, 1975.

5. Pohl, J.H. and Sarofim, A.F., 'Devolatilization and Oxidation of Coal Nitrogen', 16th Symposium (International) on Combustion, The Combustion Institute, Pittsburgh, p.491, 1976.

6. Van der Held, E.F.M., 'The Reaction Between a Surface of Solid Carbon and Oxygen', Chem.Eng.Sci., **14**, p.300, 1961.

7. Field, M.A., Gill, D.W. and Morgan, B.B., Combustion of Pulverized Coal, BCURA, Leatherhead, 1967.

8. Arthur, J.R., 'Reaction Between Carbon and Oxygen' Trans. Faraday Soc., **47**, p.164, 1951.

9. Gray, M.O. and Kimber, G.M., 'Reaction of Chemical Particles with Carbon Dioxide and Water at Temperatures up to 2800K', Nature (London), **214**, p. 797, 1967.

10. Golovina, E.S. and Khaustovich, G.P., 'The Interaction of Carbon with Carbon Dioxide and Oxygen at Temperatures up to 3000K', 8th Symp. (International) on Combustion, The Combustion Institute, p. 784, 1962.

11. Sundaresan, S. and Amundson, N.R., 'Diffusion and Reaction in a Stagnant Boundary Layer about a Carbon Particle : Effect of Water Vapour on the Pseudo-Steady-State Structure', Ind. Eng. Chem. Fundam., **19**, p.351, 1980.

12. Howard, J.B. and Essenhigh, R.H., 'Mechanisms of Ignition and Combustion of Pulverized Coal', Combustion and Flame, **9**, p.337, 1965.

13. Temmink, H.M.C. and Meulink, J., 'Operating Experiences with the TNO 2m x 1m Atmospheric Fluid Bed Boiler Facility', 3rd European Coal Utilisation Conf., Amsterdam, Oct. 1983.

PREDICTION OF THE BEHAVIOUR OF ATMOSPHERIC FLUIDIZED BED COAL COMBUSTORS

N. Selçuk
Department of Chemical Engineering
Middle East Technical University
Ankara 06531, Turkey

ABSTRACT. A mathematical model for continuous combustion of coal particles in an atmospheric fluidized bed combustor is presented. Modifications of the model for the incorporation of bubble phase reaction, rate expression, volatile combustion and various char particle combustion mechanisms are described. The accuracy of the model is tested by comparing its predictions with experimental data available on a pilot scale fluidized bed coal combustor. The effect of each modification on bed concentration profiles, carryover losses and combustion efficiencies are discussed.

1. INTRODUCTION

The development of fluidized combustion has reached the stage of industrial application and there is now an urgent need to generalize information in terms of mathematical models that give representative qualitative description of the processes of flow, combustion and mass and heat transfer in fluidized bed combustors.

In the last decade, substantial effort has been devoted to the development of comprehensive mathematical models for fluidized combustors |1-17|. These studies differ in the combinations of their assumptions as to flow patterns, char combustion mechanisms, size distribution and devolatilization of coal feed and heat and mass transfer. However, the validity of a majority of these models is questionable as no comparison of the predictions with measurements is provided |1-4, 7, 9, 11, 15, 17|. This is basically due to the incompleteness of the published data required to fix the initial conditions for the prediction procedures, or to compare with the predictions of the models so that a test of a model can be performed.

What is required for design purposes at the present time are mathematical models capable of predicting the performance of large scale continuous fluidized combustors burning coal of wide size distributions. In an attempt to achieve this objective a system model, based on two-phase theory and conservation equations for energy and chemical species has been developed |17|. Given the bed dimensions,

Y. Yürüm (ed.), New Trends in Coal Science, 495–522.

superficial air velocity, size distribution and analysis of coal feed and characteristics of inert bed material, the model predicts the concentration and temperature distributions throughout the bed, combustion efficiency, bed char loading and size distribution of particles in bed and carryover. The model has then been modified for the incorporation of bubble phase reaction, rate expression, volatile combustion and various char particle combustion mechanisms. Modified models have then been applied to the prediction of the behaviour of a pilot scale fluidized bed combustor |18-21| for which measured data had been reported previously in the literature |22|.

The purpose of the present article is to review the modifications and present the evaluation of the modified models by comparing their predictions with available measurements.

2. SYSTEM MODEL

The physical system to be considered is a fluidized bed combustor fed continuously by a coal of wide size distribution and equipped with a particle separator. Excess heat generated within the bed is assumed to be removed by means of a water cooled coil immersed into the bed.

2.1. Model Assumptions

The behaviour of the fluidized bed under consideration is described by a model based on conservation equations for energy and chemical species. The model consists of three main components: Fluid mechanics, combustion and mass transfer and heat transfer. The basic assumptions of the model components are given below.

With regard to the fluid mechanics component the following assumptions are made:

i. The bed consists of two phases, a bubble phase and an emulsion phase which contains the particles and interstitial gas. It is aggregatively fluidized and is treated by the Davidson-Harrison two-phase model |23|.

ii. The bubble phase is in plug flow and the emulsion phase is well stirred.

iii. Bubbles are of uniform size throughout the bed and are free of particles.

iv. Coal particles are fed to the bed at a constant rate and into a plane representing the base of the bed.

For the second component of the model, combustion and mass transfer, the basic assumption made is:

v. The physical properties of the gas in bubble and emulsion phases are the same and uniform throughout the bed and are based on products of complete combustion.

The sequence of events followed by a single coal particle upon entrance to bed is assumed to be as follows:

vi. Immediately upon entrance to bed, the volatile matter and moisture are released from the coal particle, leaving a char particle containing only fixed carbon and ash.

vii. The char particle reacts with the interstitial gas, the fixed carbon giving CO, and ash being removed at the same time. The ash removal is related to the rate of carbon removal at such a rate that the proportions by mass of fixed carbon and ash remain constant throughout the life of the particle, i.e., the density of burning particle is constant.

viii. Some of the CO leaving the particle surface reacts with O_2 in the gas phase to give CO_2.

ix. The reaction of CO to CO_2 is kinetically controlled.

With regard to heat transfer component of the model the following assumptions are made:

x. The bed operates adiabatically under steady state conditions.

xi. The thermal energy balances are based on convection and heat generation by the combustion reactions.

2.2. Two-Phase Modelling of Bed Behaviour

In accordance with the classical two-phase model of Davidson and Harrison |23|, volumes, volume fractions, flow rates and gas velocities in both phases, and the expanded bed height can be calculated for a specific value of equivalent bubble diameter by the following relationships:

For minimum fluidization conditions:

$$V_{mf} = H_{mf} A_o \tag{1}$$

$$q_{mf} = A_o u_{mf} \tag{2}$$

Volumetric flow rates:

$$q_f = u_o A_o \tag{3}$$

$$q_i = q_{mf} \tag{4}$$

$$q_b = \delta u_b A_o \tag{5}$$

Volumetric fraction of bubbles:

$$\delta = \frac{u_o - u_{mf}}{u_b} \tag{6}$$

where

$$u_b = u_o - u_{mf} + u_{br} \tag{7}$$

and u_{br} is the rate of rise of bubbles and is related to the bubble diameter by the following expression |23|

$$u_{br} = 0.711 (g d_b)^{1/2} \tag{8}$$

Phase volumes:

$$V_f = H_f A_o = V_e + V_b \tag{9}$$

$$V_b = \delta V_f \tag{10}$$

$$V_e = (1 - \delta)V_f = V_i + V_s \tag{11}$$

$$V_i = V_{mf} - V_s = \varepsilon_e V_{mf} \tag{12}$$

Assuming voidage in the emulsion phase is equal to the voidage at incipient fluidization, interstitial gas volume can also be calculated from

$$V_i = \varepsilon_{mf} V_{mf} \tag{13}$$

Expanded bed height:

$$H_f = \frac{H_{mf}}{(1-\delta)} \tag{14}$$

2.3. Char Particle Size Distribution

In order to obtain the size distribution and density of coal particles after devolatilisation, a coal particle of radius r_o, mass m_o and density ρ_o is considered. Following the release of volatile matter and moisture, there remains a char particle of radius r_d, mass m_d and density ρ_d. Utilising assumption (vii), densities of the coal and char particles can be related by the following expression

$$\frac{\rho_d}{\rho_o} = \frac{f_C + f_A}{(r_d/r_o)^3} \tag{15}$$

where f denotes the mass fraction in the proximate analysis of coal and subscripts C and A refer to carbon and ash respectively. Assuming that the radii of coal and char particles are related via a dimensionless swelling factor, ϕ ,

$$r_d = \phi r_o \tag{16}$$

where $\phi>1$ for swelling and $\phi<1$ for shrinking, equation (16) takes the form

$$\frac{\rho_d}{\rho_o} = \frac{f_C + f_A}{\phi^3} \tag{17}$$

Mass flow rate of char particles, F_d, can be evaluated from

$$F_d = (f_C + f_A) F_o \tag{18}$$

where F_o is the mass flow rate of coal particles.

For the estimation of size distribution of char particles, it is considered that the number flow rate of coal particles in the size range r to r+dr is the same as that of char particles i.e.

$$\frac{F_o p_o(r)dr}{\frac{4}{3}\pi \bar{r}^3 \rho_o} = \frac{F_o p_d(r)dr}{\frac{4}{3}\pi \bar{r}_d^3 \rho_d} \tag{19}$$

where p_o (r) and p_d (r) denote coal and char particle size distribution functions respectively, and the bar over the radii represent the mean size within the range. Utilising equations (16), (17) and (18), in equation (19) the size distribution function of char particles can be expressed in terms of the coal particle size distribution function as follows

$$p_d(r) = \frac{1}{\phi} p_o\left(\frac{r}{\phi}\right) \tag{20}$$

Hence, if maximum and minimum radii, r_{max} and r_{min}, ϕ and $p_o(r)$ are specified for the original coal feed, the char particle size distribution, $p_d(r)$ can be evaluated by using equation (21) and lie in the size range $\phi r_{min} \leq r \leq \phi r_{max}$.

2.4. Mass Balance on Char Particles

A steady state mass balance on char particles can be written in verbal form as |24|

$$\begin{bmatrix}\text{Char fed}\\\text{to bed}\end{bmatrix} - \begin{bmatrix}\text{Char leaving}\\\text{in overflow}\end{bmatrix} - \begin{bmatrix}\text{Char leaving}\\\text{in carryover}\end{bmatrix} + \begin{bmatrix}\text{Char shrinking}\\\text{into the interval}\\\text{from a larger size}\end{bmatrix}$$

$$- \begin{bmatrix}\text{Char shrinking}\\\text{out of the interval}\\\text{to a smaller size}\end{bmatrix} - \begin{bmatrix}\text{Char consumption}\\\text{due to shrinkage}\\\text{within the interval}\end{bmatrix} = 0 \tag{21}$$

In the size range r and r+dr, the balances takes the form

$$F_d p_d(r) - F_1 p_b(r) - F_3 p_3(r) \frac{d\{R(r)p_b(r)\}}{dr} + \frac{3M}{r} p_b(r)R(r) = 0 \tag{22}$$

where F_1 and F_3 are the mass flow rates of char particles in overflow and carryover respectively, M is the total mass of char particles in bed and R(r) denotes the rate of shrinkage of a single char particle. $p_b(r)$ and $p_3(r)$ are the size distribution functions for the char particles in bed and in carryover respectively. Assuming that there is no overflow of char particles, equation (22) can be rearranged to the following form

$$\frac{d}{dr}\left\{\frac{Mp_b(r)R(r)}{r^3}\right\} = \frac{F_d p_d(r) - F_3 p_3(r)}{r^3} \qquad (23)$$

2.5. Elutriation

The standard treatment of elutriation is to express the size distribution of the char material elutriated from the bed in terms of the size distribution of char particles in bed by putting

$$F_2 p_2(r)dr = MK(r)\ p_b(r)dr \qquad (24)$$

where K(r) is the elutriation rate constant and F_2 is the mass flow rate at which particles are elutriated from bed.

Denoting the particle separator efficiency for particles of size r by $\eta_o(r)$, the mass flow rate at which elutriated particles between r and r+dr are lost from the separator can be expressed as

$$F_3 p_3(r)dr = \{1-\eta_o(r)\}\ F_2 p_2(r)dr \qquad (25)$$

Substituting equation (24) into equation (25) and defining a net elutriation constant, E(r), yields

$$F_3 p_3(r)dr = M\ E(r)p_b(r)dr \qquad (26)$$

where

$$E(r) = \{1-\eta_o(r)\}\ K(r) \qquad (27)$$

Unless a typical functional form for $\eta_o(r)$ is obtained, equation (26) cannot be used as it stands. However, two extreme cases can be considered:

i. Complete recycle of elutriated char particles, i.e.$\eta_o(r) = 1$, thus $E(r) = 0$ for all r.
ii. No recycle of elutriated char particles indicating complete loss of elutriated material, i.e. $\eta_o(r) = 0$, thus $E(r)=K(r)$ for all r.

Deductions from mass balance on char particle: The char particles fed to the bed are in the size range $\phi r_{min} \leq r \leq \phi r_{max}$. Hence, the char particles within the bed will lie in the size range $0 \leq r \leq \phi r_{max}$, because as all char particles within the bed shrink due to combustion, those which have the longest residence time will have completely burnt, whereas the largest particles which have only just entered will still have a radius ϕr_{max}. As the bed particles are in the size range $0 \leq r \leq \phi r_{max}$, the elutriated particles will be in the same size range. Integrating equation (26) over the complete size range gives

$$F_3 = M \int_{r=0}^{r=\phi r_{max}} E(r)p_b dr \qquad (28)$$

Equation (28) can be used to evaluate F_3 once M and $p_b(r)$ have been determined. The bed size distribution function, $p_b(r)$, can be obtained by substituting equation (26) into equation (23) and takes the following form

$$p_b(r) = -\frac{F_d}{M} \frac{r^3}{R(r)I(\phi r_{max}, r)} \int_{r_i=r}^{r_i=\phi r_{max}} \frac{I(\phi r_{max}, r_i) p_d(r_i) dr_i}{r_i^3} \tag{29}$$

where

$$I(\phi r_{max}, r_i) = \exp \{-\int_{r_i=r_i}^{r_i=\phi r_{max}} \frac{E(r_i)}{R(r_i)} dr_i\} \tag{30}$$

where r_i represents dummy variable. Equation (29) can be used to calculate $p_b(r)$ once E(r), $p_d(r)$ and R(r) have been specified.

Substituting equation (29) into equation (28), the mass flow rate of carryover, F_3, can be expressed as

$$F_3 = -F_d \int_{r=0}^{r=\phi r_{max}} \frac{E(r) r^3}{R(r) I(\phi r_{max}, r)} \cdot \{ \int_{r_i=r}^{r_i=\phi r_{max}} \frac{I(\phi r_{max}, r_i) p_d(r_i)}{r_i^3} dr_i \} dr \tag{31}$$

2.6. Shrinkage Rate of a Single Char Particle

In order to be able to solve the equations developed so far, it is necessary to specify the rate of shrinkage of a single char particle, R(r), which can be written as:

$$R(r) = \frac{dr}{dt} \tag{32}$$

The functional form of R(r) depends upon the model chosen to describe combustion of a char particle of radius r. In this work three different combustion models are considered. These models and the corresponding shrinkage rates are given below.

2.6.1. <u>Diffusion controlled char combustion model with reduction reaction at the particle surface</u>. This reaction scheme involves the production of CO at the char particle surface by two heterogeneous reactions:

$$C + CO_2 \rightarrow 2CO$$
$$2C + O_2 \rightarrow 2CO$$

followed by partial oxidation of CO to CO_2 in the interstitial and bubble gases, i.e.

$$CO + 1/2\ O_2 \rightarrow CO_2$$

Assuming pseudo steady state and that species transfer occurs only by molecular diffusion the shrinkage rate of a char particle can be expressed by the following expression:

$$R(r) = \frac{dr}{dt} = -\frac{\theta}{r} \quad (33)$$

where

$$\theta = \frac{D\ M_C\ \phi^3\ (2C_{1i} + C_{3i})}{\rho_o f_C} \quad (34)$$

2.6.2. Diffusion controlled char combustion model with no reduction reaction at the particle surface. In this model the heterogeneous reaction between the char and carbon dioxide is not taken into account since this reaction is considered to be too slow to compete with the çarbon oxidation by oxygen under the conditions in fluidized bed combustors |25|.

Hence, the reaction at the surface of the particle is considered to be:

$$2C + O_2 \rightarrow 2CO$$

Assuming that surface reaction is diffusion controlled the shrinkage rate of a char particle is expressed by the following relationship:

$$R(r) = -\frac{\theta}{r}$$

where

$$\theta = \frac{2C_{1i}\ D\ M_C\ \phi^3}{\rho_o f_C} \quad (35)$$

2.6.3. Combined kinetic and diffusion controlled char combustion model with no reduction reaction at the particle surface. In the temperature range of fluidized bed combustors (between 750°C and 950°C) the resistances of gas film diffusion and chemical reaction may be both important |26|. The reactions are assumed to be the same as those in section 2.6.2. The overall reaction rate constant is given by:

$$k_c = \frac{1}{\frac{1}{k_d} + \frac{1}{k_r}} \quad (36)$$

The shrinkage rate then becomes:

$$R(r) = \frac{dr}{dt} = - \frac{2C_{1i} k_c M_C \phi^3}{\rho_o f_C} \tag{37}$$

2.7. Material Balance Equations

Assuming that the bubble phase and the interstitial gas are fully mixed at the expanded bed height, the exit concentration of any gaseous species, C_{jm}, is related to the concentration of the same species in the bubble phase at the top of the bed, C_{jbe}, and in the emulsion phase, C_{ji}, by the following expression:

$$q_f C_{jm} = q_b C_{jbe} + q_i C_{ji} \tag{38}$$

2.7.1. Balances for bubble phase. Concentration profiles in bubble phase can be determined by carrying out a steady state material balance on species j between heights z and z+dz. Using the rate expression suggested by Hottel and reported by Field et al. |26| for the homogeneous gas phase reaction, the balance can be written as:

$$u_b \frac{dC_{jb}}{dz} = (K_{be})_j (C_{ji} - C_{jb}) + \alpha_j k e^{-E/RT_b} C_{1b}^{0.3} C_{2b} C_{5b}^{0.5} \tag{39}$$

where $(K_{be})_j$ is the interchange coefficient between the bubble and emulsion phases based on the bubble volume and

$$\alpha_j = -\frac{1}{2}, -1, 1, 0, 0, 0 \text{ for } j=1, \ldots, 6 \text{ respectively.}$$

2.7.2. Balances for whole bed. Steady state material balances for the whole bed can be written for all five chemical elements (C,O,N,H,S) involved in the reactions. These balances can be expressed in the following forms:

$$C_{2m} = 2C_{1m} - \frac{2}{q_f} \left[0.21 \text{ Py} + \frac{f_C F_3}{(f_C + f_A) M_C}\right] \tag{40}$$

$$C_{3m} = \frac{1}{q_f} \left[0.42 \text{ Py} + \frac{\gamma_C f_D F_o}{M_C} + \frac{f_C F_3}{(f_C + f_A) M_C}\right] - 2C_{1m} \tag{41}$$

$$C_{4m} = \frac{1}{q_f} \left[0.79 \text{ A} + \frac{\gamma_N f_D F_o}{M_N}\right] \tag{42}$$

$$C_{5m} = \frac{1}{q_f} \left[\frac{\gamma_H f_D}{M_H} + \frac{f_W}{M_W}\right] F_o \tag{43}$$

$$C_{6m} = \frac{\gamma_S f_D F_o}{q_f M_S} \tag{44}$$

where C_{1m} to C_{6m} denote the bed exit concentrations of gaseous species O_2, CO, CO_2, N_2, H_2O and SO_2 respectively.

The complete exit gas composition cannot be determined with the equations produced so far. Equations (40) to (44) are only five equations in the six unknown exit concentrations C_{jm} (j=1,....,6). An additional equation can be found by carrying out an O_2 species balance over the whole bed and takes the following forms according to the combustion scheme assumed for char and volatiles combustion.

i) Diffusion Controlled Char Combustion Models with and Without Reduction Reaction at the Particle Surface

If the liberated volatile matter burns instantaneously and completely in the interstitial gas to give CO_2, SO_2, H_2O and N_2, O_2 species balance can be written as:

$$\underset{(1)}{q_f C_{1m}} - \underset{(2)}{q_f C_{1bo}} = - \underset{(3)}{\int_{r=0}^{r=\phi r_{max}} (4\pi D\, r\, C_{1i})\left(\frac{M\, p_b(r)\, dr}{\frac{4}{3}\pi r^3 \rho_d}\right)}$$

$$+ \underset{(4)}{F_o\left[\frac{\gamma_C f_D - f_C}{M_C} + \frac{1}{2}\,\frac{\gamma_H f_D}{M_H} + \frac{\gamma_S f_D}{M_S} - \frac{\gamma_O f_D}{M_O}\right]}$$

$$+ \underset{(5)}{\frac{1}{2}\, k\, V_i\, e^{-E/RT_i}\, C_{1i}^{0.3}\, C_{2i}\, C_{5i}^{0.5}}$$

$$+ \underset{(6)}{\frac{1}{2}\, k\, V_b\, e^{-E/R\overline{T}_b}\, \overline{C}_{1b}^{0.3}\, \overline{C}_{2b}\, \overline{C}_{5b}^{0.5}} \tag{45}$$

where (1) and (2) represent the rate of gas phase oxygen out and in respectively, (3) is the rate of consumption at particle surface, (4) corresponds to the rate of consumption by combustion of volatiles and (5) and (6) are the rates of consumption by gas phase reaction in the emulsion and bubble phases respectively. In equation (45) $\overline{C}_{jb}$ and $\overline{T}_b$ denote average concentration and temperature in bubble phase defined by:

$$\overline{C}_{jb} = \frac{\int_{z=0}^{z=H_f} C_{jb}\, dz}{H_f} \tag{46}$$

$$\overline{T}_b = \frac{\int_{z=0}^{z=H_f} T_b\, dz}{H_f} \tag{47}$$

If the volatile carbon burns instantaneously but incompletely to form CO, term (4) in equation (45) is modified as follows:

$$\begin{array}{l}\text{Rate of consumption}\\ \text{of } O_2 \text{ by combustion}\\ \text{of volatiles}\end{array} = F_o\left[\frac{1}{2}\frac{\gamma_C f_D - f_C}{M_C} + \frac{1}{2}\frac{\gamma_H f_D}{M_H} + \frac{\gamma_S f_D}{M_S} - \frac{\gamma_O f_D}{M_O}\right] \tag{48}$$

ii) Combined Kinetic and Diffusion Controlled Char Combustion Model with no Reduction Reaction at the Particle Surface
For this model term (3) in equation (45) becomes:

$$\begin{array}{l}\text{Rate of consumption}\\ \text{at particle surface}\end{array} = \int_{r=0}^{r=\phi r_{max}} (4\pi r^2 k_c C_{1i})\left(\frac{M\, p_b(r)\, dr}{\frac{4}{3}\pi r^3 \rho_d}\right) \tag{49}$$

2.8. Enthalpy Balances

2.8.1. Enthalpy balance on the bed. An additional equation containing T_i can be found by carrying out an enthalpy balance on the bed. To simplify the formulation of the balance the following additional assumptions are made:

i. Air and coal are introduced into the bed at ambient temperature.
ii. Ash removed at the grate level leaves the system at the interstitial gas temperature.
iii. Radiative heat loss from the char particles is negligible.
iv. There is no heat transfer to the cooling coil in the freeboard.
v. Specific heat of gas in bubble and emulsion phases is uniform throughout the bed.

With these assumptions the enthalpy balance on the bed takes the form:

$$F_oQ_o = [q_b\rho_g\bar{c}_g(T_{be} - T_o) + q_i\rho_g\bar{c}_g(T_i - T_o)]$$
(1) (2)

$$+ (q_bC_{2be} + q_iC_{2i})\,\Delta H_2 + F_A\bar{c}_A(T_i - T_o)$$
(3) (4)

$$+ h_{ic}A_c(T_i - T_c) + [F_3Q_d + (\Delta H_s)_d] \qquad (50)$$
(5) (6)

where (1) represents the rate of chemical enthalpy in with coal feed, (2) and (3) are the rates of sensible and chemical enthalpy out with gaseous products respectively, (4) represents the sensible enthalpy out with ash, (5) is the rate of enthalpy out through cooling coils and (6) corresponds to the rate of chemical and sensible enthalpy out with carryover.

2.8.2. Enthalpy balance on a single char particle.

2.8.2.1. Diffusion controlled char combustion model with reduction reaction at the particle surface. The enthalpy balance on a single char particle can be written as:

$$\frac{d}{dt}[m_d\bar{c}_d(T_d - T_o)] = \{4\pi D\, r\, [C_{3i}\,\Delta H_3 + 2C_{1i}\,\Delta H_1]\}$$
(1) (2)

$$- [- \frac{dm_d}{dt}\bar{c}_d(T_d - T_o) + 4\pi r^2 h_d(T_d - T_i)] \qquad (51)$$
(3) (4)

where term (1) represents the rate of increase of sensible enthalpy content, (2) is the rate at which heat is released by reaction and (3) denotes the rate at which sensible enthalpy is carried away in material removed and (4) is the rate at which heat is lost by convection.

By assuming that for the char particle Nu=2, and that devolatilization takes place at a fixed temperature T_v independent of the initial radius of an elutriated char particle, r_i, integration of Eq. (51) gives the surface temperature of the char particle as:

$$T_d = B_4 + (r/r_i)^{B_1}\,(T_v - B_4) \qquad (52)$$

where

$$B_4 = \frac{D(2C_{1i}\,\Delta H_1 + C_{3i}\,\Delta H_3) + k_g\,T_i}{k_g} \qquad (53)$$

and

$$B_1 = \frac{3\,k_g}{\rho_d \bar{c}_d \theta} \tag{54}$$

2.8.2.2. Diffusion controlled char combustion model with no reduction reaction at the particle surface. In equation (51) term (2) becomes:

$$\text{Rate at which heat is released by reaction} = 8\pi D\, r\, C_{1i}\, \Delta H_1 \tag{55}$$

The surface temperature of the char particle is given by an expression of the same form as equation (52) but with

$$B_4 = \frac{2D\, C_{1i}\, \Delta H_1 + k_g T_i}{k_g} \tag{56}$$

2.8.2.3. Combined kinetic and diffusion controlled char combustion model with no reduction reaction at the particle surface. From the enthalpy balance on a single char particle the rate of change of temperature with particle size is obtained as:

$$\frac{dT_d}{dr} = \frac{\Delta H_1 \rho_o f_C}{4\, r\, \bar{c}_d \rho_d} - \frac{k_g (T_d - T_i)\, \rho_o f_C}{8\, r^2\, \bar{c}_d k_c C_{1i}\, \rho_d} \tag{57}$$

The particle temperature can be calculated by the numerical integration of equation (57).

2.8.3. Enthalpy content of elutriated char particles.

2.8.3.1. Diffusion controlled char combustion models with and without reduction reaction at the particle surface. For these models the total sensible enthalpy content of the char material elutriated from the bed is given by the following expression:

$$(\Delta H_s)_d = F_3 \bar{c}_d (B_4 - T_o) - \Big[F_d \bar{c}_d (T_v - B_4) \int_{r=0}^{r=\phi r_{max}} \frac{E(r)\, r^{3+B_1}}{R(r) I(\phi r_{max}, r)} \Big] \cdot \Big[\int_{r_i=r}^{r_i=\phi r_{max}} \frac{I(\phi r_{max}, r_i)\, p_d(r_i)\, dr_i}{r_i^{3+B_1}} \Big]\, dr \tag{58}$$

2.8.3.2. Combined kinetic and diffusion controlled char combustion model with no reduction reaction at the particle surface. The sensible enthalpy content of the elutriated char particle can be expressed as:

$$(\Delta H_s)_d = - F_d \bar{c}_d \int_{r=0}^{r=\phi r_{max}} E(r) \cdot$$

$$\cdot \left[\int_{r_i=r}^{r_i=\phi r_{max}} \frac{r^3 I(\phi r_{max}, r_i) p_d(r_i)}{R(r) I(\phi r_{max}, r)\, r_i^3} (T_d - T_o) dr_i \right] dr \quad (59)$$

2.8.4. Enthalpy balance on bubble phase. Utilizing the assumption that the bubble phase is in plug flow, the steady state enthalpy balance for the bubble phase can be written as:

$$\rho_g \bar{c}_g u_b \frac{dT_b}{dz} = H_{be}(T_i - T_b) + \Delta H_2\, k e^{-E/RT_b}\, C_{1b}^{0.3}\, C_{2b}\, C_{5b}^{0.5} \quad (60)$$

where H_{be} is the interchange heat transfer coefficient between bubble and emulsion phases based on volume of bubbles in bed.

3. APPLICATION OF SYSTEM MODEL TO A PILOT SCALE COMBUSTOR

The input data required for the system model was taken from the data already reported in the literature |22|.

Figure 1 shows the main features of the fluidized bed and ancillaries. The combustor is constructed from stainless steel and is 1.83 m high and 0.3 m square in cross section. The bed material is an inert silica-based clinker, sized 1.68 mm to 0.25 mm. The walls of the combustor are lagged with trition Kaowool Insulation. Excess heat generated in the bed is removed by means of a 12.5 mm O.D. water cooled stainless steel coil immersed in the bed with a total area of 1000 cm^2.

Gases leaving the combustor are dedusted by two cyclones in series; there is no facility for returning the carryover back to the bed.

Steady-state experiments were carried out using-10 BSS mesh coal with varying percent excess air. Bed was operated under atmospheric pressure and bed temperature varied between 650^oC and 850^oC. Excess air levels ranged from -25 % to 40 %. Bed depth and superficial fluidization velocity were maintained constant at 60 cm and 90 cm/s respectively.

Calculations start with the evaluation of the physical parameters by using the empirical correlations listed in Table I, and evaluation of the volumes, velocities and flow rates in the bubble and emulsion phases by the two-phase theory. The size distribution of coal feed is described by the Rosin-Rammler size distribution function. Once the input data are fixed, initial estimates are made for oxygen concentration and temperature of interstitial gas and shrinkage rate is calculated. This is followed by the determination of mass flow rate of carryover by an iterative procedure. Material and enthalpy balances on bubble phase are then integrated simultaneously for the concentration profiles and temperature in that phase. Utilising the average

TABLE I. Physical Parameters

Parameter	Empirical expressions in S.I. units	Reference
u_{mf}	$\frac{1.75}{\varepsilon^3_{mf}}\left(\frac{d_p u_{mf}\rho_g}{\mu}\right)^2 + \left(\frac{150(1-\varepsilon_{mf})}{\varepsilon^3_{mf}}\right)\left(\frac{d_p u_{mf}\rho_g}{\mu}\right)$	|27|
	$= \frac{d_p^3\,\rho_g(\rho_p - \rho_g)g}{\mu^2}$	
u_{br}	$u_{br} = 0.711\,(gd_b)^{0.5}$	|23|
K_{be}	$\frac{1}{K_{be}} = \frac{1}{K_{bc}} + \frac{1}{K_{ce}}$	|27|
	$K_{bc} = 4.5\left(\frac{u_{mf}}{d_b}\right) + 10.4\,\frac{D^{1/2}}{d_b^{5/4}}$	
	$K_{cb} = 6.78\left(\frac{\varepsilon_{mf}\,D\,u_b}{d_b^3}\right)^{1/2}$	
H_{be}	$H_{be} = 4.5\left(\frac{u_{mf}\rho_g\,c_g}{d_b}\right) + 10.4\left(\frac{k_g\rho_g c_g}{d_b^{5/2}}\right)^{1/2}$	Ibid
h_{ic} (horizontal tubes)	$\frac{h_{ic}d_{ti}}{k_g} = 0.66\left(\frac{c_g\mu}{k_g}\right)^{0.3}\left[\left(\frac{d_{ti}\rho_g u_o}{\mu}\right)\left(\frac{\rho_p}{\rho_g}\right)\left(\frac{1-\varepsilon_f}{\varepsilon_f}\right)\right]^{0.44}$	Ibid
	for $\frac{d_{ti}\rho_g u_o}{\mu} < 2000$	
(vertical tubes)	$\frac{h_{ic}d_p}{k_g} = 0.01844\,C_R(1-\varepsilon_f)\left(\frac{\bar{c}_g\rho_g}{k_g}\right)^{0.43}\left(\frac{d_p\rho_g u_o}{\mu}\right)^{0.23}\left(\frac{\bar{c}_p}{\bar{c}_g}\right)\left(\frac{\rho_p}{\rho_g}\right)^{0.66}$	
	for $10^{-2} < Re < 10^2$	|28|
K^*	$\frac{K^*}{G} = A + 130\exp\left[-10.4\left(\frac{u_t}{u_o}\right)^{0.5}\left(\frac{u_{mf}}{u_o - u_{mf}}\right)^{0.25}\right]$	|29|

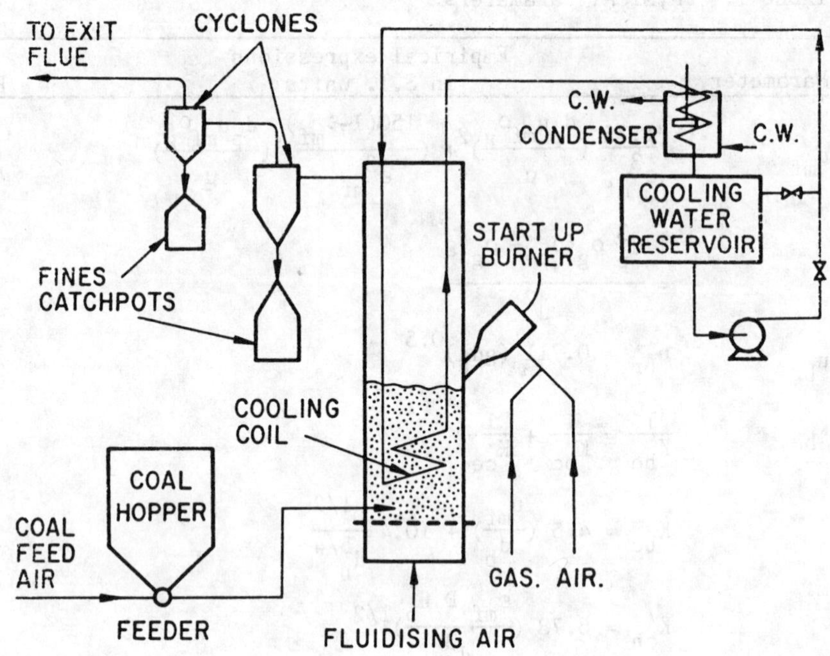

FIG. 1. PILOT SCALE FLUIDIZED COAL COMBUSTOR [22].

oxygen concentration and temperature in the bubble phase, the oxygen concentration and temperature of interstitial gas are calculated by an iterative procedure. A convergence test is then made to determine whether the differences between the calculated and estimated values satisfy the convergence criteria. If not, the initial estimates are replaced by the calculated values and the procedure is repeated until convergence is achieved. When convergence is obtained, bed concentration and temperature profiles, combustion efficiency, bed char loading and size distribution functions for bed and carryover can be evaluated. Details of the solution procedure together with the working forms of the equations representing each modified model can be found elsewhere |18, 20, 21|.

The numerical solution procedure adopted for the iterative solution of nonlinear equations is the method of false positions. It was considered that the accuracy was satisfactory when the difference between the estimated and calculated values was less than one thousandth of the estimated value for C_{li} and 1 K for T_i. The number of iterations required for convergence was of the order of 10. Integrations appearing in the model equations were carried out using Simpson's Formula. For the simultaneous solution of the differential equations, the combined two and three-point predictor-corrector method of numerical integration was adopted.

With 10 subdivisions for the numerical integration over particle size and 60 subdivisions over the bed height, the programme on Burroughs 6900 computer has a running time of about 2 minutes.

4. EVALUATION OF THE MODELS

The models which have been compared are:

a) The model derived by Selçuk et al. |17| in which the homogeneous gas phase reaction in bubble phase has been neglected-Model 1 (M1)
b) Model 1 modified for the incorporation of homogeneous gas phase reaction into the bubble phase-Model 2 (M2)
c) Model 2 modified for the incomplete combustion of volatile carbon-Model 3 (M3)
d) Model 2 applied by using the rate expression of Howard et al. |30| for the homogeneous gas phase reaction-Model 4 (M4)
e) Model 2 modified for the incorporation of incomplete combustion of volatile carbon and applied with the rate expression of Howard et al. |30| for the homogeneous gas phase reaction-Model 5 (M5)
f) Model 2 derived for diffusion controlled char combustion model with no reduction reaction at the particle surface and incomplete combustion of volatile carbon-Model 6 (M6)
g) Model 2 modified for the incorporation of combined kinetic and diffusion controlled char combustion model with no reduction reaction at the particle surface and of incomplete combustion of volatile carbon-Model 7 (M7).

4.1. Comparison of Predicted and Measured Carryover Losses

Figure 2 illustrates the effect of excess air on the measured and predicted values of carryover losses. It can be seen that for all models the variation of carryover loss with excess air follows the same trend as that of measurements, i.e., decreasing with excess air ratio. Models 1 to 5 produce the same predictions since carryover loss depends on char particle combustion mechanism which is the same for these models. However the results are underpredicted for low excess air levels and overpredicted for high excess air levels. This may be partly due to the assumed char combustion mechanism, and partly the empirical correlation used for the calculation of elutriation constant. The predictions of model-6 and model-7 are close to each other and are in better agreement with measurements. This result implies that the combustion model for single char particle with no reduction reaction at the particle surface is more realistic under the conditions prevailing in fluidized bed combustors.

4.2. Comparison of Predicted and Measured Concentration Profiles

Experimental measurements for concentration profiles are available

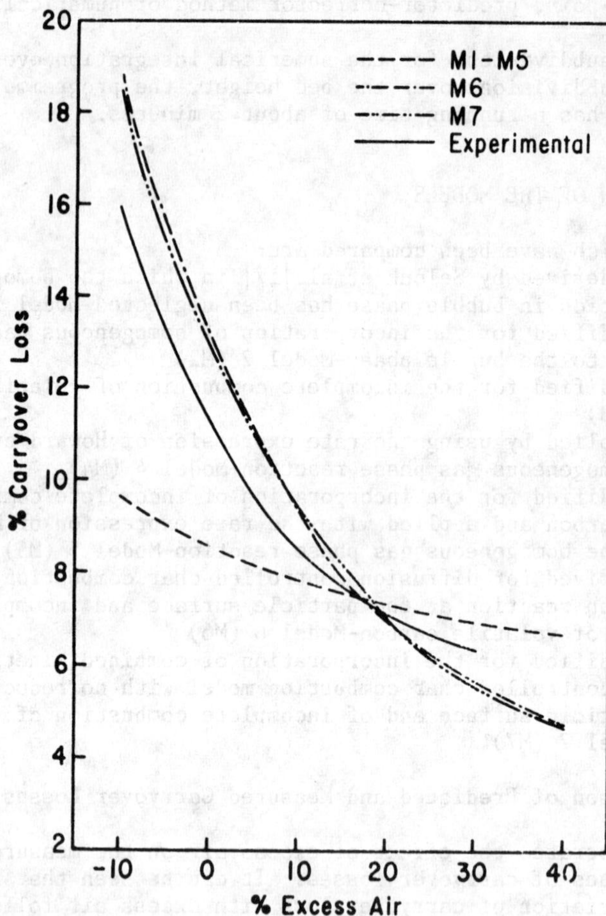

FIG. 2. VARIATION OF CARRYOVER LOSS WITH EXCESS AIR.

for -10, 0 and 10 % excess air ratios. In order to provide a comparison between the measured and predicted values, concentration profiles calculated for the same excess air percentages have been presented.

The measured profiles and the predictions of models 1 to 7 for -10 % excess air are shown in Fig.3. Model-1 predicts higher CO and O_2 and lower CO_2 concentrations compared with measured profiles. It can be seen that with model-1 the percentage of carryover loss is underestimated for -10 % excess air (see Fig.2). Therefore lower values for CO_2 concentrations can be considered to be the result of oxygen deficiency which decreases the rate of oxidation of CO to CO_2. Predictions of models 2 to 5 are in fairly good agreement with experimental measurements. For these models, although the carryover loss is also underestimated, the occurrence of homogeneous reaction in the bubble phase increases the production of CO_2. However as CO_2

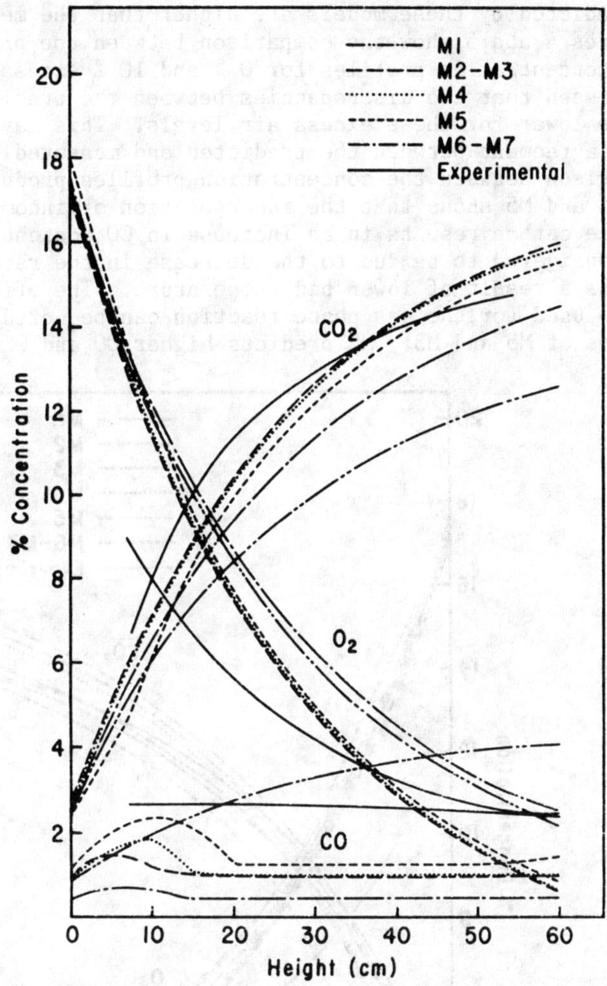

FIG. 3. BED CONCENTRATION PROFILES FOR -10 % EXCESS AIR.

production increases, bed temperature and hence the rate of homogeneous gas phase reaction also increase which results in the underprediction of CO concentration profiles. It can be seen that while experimental CO concentration profile shows a maximum in the middle of the bed, model-1 exhibits a steady rise along the bed and models 2 to 7 give a peak at the bottom of the bed and then remain at a nearly constant value along the bed height. Predicted O_2 concentrations by models 2 to 7 are higher than experimental results at the lower parts of the bed and lower at the upper parts of the bed. On the overall the concentration profiles predicted by models 2,3,4 and 5 show better agreement with experimental results. Model-6 and model-7 underpredict CO_2 and CO concentrations and overpredict O_2 concentrations, since the carryover losses predicted by these models are higher than the measured values.

Figures 4 and 5 show the comparison between the predicted and measured concentration profiles for 0 % and 10 % excess air respectively. It can be seen that the discrepancies between the predicted and measured values are lower for these excess air levels. This may be the result of better agreement between the predicted and measured carryover losses. The comparison between the concentration profiles produced by M2 and M3, and M4 and M5 shows that the incorporation of incomplete combustion of volatile carbon results in an increase in CO content as expected. This is considered to be due to the decrease in the rate of gas phase reaction as a result of lower bed temperature. The effect of rate expression used for the gas phase reaction can be noted by comparing the predictions of M5 and M3. M5 predicts higher CO and O_2 and lower CO_2

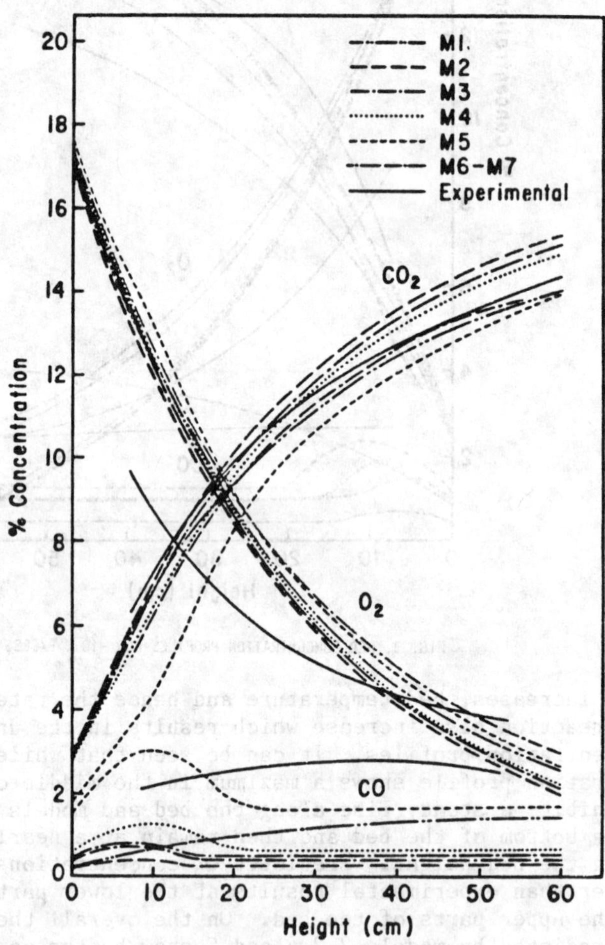

FIG. 4. BED CONCENTRATION PROFILES FOR 0 % EXCESS AIR.

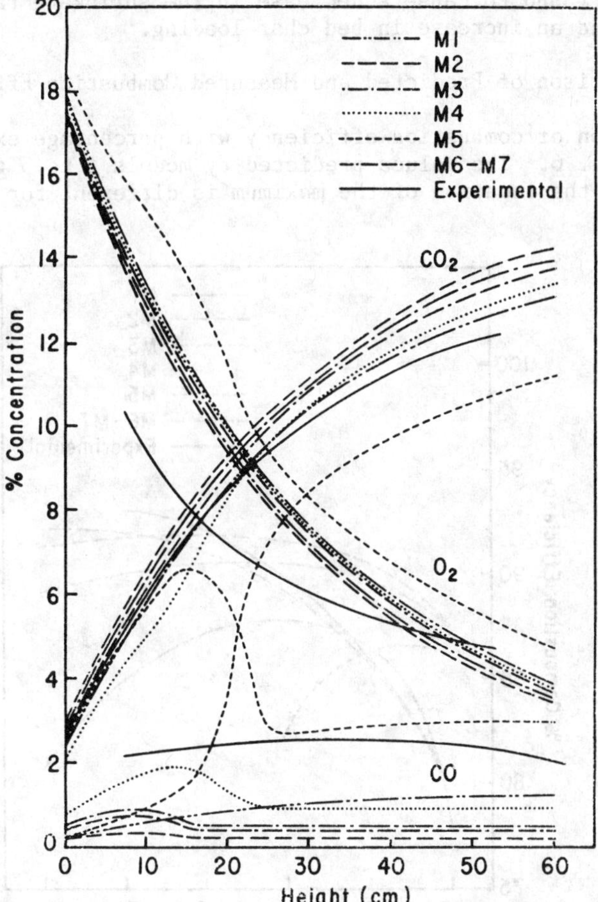

FIG. 5. BED CONCENTRATION PROFILES FOR 10 % EXCESS AIR.

concentrations throughout the bed than those predicted by M3. The reason for this behavior is that the activation energy in the rate expression of Howard et al. is about two times that in the rate expression of Hottel. Therefore the rate of homogeneous reaction is lower for M5. The discrepancies increase with excess air, as the interstitial gas temperature and hence the rate of reaction decrease with increasing excess air levels.

Figures 3,4 and 5 also show that M7 which incorporates combined kinetic and diffusion controlled char combustion model with no reduction reaction at the particle surface gives the same concentration profiles as those obtained from M6. This can be due to the fact that both models utilize the same reactions for char combustion. However, the incorporation of combined kinetic and diffusion controlled char reaction is found to cause a decrease in the shrinkage rates of particles and an increase in bed char loading.

4.3. Comparison of Predicted and Measured Combustion Efficiencies

The variation of combustion efficiency with percentage excess air is shown in Fig. 6. The values predicted by models 1 to 5 go through a maximum but the location of the maximum is different for each model.

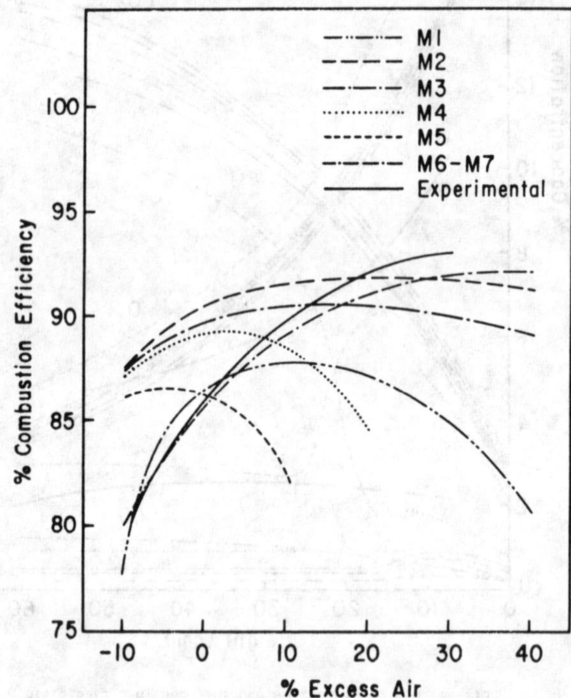

FIG. 6. VARIATION OF COMBUSTION EFFICIENCY WITH EXCESS AIR.

Up to the maximum point, the carryover loss is the dominant factor in the determination of combustion efficiency but after the maximum point the unburnt CO loss becomes the dominant factor. For model-6 and model-7 combustion efficiency increases with excess air. This can be explained by referring to the variation of carryover loss with excess air. The rate of decrease of carryover loss with excess air is higher for models 6 and 7 than those for other models. Therefore the combustion efficiency is controlled by the carryover loss over the whole range of excess air ratios. The good agreement obtained between the predictions of model 6 and 7 and the measurements for the variation of combustion efficiencies with excess air is another evidence implying that reduction reaction does not take place at the char particle surfaces in fluidized bed combustors.

5. CONCLUDING REMARKS

A mathematical model of continuously fed fluidized combustors burning coal of wide size distribution is presented. Modifications of the model for the incorporation of bubble phase reaction, rate expression, volatile combustion and various char particle combustion mechanisms are discussed. Application of the modified models to the prediction of the behaviour of a pilot scale coal fired fluidized combustor is reviewed. Evaluation of the models by comparing their predictions with available measurements is presented.

On the basis of comparisons between predicted and measured values the following conclusions are reached:

i. The homogeneous gas phase reaction in the bubble phase should not be neglected, otherwise the combustion efficiency of the bed is underpredicted.

ii. With regard to CO emissions, incomplete combustion of volatile carbon is more satisfactory than complete combustion of it.

iii. Comparison of the predictions of the models which use different reaction rate expressions for the homogeneous gas phase reaction indicates that reliable reaction rate expressions are very important for quantitative predictions.

iv. Carryover losses and combustion efficiencies predicted by the models which incorporate the single char particle combustion mechanisms with no reduction reaction at the particle surface are in better agreement with measured values implying that these char combustion mechanisms are more realistic under the conditions prevailing in fluidized bed combustors.

v. The effect of the kinetics of the heterogeneous reaction between the char and oxygen is not negligible, since the char loading of the bed is affected considerably by the incorporation of combined kinetic and diffusion controlled char combustion mechanism into the system model.

vi. The system models utilizing the rate expression of Hottel for homogeneous reaction in both phases, involving incomplete combustion of volatile carbon and char particle combustion mechanism with no reduction reaction at the particle surface are found to be in reasonable agreement with the limited experimental data available. These models should prove useful to design engineers in evaluating the effects of variation of excess air and recycle on the concentration and temperature profiles, combustion efficiency, and size distribution of particles in bed and carryover in fluidized coal combustors. The application of the models is however limited to situations in which uniform bubble size may be considered and well mixed and plug flow may be assumed in emulsion and bubble phases respectively.

NOMENCLATURE

A	: Air flow rate to bed, gmol/s
A_c	: Heat transfer area of immersed coil, cm^2
A_o	: Cross-sectional area of the bed, cm^2
$\overline{c}$	: Average specific heat, cal/gK
C	: Concentration, g mol/cm^3
$\overline{C}$	: Average concentration, g mol/cm^3
d_b	: Equivalent bubble diameter, cm
d_p	: Particle diameter, cm
d_{ti}	: Tube diameter, cm
D	: Diffusion coefficient, cm^2/s
$E(r)$	: Net elutriation rate constant, s^{-1}
E	: Activation energy, cal/g mol
f	: Mass fraction in proximate analysis of coal
F_d, F_o, F_3, F_1	: Mass flow rate of char, coal, carryover and overflow respectively, g/s
g	: Gravitational acceleration, cm/s^2
G	: Mass flux of air, $g/cm^2 s$
h_d	: Heat transfer coefficient between particle and gas, $cal/cm^2 s$ K
h_{ic}	: Heat transfer coefficient between bed and coil, $cal/cm^2 s$ K
H	: Bed height, cm
H_{be}, H_{bc}	: Heat transfer coefficient between bubble and emulsion and bubble cloud phases based on the bubble volume respectively, $cal/cm^3 s$ K
k	: Conductivity, $cal/cm^2 s$ K
	: Reaction rate constant for homogeneous gas phase reaction
k_d, k_r, k_c	: Diffusional, surface reaction and overall reaction rate coefficients respectively, cm/s
K	: Elutriation rate constant, s^{-1}
K^*	: Specific elutriation rate constant, $g/cm^2 s$
K_{be}, K_{bc}, K_{ce}	: Mass transfer coefficient between bubble and emulsion, bubble and cloud, cloud and emulsion based on the bubble volume respectively, s^{-1}
m	: Mass of a particles, g
M	: Mass of char particles in bed, g
	: Molecular weight, g/g mol
$p_b(r), p_o(r), p_3(r), p_d(r)$	: Size distribution functions for bed, feed, carryover and char respectively
P	: Excess air ratio
q	: Volumetric flow rate, cm^3/s
	: Rate of removal of carbon per unit external surface area of a particle, $g/cm^2 s$
Q	: Net calorific value, cal/g
r	: Particle radius, cm
$R(r)$	: Shrinkage rate of a particle, cm/s

T : Temperature, K
u : Velocity, cm/s
V : Volume, cm^3
y : Theoretical air requirement, g mol/s
z : Variable representing distance from the base of the bed, cm.

Greek Letters

α_b : Dimensionless mass transfer coefficient between phases
γ : Mass fraction in ultimate analysis of coal
δ : Volume fraction of bubbles in bed
$\Delta H_1, \Delta H_2, \Delta H_3$: Heats of reaction for $C + 1/2\ O_2 \rightarrow CO$, $CO + 1/2\ O_2 \rightarrow CO_2$, $C + CO_2 \rightarrow 2CO$, respectively, cal/g mol
ε_f : Bed voidage fraction
ε_{mf} : Voidage fraction at minimum fluidization
μ : Viscosity, poise
$\eta_o(r)$: Particle separator efficiency
η_c : Combustion efficiency
ρ : Density, g/cm^3
ϕ : Coal swelling factor

Subscripts

A : Ash
b : Bubble phase
be : Bubble phase exit
bo : Bubble phase inlet
br : Bubble rise
c : Surface of immersed coil
C : Carbon
d : Char
D : Dry coal
e : Emulsion phase
f : Operating condition
g : Gas
H : Hydrogen
i : Interstitial gas
j : Component
m : Exit gas
mf : Minimum fluidization
N : Nitrogen
o : Bed inlet
O : Oxygen
p : Particle
s : Solids
S : Sulfur
t : Terminal
v : Devolatization
W : Water

REFERENCES

1. Highley, J. and Merrick, D., 'The Effect of the Spacing Between Solid Feed Points on the Performances of a Large Fluidised Bed Reactor', A.I.Ch.E. Symposium Series, **67**, No.116, p.219, 1971.

2. Becker, H.A., Beer, J.M. and Gibbs, B.M., 'A Model for Fluidised Bed Combustion of Coal', Inst. of Fuel Symposium Series, No.1, A1-1, 1975.

3. Gibbs, B.M., 'A Mechanistic Model for Predicting the Performance of a Fluidised Bed Coal Combustor', Inst. of Fuel Symposium Series, No.1, A5-1, 1975.

4. Gordon, A.L. and Amundson, N.R., 'Modelling of Fluidised Bed Reactors', Chem.Eng.Sci., **31**, p.1163, 1976.

5. Chen, T.P. and Saxena, S.C., 'Mathematical Modelling of Coal Combustion in Fluidised Beds with Sulphur Emission Control by Limestone or Dolomite', Fuel, **56**, p.401, 1977.

6. Beer, J.M.,Baron, R.E., Borghi, G., Hodges, J.L., and Sarofim, A.F., 'A Model of Coal Combustion in Fluidised Bed Combustors', Proceedings of 5th Int.Conf. on FBC., **3**, No.5, p.437, 1977.

7. Gordon, A.L., Caram, H.S., and Amundson, N.R., 'Modelling of Fluidised Bed Reactors-V: Combustion of Carbon Particles-an Extension', Chem.Eng.Sci., **33**, p.713, 1978.

8. Baron, R.E., Hodges, J.L., and Sarofim, A.F., A.I.Ch.E. Symposium Series, **74**, No.176, p.120, 1978.

9. Horio, M., and Wen, C.Y., 'Simulation of Fluidized Bed Combustors: Part I. Combustion Efficiency and Temperature Profile', A.I.Ch.E. Symposium Series, **74**, No. 176, p.101, 1978.

10. Rajan, R., Krishnan, R., and Wen, C.Y., 'Simulation of Fluidized Bed Combustors: Part II. Coal Devolatilization and Sulfur Oxides Retention', A.I.Ch.E. Symposium Series, **74**, No.176, p.112, 1978.

11. Leung, L.S., and Smith, I.W., 'The Role of Fuel Reactivity in Fluidized Bed Combustion', Fuel, **58**, p.354, 1979.

12. Fan, L.T., Tojo, K., and Chang, C.C., 'Modelling of Shallow Fluidized Bed Combustion of Coal Particles', Ind.Eng.Chem.Process Des.Dev., **18**, No.2, p.333, 1979.

13. Rajan, R.R., and Wen, C.Y., 'A Comprehensive Model for Fluidized Bed Coal Combustors', A.I.Ch.E. Journal, **26**, No.4, p.642, 1980.

14. Bukur, D.B., and Amundson, N.R., 'Fluidized Bed Char Combustion Diffusion Limited Models', Chem.Eng.Sci., **36**, p.1239, 1981.

15. Park, D., Levenspiel, O., and Fitzgerald, T.J., 'Plume Model for Large Particle Fluidized Bed Combustors', Fuel, **60**, p.295, 1981.

16. Bukur, D., and Amundson, N.R., 'Fluidized Bed Char Combustion Kinetic Models', Chem.Eng.Sci., **37**, No.1, p.17, 1982.

17. Selçuk, N., Sivrioğlu, Ü., and Siddall, R.G., 'Mathematical Modelling of Coal-Fired Fluidized Bed Combustors', Alternative Energy Sources III (T.N. Veziroğlu, Ed.), **6**, Hydrocarbon Technology, p.231, Hemisphere, 1983.

18. Selçuk, N., and Özkan, Ü., 'Testing of a Model for Fluidized Bed Coal Combustors', Proc. of 20th Intern.Symp.on Combustion. The Combustion Institute, Pittsburgh, p.1471, 1985.

19. Selçuk, N., and Pekyılmaz, A., 'Modelling of Coal-Fired Fluidised Bed Combustors-Effect of Bubble Phase Reaction', Preprints of 3. Intern. Fluidised Conference, Inst. of Energy, London, DISC/28/241, 1984.

20. Selçuk, N., and Pekyılmaz, A., 'Testing of a Model for Fluidised Bed Coal Combustors-Effect of Rate Expression and Volatile Combustion' Alternative Energy Sources VII (T.N. Veziroğlu Ed.), **5**, Hydrocarbons/Energy Transfer, Hemisphere, p.337, 1987.

21. Selçuk, N., and Pekyılmaz, A., 'Testing of a Model for Fluidised Bed Coal Combustors-Effect of Char Combustion Model', 21st.Symp. (Intern.) on Combustion, The Combustion Institute, Pittsburgh, 1986 (in press).

22. Gibbs, B.M., Pereira, F.J., and Beer, J.M., 'Coal Combustion and NO Formation in an Experimental Fluidized Bed', Inst. of Fuel Symposium Series No.1, D6-1, 1975.

23. Davidson, J.F. and Harrison, D., Fluidised Particles, Cambridge University Press, 1963.

24. Levenspiel, O., Kunii, D., and Fitzgerald, T., 'The Proceeding of Solids of Changing Size in Bubbling Fluidized Beds', Powder Technology, **2**, p.87, 1968.

25. Dutta, S., Wen C.Y., and Bett, R.J., 'Reactivity of Coal and Char. 1. in Carbon Dioxide Atmosphere', Ind.Eng.Chem.Process Des.Dev., **16**, No.1, p.20, 1977.

26. Field, M.A., Gill, D.W., and Hawksley, P.G.W., Combustion of Pulverised Coal, The British Coal Utilization Research Association, Leatherhead, 1967.

27. Kunii, D., and Levenspiel, O., Fluidization Engineering, John Wiley and Sons, 1969.

28. Wender, L., and Cooper, G.T., 'Heat Transfer Between Fluidized-Solids Beds and Boundary Surfaces-Correlation of Data', A.I.Ch.E. Journal, 4, No.1, p.15, 1958.

29. Merrick, D., and Highley, J., 'Particle Size Reduction and Elutriation in a Fluidized Bed Process', A.I.Ch.E. Symposium Series, 70, No.137, p.366, 1973.

30. Howard, J.B., Williams, G.C., and Fine, D.H., 'Kinetics of Carbon Monoxide Oxidation in Postflame Gases', Proc. 14th Symp. (Int) on Combustion, The Combustion Institute, Pittsburgh, p.975, 1973.

SUBJECT INDEX